Microorganisms for Sustainability

Volume 46

Microorganisms have been in existence since the origin of life on earth and can survive the most extreme habitats or conditions on earth. Microorganisms are involved in regulating biogeochemical cycles, maintaining plant and animal health, and sustaining the global food chain. Moreover, they play crucial roles in addressing the challenges of climate change and achieving the targets of Sustainable Development Goals (SDGs).

This multidisciplinary book series captures the role of microbes towards building a sustainable world, while encompassing cutting-edge technologies and current needs across various fields such as agriculture sustainability, bioremediation, restoration of degraded habitats and wastelands, and food security. Additionally, this series explores microbial applications in industries, and building their utilization in clean and green energy solutions. Furthermore, themes like microbial secondary metabolites, extremophilic microbes and modern omics, including next generation sequencing and metagenomics, are also covered in this series.

With contributions from researchers across the globe, this series addresses the important call of 'One Planet-One Health-One Future'. It comprises a collection of diverse volumes that provides insights for scientists, young researchers, educators and decision-makers in the government, private sector, and non-governmental organizations, empowering their efforts to achieve the global goals.

The series invites, evaluates, and accepts book proposals to ensure a diverse, inclusive, and evolving program. The final decision regarding acceptance rests with the series editor.

Peer-review

This book series follows a stratified review process. Proposals for individual volumes are reviewed by the series editor and then the editorial board members. On a case-to-case basis, external reviewers are also invited for further evaluation of the book proposal. Review of the chapters is the responsibility of the volume editor(s). A manuscript submission platform has been recently made available to the authors, volume editors and the series editor.

Naveen Kumar Arora • Piyush Pandey
Jitendra Mishra
Editors

Environmental Sustainability and Biotechnology: Opportunities and Challenges

Springer

Editors
Naveen Kumar Arora
Department of Environmental Science
School of Earth and Environmental Sciences
Babasaheb Bhimrao Ambedkar University
Lucknow, UP, India

Piyush Pandey
Department of Microbiology
Assam University
Silchar, India

Jitendra Mishra
Department of Microbiology
Amity University Punjab
Mohali, India

ISSN 2512-1901 ISSN 2512-1898 (electronic)
Microorganisms for Sustainability
ISBN 978-981-96-9024-4 ISBN 978-981-96-9025-1 (eBook)
https://doi.org/10.1007/978-981-96-9025-1

This Springer imprint is published by the registered company Springer Nature Singapore Pte Ltd.
The registered company address is: 152 Beach Road, #21-01/04 Gateway East, Singapore 189721, Singapore

Preface

In *Environmental Sustainability and Biotechnology: Opportunities and Challenges* (part of the groundbreaking Microorganisms for Sustainability series), visionary scientists and thought leaders unite to unravel the extraordinary potential of microbial innovation in steering humanity toward the United Nations' Sustainable Development Goals (SDGs). Born from the dynamic exchange of ideas at the *International Conference on Environmental Sustainability and Biotechnology* (Goa, India, 16–19 November 2022), organized by the Society for Environmental Sustainability, this volume pulses with cutting-edge insights as contributors translate their pioneering research into actionable strategies. Bridging the microscopic world of microbes to the macroscopic challenges of planetary health, this book charts a bold course for harnessing biotechnology to heal ecosystems, reimagine waste, and forge a future where sustainability and scientific ingenuity thrive in tandem.

Spanning 12 meticulously curated chapters across 3 thematic parts, this volume captures the breadth and depth of contemporary microbial research at the forefront of sustainability.

Part I: Foundations of Microbial Biotechnology in Sustainability

The opening chapter, "Microbes for a Sustainable World: Pioneering the SDGs Through Microbial Ingenuity," sets a broad and compelling context. This chapter connects microbial innovations directly to the global SDG framework, offering readers a foundation for understanding how microbial processes intersect with pressing societal goals. By highlighting roles in food production, clean water access, waste recycling, and climate resilience, this introductory chapter serves as a thematic gateway to the rest of the volume.

Part II: Sustainable Agriculture and Biotechnological Solutions

Sustainable food production and soil health underpin ecological and economic stability. In this section, the authors explore how microbes and bio-based solutions can rejuvenate agricultural ecosystems.

Chapter 2, "Microorganisms for Sustainable Agriculture and Food Security," delves into the essential roles played by plant growth-promoting microbes as biofertilizers and biopesticides. The chapter emphasizes microbial strategies that enhance soil fertility, reduce chemical dependency, and promote food security in a sustainable manner.

Chapter 3, "Biochar-Mediated Soil Amendment for Sustainable Agriculture," investigates the synergetic interaction between biochar and soil microbes. The use of biochar to improve soil physical properties, combined with microbial nutrient cycling, promises sustainable intensification of agriculture.

Chapter 4, "Enhancing Strawberry Production in Different Systems: The Role of Nutrient Solutions and Biofertilizers," presents case studies where microbial biofertilizers combined with nutrient management strategies significantly enhanced strawberry yield and quality across cultivation systems.

Chapter 5, "Salicylic Acid in Sustainable Agriculture by Mitigating Salt Stress: Impact on Membrane Lipids," explores how salicylic acid, together with microbial partners, mitigates salt-induced stress at the cellular level, reconfiguring lipid membranes to sustain plant growth.

Chapter 6, "Bioprospecting of Plant Growth Promoting Fungi (PGPF) for Remediating Plant Stress," highlights the isolation and characterization of fungal bioinoculants that improve plant tolerance to abiotic stresses such as drought, salinity, and oxidative stress.

Part III: Industrial Biotechnology and Environmental Remediation

This section explores the transformative role of industrial biotechnology in addressing environmental degradation, utilizing microbial systems, enzymatic processes, and innovative biocatalysts to detoxify pollutants, remediate contaminated ecosystems, and pioneer waste-to-resource paradigms. It sheds light on the synergy between sustainable industrial practices and ecological restoration, while confronting challenges such as scalability, site-specific complexities, and the integration of circular economy principles.

Chapter 7, "Microbially Derived Chitinases: Multifunctional Enzymes for Biocontrol, Bioremediation, and Sustainable Industrial Applications," examines the dual role of chitinases in pathogen suppression and biopolymer degradation, presenting broad applications in agriculture and waste treatment.

Chapter 8, "Unveiling Nature's Bacterial Solution to Radioactive Waste," presents an exploration of bacterial strains capable of transforming and immobilizing radionuclides, offering a biological solution to one of the world's most daunting waste management challenges.

Chapter 9, "Bacterial Remediation of Radioactive Waste: Current Understanding on Mechanistic Insights and Future Direction," delves into molecular mechanisms, offering insights into how bacteria adapt to and detoxify radioactive environments.

Chapter 10, "Food Waste Biotransformation: Advancing Enzyme Production Through Microbial Processes," details how food industry waste streams can serve as substrates for high-value enzyme production, fostering circular economy models.

Chapter 11, "Pyomelanin Biosynthesis by *Pseudomonas*: A Sustainable Resource for Cosmetic Formulations," examines the biosynthesis of pyomelanin, a promising bioactive pigment with antioxidant and UV-protective properties, highlighting its commercial potential in eco-friendly cosmetic formulations.

Chapter 12, "Proteomics in Unravelling Secondary Metabolite Synthesis in Plants: Exploring the Microbial Connection," uses proteomics tools to reveal how microbial symbiosis can enhance the biosynthesis of valuable secondary metabolites, with implications for agriculture, nutrition, and medicine.

By illuminating the diverse capacities of microorganisms to support sustainable systems, this book reinforces the critical role of microbiology in achieving global development targets. From soil remediation to renewable bio-products and plant resilience, the microbial world offers a treasure trove of solutions that are as sophisticated as they are sustainable. We hope that this volume inspires researchers, innovators, policymakers, and students to further harness microbial ingenuity for a resilient, equitable, and sustainable world. The editors thank all the contributing authors and hope that the book will be useful for the readers in getting an in-depth knowledge on the role and utilization of microorganisms in diverse ways.

Lucknow, UP, India Naveen Kumar Arora
Silchar, India Piyush Pandey
Mohali, India Jitendra Mishra

Contents

Series Editor and Editors

About the Series Editor

Naveen Kumar Arora, PhD in Microbiology, Fellow of International Society of Environmental Botanists (FISEB), is a Professor in the Department of Environmental Science, Babasaheb Bhimrao Ambedkar University (A Central University), India. He is a renowned researcher in the field of environmental microbiology and biotechnology. His specific area of research is plant-microbe interactions and sustainable agriculture. He has more than 150 research articles published in premium international journals and several articles published in magazines and dailies. He is an editor of 54 books, published by Springer Nature. He is the series editor of book series, Microorganisms for Sustainability, and he is the editor-in-chief of the journal, *Environmental Sustainability*, both published by Springer Nature.

About the Editors

Naveen Kumar Arora, PhD in Microbiology, Fellow of International Society of Environmental Botanists (FISEB), Professor in Department of Environmental Science, Former Dean, School of Earth and Environmental Sciences, Former Head, Department of Environmental Science, Former Head, Department of Environmental Microbiology and President of Institutions Innovation Council, Babasaheb Bhimrao Ambedkar University (A Central University), Lucknow, Uttar Pradesh, India, is a renowned researcher in the field of Environmental Microbiology and Biotechnology.

His specific area of research is plant-microbe interactions and sustainable agriculture. Before joining Babasaheb Bhimrao Ambedkar University, he was Head, Department of Microbiology, Chhatrapati Shahu Ji Maharaj University, Kanpur, Uttar Pradesh, India, from 2006 to 2011. He has more than 150 research articles published in premium international journals and several articles published in magazines and dailies. He is editor of 54 books, published by Springer Nature. He is the Series Editor of the book series Microorganisms for Sustainability published by Springer Nature and Editor of the journal *Plant Pathology* published by Wiley. He is the Editor-in-Chief of the journal *Environmental Sustainability* published by Springer Nature. He is also the member of several national and international societies, and in editorial board of four journals. As per the list released by Stanford University in collaboration with Elsevier, he is among the top 2% most cited scientists of the world for four years in a row. He has delivered lectures in conferences and seminars around the globe, some of which include Lead Lectures at University of Lisbon, Portugal; SIIRT University, Turkey; University of Tashkent, Uzbekistan; Ibn Zohr University, Agadir, Morocco; IUSS Conference at Philadelphia, USA; conferences at Beijing (China) and Hanoi (Vietnam); and several conferences in India. He has been advisor to 126 postgraduate and 22 doctoral students. He has received awards for excellence in research by the Honorable Governor of Uttar Pradesh, Asian PGPR Society, Vijnana Bharti, Samagra Vikas Welfare Society, and Research & Academic Excellence Awards in the discipline of Life Sciences by Babasaheb Bhimrao Ambedkar University (A Central University), Lucknow, Uttar Pradesh, India. He is also President of Society for Conservation of Wildlife and is Secretary, Society for Environmental Sustainability (website: www.sesindia.org).

Piyush Pandey is working at the Department of Microbiology, Assam University, India. He has 24 years of academic and research experience. His research group works on plant-associated bacteria, with their potential applications in agriculture and environment. He has published more than 160 papers in journals of high impact, 4 books, and undertaken 16 research grants. He has been awarded with "DBT-Unit for Excellence award" for working on plant-rhizosphere mediated bioremediation of polyaromatic hydrocarbons. Also, he has been "Overseas Associate" to Agroscope, Switzerland, SICI-SMP fellow to York University, Canada, and had visited Shizuoka University, Japan. He is a Fellow of International Society of Environmental Botanists, member of several scientific societies, and has editorial role in reputed journals. He has been extensively working on soil and plant microbiomes, rhizoremediation of environmental pollutants, and environmental genomics. He has been awarded from Springer, Society for Environmental Sustainability, and Asian PGPR Society for his research contributions. Seventeen students have been awarded PhD under his supervision. He has two Indian patents granted.

Jitendra Mishra is Assistant Professor of Microbiology at Amity University Punjab, India. A globally recognized scientist, he ranked among the top 2% of microbiologists worldwide (Elsevier, 2024). He previously coordinated a DST-SEED-funded project (2020–2023) at Babasaheb Bhimrao Ambedkar University, spearheading R&D in microbial technologies. With prior expertise as a DST Research Associate, he specializes in biocontrol agent development, including strain selection, carrier formulation, and quality control. His research focuses on microbial secondary metabolites, bioinoculants, and plant growth-promoting rhizobacteria (PGPR), leveraging fluorescent pseudomonads for eco-friendly phytopathogen management in saline soils. An active academic, he mentors postgraduate students, holds a patent, and has published extensively in reputed journals. He has authored three books and serves as an editor for Frontiers and Springer-Nature journals. His interdisciplinary work bridges microbial innovation and sustainable agriculture, addressing global agricultural challenges through cutting-edge biotechnological solutions.

Part I
Foundations of Microbial Biotechnology in Sustainability

Theoretical and Conceptual Groundwork

Chapter 1
Microbes for a Sustainable World: Pioneering the SDGs Through Microbial Ingenuity

Nandita Das, Ankita Das, Dona Mazumder, Bhrigu Bhuyan, Sourav Debnath, Rajarshi Chowdhury, Rajnanda Thaosen, and Piyush Pandey

Abstract The Sustainable Development Goals (SDGs) include a broad spectrum of global challenges, ranging from poverty eradication and reducing inequalities to addressing climate change, ensuring access to clean energy, promoting responsible consumer behaviour and establishing sustainable practices in terrestrial and marine ecosystems. Microorganisms, with their remarkable adaptability and growth capabilities in diverse environments, have emerged as important components for addressing these social concerns and directly contributing to the achievement of the SDGs. Primarily, the pathogenic abilities of microorganisms have been the focus for many centuries; however, it is essential to recognise the vast array of beneficial roles that microbes play in the sustainability of the planet. Therefore, a comprehensive and scientific exploration of the microbial domain is necessary, as it holds the potential to significantly contribute to the attainment of many SDGs. This chapter provides a comprehensive understanding of the capabilities of microorganisms and recommends innovative solutions for global challenges such as bioenergy production, bioremediation of polluted ecosystems, sustainable agriculture and promoting ecosystems. The objective of this chapter is to investigate the intricate interactions of microbial communities in terrestrial and aquatic ecosystems, which are crucial for maintaining ecological balance and biodiversity. Further, the role of microbes in nutrient cycling, soil health and carbon sequestration is highlighted. Harnessing the unique metabolic capabilities of microorganisms can lead to the development of sustainable agricultural practices that rely on reduced chemical inputs, improved nutrient uptake and enhanced crop resilience to environmental stresses for a transition to a circular and sustainable economy.

Keywords Sustainable Development Goals · Microbiome · Green Technology · Bioenergy · Bioeconomy

---

N. Das · A. Das · D. Mazumder · B. Bhuyan · S. Debnath · R. Chowdhury · R. Thaosen
P. Pandey (✉)
Soil and Environment Microbiology Laboratory, Department of Microbiology, Assam University, Silchar, Assam, India

N. K. Arora et al. (eds.), *Environmental Sustainability and Biotechnology: Opportunities and Challenges*, Microorganisms for Sustainability 46,
https://doi.org/10.1007/978-981-96-9025-1_1

1.1 Introduction

Indiscriminate use of chemicals, reliance on non-renewable energy sources and uncontrolled generation of waste in various industrial processes poses a significant threat to the environment's sustainability. It is now imperative for the world to assume greater responsibility and embrace sustainable practices, cleaner production techniques and green technologies. By doing so, we can ensure the preservation of Earth's ecology for coming generations (Leal Filho et al. 2019). In recent years, there has been a growing global focus on the concept of sustainable development as societies grapple with urgent environmental and societal challenges. To foster collective efforts in this direction, 193 countries agreed upon the 17 Sustainable Development Goals (SDGs), a UN-sponsored initiative aimed at achieving sustainable economic development worldwide (Fund 2015). The SDGs aim to achieve economic and societal progress without environmental harm through solutions based on people, planet, prosperity, peace and partnerships (Urata et al. 2023).

Microorganisms, including bacteria, fungi, viruses and other microscopic life forms, play an essential role in our planet's ecosystems. They are ubiquitous, existing in each component of the biosphere, and their influence reaches far beyond their minuscule size. From the depths of the ocean to the soil beneath our feet and even within our own bodies, microorganisms are involved in biochemical processes (Prussin and Marr 2015). Microorganisms as pathogens that cause diseases tend to receive significant attention, so it is crucial to recognise the multitude of positive roles that microbes play in the environment. Therefore, it is crucial to thoroughly investigate the microbial world in order to fully comprehend its substantial contributions towards sustainable development (Akinsemolu 2018). In the past few decades, the microbial kingdom has been explored for its potential in every field, and the success stories are plenty. Microbial biotechnology and the discovery of novel antimicrobial agents offer promising avenues for combating infectious diseases and antibiotic resistance (Mohr 2016). In agriculture and food production, microorganisms offer innovative solutions for enhancing crop yields, improving soil fertility and reducing the use of chemical fertilisers and pesticides (de Araujo Avila et al. 2021). Further, microbes are significant in mitigating climate change as they serve as nature's ultimate recycling bioremediators and have roles in pollution abatement and the restoration of degraded ecosystems (Singh et al. 2010; Arora 2019; Pande et al. 2020). Their potential exists in harnessing renewable energy sources, such as biofuels and microbial fuel cells, thus paving the way towards a greener and more sustainable energy (Joshi et al. 2017; Jenol et al. 2019).

The primary objective of this chapter is to examine and evaluate the various pertinent aspects of microorganisms that need consideration to achieve the SDGs. The chapter aims to investigate and emphasise the diverse roles played by microorganisms in addressing the global environmental challenges outlined in the SDGs. By conducting a comprehensive analysis of the scientific literature, this chapter seeks to provide a comprehensive understanding of how microbial communities play crucial roles in environmental processes, public health, agriculture, biotechnology, energy production and pollution reduction. By analysing the specific connections between microorganisms and each of the SDGs, the study intends to highlight the potential of microbial applications in advancing the goals of sustainable development, ultimately contributing to a more comprehensive and well-informed approach to addressing complex global issues.

1.2 Needs and Challenges of Sustainable Development Goals (SDGs)

Sustainable development often implies a future devoid of any environmental issues, but this is not sustainable due to degradation in biospheres, climate change, desertification and water quality. However, human well-being is also affected by many challenges, like hunger, regional conflicts, poverty, human trafficking and healthcare costs (Timmis et al. 2017a, b). Rapid population growth and urbanisation increase food and energy demands, causing excessive land use and soil issues like microbial decline and organism displacement (Furtak and Gajda 2018). Excessive chemical use, non-renewable energy and unregulated waste generation threaten environmental sustainability. It is crucial to adopt sustainable practices, cleaner production methods and green technologies for future generations (Akinsemolu 2018). The 2030 Agenda for Sustainable Development was unanimously embraced by every 193-member state of the United Nations in 2015. It serves as a collective framework for promoting peace and prosperity for present and future generations on our planet. The SDGs necessitate urgent and collective action from all nations, irrespective of their developmental status, fostering a collaborative global partnership. These goals acknowledge the interconnectedness of poverty eradication, improvements in health and education, reduction of inequality and stimulation of economic growth, while simultaneously addressing climate change and the preservation of our oceans and forests (Olabi et al. 2023). The SDGs strive to create solutions that foster economic and societal progress without causing harm to the environment (Akinsemolu 2018). Countries are facing challenges like population growth, communication systems, limited resources, policies and funding to meet the goals. These shortcomings highlight the urgent need to accelerate microbiological initiatives in order to achieve these goals (Sutherland et al. 2021).

1.3 Role of Microbes Achieving Sustainable Goals

1.3.1 Microorganisms in Mitigation of Poverty (SDG 1)

According to the World Health Organization, there is a correlation between poverty and the decline of various vital human attributes, such as health. Poverty deprives individuals of access to basic necessities, leaving them malnourished and prone to disability and illness. Disabilities and diseases reduce earnings, productivity and quality of life (World Health Organization 2017). In this scenario, microorganisms have both positive and negative effects. Infectious diseases, malnutrition and child-bearing difficulties cause 52% of third-world mortality. These results shed light on how microbial diseases affect the health of people who do not have access to food, health care or good sanitation (Irwin et al. 2018). Awareness programmes on sanitation and knowledge of pathogenic microorganisms having antimicrobial resistance such as MRSA, *Escherichia coli, Streptococcus*, etc. can greatly help in this regard (Akinsemolu 2018).

Across the world, traditional food preparation frequently uses the microbe-aided fermentation process. The application of this knowledge for the development of small-scale businesses will greatly help in creating employment and contribute to the country's economics (Nikoli and Matsi 2014). Microbial phytase hydrolyses almost 90% of the phytate from maize, cassava, soy beans, cowpeas, sorghum and cocoyam during fermentation. Fermentation improves vitamin B content and protein quality while lengthening the product's shelf life. Thus, traditional cuisine in third-world diets can reduce malnutrition (Friedman 2013). Income generation through the production of *Gobar Gas* using methane-producing organisms like *Methylocystis*, *Methylosinus, Methanomonas, Methanobacter* and *Methylomonas* is a predominant resource contributing to the circular economy. Production of traditional Nigerian foods such as Ogi, Garri and Pito, as well as pickles and beverages through the use of lactic acid bacteria and yeast, also contributes to income generation and employment boosting (Rainieri and Zambonelli 2009; Bhutto et al. 2011; Adesulu and Awojobi 2014). Tequila and mescal originate from Mexico, which are now worldwide known beverages and contribute greatly to employment generation (Ramírez-Guzmán et al. 2019). Similarly, African traditional beverages such as Bushera, Borde, Gowé, Kunun-zaki, Mahewu and Malwa are produced on an industrial scale (Aka et al. 2014). Traditional understanding does not include understanding of microbiological methods, which involve plant or animal inoculum. Thus, harnessing the significance of microbial diversity in the production of fermented foods, which contribute to one-third of the world's diet, will greatly contribute towards attaining sustainable agendas by reducing poverty and providing food security (Ojeda-Linares et al. 2021).

1.3.2 Microorganisms in Promoting Food Security to Promote Zero Hunger (SDG 2)

Despite the improved production of food, it is still insufficient to sustain for future generations. In 2050, the global population is expected to reach 9.7 billion (Chakraborty et al. 2013). Thus, proper agricultural production strategies are being considered to achieve increased, sustainable food production to feed the globe within the next few decades. Investments in small-scale industries such as fisheries, dairy and crop industries can contribute to achieving hunger needs if executed properly (Azadi et al. 2023). Microorganisms, such as bacteria and fungi, are essential for maintaining soil health and facilitate nutrient cycling to ensure the fertility of agricultural soils, enabling the production of healthy crops (Coonan et al. 2020). Leguminous plants, such as peas, beans and soybeans, form symbiotic relationships with these bacteria, allowing them to acquire nitrogen and reduce the need for synthetic fertilisers (Favero et al. 2021; Ciampitti et al. 2021; Gungaabayar et al. 2022). Rhizobacteria and some endophytic bacteria alleviate salt stress and exhibit drought tolerance (El-Sayed et al. 2014; Vurukonda et al. 2016), produce phytohormones, siderophores, osmolytes, antioxidants and volatile compounds, which improve crops growth and yield (Camaille et al. 2021; Naseem et al. 2018). With the presence of all these properties, microbes can be used for green technology to restore unfertile agricultural lands and enhance the yield production of crops.

Addressing a global issue such as hunger necessitates not only enhancing productivity but also improving the availability of food for everyone. Probiotics can maintain nutrients in fermented foods like yoghurt and sauerkraut, decreasing food waste with the increasing shelf-life of the products (Chugh and Kamal-Eldin 2020). These beneficial bacteria also secrete various enzymes industrially, including amylase produced by *Aeromonas, Bacillus, Bacteridaceae* and *Lactobacillus* as well as protease and cellulase enzymes produced by *B. subtilis*, *L. plantarum* and *Staphylococcus* sp. (El-Saadony et al. 2021). Further, probiotics can be used in livestock and aquaculture to improve animal health, enhance nutrient absorption and reduce the need for antibiotics, ultimately promoting sustainable food production (Abdel-Latif et al. 2022; Sharma et al. 2020; Peñalosa-Martinell et al. 2021). The biofertiliser market generates around $495 million per annum. In the late 90s, developing countries like India implemented schemes like the "Green revolution" to increase wheat and rice production (Antar et al. 2021). India's "Vision 2030" focuses on food security and reducing hunger through microorganism-based initiatives. Nigeria's "Green Alternative" could support the economy, create employment and contribute to reducing hunger and achieving the second SDG (Gil et al. 2019).

1.3.3 Microorganisms in Sustaining Health and Well-Being (SDG 3)

Microorganisms have a profound impact on human health through the interaction with human microbiome. Gut microbiome consisting of trillions of bacteria influences the digestion, nutrient absorption, immune system development, vitamin synthesis and also disease prevention (De Filippis et al. 2020; Shreiner et al. 2015; Kau et al. 2011). One of the major contributions of the microorganisms is in the development of vaccines. The Global Polio Eradication Initiative's polio eradication program results in the reduction in polio cases by 99% (Lima et al. 2021), while Taiwan's Hepatitis B vaccine campaign reduced liver cancer rates to 0.1% in the early 2000s (Heck et al. 2022). This aligns with SDG 3 which focuses on preventing and treating these disease outbreaks. In addition to disease prevention, gut microbiome also balances and regulates many non-communicable diseases like obesity, diabetes, cardiovascular diseases and cancers. Dysbiosis refers to an imbalance or disruption in the gut microbiota, which can be caused by incorrect signalling between the immune system and the microbiota. (Hosseinkhani et al. 2021). Recent studies reveal the gut microbiome associated with the gut–brain axis plays a significant role in developing innate immunity, supporting mental well-being and help maintain the overall health (Shoubridge et al. 2022).

Microorganisms are also intertwined with the global health threat of antimicrobial resistance (AMR), caused by the misuse and overuse of antibiotics and making it challenging to treat infections effectively (Dadgostar 2019). Maternal and child health outcomes are impacted as AMR compromises the effectiveness of antibiotics in preventing and treating infections during pregnancy and childbirth (Rallis et al. 2023). Moreover, non-communicable diseases are also affected due to the limited availability of effective antibiotics (Kelley 2011). Probiotics also help in the restoration of healthy gut microbiota after antibiotic therapy, thus helping in combating AMR (Kumar et al. 2020). Addressing AMR is crucial to achieving SDG 3, and The Global Action Plan on Antimicrobial Resistance (AMR) is a WHO-designed framework aiming to address global antimicrobial resistance by enhancing knowledge, boosting surveillance, reducing infection incidence, optimising antimicrobial agent use and guaranteeing sustainable funding (Vekemans et al. 2021).

The One Health concept emphasises the interconnectedness of human, animal and environmental health. Microorganisms, including bacteria, viruses, fungi and parasites, play a crucial role in maintaining the ecosystem balance and supporting human and animal health. They can cause infectious diseases like malaria, tuberculosis, influenza and COVID-19, leading to significant morbidity and mortality. SDG 3 focuses on understanding transmission dynamics, developing vaccines and implementing public health measures to prevent diseases, while addressing the livestock health and biosecurity are crucial components of One Health (Muluneh 2021) (Fig. 1.1).

1.3.4 Microorganisms in Quality Education (SDG 4)

The most significant method for attaining sustainable development is education. Therefore, the fourth goal of SDG emphasises on providing excellent education among the youth, removal of the inequalities in access to education, expanding scholarships and also promotes opportunities for continuing further education (United Nations 2016). In the modern period, microorganisms play a vital role in the daily lives of humans. Through teaching, research and innovation, the significant role of bacteria on humanity can be understood. In order to help the students understand how microbiology and biotechnology are potentially used for improvising livelihoods and environmental security, efforts are made to establish new

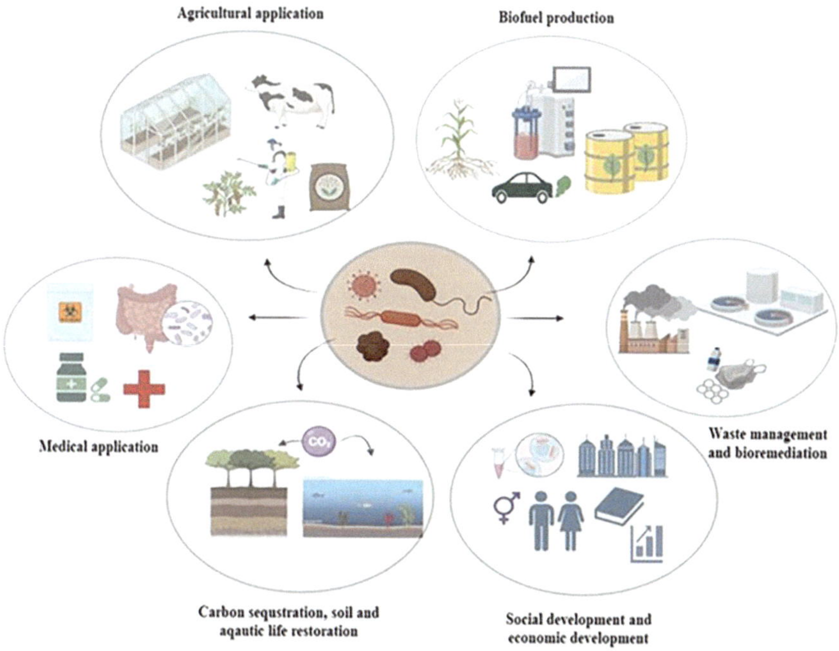

Fig. 1.1 Role of microbes in different fields of sustainable developmental goals

teaching methods for a variety of technologies such as environmental technology, food biotechnology, fermentation technology and immunology (Simonneaux 2000). In each of the abovementioned techniques, microorganisms play a pivotal role, and education regarding the role of microorganisms in our environment is of utmost necessity. Food fermentation is the most economical method to produce and preserve food, and it is the most ancient method that all cultures have used (Rizo et al. 2020). Microorganisms like *Lactobacillus, Saccharomyces cerevisiae, Leuconostoc,* etc. are extensively used in food industry and fermentation technology, and enriching the youths with the knowledge regarding the intake of these microbes that work as probiotics for human is important. Moreover, quality education to the farmers and skilled youths regarding the implementation of soil bacteria like *Rhizobium, Pseudomonas, Azotobacter,* etc. as compost in the agricultural fields should be given priority. Earlier, in a report by Milandri (2004), in Italy, it was demonstrated that around 60% of children consider microorganisms to be dangerous, and only 25% acknowledge their beneficial function. It has been suggested that a better educational system might contribute to a lower chance of contracting a harmful illness by a pathogenic bacterium. For example, *Salmonella typhi* that causes typhoid and *Vibrio cholerae* causing cholera impose a negative effect on mankind. However, there are already numerous reports which indicate that students who have taken microbiology as a subject have understood the importance of microbes (Jones et al. 2013). Those students can come forward and educate others about the positive and negative impacts of the microorganisms in our daily life. In one of the published research works (Mafra et al. 2015), it was mentioned that children's ability to maintain hygiene has been significantly improved in primary school after learning about microorganisms. Furthermore, promoting breastfeeding education among the new mothers and creating awareness among them help to strengthen the child's immune system with the assistance of *Bacteroides* and *Bifidobacterium* that provide the capacity to fight sickness (O'Sullivan et al. 2015). Foremostly, it is essential to educate the students from primary levels and teach them the benefits of microbiological education by adding such types of subjects in their curriculum and also train the educators in the best ways to establish the course (Barberan et al. 2016).

1.3.5 Achieve Gender Equality and Empower All Women and Girls (SDG 5)

Highlighted by UNESCO and the United Nations, the significance of the intersection between science and gender equality is essential for attaining sustainable development. Despite the growing awareness of women's rights in recent years, a 2018 UN report highlighted the disproportionate challenges faced by women and girls in accessing basic necessities, experiencing violence and facing gender disparities (Pereira et al. 2022). In 2016, women made up less than 30% of researchers globally, as reported by the UNESCO Institute for Statistics. Additionally, only 35% of research grants given by the National Institutes of Health (NIH) in 2020 were awarded to women. In order to shift conventional perspectives, it is essential to advocate for gender equality, challenge stereotypes and support girls and women in pursuing careers in STEM (Science, Technology, Engineering and Mathematics) fields (Frize et al. 2022). Many of the food sources worldwide come from farms that are 2 hectares or smaller, often run by women who work together to support their families. Understanding microbial sciences and their impact on the environment and cultivation can help small farmers increase their income. When it comes to agriculture, women can learn about the significance of soil microflora and how beneficial microbes such as *Azospirillum*, *Bradyrhizobium*, *Burkholderia*, *Pseudomonas*, *Rhizobium* and *Trichoderma* can enhance crop yields. Further, education is vital for health, and people with the knowledge of microorganisms are usually more concerned about hygiene (Gamage et al. 2023). Thus, encouraging women to pursue education will benefit their offspring, families and society as a whole. As previous studies showed the gut microbiota and its impact on health and disease advances, the vaginal microbiota still lags behind in research (Bach et al. 2021). Within this focus, Jacques Ravel and his team present a comprehensive summary of the most recent research on the human vaginal microbiota. They emphasise the necessity of creating models for in-depth studies due to the variations in vaginal communities among current animal models. In a recent Comment article, Pawel Łaniewski and Melissa Herbst-Kralovetz explore the relationship between the vaginal microbiota and menopause, a topic that impacts a significant portion of the global population yet remains lacking in research and funding (France et al. 2022). Despite these obstacles, women have been pioneers in the field of microbiology. Jane Hinton is widely recognised for her role in creating Mueller-Hinton agar in 1941 (Munson 2020). Elizabeth Lee Hazen and Rachel Fuller Brown are credited with the discovery of the antifungal nystatin in the 1950s (Larkin et al. 2019). Fanny Hesse's proposal to substitute agar for gelatin in the late nineteenth century brought about a significant change in culturing microorganisms (Nussinovitch and Nussinovitch 1997). Recently, Jennifer Doudna and Emmanuelle Charpentier, Nobel laureates, were acknowledged for their pioneering efforts in creating the CRISPR-Cas9 genome editing tools, which have transformed the field of science (Ahmad et al. 2024).

1.3.6 Microorganisms in Maintaining Clean Water and Sanitation (SDG 6)

SDG 6 aims to ensure access to clean water and adequate sanitation for all people. Based on data by Centres for Disease Control and Prevention (CDC), it has been determined that approximately 780 million people are deprived of adequate access to a suitable water source. Furthermore, an estimated 2.5 million individuals are found to be lacking improved sanitation facilities. Microorganisms break down organic matter, remove pollutants and decompose hazardous compounds in wastewater, making it safer to release into the environment (Fufă et al. 2017). Efficient wastewater treatment is essential for maintaining clean water sources and preventing waterborne diseases. On the other hand, pathogenic bacteria like *Escherichia coli*, *Salmonella*, *Shigella*, etc. can contaminate water sources and cause waterborne diseases (Ahmed et al. 2020). Certain protozoa are determinants of water quality by serving as indicators of faecal contamination such as *Cryptosporidium* and *Giardia* (Ligda et al. 2020). Many species of bacteria and fungi have the ability to degrade pollutants and toxin compounds present in water, such as oil spills or industrial waste (Pandey and Fulekar 2012). Furthermore, probiotics in water treatment can maintain microbial equilibrium and reduce chemical interventions (Ren et al. 2021; Jóźwiakowski et al. 2009). Economical green technologies such as activated sludge and filter are commonly employed for the treatment of municipal wastewater.

The implementation of measures to prevent water-related diseases and improve sanitation has the potential to result in annual cost savings of $7 billion. Furthermore, these efforts can also contribute to a significant increase in the gross domestic product (GDP) of developing nations, with an estimated growth rate of 89%. Nationwide sanitation projects in India, such as Clean India Campaign—Nirmal Bharat Abhiyan, Total Sanitation Campaign (TSC) and Swachh Bharat Abhiyan—Clean India Mission, are employed to create awareness and generate clean environment. Moreover, in such schemes, 33% of the spots are reserved for women, thereby addressing gender inequality (Routray et al. 2017). The RAIN project, which is supported by the Coca-Cola Company and the Imo State Drinking Water Supply and Sanitation Project in Nigeria, has as its primary objective the provision of clean drinking water and an adequate supply of water to the people of Africa and Nigeria, respectively (Akinsemolu 2018).

1.3.7 Microorganisms in Affordable and Clean Energy (SDG 7)

Affordable and clean energy is one of the sustainable development goals of the 17 such goals adopted by the United Member States. Recent problems that have emerged with the rise in population are the increase in the demand of energy. Fossil fuels like coal, petroleum and natural gas serve as the major source of energy generation; however, this has several negative impacts on the environment. Their use intensifies the generation of greenhouse gases such as carbon dioxide, contributing to problems such as global warming (Al-Ghussain 2019). Fossil fuels are also non-renewable energy resources and as such are being depleted at a fast rate (Holechek et al. 2022). In order to tackle these problems, several researches are going on in various fields for the production of clean and renewable energy. Researches have also applied the use of microbes like bacteria, fungi, algae, etc. for generating energy (Pandya et al. 2024). Moreover, using microorganisms also have several advantages because several substrates which are easily available and cheap can be used for the generation of energy like wastewater from municipal waste and industrial waste, agricultural waste, contaminated soil, etc. (Majidian et al. 2018; Chaurasia et al. 2020). Some of the techniques where microbes are incorporated for the generation of energy are through microbial fuel cells (MFCs), microbial electrolysis cells, generation of biofuels, carbon capture and utilisation. These techniques are at various stages of development, and research is still going on for improvement and efficiency. MFCs are one of bio-electrochemical devices that use an electrochemically active microorganism as a biocatalyst for the production of electricity (Du et al. 2007; Logan 2009; Wang et al. 2015; Dwivedi et al. 2022). Electrical current produced by bacteria was observed by Potter in 1911. A wide variety of microorganisms belonging to archaea, bacteria and fungi have been identified that are capable of generating electricity. A variety of microbial phyla have exoelectrogenic members, including *Proteobacteria, Firmicutes, Acidobacteria, Actinobacteria, Eumycota* and *Chlorophyta*. Microorganisms such as *Geobacter, Shewanella, Pseudomonas* and *Rhodoferax* are the most studied organisms in this regard (Li et al. 2018).

Microbial electrolysis cell (MEC) is another such technique where the cell consists of a microbial anode and a hydrogen evolution cathode. The electrolyte in the anodic compartment must contain microorganisms and the culture medium necessary for their growth. Some microorganisms spontaneously colonise the anode surface and form an electroactive biofilm, which acts as the electrocatalyst (Rousseau et al. 2020; Koul et al. 2022). Different types of substrates are used for different microorganisms. For example, *Aeromonas hydrophila, Thermincola sp, Geothrix fermentans* and *Gluconobacter oxydans* use the substrate acetate; for *Shewanella putrefaciens* and *Shewanella oneidensis,* the substrate is acetate lactate, while for *Klebsiella pneumoniae, Rhodoferax ferrireducens* and *Enterobacter cloacae,* the substrate is acetate glucose (Sonawane et al. 2022). Biofuels are referred to as any solid, liquid or gaseous fuels that are predominantly produced from biomass (Demirbas et al. 2011). Liquid and gaseous biofuels have become more attractive recently because of their environmental benefits. Biofuels are non-polluting, locally available, accessible, sustainable and reliable fuels obtained from renewable sources. A variety of waste can be used as substrate for the production of biofuels, such as plant waste and agricultural waste (Das et al. 2024a, Silva et al. 2014). Some of the most common used biofuels include bioethanol, biodiesel, biohydrogen, butanol and biomethane. Microorganisms that play a crucial role in biofuel production belong to bacteria, fungi and algae like *Penicillium* sp., *S. cerevisiae, Nannochloropsis* sp., *Schizochytrium* sp., *Botryococcus* sp., *Streptomyces* sp., *Rhodococcus* sp. and *Nocardia* sp. (Bardhan et al. 2019). The choice of microbial host is generally dictated by several factors, including the source of carbon (such as cellulose, lignin, methane and carbon dioxide), the toxicity of the substrate (or any impurities in it) or the fuel itself and the processing conditions needed to transform the substrate and/or produce the fuel (Keasling et al. 2021). Carbon capture and storage (CCS) and carbon capture and utilisation (CCU) are technologies that capture emissions of CO_2 from point sources (e.g. industrial operations and power plants) for storage to prevent them from being released into the atmosphere (Markewitz et al. 2012). Bacteria have many benefits over these species, including a considerably quicker rapid growth rate and life cycle, the capacity to exist in a culture of high density and the ability to be genetically engineered more readily (Bharti et al. 2014). With the help of microbes, CO_2 can be converted to bio-based products and used for the generation of energy. Some of these techniques are still at an early age and are yet to fully develop, the major obstacle being the production cost for this bioenergy using microbes. Crucial steps need to be taken to reduce the cost of production and commercialisation, to increase efficiency. If these are made affordable for general use, it will in turn help in dealing with the problems caused by the use of fossil fuels and will lead towards the goal of affordable and clean energy.

1.3.8 Microbes in Decent Work and Economic Growth (SDG 8)

Due to inadequate investment and a lack of development initiatives brought about by uneven progress around the globe, particularly in developing nations, there are significant issues with the lack of acceptable work prospects. Therefore, SDG 8 aims to reduce the unemployment rate by raising labour productivity and promote economic growth. However, due to the COVID-19 pandemic, the world has witnessed global economic fallout, mainly in the production sector, together with the loss of labour and capital stock (Hughes et al. 2021). As per the reports submitted by the UN in 2020, the rate of income in the developing countries reduced by 15%, and around 255 million people lost their jobs, out of which 54% were women, from all over the world due to the COVID-19 crisis. The COVID-19 pandemic has shifted the paradigm of the world towards a green transition, as observed by the UN in the year 2020. The adoption of 'green technologies' with the help of microbes could provide better global sustainability. To be precise, the microorganisms like *Favolaschia calocera*, *Sanghuangporus* spp., *Coprinus comatus* and *Aspergillus terreus* that are used for the production of antibiotics in the medicine industry allow for a sustainable and cost-effective supply of active pharmaceutical ingredients that benefit the healthcare and also generate employment (Khan 2019). Additionally, biogas can contribute positively to economic growth as it is known as green self-employment in which microorganisms such as *Saccharomyces cerevisiae*, *Clostridium phytofermentans*, *Ruminococcus* sp., *Ethanoigenens harbinense*, *Dunaliella tertiolecta* and *Nannochloropsis oculate* are involved (Khan et al. 2015). In addition, the introduction of biogas as a domestic fuel could positively affect the quality of rural life, as mentioned by Patinvoh et al. (2019), the development of such efficient processing technologies can be another source of employment to the people. Comparably, in emerging nations, small-scale vendors of fermented goods—such as tempeh (Indonesia), pitha, dosa, appam (India), fufu (Africa) and other regional foods—have created jobs that range from 'for family' to 'large-scale' enterprises (Okafor 1992). Timmis et al. (2017a, b) suggested microbial biotechnology is a key knowledge-based bio-business that has applications in almost all aspects of life and society and will be a major driver of enterprise and employment creation in the future and hence significantly contributes to SDG 8. For instance, an alternative for the use of chemically synthetised insecticides is the employment of bioinsecticides, which include bacteria, mainly *Bacillus thuringiensis*, and fungi which are non-toxic and highly specific for certain pests (Soberón et al. 2023). The utilisation of bioinsecticides would not only enhance the productivity of the crops but can also become a source of income to the farmers through their knowledge of the manufacturing process (Khan 2019; Debnath et al. 2022).

Furthermore, the post-pandemic SDG actions adopted a systematic framework of 'Classification–Coordination–Collaboration'. This approach provided a practicable way to achieve the goals of SDGs in various countries, comprising the necessary targets, and with the execution of the essential steps, the targets can also be reached. Therefore, strategies to accelerate the implementation of the SDGs should be prioritised over different periods of time (Fu et al. 2024). According to the International Labour Organization (ILO) report of 2018, the focus on recycling, reuse and repair has been emphasised, which would significantly create job opportunities. The report suggests that by 2030, approximately 45 million workers could be employed in the green technology sector, while an additional 50 million jobs could be generated in repair and remanufacturing services. Notably, investing in the recycling infrastructure and promoting circular economy practices not only create jobs but also recycling and reusing materials reduce the need for resource extraction and energy-intensive manufacturing processes, thereby contributing to the process of sustainable development (Zhao et al. 2022).

1.3.9 Microbes in Industry Infrastructure for Sustainable Development Goals (SDG 9)

Sustainable industrial expansion and technological advancement are crucial for long-term global development (Akinsemolu 2018). Traditionally, microbes are employed in different industrial processes such as the production of antibiotics, biofertilisers, yogurt, etc. (Rehman et al. 2020; Marhons et al. 2023; Isabel et al. 2024). Microbes have a critical role in regulating numerous biogeochemical cycles (Das et al. 2025). This metabolic versatility may be potentially used to minimise the influence of pollutants, climatic conditions and other environmental factors, which are included in the SDGs (Kaksonen 2018; Jansson 2023). Industrialisation and urbanisation have produced hazardous wastewater, which causes adverse effects on humans and the surrounding environment if disposed off with inadequate treatment (Sharma et al. 2021a, b, c; Lin et al. 2022). Wastewater treatment is an expensive and energy-intensive procedure, due not just to the high concentrations of contaminants but also to the sheer volume involved (Silva 2023). Microbes are resistant to the harmful effects of metals and contribute considerably to resource recycling and energy flow within ecosystems (Zhou et al. 2023). Bacterial genera such as *Micrococcus*, *Microbacterium*, *Arthrobacter*, *Shinella*, *Serratia, Bacillus,* etc. were reported to be associated with wastewater management. Microbial processes like trickling filters, activated sludge and anaerobic digestion are frequently used in industrial wastewater treatment plants, which metabolise the organic materials, nitrogen and phosphorus found in wastewater, cleansing it before disposal through several processes such as oxidation–reduction, biomineralisation, biosorption processes, bioaccumulation, bioleaching, biosurfactant technology, bioprecipitation and biovolatilisation (Sharma et al. 2021a, b, c; Singh et al. 2022). Recently, MFCs have been a promising renewable energy option because they use microorganisms to directly convert chemical energy from organic waste into electrical energy (Erenler et al. 2024). MFCs have the potential to produce sustainable energy from organic sources while simultaneously offering excellent wastewater treatment (Jalili et al. 2024).

Hydrogen stands out as an environmentally friendly and sustainable choice for future biofuel use due to its absence of carbon emissions, non-toxic nature and high energy conversion efficiency (Sharma et al. 2021a, b, c). Biohydrogen generation from microorganisms, specifically photosynthetic bacteria, cyanobacteria, fermentative bacteria and green microalgae, via biological metabolism, has recently gained substantial interest due to its sustainability and ecologically benign potential (Woon et al. 2023). Several parameters such as pH, light intensity, nutrients and temperature could enhance the microbial production of biohydrogen (El Bari et al. 2022).

Additionally, microbes are utilised in the process of biodegradation to transform numerous harmful compounds, such as hydrocarbons, heavy metals and pesticides, into less hazardous substances (Kotoky and Pandey 2020; Malla et al. 2022). Microbes capable of breaking down hazardous substances have been isolated from soil, seawater, compost, crude oil-contaminated sites, activated sludge, etc. (Montazer et al. 2020). For example, bacteria isolated from crude oil-contaminated sites could degrade highly recalcitrant asphaltene (Das et al. 2023a, Das et al. 2024b). These microbes can be introduced into deficient environments that lack beneficial strains and transform different toxic elements from the surroundings (Maslennikova et al. 2022; Miglani et al. 2022). However, specific strains will thrive and propagate only under conditions favourable to their growth. This application is especially crucial for industrial organisations that deal with chemical contamination, hazardous waste and oil spills.

Machine learning (ML) has caused a revolution in the modern era. It has the potential qualities to improve climate change models (Chen et al. 2023). An exciting possibility is to use machine learning (ML) to gather microbiome data from many habitats. If ML identifies an imbalance in a certain category of microbial functional groups, inoculation of microbes may be considered to reintroduce the beneficial bacteria and their important functions into a specific environment (Jansson 2023). Furthermore, microorganisms offer opportunities for small-scale industries, job creation and economic development (Akinsemolu 2018). Some more examples of effective research and industrialisation are tofu, butanol, methanol, citric acid and glycerol and fermented food production (Rossi et al. 2016; Clomburg et al. 2017). Microbial products such as antibiotics, ethanol, enzymes (e.g. beta-glucosidase), amino acids and vitamins have proven successful on a wide scale (Singh et al. 2017). Microbes such as *Rhizobium* and *Pseudomonas* are utilised as biofertilisers to increase agricultural and industrial productivity (Seenivasagan and Babalola 2021); genetically modified *E. coli* is utilised to produce an enormous amount of insulin (Baeshen et al. 2014); and *S. cerevisiae* for fermented beverages (Walker and Stewart 2016). In broad terms, microorganisms act as adaptable agents of transformation in industrial infrastructure, providing long-term solutions to environmental complications, increasing resource efficiency and pushing innovation across a wide range of industries. Their incorporation into industrial processes emphasises the significance of using nature's microbial diversity to develop robust and environmentally friendly infrastructure systems for a sustainable future.

1.3.10 Microbes in Reducing Inequalities (SDG 10)

Microbiologists and microbiological organisations can help reduce income disparities by using microbial products from different regions. For instance, Yakult is a widely recognised brand that generates substantial revenue in developed nations. Thus, a strong collaboration between experts in microbiology and bio-investors from both developed and developing nations has the potential to address income disparities (Fagunwa and Olanbiwoninu 2020). Moreover, globalisation is crucial for ensuring a uniform healthcare system for all individuals worldwide. Enhancing global trade can enhance the economy of nations that are underdeveloped or developing and offer a source of income for local residents. Microorganisms are ubiquitous and hold pivotal significance in diverse societal contexts such as agriculture, industry, environmental preservation, confidentiality, healthcare and the management of biological resources. The innovative idea of "microbes and social equity" is based on the understanding that we depend on the microorganisms residing in or on us and our surroundings to offer essential ecosystem services for development and waste management (Mittelman and Hanaway 2012).

Our microbiome, the combined community of microorganisms, is susceptible to influences from various aspects of our lifestyle, encompassing factors like diet, hygiene, health condition and others. However, numerous individuals face limitations imposed by social, economic, medical or political constraints, constraining the choices available to them that might affect their health. Numerous clinical conditions or diseases in humans have been identified as interconnected with the condition of the human microbiome (Aggarwal et al. 2022). It is recognised that widespread social inequities can significantly influence the occurrence, severity and fatality of certain diseases or conditions. When barriers in obtaining a nourishing diet and healthcare are imposed by social disparities, these inequalities can also impact the human microbiome, potentially leading to diminished or impaired microbial functionality. The vitality of robust ecosystems depends on microbial communities for the provision of clean water, fresh air and nourishing foods (Amato et al. 2021). Likewise, the well-being of human microbiomes flourishes when these essential resources are conserved. Ensuring access to these resources is a fundamental human right. Regrettably, both these resources and beneficial microbiomes are not distributed evenly. Promoting the health of microbiomes, whether in the human gut or soil is required for addressing social disparities (Ogunrinola et al. 2020).

1.3.11 Microorganisms in Sustainable Cities and Communities' Development (SDG 11)

The rapid urbanisation of recent years has had a large negative impact on the surroundings. In 2015, about 54% of the human population lived in urban regions, and this expansion is predicted as the population grows (United Nations reports 2022). The increasing need for infrastructure and services in urban areas has led to various environmental challenges, such as air pollution and uncontrolled urban sprawl. These issues have increased vulnerability to natural hazards (Escap 2018). Furthermore, in the last decade, urban areas expanded faster, reflecting the environmental consequences of urbanisation and attractive urban development. Because of their roles in waste management, soil fertility, pollution prevention, renewable energy, green resources and public health, bacteria make a significant contribution to the growth of sustainable cities and communities. An effective solid waste disposal system is crucial for resilient and sustainable society development, mitigating issues like pipe blockages, flooding and waterborne disease transmission. Microbes can be used to clean up contaminated brownfields in urban areas and former industrial sites (Rodríguez-Espinosa et al. 2021). Additionally, microorganisms are used in wastewater treatment facilities to purify wastewater and make it suitable for release into the surroundings by means of decreasing and removing organic contaminants, pathogens and other heavy metals (Saravanan et al. 2021). Thus, the usage of bioremediation in urban areas may improve the quality of soil, water and air due to its low price range and eco-friendly nature (Chen et al. 2015). Advances in the biosynthesis of bioplastic metabolic pathways in microbes may offer an alternative to conventional plastic pollution, aligning with the circular bioeconomy (Sarah Marshall 2022). Microbial bioconversion is a sustainable strategy for effective solid waste management, converting waste into valuable resources. German company Weltec Biopower was awarded a 2020 contract to build a 7-megawatt food waste biogas plant in Gyeonggi Province, aiming to increase Korea's renewable energy generation in line with the Green New Deal. This project also includes a hydrogen fuel station in Chungju (Sarah Marshall 2022). Microorganisms are necessary for the successful functioning of green infrastructure, which includes elements such as green roofing, living walls, rain forests and urban wetlands. Recently, *Bacillus subtilis* has been used as a substitute for water in the concrete mixture and as a surface treatment gel to increase the weak tensile and flexural strengths of concrete (Karimi and Mostofinejad 2020). These microbial communities contribute to the overall biodiversity and ecological balance of urban ecosystems and other wildlife (Mayrand and Clergeau 2018). Architects and scientists are integrating microbiome study and bio-informed design to improve health safety, building resilience and sustainability. This involves reducing the biodeterioration of materials, soil–plant frameworks, waste collection and appropriate treatment systems (Fagunwa and Olanbiwoninu 2020; Yang et al. 2023).

1.3.12 Microbes in Responsible Consumption and Production (SDG 12)

The full integration of environmentally friendly production and consumption frameworks is necessary for sectoral and national strategies. This involves practising green practices, acting responsibly as a customer, using sustainability and services and appropriate handling of hazardous wastes and chemical substances (United Nation reports 2019). By 2050, 9.7 billion people will require natural resources for life on this planet. Less than 3% of Earth's freshwater sources are suitable for drinking, and 1 billion lack access. Thus, SDG 12 emphasises maximising natural resource utilisation and transitioning to alternative energy sources (Hoballah and Averous 2015; Arora and Mishra 2023). Microorganisms have a long history of contribution to achieve these goals by synthesising pharmaceutical compounds, reducing waste, promoting sustainable agriculture and deriving innovative biotechnological solutions. In addition, the implementation of responsible and sustainable production technologies that involve the utilisation of microorganisms represents a significant approach towards attaining SDG 12. Bacteria *Komagataeibacter xylinus* offers environmentally conscious production through sustainable microbial facilities, including cellulases, ethanol and bacterial cellulose. Furthermore, bacterial cellulose can be utilised for the production of Li-ion batteries and biofuel to provide a cleaner and sustainable energy resource. Microorganisms, like bacteria and mycorrhizal fungi, significantly impact soil biology, enhancing crop yields, reducing erosion, providing nutrients, fixing nitrogen and supporting sustainable agriculture practices without compromising ecological health (Kumar et al. 2022). Furthermore, one of the most critical aspects of achieving SDG 12 is ethical and clean energy technologies using microorganisms, such as aerobic digestion, for renewable energy production and biogas production (Tauseef et al. 2013). Biogas made from methane may be stored and used as a renewable fuel for heating, cooking, transportation and generating energy (Scarlat et al. 2018; Pham 2018). Moreover, the utilisation of microalgae cultivated in ponds or photobioreactors has been explored for the purpose of biomass and biofuel generation (Zhu et al. 2014). Microorganism-derived nanoparticles are used in various industries, including antimicrobials, agriculture, medicine, environment, cosmetics, waste control, electronics and building materials. Understanding the microbial role and addressing their responsibilities are crucial for a sustainable future (Mohanty and Swain 2018). Public awareness is crucial for promoting sustainable consumption and understanding microbes' role in environmental preservation. Practices like personal hygiene, cleanliness, self-medicating, waste segregation and reducing fossil fuel reliance benefit individuals and citizens, fostering societal responsibility.

1.3.13 Microbes in Climate Action (SDG 13)

The mitigation of the effects of global environmental changes resulting from rising greenhouse gas (GHG) emissions is one of the main issues of the twenty-first century. Methane (CH_4), nitrous oxide (N_2O) and other remaining amounts of trace gases account for 24, 10 and 3% of all greenhouse gas emissions, respectively. Carbon dioxide (CO_2) holds a major contribution to greenhouse gas emissions, accounting for 63% of total emissions (Ravindranath et al. 2006; IPCC 2007). According to Crowther et al. (2015), the amount of atmospheric carbon is seven to nine times greater that is produced by soil microbial decomposition than the amount released annually due to anthropogenic emissions globally. In reality, microbes emit 5.5×10^{12} kg (55 billion tons) of carbon dioxide in soil annually through the process of plant organic matter breakdown, which is eight times more than what humans release (Dutta and Dutta 2016). This has a significant impact on the carbon cycle and climate feedback, which is the foundation for Intergovernmental Panel's

Climate Change (IPCC) projections. As per the published reports of Crowther et al. (2015), soil biotic interactions are significant because they play a critical role in moderating soil microbial feedbacks to climate change, yet the breakdown of organic matter by microorganisms is extremely susceptible to changes in the climate. Compared to CO_2, the global emission of CH_4 is more directly influenced by microbes (Singh et al. 2010). Subsequently, microbes account for over 85% of the global methane emissions (Zimmerman and Labonte 2015). In this regard, methane emission from spongy bog soils warrants special attention as a huge number of microbial genes and proteins are involved in it. Methylcoenzyme M reductase, the enzyme which is responsible for converting carbon dioxide to methane, is a good example of this. Increased methane emissions from many of these local bacteria could be caused by warming trends (Perner et al. 2022). Approximately 60% of methane produced worldwide is consumed by microorganisms (Zimmerman and Labonte 2015). Methane-consuming microbes are actually found to occur in both the sea and on land, and also, they have the ability to remove atmospheric methane even at trace amounts (Zimmerman and Labonte 2015). Therefore, bacteria have a role in regulating climate change by reducing methane emissions. The natural prairie grasslands bacteria are especially worth mentioning in this context because they significantly contribute to the stabilisation of the carbon flux. These bacteria are impacted by modified precipitation patterns brought about by climate change. Thus, altering the storage of carbon in the soil could have a significant impact on the carbon balance in these ecosystems (Weiman 2015).

Aquatic environments are home to a wide variety of bacterial and fungal species that cycle carbon. Carbon conversion can occur anaerobically even in mostly oxygen-free environments, such as the deep mud of ponds, lakes and other water bodies. In the presence or absence of oxygen, both forms of conversion can occur. The process of algae participation is aerobic. Fermentation is the process by which bacteria cycle carbon molecules in anaerobic environments to produce energy. It is possible for other microbes to take part in the carbon cycle. For instance, green or purple sulfur bacteria can break down carbon compounds using the energy they obtain from a substance breakdown known as hydrogen sulfide. *Thiobacillus ferrooxidans* and other bacteria use the energy released when an electron is taken out of molecules containing iron to transform carbon. Only microbes can break down carbon anaerobically. Many different bacteria, including *Clostridium butyricum*, *Syntrophomonas* sp. and *Bacteroides succinogenes*, are involved in this process of degradation. The interspecies hydrogen transfer that results from this bacterial cooperation is ultimately what releases the majority of carbon dioxide and methane into the atmosphere (Abatenh et al. 2018) (Table 1.1).

1.3.14 Life Below Water (SDG 14)

The oceans, seas and other water bodies are the most important parts of ecosystems, whose quality is being degraded over time due to both natural and human activities. This raises a critical concern about how to mitigate these problems sustainably.

Table 1.1 Microorganisms directly involved in achieving sustainable development goals (SDGs)

Sustainable Development Goals (SDGs)	Objectives	Types of processes	Reported microorganism involved	Mechanism	References
SDG 1	*No Poverty* (End poverty in all its forms everywhere)	Livelihood opportunities	Lactic acid bacteria and yeast	Production and preservation of traditional food like Ogi, Garri, Pito, Iru and Vinegar	Obafemi et al. (2022)
		Promotion of good health	Probiotics	Probiotics can contribute to addressing poverty by improving health and reducing healthcare expenses	Lenoir-Wijnkoop et al. (2019)
SDG 2	*Zero Hunger* (End hunger, achieve food security and improved nutrition and promote sustainable agriculture)	Preservation of food for longer period of time	Pickles (*Leuconostoc mesenteroides, Pediococcus cerevisiae*); Cheese (*Penicillium roqueforti*); Yogurt (*L. delbruekii, Streptococcus thermophilus*); Bread (Saccharomyces cerevisiae)	Fermentation	Ghnimi and Guizani (2018, Chávez et al. (2023, Kok and Hutkins (2018), Lahue et al. (2020)
		Enhancement of soil fertility	*Rhizobium leguminosarum, Azospirillum* sp.	Enhance soil fertility by fixing atmospheric nitrogen to soil	Fukami et al. (2018), Pastor-Bueis et al. (2019)
		Biocontrol agents	Endophytic bacteria are mainly associated with it. *P. flourescens,* *B.subtilis,* *B. thuringiensis,* *Burkholderia anthina*	They counteract the pathogens by producing various lytic enzymes, antimicrobial compounds, endotoxins, siderophores and also by inducing systemic response	Milner et al. (2019), Saber et al. (2015), Villarreal-Delgado et al. (2018), de Los Santos-Villalobos et al. (2012)
		Reduce disease risk in livestock	*Lactobacillus, Enterococcus, Lactococcus, Streptococcus, Bifidobacteria*	These bacteria act as probiotics and thus improve the animal's microbiota and counter pathogenic microbes	Lambo et al. (2021), Kimoto-Nira et al. (2007), Kulkarni et al. (2022)

(continued)

Table 1.1 (continued)

Sustainable Development Goals (SDGs)	Objectives	Types of processes	Reported microorganism involved	Mechanism	References
SDG 3	*Good Health and Well-being* (Ensure healthy lives and promote well-being for all at all ages)	Development and restoration of healthy gut microbiome among individuals	Lactic acid bacteria, *Enterococci, Propionibacteria, Bifidobacterial, S. boulardii, Bacillus* spp.	They modulate intestinal inflammation, immune response, supress allergic response and secrete many extracellular enzymes	Gareau et al., 2010, Ohno et al. (2005), Zhang et al. (2016)
		Probiotics and prebiotics	*L. acidophilus, S. boulardii, Aspergillus oryzae, A. niger, S. carisbergensis*	Probiotics and prebiotics enhance host health by bolstering resistance to pathogenic bacteria colonisation and strengthening immune responses, resulting in the reduced presence of harmful pathogens	Yoha et al. (2021), Bull et al. (2013)
		Development and enhancement in production of antibiotics	Penicillin (*Penicillium notatum*), Chloramphenicol (*Streptomyces venezuelae*)	Antibiotic functions by antibiosis. The strains were developed by genetic engineering to enhance the production of antibiotics industrially	Sawant and Vamkudoth (2022), Zong et al. (2022)
SDG 4	*Quality Education* (Ensure inclusive and equitable quality education and promote lifelong learning opportunities for all)	Involvement of microbiology in school syllabus	*Agrobacterium, Rhizobium*	Educating students about microbes, their impact on health, hygiene, agriculture, biodiversity and how to apply this knowledge can be done	Chen and Feng (2021)

SDG 5	*Gender Equality* (Achieve gender equality and empower all women and girls)	Sustainable agriculture practice training	*Rhizobium* sp., *Pararhizobium* sp. Etc.	Empowering women with knowledge and access to sustainable agricultural practices that harness beneficial microbes can contribute to increased agricultural productivity and economic empowerment	Debnath et al. (2023)
		Promotion of women's health through microbial innovations in reproductive healthcare	Probiotics, *Lactobacillus*	Positively impact women's reproductive health by maintaining a balanced microbiome	Reid (2017)
SDG 6	*Clean Water and Sanitation* (Ensure availability and sustainable management of water and sanitation for all)	Wastewater treatment	*Desulfovibrio, Flavobacterium, Geotrichum*	They degrade recalcitrant xenobiotic and organic matter into non-toxic simple compounds	Diao et al. (2023), Li et al. (2022), Kamilari et al. (2023)
		Indicator organism	*Escherichia coli*, *Cryptosporidium, Giardia*	The presence of these organisms indicates poor water quality and indicates the needs for treatments	Ligda et al. (2020), Khan and Gupta (2020)
		Composting toilets	Moisture activated aerobic microorganisms and dried enzymes	Decompose organic matters of human waste and converts it to compost-like material	Oishi et al. (2023)

(continued)

Table 1.1 (continued)

Sustainable Development Goals (SDGs)	Objectives	Types of processes	Reported microorganism involved	Mechanism	References
SDG 7	*Affordable and Clean Energy* (Ensure access to affordable, reliable, sustainable and modern energy for all)	Bioethanol	*Clostridium ljungdahlii*	Complex starch was hydrolysed first and then converted to ethanol	Richter et al. (2016)
		Biogas	*Methanobacterium* spp.	Digestion of organic matter in the absence of oxygen	Li et al. (2021)
		Biodiesel	*Rhodococcus opacus, Cryptococcus curvatus*	Facilitate production by transesterification and esterification	Palmer and Brigham (2016), Leiva-Candia et al. (2014)
		Microbial fuel cell	*G. sulfurreducens, G. metallireducens P. aregunosa, Shewanella oneidensis*	Transfer electrons from organic matter to an electrode	Cusick et al. (2010), Dai et al. (2021)
		Microbe-enhanced oil recovery	*Bacillus* sp. *Pseudomonas* sp	Release crude oil from trapped zones of oil reservoir by reducing surface tension, wettability alteration and degradation of heavy oil	Das et al. (2025), Patel et al. (2015)
SDG 8	*Decent Work and Economic Growth* (Promote sustained, inclusive and sustainable economic growth, full and productive employment and decent work for all)	Industrial processes and employment generation	*Pseudomonas* sp., *Rhobium* sp., *Bacillus* sp.,	Biogas stove manufacturer providing jobs in Sub-Sahara Africa Denmark's biogas production policy plays a vital role in advancing the country's bioenergy strategy and efforts to reduce carbon emissions	Tumwesige et al. (2014), Raven and Gregersen (2007)
		Innovation and research	*Pseudomonas* sp., *Rhobium* sp., *Bacillus* sp.,	Penicillium reported to produce the first antibiotic Development of vaccines Designing of drugs	Akinsemolu (2018), Plotkin and Plotkin (2011)

SDG 9	*Industry, Innovation and Infrastructure* (Build resilient infrastructure, promote inclusive and sustainable industrialisation and foster innovation)	Biofertilisers	Nitragin *Gold (Rhizobium)*, Nodulator (*Bradyrhizobium japonicum*), Legume Fix (*Rhizobium* sp., *Bradyrhizobium japonicum*),	These biofertilisers fix nitrogen	García-Fraile et al. (2017), Aloo et al. (2022)
		Microbial products	Vitamin B12 (*Propionibacterium shermanii*), Biotin (*Serratia marcescens, B. sphaericus*), Riboflavin (*Ashbya gossypii, Bacillus* sp.)		Tanaka et al. (2017), Acevedo-Rocha et al. (2019), Averianova et al. (2020)
SDG 10	*Reduced Inequalities* (Reduce inequality within and among countries)	Health equity	*Streptococcus pyogenes* and heat-killed *Serratia marcescens* for cancer treatment	Microbial-related health issues, such as infectious diseases, and promoting access to microbial-based medicines, contributing to reducing health inequalities can be addressed	Sawant et al. (2020), Vargason and Anselmo (2018)
		Biodiversity and environmental justice	*Rhizobium*, etc.	Preserving microbial biodiversity through sustainable land management practices promotes environmental justice by ensuring equitable access to ecosystem services for all	Martin et al. (2013)

(continued)

Table 1.1 (continued)

Sustainable Development Goals (SDGs)	Objectives	Types of processes	Reported microorganism involved	Mechanism	References
SDG 11	*Sustainable Cities and Communities* (Make cities and human settlements inclusive, safe, resilient and sustainable)	Planning of urban cities by planting trees	*Azadirachta indica* plant with rhizobacteria *P. aeruginosa and Gordonia amicalis* facilitates degradation of crude oil	More trees can be planted on both polluted and non-polluted places as many trees are associated with phytoremediation	Bhuyan and Pandey (2022)
		Solid waste management	*Pseudomonas*, *Bacillus*, *Thermoactinomycetes*, *Microbispora* and *Actinobifida*	Proper sorting of biodegradable and non-biodegradable waste and its proper planning of reuse and disposal	Akinsemolu (2018)

SDG 12	*Responsible Consumption and Production* (Ensure sustainable consumption and production pattern)	Sustainable agriculture	Nitrogen-fixing bacteria and mycorrhizal fungi	They enhance nutrient availability, promote plant growth, reduce the need for synthetic fertilisers, suppress plant diseases and enhancing soil fertility, leading to more sustainable and productive agricultural systems.	Pathania et al. (2020)
		Sustainable consumption		Microorganisms are involved in the production of probiotics and fermented foods, which contribute to healthier and more sustainable dietary choices. They can promote better nutrition, reduce food waste and enhance food security	Rumpala (2011)
		Resource efficiency	Microbes used in industrial processes (*Lactobacillus* sp., *Bacillus* sp.)	Microbial processes can offer more sustainable alternatives to conventional production methods, reducing resource consumption and environmental impacts	Sharma et al. (2020)

(continued)

Table 1.1 (continued)

Sustainable Development Goals (SDGs)	Objectives	Types of processes	Reported microorganism involved	Mechanism	References
SDG 13	*Climate Action* (Take urgent action to combat climate change and its impacts)	Carbon cycling	Mainly by aerobic photoheterotrophic bacteria and cyanobacteria	They can capture CO2 through photosynthesis and convert it into organic compounds, thus mitigating greenhouse gas emissions and helping to regulate the carbon balance	Kolber et al. (2001, Sánchez-Baracaldo et al. (2022)
		Methane regulation	Methane-consuming archaea	Some microorganisms consume methane as a source of energy, reducing its release into the atmosphere and mitigating its impact on climate change	Nariya and Kalyuzhnaya (2019)
		Climate-resilient agriculture	*Pseudomonas* sp., *Enterobacter hormaechei*, *Agrobacterium* sp., *Ochrobactrum* sp.	Certain beneficial soil microorganisms can enhance plant tolerance to environmental stresses, such as drought and salinity, by improving nutrient uptake and enhancing plant growth	Niu et al. (2018), Mishra et al. (2017)

SDG 14	*Life Below Water* (Conserve and sustainably use the oceans, seas and marine resources for sustainable development)	Marine food web	Phytoplankton	They are primary producers and convert sunlight and nutrients into organic matter	Trombetta et al. (2020)
		Coral reef health	Symbiotic algae (Zooxanthellae)	They provide corals with essential nutrients and involve in calcification, which helps corals build their calcium carbonate skeletons	Titlyanov and Titlyanova (2020)
		Nutrient cycling	Diatom, *Trichodesmium*	They break down organic matter, releasing nutrients back into water	Pajares and Ramos (2019)
SDG 15	*Life on Land* (Protect, restore and promote sustainable use of terrestrial ecosystems, sustainably manage forests, combat desertification and halt and reverse land degradation and halt biodiversity loss)	Restoration of degraded ecosystems	Overall soil microbiome	They contribute to the process of ecological succession by facilitating soil formation, nutrient cycling and the establishment of pioneer plant species	Coban et al. (2022)
		Decomposition and detoxification	*Actinobacteria, Proteobacteria, Acidobacteria, Verrucomicrobia*	They break down complex organic compounds, returning them to the soil and completing the nutrient cycle	Purahong et al. (2015), Yarwood (2018)
		Disease suppression	*Pseudomonas* sp.	They protect plant health and reduce the need for chemical pesticides	Tao et al. (2020)

(continued)

Table 1.1 (continued)

Sustainable Development Goals (SDGs)	Objectives	Types of processes	Reported microorganism involved	Mechanism	References
SDG 16	*Peace, Justice and Strong institutions* (Promote peaceful and inclusive societies for sustainable development, provide access to justice for all and build effective, accountable and inclusive institutions at all levels)	Health and disease prevention	*E. coli* (indicator microorganism)	Monitoring microbial indicators in water and food helps ensure public health and prevent the spread of waterborne diseases	Gerba (2009)
		Bioremediation and environmental justice	*Pseudomonas* sp., *Mycobacterium* sp., *Bacillus* sp., *Acinetobacter* sp. Etc.	This is particularly relevant for addressing environmental injustices caused by pollution, supporting SDG 16's goal of building accountable and inclusive institutions that ensure justice for affected communities	Das et al. (2023b)
		Global collaboration (microbial research networks)		International collaboration in microbial research fosters cooperation between institutions and nations, promoting the exchange of knowledge and resources	
SDG 17	*Partnerships for the Goals* (Strengthen the means of implementation and revitalise the Global Partnership for Sustainable Development)	Research and innovation	Agriculturally important microbes	Microbes are critical in biotechnological research, contributing to innovations in medicine, agriculture and industry	
		Sharing microbial technologies		International collaborations in microbial research facilitate the transfer of technologies that can address various challenges	

SDG 14 focuses on "Life Below Water", aiming to conserve and sustainably use the oceans, seas and marine resources. Microorganisms play a crucial role in supporting and maintaining the health of marine ecosystems, contributing significantly to the achievement of SDG 14 (Akinsemolu 2018). Microscopic algae, such as diatoms and dinoflagellates, are crucial for marine ecosystems as they contribute to primary production through photosynthesis (Larkum et al. 2020). Bacteria in the water column help decompose organic matter, recycling nutrients essential for the growth of phytoplankton. This microbial activity maintains the balance of nutrients and supports the marine food web (Boyd and Boyd 2020). Coral reefs, vital for marine biodiversity, rely on a symbiotic relationship between corals and microorganisms, particularly dinoflagellates known as zooxanthellae. These photosynthetic organisms provide energy to corals, supporting their growth and resilience. Bacteria associated with corals contribute to nutrient cycling, disease resistance and overall coral health (Nelson et al. 2023). Understanding and preserving the coral microbiome are essential for the conservation of coral reef ecosystems (Marangon et al. 2021). For example, in the microbiome of coral reef demosponges, the abundant microbial community includes Proteobacteria, *Chloroflexota, Actinobacteriota, Acidobacteriota, Gemmatimonadota, Latescibacterota, Cyanobacteria, Bacteroidota, Poribacteria, Dadabacteria, Nitrospirota, Planctomycetota, Bdellovibrionota, Patescibacteria, Spirochaetota, Nitrospinota, Myxococcota, Entotheonella* and *Nitrososphaeria* (Robbins et al. 2021). The bacteria belonging to these communities are widely reported in maintaining coral reel health, nutrient cycling and decomposition of organic pollutants (Oyetibo et al. 2021). A study reported the *Staphylococcus* profile in Ganga River, India, which showed that 90% of *Staphylococcus* are resistant to several antibiotics (Sood et al. 2014).

Annually, a range of 1.5 to ten million tons of petroleum hydrocarbons is released into seawater, leading to alterations in the physical and chemical attributes as well as the viscosity of seawater, and these changes pose a threat to the marine environment (Hazaimeh and Ahmed 2021). In the past decade alone, the global spillage exceeded one billion gallons of oil, with an annual influx of six million tonnes into the oceans. For example, prominent oil spills like the 1989 Exxon Valdez incident and the more recent British Petroleum Deepwater Horizon (2010), East China Sea (2018) and Southeast Coast of Mauritius (2020) spills have gained worldwide attention owing to their disastrous consequences (Abdul-Hamid et al. 2013; Hazaimeh and Ahmed 2021). Microbes play an important role in breaking down pollutants in coastal areas and estuaries. Bacteria and other microorganisms can be harnessed in bioremediation processes to treat wastewater before it enters the oceans, helping to mitigate the impact of human activities on marine ecosystems (Dai et al. 2023). For example, microorganisms isolated from the marine environment, including *Exiguobacterium* sp., *Haererehalobacter* sp., *Photobacterium* sp., *Ruegeria* sp., *Oceanobacillus* sp., *Nesiotobacter* sp., *Acinetobacter* sp. and *Enterobacter* sp., exhibit the ability to thrive across various temperatures (25–40 °C), withstand a broad pH range (6–10), endure high NaCl concentrations (10%) and have different metabolic pathways (Kumar et al. 2019). Microorganisms in the rhizosphere of mangrove trees contribute to nutrient cycling and enhance the trees' ability to

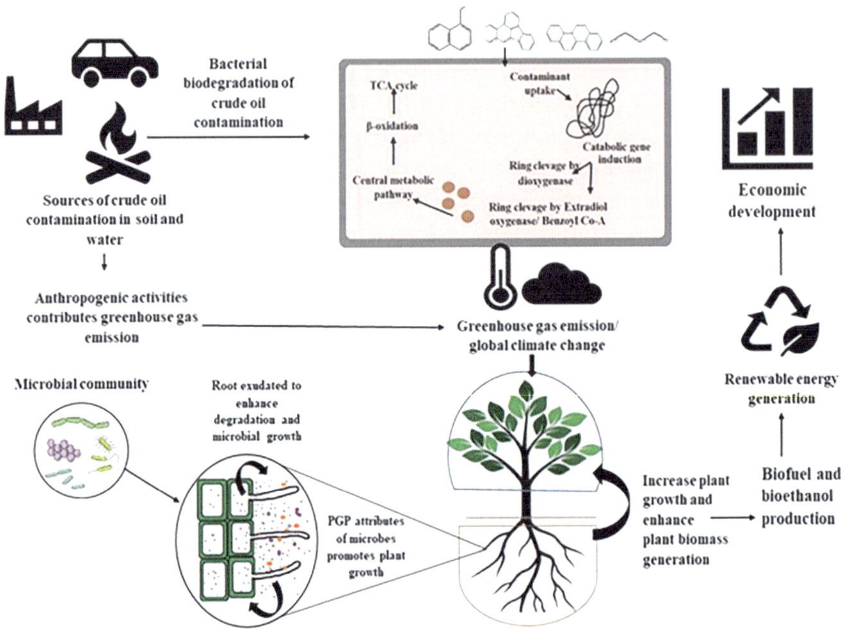

Fig. 1.2 Role of microbes in the degradation of contaminants and enhancement of plant growth

tolerate saline conditions. This microbial interaction is vital for the health and sustainability of mangrove ecosystems (Helfer and Hassenrück 2021). *Pleurocapsa*, *Marinomonas*, *Tunicatimonas*, *Erythrobacter*, *Altererythrobacte*, *Halomonas*, *Lewinella* and *Rubrivirga* reported from rhizospheres' microbiome of *Avicenia alba* and *Sonneratia alba* are involved in nutrient cycling and also promote the survivability of hosts in harsh environments (Wainwright et al. 2023). Common microbiome of mangrove rhizosphere includes Proteobacteria, Firmicutes and Bacteroidetes where sulphate reducers and methanogens are the highest (Alzubaidy et al. 2016). Thus, for sustainable ocean management, it is essential to comprehend and utilise the contributions that microbes make to life below the surface of the ocean. We can improve the resilience and health of marine ecosystems by incorporating microbial considerations into conservation and management measures, which will ultimately help achieve SDG 14 (Fig. 1.2).

1.3.15 Life on Land (SDG 15)

The achievement of SDG 15 is crucial for addressing urgent environmental problems related to terrestrial ecosystems. The goal focuses on biodiversity conservation, combating desertification, halting deforestation, promoting sustainable land use and mitigating climate change. By emphasising sustainable practices in agriculture, forestry and land management, SDG 15 aims to ensure the resilience of ecosystems, maintain vital ecosystem services and preserve biodiversity (Krauss 2022). Microorganisms play an important role in supporting life on land through various ecological processes. Microbes are the crucial component for sustainable agriculture and land use (Arora 2023). Mycorrhizal fungi form symbiotic relationships with plant roots, which enhance nutrient uptake by plants, especially phosphorus and nitrogen. This contributes to soil fertility, improves plant health and supports sustainable agriculture (Devi et al. 2021). Certain bacteria, like rhizobia, forming nodules on the roots of leguminous plants, fix atmospheric nitrogen, which can be utilised by plants (Lindström and Mousavi 2020). This biological nitrogen fixation is crucial for maintaining soil fertility and reducing the need for synthetic fertilisers. Some common microorganisms involved in the commercial production of biofertilisers are *Rhizobium* sp., *Azotobacter* sp., *Advenella mimigardefordensis*, *Bacillus amyloliquefaciens*, *Pseudomonas aeruginosa,* etc. (Ibáñez et al. 2023). Some bacteria in the rhizosphere promote plant growth by producing growth hormones, solubilising nutrients and protecting plants from pathogens (Basu et al. 2021). Certain bacterial species such as *Stenotrophomonas maltophilia* act as plant probiotic and are also involved in the composting of agricultural residues, which thereby facilitate land restoration along with plant growth promotion (Nevita et al. 2018). Crop rotation is a crucial practice for the proper utilisation of agricultural lands. However, nutrient loss occurs in the process, leaving the land barren in alternate seasons. For instance, applying *Pararhizobium giardinii* and *Ochrobactrum* sp. treatments to lands used for rice fallow cultivation can boost yields and enhance the abundance of plant growth-promoting rhizobacteria (PGPR) in the soil (Debnath et al. 2023). These microbes contribute to sustainable agriculture practices and enhance crop yields.

Microorganisms play a key role in breaking down dead plant and animal matter. Bacteria and fungi decompose organic materials and return essential nutrients to the soil and facilitate the recycling of organic matter (Raza et al. 2023). Both bacteria and fungi have the ability to break down hydrocarbons found in oil and other pollutants, can be harnessed in bioremediation efforts to clean up contaminated soils and restore ecosystems affected by oil spills (Kotoky et al. 2018). Some common bacterial examples are *Pseudomonas* sp., *Mycobacterium* sp., *Bacillus* sp., *Acinetobacter* sp. etc. (Das et al. 2023a, b, c); and certain fungi involved are *Aspergillus* sp., *Penicillium* sp., *Neurospora* sp., *Phanerochaete* sp., etc. (Li et al. 2020). Fungi from *Mycenaceae* and *Marasmiaceae* families of *Basidiomycota,* and *Rhytismataceae* and *Xylariaceae* families of *Ascomycotaare,* are reported to be involved in breaking down leaf litter and other organic materials in forest ecosystems (Osono 2020). This decomposition process influences nutrient cycling and supports the overall health and biodiversity of forests. In anaerobic conditions, certain microbial species, such as methanogenic archaea (*Methanobacterium* sp., *Methanoculleus* sp., *Methanobrevibacter* sp. etc.), play a role in the breakdown of organic waste in landfills, producing methane gas. While methane is a potent greenhouse gas, its capture and utilisation can contribute to sustainable waste management practices (Das et al.

2023b, c). Understanding and harnessing the microbial contributions to life on land are essential for sustainable land management and biodiversity conservation. Integrating microbial considerations into land-use planning and conservation strategies can enhance the resilience and health of terrestrial ecosystems, ultimately contributing to the achievement of SDG 15.

1.3.16 Role of Microbes in Peace, Justice and Strong Institutions (SDG 16)

SDG 16 aims to advance peaceful, inclusive societies where institutions are powerful and effective and maintain justice (United Nations 2019). Since this objective mostly requires legislation and administration, microbes do not directly contribute to its achievement. The phenomenon of bioterrorism, which involves the deliberate utilisation of live organisms such as diseases or their byproducts, such as poisons, with the intention of causing harm to human populations, has emerged as a significant global concern. This concern is rooted in the limited level of tolerance exhibited by nations towards one another. Additionally, several bacteria that are known for spreading epidemics of fulminant sickness, such *Francisella tularensis, Streptococcus pyogenes, Bacillus anthracis* and *Neisseria meningitidis*, are regarded as being extremely harmful biological components (Casadevall and Pirofski 2004). However, through microbial forensics and counter-bioterrorism, microbiology might indirectly contribute to this SDG goal (Keim 2003). Identification of bacteria at various phases of a body's decomposition is made possible by this microbial clock (Castro and De Ungria 2022). However, microbiological forensics is not yet a particularly well-established field; thus, study and education are needed to prepare the officials to use these techniques to maintain biological warfare and peace. Poverty, poor food availability and illiteracy directly impact children's growth and development. According to Chilton and Rose (2009), food insecurity still affects over 11% of American households, which is identified as a significant determinant of children's mental as well as physical health. As was mentioned previously, microorganisms can offer small-scale establishments a way to generate income. Having a secure income source can make individuals feel pleased, and young brains are especially susceptible to turning to violence (Fagunwa and Olanbiwoninu 2020). Microbes can aid in achieving SDG 16 and ultimately aid in the development of society and the predominance of peace and justice by preventing bioterrorism, supplying nutrition, improving environmental conditions, implementing green technology and enhancing national and international infrastructure (Mohamad Taghvaee et al. 2023).

1.3.17 Role of Microbes in Global Partnership for Sustainable Development (SDG 17)

Microbes promote sustainable technologies in developing countries, strengthening science, technology and innovation capacity through knowledge exchange and collaboration, achieving SDG 17 goals through data sharing, data evacuation and microbial resource sharing (Pearce et al. 2020). It can also build or re-energise international partnerships for sustainable development through the use of microbial technology (Timmis et al. 2017a, b), encourage and develop productive relationships between the state, corporate sector and civil society, based on their resources and expertise. Agriculture, medical research, biofuels and fermented foods are the most crucial areas that demand rapid attention on the use of microorganisms. More open commerce, flexible policies and research exchanges are essential for all nations (Olabi et al. 2023). Global partnerships are essential for disease surveillance and control, as well as for responding to infectious disease outbreaks caused by microbial pathogens. The antimicrobial efficacy has to be acknowledged as a fundamentally significant global public good and controlled appropriately in order to prevent an AMR situation (Jasovský et al. 2016). Through globalisation and the removal of geographical obstacles, the advantages of microorganisms and microbial technologies can be made accessible to the general public. Political will is needed to support technology implementation and capacity development for low-income communities and high-income nations. Social justice and equality principles should be reinforced through green technology. Organisations like the North American Free Trade Agreement, European Union, South African Development Committee, Association of Southeast Nations, Arab Organisation for Agricultural Development and Asia-Pacific Economic Cooperation are working towards progress (DaSilva 1998).

1.4 Recommendations

Sustainable goals require collaboration, education and cooperation among microbiologists and colleagues to achieve desired results.

1. The SDG framework should emphasise the importance of microbes in enhancing the living organism well-being, providing essential nutrients, preventing diseases and performing ecological functions. It should encourage collaboration between government, industry, civil society and academia to tackle interconnected challenges in sustainable development.
2. Establishing policies for social and environmental justice requires establishing foundations and requirements for microbial biotechnology. This strategy enhances industrial infrastructure, employment rates and GDP by promoting microorganisms as biofertilisers, improving crop output and addressing soil quality degradation.

3. Data monitoring and accounting should be strengthened for better policy decisions, measuring the social–environmental impact and promoting sustainable development through research networks, training programs and knowledge-sharing mechanisms.
4. Researchers should prioritise microbial applications in medicine, electricity generation, biofuel production, bio-detector, fermentation, agriculture and biocontrol, fostering collaboration among biologists, environmental scientists, engineers and social scientists.
5. The framework should explore innovation, promote partnerships, mobilise microbial resources for sustainable development, raise awareness, empower stakeholders and promote global microbial culture-sharing. Budapest's International Microorganism Deposit System should be improved for the reliable commercialisation of microbial cultures and biological materials.
6. Microbial diversity should be prioritised for climate resilience and adaptations, promoting renewable energy, sustainable land and water management and infrastructure development.

1.5 Conclusions

In order to achieve a sustainable ecosystem, various microorganisms play a crucial role in mitigating global environmental issues and promoting the development of green production technologies. Additionally, they are well recognised as an exceptional source of renewable energy. One significant achievement with the potential to catalyse the 'microbial revolution' involves the development of technology capable of employing microbial manufacturing methods. This advancement is noteworthy, as microorganisms possess the ability to mitigate the transmission of infectious diseases, safeguard food security, conserve biodiversity in terrestrial and marine ecosystems, harness renewable energy sources such as biofuels for commercial purposes and achieve sustainable outcomes. Microorganisms exhibit a notable capacity for swift proliferation and possess the advantage of being economically feasible to cultivate, rendering them a promising candidate as a sustainable food resource that has the potential to make substantial contributions towards alleviating issues of food scarcity and socioeconomic deprivation. These organisms possess the ability to produce significant amounts of bioactive compounds, therefore playing a crucial role in enhancing human well-being. In addition, microorganisms possess the capacity to contribute to the mitigation of pollution. The use of these technologies has the potential to not only generate employment opportunities for economically disadvantaged populations on a worldwide scale but also enhance productivity, thereby fostering economic growth and wealth. It is imperative to prioritise the enhancement of domestic policies and the establishment of international collaborations on a worldwide level. Greater emphasis should be placed on implementing rational strategies and fostering collaborative alliances in order to ultimately achieve sustainable outcomes, adopt cleaner manufacturing techniques and cultivate a robust and sustainable ecosystem.

Author Contributions *Draft preparation:* writing—original draft preparation [ND], formal analysis and investigation: [AD], [DM] [BB] [SD] [RT] and [RC]. *Conceptualisation:* [PP], writing—review and editing: [PP], supervision: [PP].

Funding and Acknowledgements ND, BB and PiyushP gratefully acknowledge the financial support received from Department of Biotechnology, Ministry of Science and Technology, Government of India. AD and DM are grateful to DST, Government of India, to receive INSPIRE fellowship.

Conflict of Interest There are no conflicts of interests or competing interests with respect to this manuscript.

References

Abatenh E, Gizaw B, Tsegaye Z, Tefera G (2018) Microbial function on climate change – a review. Open J Environ Biol 3(1):001–007. https://doi.org/10.17352/ojeb.000008

Abdel-Latif HM, Yilmaz E, Dawood MA, Ringø E, Ahmadifar E, Yilmaz S (2022) Shrimp vibriosis and possible control measures using probiotics, postbiotics, prebiotics, and synbiotics: a review. Aquaculture 551:737951

Abdul-Hamid O, Agoawike A, Odulaja A (2013) The OPEC annual statistical bulletin. OPEC, Vienna

Acevedo-Rocha CG, Gronenberg LS, Mack M, Commichau FM, Genee HJ (2019) Microbial cell factories for the sustainable manufacturing of B vitamins. Curr Opin Biotechnol 56:18–29

Adesulu AT, Awojobi KO (2014) Enhancing sustainable development through indigenous fermented food products in Nigeria. Afr J Microbiol Res 8:1338e1343. https://doi.org/10.5897/AJMR2013.5439

Aggarwal N, Kitano S, Puah GRY, Kittelmann S, Hwang IY, Chang MW (2022) Microbiome and human health: current understanding, engineering, and enabling technologies. Chem Rev 123(1):31–72

Ahmad A, Arif A, Munir A (2024) CRISPR technology commercialization and biosafety. In: Global regulatory outlook for CRISPRized plants. Academic Press, pp 461–514

Ahmed J, Wong LP, Chua YP, Channa N, Mahar RB, Yasmin A, VanDerslice JA, Garn JV (2020) Quantitative microbial risk assessment of drinking water quality to predict the risk of waterborne diseases in primary-school children. Int J Environ Res Public Health 17:2774

Aka S, Konan G, Fokou G, Dje KM, Bonfoh B (2014) Review on African traditional cereal beverages. Am J Res Commun 2(5):103–153

Akinsemolu AA (2018) The role of microorganisms in achieving the sustainable development goals. J Clean Prod 182:139–155. https://doi.org/10.1016/j.jclepro.2018.02.081

Al-Ghussain L (2019) Global warming: review on driving forces and mitigation. Environ Prog Sustain Energy 38(1):13–21

Aloo BN, Tripathi V, Makumba BA, Mbega ER (2022) Plant growth-promoting rhizobacterial biofertilizers for crop production: the past, present, and future. Front Plant Sci 13:1002448

Alzubaidy H, Essack M, Malas TB, Bokhari A, Motwalli O, Kamanu FK, Jamhor SA, Mokhtar NA, Antunes A, Simões MF (2016) Rhizosphere microbiome metagenomics of gray mangroves (Avicennia marina) in the Red Sea. Gene 576(2):626–636

Amato KR, Arrieta MC, Azad MB, Bailey MT, Broussard JL, Bruggeling CE et al (2021) The human gut microbiome and health inequities. Proc Natl Acad Sci 118(25):e2017947118

Antar M, Lyu D, Nazari M, Shah A, Zhou X, Smith DL (2021) Biomass for a sustainable bioeconomy: an overview of world biomass production and utilization. Renew Sust Energ Rev 139:110691

Arora NK (2019) Impact of climate change on agriculture production and its sustainable solutions. Environ Sustain 2:95–96

Arora P (2023) Life below land: the need for a new sustainable development goal. Front Environ Sci 11:1215282. https://doi.org/10.3389/fenvs.2023.1215282

Arora NK, Mishra I (2023) Responsible consumption and production: a roadmap to sustainable development. Environ Sustain 6:1–6

Averianova LA, Balabanova LA, Son OM, Podvolotskaya AB, Tekutyeva LA (2020) Production of vitamin B2 (riboflavin) by microorganisms: an overview. Front Bioeng Biotechnol 8:570828

Azadi H, Ghazali S, Ghorbani M, Tan R, Witlox F (2023) Contribution of small-scale farmers to global food security: a meta-analysis. J Sci Food Agric 103(6):2715–2726

Bach LL, Ram A, Ijaz UZ, Evans TJ, Lindström J (2021) A longitudinal study of the human oropharynx microbiota over time reveals a common core and significant variations with self-reported disease. Front Microbiol 11:573969

Baeshen NA, Baeshen MN, Sheikh A, Bora RS, Ahmed MMM, Ramadan HA et al (2014) Cell factories for insulin production. Microb Cell Factories 13(1):1–9. https://doi.org/10.1186/s12934-014-0141-0

Barberan A, Hammer TJ, Madden AA, Fierer N (2016) Microbes should be central to ecological education and outreach. J Microbiol Biol Educ 17:23–28

Bardhan P, Gupta K, Mandal M (2019) Microbes as bio-resource for sustainable production of biofuels and other bioenergy products. In: New and future developments in microbial biotechnology and bioengineering. Elsevier, pp 205–222

Basu A, Prasad P, Das SN, Kalam S, Sayyed R, Reddy M, El Enshasy H (2021) Plant growth promoting rhizobacteria (PGPR) as green bioinoculants: recent developments, constraints, and prospects. Sustainability 13(3):1140

Bharti RK, Srivastava S, Thakur IS (2014) Production and characterization of biodiesel from carbon dioxide concentrating chemolithotrophic bacteria, Serratia sp. ISTD04. Bioresour Technol 153:189–197

Bhutto AW, Bazmi AA, Zahedi G (2011) Greener energy: issues and challenges for Pakistan—Biomass energy prospective. Renew Sust Energ Rev 15(6):3207–3219

Bhuyan B, Pandey P (2022) Remediation of petroleum hydrocarbon contaminated soil using hydrocarbonoclastic rhizobacteria, applied through Azadirachta indica rhizosphere. Int J Phytoremediation:1–11

Boyd CE, Boyd CE (2020) Microorganisms and water quality. In: Water quality: an introduction, pp 233–267

Bull M, Plummer S, Marchesi J, Mahenthiralingam E (2013) The life history of Lactobacillus acidophilus as a probiotic: a tale of revisionary taxonomy, misidentification and commercial success. FEMS Microbiol Lett 349:77–87

Camaille M, Fabre N, Clément C, Ait Barka E (2021) Advances in wheat physiology in response to drought and the role of plant growth promoting rhizobacteria to trigger drought tolerance. Microorganisms 9:687

Casadevall A, Pirofski LA (2004) The weapon potential of a microbe. Trends Microbiol 12(6):259–263

Castro AE, De Ungria MCA (2022) Methods used in microbial forensics and epidemiological investigations for stronger health systems. Forensic Sci Res 7(4):650–661

Chakraborty U, Chakraborty B, Chakraborty A, Dey P (2013) Water stress amelioration and plant growth promotion in wheat plants by osmotic stress tolerant bacteria. World J Microbiol Biotechnol 29:789–803

Chaurasia D, Boudh S, Singh JS, Chaturvedi P (2020) Microbes and microbial enzymes as a sustainable energy source for biofuel production. In: Microbial services in restoration ecology. Elsevier, pp 207–217

Chávez R, Vaca I, García-Estrada C (2023) Secondary metabolites produced by the blue-cheese ripening mold Penicillium roqueforti; biosynthesis and regulation mechanisms. J Fungi 9:459

Chen H, Feng Y (2021) Moving forward to the future: a review of microbial food safety education in China. Foodborne Pathog Dis 18(8):547–566

Chen M, Xu P, Zeng G, Yang C, Huang D, Zhang J (2015) Bioremediation of soils contaminated with polycyclic aromatic hydrocarbons, petroleum, pesticides, chlorophenols and heavy metals by composting: applications, microbes and future research needs. Biotechnol Adv 33(6):745–755

Chen L, Chen Z, Zhang Y, Liu Y, Osman AI, Farghali M et al (2023) Artificial intelligence-based solutions for climate change: a review. Environ Chem Lett:1–33. https://doi.org/10.1007/s10311-023-01617-y

Chilton M, Rose D (2009) A rights-based approach to food insecurity in the United States. American Journal of Public Health 99(7):1203–1211

Chugh B, Kamal-Eldin A (2020) Bioactive compounds produced by probiotics in food products. Curr Opin Food Sci 32:76–82

Ciampitti IA, de Borja Reis AF, Córdova SC, Castellano MJ, Archontoulis SV, Correndo AA, Antunes De Almeida LF, Moro Rosso LH (2021) Revisiting biological nitrogen fixation dynamics in soybeans. Front Plant Sci 12:727021

Clomburg JM, Crumbley AM, Gonzalez R (2017) Industrial biomanufacturing: the future of chemical production. Science 355(6320):aag0804. https://doi.org/10.1126/science.aag0804

Coban O, De Deyn GB, van der Ploeg M (2022) Soil microbiota as game-changers in restoration of degraded lands. Science 375:abe0725

Coonan EC, Kirkby CA, Kirkegaard JA, Amidy MR, Strong CL, Richardson AE (2020) Microorganisms and nutrient stoichiometry as mediators of soil organic matter dynamics. Nutr Cycl Agroecosyst 117:273–298

Crowther TW, Thomas SM, Maynard DS, Baldrian P, Covey K, Frey SD, van Diepen LTA, Bradford MA (2015) Biotic interactions mediate soil microbial feedbacks to climate change. Proc Natl Acad Sci 112(22):7033–7038

Cusick RD, Kiely PD, Logan BE (2010) A monetary comparison of energy recovered from microbial fuel cells and microbial electrolysis cells fed winery or domestic wastewaters. Int J Hydrog Energy 35:8855–8861

Da Silva TL, Gouveia L, Reis A (2014) Integrated microbial processes for biofuels and high value-added products: the way to improve the cost effectiveness of biofuel production. Appl Microbiol Biotechnol 98:1043–1053

Dadgostar P (2019) Antimicrobial resistance: implications and costs. Infect Drug Resist 12:3903–3910

Dai HN, Nguyen T-AD, Loan-Phung My LE, Van Tran M, Lan T-H, Wang C-T (2021) Power generation of Shewanella oneidensis MR-1 microbial fuel cells in bamboo fermentation effluent. Int J Hydrog Energy 46:16612–16621

Dai X, Lv J, Fu P, Guo S (2023) Microbial remediation of oil-contaminated shorelines: a review. Environ Sci Pollut Res 30(41):93491–93518

Das S, Das N, Choure K, Pandey P (2023a) Biodegradation of asphaltene by lipopeptide-biosurfactant producing hydrocarbonoclastic, crude oil degrading Bacillus spp. Bioresour Technol 382:129198. https://doi.org/10.1016/j.biortech.2023.129198

Das N, Das A, Das S, Bhatawadekar V, Pandey P, Choure K, Damare S, Pandey P (2023b) Petroleum hydrocarbon catabolic pathways as targets for metabolic engineering strategies for enhanced bioremediation of crude-oil-contaminated environments. Fermentation 9(2):196

Das A, Das S, Das N, Pandey P, Ingti B, Panchenko V, Bolshev V, Kovalev A, Pandey P (2023c) Advancements and innovations in harnessing microbial processes for enhanced biogas production from waste materials. Agriculture 13(9):1689

Das N, Maheshwari DK, Pandey P (2024a) Energy crop-based rhizoremediation and lignocellulosic biomass production as sustainable bioeconomy-driven solution for biofuel generation and waste mitigation. Progress Energy Combustion Sci 103:101161

Das A, Das N, Rajkumari J, Pandey P, Pandey P (2024b) Exploring the bioremediation potential of Bacillus spp. for sustainable mitigation of hydrocarbon contaminants. Environ Sust 7(2):135–156

Das A, Das N, Pandey P, Pandey P (2025) Microbial enhanced oil recovery: process perspectives, challenges, and advanced technologies for its efficient applications and feasibility. Archives Microbiol 207(5):1–43

DaSilva E (1998) Biotechnology: developing countries and globalization. World J Microbiol Biotechnol 14:463–486

de Araujo Avila GM, Gabardo G, Clock DC, de Lima Junior OS (2021) Use of efficient microorganisms in agriculture. Res Soc Dev 10(8):e40610817515

De Filippis F, Pasolli E, Ercolini D (2020) The food-gut axis: lactic acid bacteria and their link to food, the gut microbiome and human health. FEMS Microbiol Rev 44:454–489

de Los Santos-Villalobos S, Barrera-Galicia GC, Miranda-Salcedo MA, Peña-Cabriales JJ (2012) Burkholderia cepacia XXVI siderophore with biocontrol capacity against Colletotrichum gloeosporioides. World J Microbiol Biotechnol 28:2615–2623

Debnath S, Das N, Maheshwari DK, Pandey P (2022) Interactions of rhizobia rhizobia with non-leguminous plants: a molecular ecology perspective for enhanced plant growth. In: Nitrogen fixing bacteria: sustainable growth of non-legumes. Springer Nature Singapore, Singapore, pp 23–64

Debnath S, Das A, Maheshwari D, Pandey P (2023) Treatment with atypical rhizobia, pararhizobium giardinii and ochrobactrum sp. modulate the rhizospheric bacterial community, and enhances lens culinaris growth in fallow-soil. Microbiol Res 267:127255

Demirbas MF, Balat M, Balat H (2011) Biowastes-to-biofuels. Energy Convers Manag 52(4):1815–1828

Devi SH, Bhupenchandra I, Sinyorita S, Chongtham S, Devi EL (2021) Mycorrhizal fungi and sustainable agriculture. In: Nitrogen in agriculture-physiological, agricultural and ecological aspects. IntechOpen

Diao C, Ye W, Yan J, Hao T, Huang L, Chen Y, Long J, Xiao T, Zhang H (2023) Application of microbial sulfate-reduction process for sulfate-laden wastewater treatment: a review. J Water Process Eng 52:103537

Du Z, Li H, Gu T (2007) A state of the art review on microbial fuel cells: a promising technology for wastewater treatment and bioenergy. Biotechnol Adv 25(5):464–482

Dutta H, Dutta A (2016) The microbial aspect of climate change. Energy Ecol Environ 1:209–232

Dwivedi KA, Huang SJ, Wang CT, Kumar S (2022) Fundamental understanding of microbial fuel cell technology: recent development and challenges. Chemosphere 288:132446

El Bari H, Lahboubi N, Habchi S, Rachidi S, Bayssi O, Nabil N et al (2022) Biohydrogen production from fermentation of organic waste, storage and applications. Cleaner Waste Syst:100043. https://doi.org/10.1016/j.clwas.2022.100043

El-Saadony MT, Alagawany M, Patra AK, Kar I, Tiwari R, Dawood MA, Dhama K, Abdel-Latif HM (2021) The functionality of probiotics in aquaculture: an overview. Fish Shellfish Immunol 117:36–52

El-Sayed WS, Akhkha A, El-Naggar MY, Elbadry M (2014) In vitro antagonistic activity, plant growth promoting traits and phylogenetic affiliation of rhizobacteria associated with wild plants grown in arid soil. Front Microbiol 5:651

Erenler SA, Unver T, Ozaslan BF, Koytepe S, Sezer S (2024) Development of microbial chondroitin sulfate-based proton exchange membranes for microbial fuel cells. Fuel 363:130976. https://doi.org/10.1016/j.fuel.2024.130976

Escap U (2018) SDG 11: sustainable cities and communities: make cities and human settlements inclusive, safe, resilient and sustainable

Fagunwa OE, Olanbiwoninu AA (2020) Accelerating the sustainable development goals through microbiology: some efforts and opportunities. Access Microbiol 2(5):e000112

Favero VO, Carvalho RH, Motta VM, Leite ABC, Coelho MRR, Xavier GR, Rumjanek NG, Urquiaga S (2021) Bradyrhizobium as the only rhizobial inhabitant of mung bean (Vigna radiata) nodules in tropical soils: a strategy based on microbiome for improving biological nitrogen fixation using bio-products. Front Plant Sci 11:602645

France M, Alizadeh M, Brown S, Ma B, Ravel J (2022) Towards a deeper understanding of the vaginal microbiota. Nat Microbiol 7(3):367–378

Friedman M (2013) Nutritional and toxicological consequences of food processing. Springer Science & Business Media

Frize M, Tsapaki V, Lhotska L, da Silva AMM, Ibrahim F, Bezak E et al (2022) Women in medical physics and biomedical engineering: past, present and future. Heal Technol 12(3):655–662

Fu H, Fu L, Dávid LD, Zhong Q, Zhu K (2024) Bridging gaps towards the 2030 agenda: a data-driven comparative analysis of government and public engagement in China towards achieving sustainable development goals. Land 13(6):818

Fufă MOM, Popescu RC, Grumezescu AM, Holban AM (2017) Microorganisms: new trends in environment-friendly and energy-saving water purification. In: Water Purification. Elsevier

Fukami J, Cerezini P, Hungria M (2018) Azospirillum: benefits that go far beyond biological nitrogen fixation. Amb Express 8:73

Fund S (2015) Sustainable development goals. Available at this link: https://www.un.org/sustainable development/inequality

Furtak K, Gajda AM (2018) Activity and variety of soil microorganisms depending on the diversity of the soil tillage system. Sustain Agroecosyst:45. https://doi.org/10.5772/intechopen.72966

Gamage A, Gangahagedara R, Gamage J, Jayasinghe N, Kodikara N, Suraweera P, Merah O (2023) Role of organic farming for achieving sustainability in agriculture. Farming System 1(1):100005

García-Fraile P, Menéndez E, Celador-Lera L, Díez-Méndez A, Jiménez-Gómez A, Marcos-García M, Cruz-González XA, Martínez-Hidalgo P, Mateos PF, Rivas R (2017) Bacterial probiotics: a truly green revolution. In: Probiotics and plant health, pp 131–162

Gareau MG, Sherman PM, Walker WA (2010) Probiotics and the gut microbiota in intestinal health and disease. Nat Rev Gastroenterol Hepatol 7:503–514

Gerba CP (2009) Indicator microorganisms. In: Environmental microbiology. Elsevier, pp 485–499

Ghnimi S, Guizani N (2018) Vegetable fermentation and pickling. In: Handbook of vegetables and vegetable processing, pp 407–427

Gil JDB, Reidsma P, Giller K, Todman L, Whitmore A, van Ittersum M (2019) Sustainable development goal 2: improved targets and indicators for agriculture and food security. Ambio 48(7):685–698

Gungaabayar A, Jha A, Warkentin T, Knight D, Penner G, Biligetu B (2022) Forage yield and biological nitrogen fixation of pea–cereal intercrops for hay production. Agron J 115(2): 607–619

Hazaimeh MD, Ahmed ES (2021) Bioremediation perspectives and progress in petroleum pollution in the marine environment: a review. Environ Sci Pollut Res 28(39):54238–54259

Heck JE, Wu CK, Huang X, Chew KW, Tong M, Federman N et al (2022) Cohort study of familial viral hepatitis and risks of paediatric cancers. Int J Epidemiol 51(2):448–457

Helfer V, Hassenrück C (2021) Microbial communities in mangrove sediments. In: Dynamic sedimentary environments of mangrove coasts, pp 141–175

Hoballah A, Averous S (2015) Ensure sustainable consumption and production patterns. UN Chron 51(4):28–29

Holechek JL, Geli HM, Sawalhah MN, Valdez R (2022) A global assessment: can renewable energy replace fossil fuels by 2050? Sustainability 14(8):4792

Hosseinkhani F, Heinken A, Thiele I, Lindenburg P, Harms A, Hankemeier T (2021) The contribution of gut bacterial metabolites in the human immune signaling pathway of non-communicable diseases. Gut Microbes 13:1882927

Hughes BB, Hanna T, McNeil K, Bohl DK, Moyer JD, Abidoye B, Felix J, Kapto S, Patterson L (2021) Pursuing the Sustainable Development Goals in a World Reshaped by COVID-19. Pardee Center for International Futures and United Nations Development Programme, Denver and New York

Ibáñez A, Garrido-Chamorro S, Vasco-Cárdenas MF, Barreiro C (2023) From lab to field: biofertilizers in the 21st century. Horticulturae 9(12):1306

IPCC (2007) Climate change IPCC Working Group II. Switzerland, Geneva

Irwin EG, Culligan PJ, Fischer-Kowalski M, Law KL, Murtugudde R, Pfirman S (2018) Bridging barriers to advance global sustainability. Nat Sustain 1(7):324–326

Isabel JB, Balamurugan A, Devi PR, Periyasamy S (2024) Chitosan-encapsulated microbial biofertilizer: a breakthrough for enhanced tomato crop productivity. Int J Biol Macromol 260:129462. https://doi.org/10.1016/j.ijbiomac.2024.129462

Jalili P, Ala A, Nazari P, Jalili B, Ganji DD (2024) A comprehensive review of microbial fuel cells considering materials, methods, structures, and microorganisms. Heliyon. https://doi.org/10.1016/j.heliyon.2024.e25439

Jansson JK (2023) Microorganisms, climate change, and the Sustainable Development Goals: progress and challenges. Nat Rev Microbiol 21(10):622–623. https://doi.org/10.1038/s41579-023-00953-8

Jasovský D, Littmann J, Zorzet A, Cars O (2016) Antimicrobial resistance—a threat to the world's sustainable development. Ups J Med Sci 121(3):159–164

Jenol MA, Ibrahim MF, Kamal Bahrin E, Kim SW, Abd-Aziz S (2019) Direct bioelectricity generation from sago hampas by Clostridium beijerinckii SR1 using microbial fuel cell. Molecules 24:2397

Jones G, Gardner GE, Lee T, Poland K, Robert S (2013) The impact of microbiology instruction on students' perceptions of risks related to microbial illness. Int J Sci Educ pp 199–213 https://doi.org/10.1080/21548455.2012.684434

Joshi G, Pandey JK, Rana S, Rawat DS (2017) Challenges and opportunities for the application of biofuel. Renew Sust Energ Rev 79:850–866

Jóźwiakowski K, Czernaś K, Szczurowska A (2009) Preliminary results of studies on the purification of water in a pond using the SCD Probiotics technology. Ecohydrol Hydrobiol 9:307–312

Kaksonen AH (2018) Editorial for special issue "Microorganisms for Environmental and Industrial Applications". Microorganisms 6(3):62. https://doi.org/10.3390/microorganisms6030062

Kamilari E, Stanton C, Reen FJ, Ross RP (2023) Uncovering the biotechnological importance of Geotrichum candidum. Foods 12:1124

Karimi N, Mostofinejad D (2020) Bacillus subtilis bacteria used in fiber reinforced concrete and their effects on concrete penetrability. Constr Build Mater 230:117051

Kau AL, Ahern PP, Griffin NW, Goodman AL, Gordon JI (2011) Human nutrition, the gut microbiome and the immune system. Nature 474:327–336

Keasling J, Garcia Martin H, Lee TS, Mukhopadhyay A, Singer SW, Sundstrom E (2021) Microbial production of advanced biofuels. Nat Rev Microbiol 19(11):701–715

Keim P (2003) Microbial forensics: a scientific assessment (No. DOEER63403). American Society for Microbiology (US)

Kelley PW (2011) Antimicrobial resistance in the age of noncommunicable diseases. Rev Panam Salud Publica 30:515–518

Khan N (2019) Microbes role in the economic development of a world. SSRN

Khan FM, Gupta R (2020) Escherichia coli (E. coli) as an indicator of fecal contamination in groundwater: a review. Sustainable Development of Water and Environment: proceedings of the ICSDWE2020, pp 225–235

Khan S, Paliwal V, Pandey VV, Kumar V (2015) Biomass as renewable energy. Int Adv Res J Sci Eng Technol 2:301–304

Kimoto-Nira H, Mizumachi K, Nomura M, Kobayashi M, Fujita Y, Okamoto T, Suzuki I, Tsuji NM, KURISAKI J.-i. Ohmomo S (2007) Lactococcus sp. as potential probiotic lactic acid bacteria. Japan Agricultural Research Quarterly: JARQ 41:181–189

Kok CR, Hutkins R (2018) Yogurt and other fermented foods as sources of health-promoting bacteria. Nutr Rev 76:4–15

Kolber ZS, Gerald F, Plumley, Lang AS, Beatty JT, Blankenship RE, VanDover CL, Vetriani C, Koblizek M, Rathgeber C (2001) Contribution of aerobic photoheterotrophic bacteria to the carbon cycle in the ocean. Science 292:2492–2495

Kotoky R, Pandey P (2020) Rhizosphere assisted biodegradation of benzo (a) pyrene by cadmium resistant plant-probiotic Serratia marcescens S2I7, and its genomic traits. Sci Rep 10(1):5279. https://doi.org/10.1038/s41598-020-62285-4

Kotoky R, Rajkumari J, Pandey P (2018) The rhizosphere microbiome: significance in rhizoremediation of polyaromatic hydrocarbon contaminated soil. J Environ Manag 217:858–870

Koul Y, Devda V, Varjani S, Guo W, Ngo HH, Taherzadeh MJ, Chang JS, Wong JWC, Bilal M, Kim SH, Bui XT, Parra-Saldívar R (2022) Microbial electrolysis: a promising approach for treatment and resource recovery from industrial wastewater. Bioengineered 13(4):8115–8134

Krauss JE (2022) Unpacking SDG 15, its targets and indicators: tracing ideas of conservation. Globalizations 19(8):1179–1194

Kulkarni NA, Chethan H, Srivastava R, Gabbur AB (2022) Role of probiotics in ruminant nutrition as natural modulators of health and productivity of animals in tropical countries: an overview. Tropical Animal Health and Production 54:110

Kumar AG, Rajan NN, Kirubagaran R, Dharani G (2019) Biodegradation of crude oil using self-immobilized hydrocarbonoclastic deep sea bacterial consortium. Mar Pollut Bull 146:741–750

Kumar R, Sood U, Gupta V, Singh M, Scaria J, Lal R (2020) Recent advancements in the development of modern probiotics for restoring human gut microbiome dysbiosis. Indian journal of microbiology 60:12–25. https://doi.org/10.1007/s12088-019-00808-y

Kumar S, Sindhu SS, Kumar R (2022) Biofertilizers: an ecofriendly technology for nutrient recycling and environmental sustainability. Curr Res Microbial Sci 3:100094

Lahue C, Madden AA, Dunn RR, Smukowski HC (2020) History and domestication of Saccharomyces cerevisiae in bread baking. Frontiers in genetics 11:584718

Lambo MT, Chang X, Liu D (2021) The recent trend in the use of multistrain probiotics in livestock production: an overview. Animals 11:2805

Larkin EL, Ali AAL, Swindell K (2019) History of antifungals. In: Antifungal therapy, Second Edition. CRC Press, pp 1–10

Larkum AW, Grossman AR, Raven JA (2020) Photosynthesis in algae: biochemical and physiological mechanisms. Springer

Leal Filho W, Tripathi SK, Andrade Guerra JBSOD, Giné-Garriga R, Orlovic Lovren V, Willats J (2019) Using the sustainable development goals towards a better understanding of sustainability challenges. Int J Sustain Dev World Ecol 26(2):179–190

Leiva-Candia D, Pinzi S, Redel-Macías M, Koutinas A, Webb C, Dorado M (2014) The potential for agro-industrial waste utilization using oleaginous yeast for the production of biodiesel. Fuel 123:33–42

Lenoir-Wijnkoop I, Merenstein D, Korchagina D, Broholm C, Sanders ME, Tancredi D (2019) Probiotics reduce health care cost and societal impact of flu-like respiratory tract infections in the USA: an economic modeling study. Front Pharmacol 10:980

Li M, Zhou M, Tian X, Tan C, McDaniel CT, Hassett DJ, Gu T (2018) Microbial fuel cell (MFC) power performance improvement through enhanced microbial electrogenicity. Biotechnol Adv 36(4):1316–1327

Li Q, Liu J, Gadd GM (2020) Fungal bioremediation of soil co-contaminated with petroleum hydrocarbons and toxic metals. Appl Microbiol Biotechnol 104:8999–9008

Li Y, Meng Z, Xu Y, Shi Q, Ma Y, Aung M, Cheng Y, Zhu W (2021) Interactions between anaerobic fungi and methanogens in the rumen and their biotechnological potential in biogas production from lignocellulosic materials. Microorganisms 9:190

Li S, Qu W, Chang H, Li J, Ho S-H (2022) Microalgae-driven swine wastewater biotreatment: Nutrient recovery, key microbial community and current challenges. J Hazard Mater 440:129785

Ligda P, Claerebout E, Kostopoulou D, Zdragas A, Casaert S, Robertson LJ, Sotiraki S (2020) Cryptosporidium and Giardia in surface water and drinking water: animal sources and towards the use of a machine-learning approach as a tool for predicting contamination. Environ Pollut 264:114766

Lima ES, Romero EC, Granato CFH (2021). Current polio status in the world. Jornal Brasileiro de Patologia e Medicina Laboratorial, 57

Lin L, Yang H, Xu X (2022) Effects of water pollution on human health and disease heterogeneity: a review. Front Environ Sci 10:880246. https://doi.org/10.3389/fenvs.2022.880246

Lindström K, Mousavi SA (2020) Effectiveness of nitrogen fixation in rhizobia. Microb Biotechnol 13(5):1314–1335

Logan BE (2009) Exoelectrogenic bacteria that power microbial fuel cells. Nat Rev Microbiol 7(5):375–381

Mafra P, Lima N, Carvalho GS (2015) Experimental activities in primary school to learn about microbes in an oral health education context. J Biol Educ 49:190–203

Majidian P, Tabatabaei M, Zeinolabedini M, Naghshbandi MP, Chisti Y (2018) Metabolic engineering of microorganisms for biofuel production. Renew Sust Energ Rev 82:3863–3885

Malla MA, Dubey A, Raj A, Kumar A, Upadhyay N, Yadav S (2022) Emerging frontiers in microbe-mediated pesticide remediation: unveiling role of omics and In silico approaches in engineered environment. Environ Pollut 299:118851. https://doi.org/10.1016/j.envpol.2022.118851

Marangon E, Laffy PW, Bourne DG, Webster NS (2021) Microbiome-mediated mechanisms contributing to the environmental tolerance of reef invertebrate species. Mar Biol 168(6):89

Marhons Š, Hyršlová I, Stetsenko V, Jablonská E, Veselý M, Míchová H et al (2023) Properties of yoghurt treated with microbial transglutaminase and exopolysaccharides. Int Dairy J 144:105701. https://doi.org/10.1016/j.idairyj.2023.105701

Markewitz P, Kuckshinrichs W, Leitner W, Linssen J, Zapp P, Bongartz R et al (2012) Worldwide innovations in the development of carbon capture technologies and the utilization of CO 2. Energy Environ Sci 5(6):7281–7305

Martin A, McGuire S, Sullivan S (2013) Global environmental justice and biodiversity conservation. Geogr J 179(2):122–131

Maslennikova D, Nasyrova K, Chubukova O, Akimova E, Baymiev A, Blagova D et al (2022) Effects of Rhizobium leguminosarum Thy2 on the growth and tolerance to cadmium stress of wheat plants. Life 12(10):1675. https://doi.org/10.3390/life12101675

Mayrand F, Clergeau P (2018) Green roofs and green walls for biodiversity conservation: a contribution to urban connectivity? Sustainability 10(4):985

Miglani R, Parveen N, Kumar A, Ansari MA, Khanna S, Rawat G et al (2022) Degradation of xenobiotic pollutants: an environmentally sustainable approach. Metabolites 12(9):818. https://doi.org/10.3390/metabo12090818

Milner J, Silo-Suh L, Goodman RM, Handelsman J (2019) Antibiosis and beyond: genetic diversity, microbial communities, and biological control. In: Ecological interactions and biological control. CRC Press

Milandri M (2004) Children's views of microbes: current beliefs about bacteria in Italian grade school children. Pediatr Infect Dis J 23:1077–1080

Mishra SK, Khan MH, Misra S, Dixit VK, Khare P, Srivastava S, Chauhan PS (2017) Characterisation of Pseudomonas spp. and Ochrobactrum sp. isolated from volcanic soil. Antonie Van Leeuwenhoek 110:253–270

Mittelman M, Hanaway P (2012) Globalization of healthcare. Glob Adv Health Med 1(2):5–7. https://doi.org/10.7453/gahmj.2012.1.2.001. PMID: 24278809; PMCID: PMC3833491

Mohamad Taghvaee V, Assari Arani A, Nodehi M, Khodaparast Shirazi J, Agheli L, Neshat Ghojogh HM et al (2023) Sustainable development goals: transportation, health and public policy. Rev Econ Polit Sci 8(2):134–161

Mohanty S, Swain CK (2018) Role of microbes in climate smart agriculture. In: Microorganisms for green revolution: volume 2: microbes for sustainable agro-ecosystem, pp 129–140

Mohr KI (2016) History of antibiotics research. In: How to overcome the antibiotic crisis: facts, challenges, technologies and future perspectives, pp 237–272

Montazer Z, Habibi Najafi MB, Levin DB (2020) Challenges with verifying microbial degradation of polyethylene. Polymers 12(1):123. https://doi.org/10.3390/polym12010123

Muluneh MG (2021) Impact of climate change on biodiversity and food security: a global perspective—a review article. Agric Food Secur 10(1):1–25

Munson E (2020) Biographical feature: William A. Hinton, MD. J Clin Microbiol 58(11):10–1128

Nariya S, Kalyuzhnaya MG (2019) Diversity, physiology, and biotechnological potential of halo (alkali) philic methane-consuming bacteria. In: Methanotrophs: microbiology fundamentals and biotechnological applications. Springer

Naseem H, Ahsan M, Shahid MA, Khan N (2018) Exopolysaccharides producing rhizobacteria and their role in plant growth and drought tolerance. J Basic Microbiol 58:1009–1022

Nelson CE, Wegley Kelly L, Haas AF (2023) Microbial interactions with dissolved organic matter are central to coral reef ecosystem function and resilience. Annu Rev Mar Sci 15:431–460

Nevita T, Sharma OG, Pandey P (2018) Composting of rice-residues using lignocellulolytic plant-probiotic Stenotrophomonas maltophilia, and its evaluation for growth enhancement of Oryza sativa L. Environ Sustain 1:185–196

Nikoli T, Matsi T (2014) Evaluation of certain Ni soil tests for an initial estimation of Ni sufficiency critical levels. J Plant Nutr Soil Sci 177(4):596–603

Niu X, Song L, Xiao Y, Ge W (2018) Drought-tolerant plant growth-promoting rhizobacteria associated with foxtail millet in a semi-arid agroecosystem and their potential in alleviating drought stress. Front Microbiol 8:2580

Nussinovitch A, Nussinovitch A (1997) Agar. In: Hydrocolloid applications: gum technology in the food and other industries, pp 1–18

O'Sullivan A, Farver M, Smilowitz JT (2015) The influence of early infant-feeding practices on the intestinal microbiome and body composition in infants. Nutr Metab Insights 8:1–9

Obafemi YD, Oranusi SU, Ajanaku KO, Akinduti PA, Leech J, Cotter PD (2022) African fermented foods: overview, emerging benefits, and novel approaches to microbiome profiling. npj Sci Food 6:15

Ogunrinola GA, Oyewale JO, Oshamika OO, Olasehinde GI (2020) The human microbiome and its impacts on health. Int J Microbiol 2020:8045646

Ohno H, Tsunemine S, Isa Y, Shimakawa M, Yamamura H (2005) Oral administration of Bifidobacterium bifidum G9-1 suppresses total and antigen specific immunoglobulin E production in mice. Biol Pharm Bull 28:1462–1466

Oishi W, Vinnerås B, Sano D (2023) Resource recovery technologies as microbial risk barriers: towards safe use of excreta in agriculture based on hazard analysis and critical control point. Environ Sci: Water Res Technol 9:1008–1029

Ojeda-Linares C, Álvarez-Ríos GD, Figueredo-Urbina CJ, Islas LA, Lappe-Oliveras P, Nabhan GP et al (2021) Traditional fermented beverages of Mexico: a biocultural unseen foodscape. Foods 10(10):2390

Okafor N (1992) Commercialization of fermented foods in Sub-Saharan Africa. Applications of Biotechnology to Fermented Foods: report of an Ad Hoc Panel of the Board on Science and Technology for International Development. In: National Research Council (US) Panel on the applications of biotechnology to traditional fermented foods. national Academy Press, Washington, DC

Olabi AG, Shehata N, Sayed ET, Rodriguez C, Anyanwu RC, Russell C, Abdelkareem MA (2023) Role of microalgae in achieving sustainable development goals and circular economy. Sci Total Environ 854:158689

Osono T (2020) Functional diversity of ligninolytic fungi associated with leaf litter decomposition. Ecol Res 35(1):30–43

Oyetibo GO, Ige OO, Obinani PK, Amund OO (2021) Ecological risk potentials of petroleum hydrocarbons and heavy metals shape the bacterial communities of marine hydrosphere at Atlantic Ocean, Atlas Cove. Nigeria J Environ Manag 289:112563

Pandya RS, Kaur T, Bhattacharya R, Bose D, Saraf D (2024) Harnessing microorganisms for bioenergy with Microbial Fuel Cells: Powering the future. Water-Energy Nexus 7:1–12
Pajares S, Ramos R (2019) Processes and microorganisms involved in the marine nitrogen cycle: knowledge and gaps. Front Mar Sci 6:739
Palmer JD, Brigham CJ (2016) Feasibility of triacylglycerol production for biodiesel, utilizing Rhodococcus opacus as a biocatalyst and fishery waste as feedstock. Renew Sust Energ Rev 56:922–928
Pande V, Pandey SC, Sati D, Pande V, Samant M (2020) Bioremediation: an emerging effective approach towards environment restoration. Environ Sustain 3:91–103
Pandey B, Fulekar M (2012) Bioremediation technology: a new horizon for environmental clean-up. Biol Med 4:51
Pastor-Bueis R, Sánchez-Cañizares C, James EK, González-Andrés F (2019) Formulation of a highly effective inoculant for common bean based on an autochthonous elite strain of Rhizobium leguminosarum bv. phaseoli, and genomic-based insights into its agronomic performance. Front Microbiol 10:2724
Patel J, Borgohain S, Kumar M, Rangarajan V, Somasundaran P, Sen R (2015) Recent developments in microbial enhanced oil recovery. Renew Sust Energ Rev 52:1539–1558
Pathania P, Rajta A, Singh PC, Bhatia R (2020) Role of plant growth-promoting bacteria in sustainable agriculture. Biocatal Agric Biotechnol 30:101842
Patinvoh RJ, Taherzadeh MJ (2019) Challenges of biogas implementation in developing countries. Curr Opin Environ Sci Health 12:30–37
Pearce TR, Antonelli A, Brearley FQ, Couch C, Campostrini Forzza R, Gonçalves SC et al (2020) International collaboration between collections-based institutes for halting biodiversity loss and unlocking the useful properties of plants and fungi. Plants People Planet 2(5):515–534
Peñalosa-Martinell D, Araneda-Padilla M, Dumas S, Martinez-Díaz S, Vela-Magaña M (2021) The use of probiotics in larval whiteleg shrimp (Litopenaeus vannamei) production: a marginal analysis of bioeconomic feasibility. Aquac Res 52:943–951
Pereira VC, Silva SN, Carvalho VK, Zanghelini F, Barreto, JO (2022) Strategies for the implementation of clinical practice guidelines in public health: an overview of systematic reviews. Health Res Policy Systems 20(1):13
Perner M, Wallmann K, Adam-Beyer N, Hepach H, Laufer-Meiser K, Böhnke S, ... & Scholz F (2022) Environmental changes affect the microbial release of hydrogen sulfide and methane from sediments at Boknis Eck (SW Baltic Sea). Frontiers Microbiol 13:1096062
Pham HT (2018) Biosensors based on lithotrophic microbial fuel cells in relation to heterotrophic counterparts: research progress, challenges, and opportunities. AIMS Microbiol 4(3):567
Plotkin SA, Plotkin SL (2011) The development of vaccines: how the past led to the future. Nat Rev Microbiol 9:889–893
Potter MC (1911) Electrical effects accompanying the decomposition of organic compounds. Proc Royal Soc London B 84(571):260–276
Prussin AJ, Marr LC (2015) Sources of airborne microorganisms in the built environment. Microbiome 3:1–10
Purahong W, Kapturska D, Pecyna MJ, Jariyavidyanont K, Kaunzner J, Juncheed K, Uengwetwanit T, Rudloff R, Schulz E, Hofrichter M (2015) Effects of forest management practices in temperate beech forests on bacterial and fungal communities involved in leaf litter degradation. Microb Ecol 69:905–913
Rainieri S, Zambonelli C (2009) Organisms associated with acetic acid bacteria in vinegar production. In: Vinegars of the world, pp 73–95. https://doi.org/10.1007/978-88-470-0866-3_5
Rallis D, Giapros V, Serbis A, Kosmeri C, Baltogianni M (2023) Fighting antimicrobial resistance in neonatal intensive care units: rational use of antibiotics in neonatal sepsis. Antibiotics 12:508
Ramírez-Guzmán KN, Torres-León C, Martinez-Medina GA, de la Rosa O, Hernández-Almanza A, Alvarez-Perez OB et al (2019) Traditional fermented beverages in Mexico. In: Fermented beverages. Woodhead Publishing, pp 605–635

Raven RP, Gregersen KH (2007) Biogas plants in Denmark: successes and setbacks. Renew Sust Energ Rev 11:116–132

Ravindranath NH, Joshi NV, Sukumar R (2006) Curr Sci 90(3):354–361

Raza T, Qadir MF, Khan KS, Eash NS, Yousuf M, Chatterjee S, Manzoor R, ur Rehman, S., & Oetting, J. N. (2023) Unrevealing the potential of microbes in decomposition of organic matter and release of carbon in the ecosystem. J Environ Manag 344:118529

Rehman K, Niaz S, Tahir A, Akash MSH (2020) Microorganisms and antibiotic production. In: Antibiotics and antimicrobial resistance genes in the environment. Elsevier, pp 1–6. https://doi.org/10.1016/B978-0-12-818882-8.00001-2

Reid G (2017) The development of probiotics for women's health. Can J Microbiol 63(4):269–277

Richter H, Molitor B, Wei H, Chen W, Aristilde L, Angenent L (2016) Ethanol production in syngas-fermenting Clostridium ljungdahlii is controlled by thermodynamics rather than by enzyme expression. Energy Environ Sci 9:2392–2399

Ren W, Wu H, Guo C, Xue B, Long H, Zhang X, Cai X, Huang A, Xie Z (2021) Multi-strain tropical Bacillus spp. as a potential probiotic biocontrol agent for large-scale enhancement of mariculture water quality. Frontiers Microbiol 12:699378

Rizo J, Guillén D, Farrés A, Díaz-Ruiz G, Sánchez S (2020) Omics in traditional vegetable fermented foods and beverages. Crit Rev Food Sci Nutr 60(5):791–809

Robbins S, Song W, Engelberts J, Glasl B, Slaby BM, Boyd J, Marangon E, Botté E, Laffy P, Thomas T (2021) A genomic view of the microbiome of coral reef demosponges. ISME J 15(6):1641–1654

Rodríguez-Espinosa T, Navarro-Pedreño J, Gómez-Lucas I, Jordán-Vidal MM, Bech-Borras J, Zorpas AA (2021) Urban areas, human health and technosols for the green deal. Environ Geochem Health 43(12):5065–5086

Rossi F, Felis GE, Martinelli A, Calcavecchia B, Torriani S (2016) Microbiological characteristics of fresh tofu produced in small industrial scale and identification of specific spoiling microorganisms (SSO). LWT 70:280–285. https://doi.org/10.1016/j.lwt.2016.02.057

Rousseau R, Etcheverry L, Roubaud E, Basséguy R, Délia ML, Bergel A (2020) Microbial electrolysis cell (MEC): strengths, weaknesses and research needs from electrochemical engineering standpoint. Appl Energy 257:113938

Routray P, Torondel B, Clasen T, Schmidt WP (2017) Women's role in sanitation decision making in rural coastal Odisha, India. PloS one 12:e0178042

Rumpala Y (2011) "Sustainable consumption" as a new phase in a governmentalization of consumption. Theory and Society 40:669–699

Saber WI, Ghoneem KM, Al-Askar AA, Rashad YM, Ali AA, Rashad EM (2015) Chitinase production by Bacillus subtilis ATCC 11774 and its effect on biocontrol of Rhizoctonia diseases of potato. Acta Biol Hung 66:436–448

Sánchez-Baracaldo P, Bianchini G, Wilson JD, Knoll AH (2022) Cyanobacteria and biogeochemical cycles through Earth history. Trends Microbiol 30:143–157

Sarah Marshall (2022) South Korea's food waste system is a model for developed nations. https://keia.org/the-peninsula/south-koreas-food-waste-system-is-a-model-for-developed-nations/

Saravanan A, Kumar PS, Jeevanantham S, Karishma S, Tajsabreen B, Yaashikaa PR, Reshma B (2021) Effective water/wastewater treatment methodologies for toxic pollutants removal: processes and applications towards sustainable development. Chemosphere 280:130595

Sawant AM, Vamkudoth KR (2022) Biosynthetic process and strain improvement approaches for industrial penicillin production. Biotechnol Lett:1–14

Sawant SS, Patil SM, Gupta V, Kunda NK (2020) Microbes as medicines: harnessing the power of bacteria in advancing cancer treatment. Int J Mol Sci 21(20):7575

Scarlat N, Dallemand JF, Fahl F (2018) Biogas: developments and perspectives in Europe. Renew Energy 129:457–472. Science. aag0804

Seenivasagan R, Babalola OO (2021) Utilization of microbial consortia as biofertilizers and biopesticides for the production of feasible agricultural product. Biology 10(11):1111. https://doi.org/10.3390/biology10111111

Sharma R, Garg P, Kumar P, Bhatia SK, Kulshrestha S (2020) Microbial fermentation and its role in quality improvement of fermented foods. Fermentation 6:106
Sharma HB, Panigrahi S, Dubey BK (2021a) Food waste hydrothermal carbonization: study on the effects of reaction severities, pelletization and framework development using approaches of the circular economy. Bioresour Technol 333:125187
Sharma P, Dutta D, Udayan A, Kumar S (2021b) Industrial wastewater purification through metal pollution reduction employing microbes and magnetic nanocomposites. J Environ Chem Eng 9(6):10667. https://doi.org/10.1016/j.jece.2021.106673
Sharma S, Agarwal S, Jain A (2021c) Significance of hydrogen as economic and environmentally friendly fuel. Energies 14(21):7389. https://doi.org/10.3390/en14217389
Shoubridge AP, Choo JM, Martin AM, Keating DJ, Wong M-L, Licinio J, Rogers GB (2022) The gut microbiome and mental health: advances in research and emerging priorities. Mol Psychiatry 27:1908–1919
Shreiner AB, Kao JY, Young VB (2015) The gut microbiome in health and in disease. Curr Opin Gastroenterol 31:69
Silva JA (2023) Wastewater treatment and reuse for sustainable water resources management: a systematic literature review. Sustainability 15(14):10940. https://doi.org/10.3390/su151410940
Simonneaux L (2000) A study of pupils' conceptions and reasoning in connection with "microbes", as a contribution to research in biotechnology education. Int J Sci Educ 22:619e644
Singh BK, Bardgett RD, Smith P, Reay DS (2010) Microorganisms and climate change: terrestrial feedbacks and mitigation options. Nat Rev Microbiol 8:779–790
Singh R, Kumar M, Mittal A, Mehta PK (2017) Microbial metabolites in nutrition, healthcare and agriculture. 3 Biotech 7:1–14. https://doi.org/10.1007/s13205-016-0586-4
Singh D, Goswami RK, Agrawal K, Chaturvedi V, Verma P (2022) Bio-inspired remediation of wastewater: a contemporary approach for environmental clean-up. Curr Res Green Sustain Chem 5:100261. https://doi.org/10.1016/j.crgsc.2022.100261
Soberón M, Bravo A, Blanco CA (2023) Strategies to reduce insecticide use in agricultural production. In: Ferranti P (ed) Sustainable food science: a comprehensive approach, vol 4. Elsevier, Netherlands, pp 507–514
Sood A, Pandey P, Bisht S, Sharma S (2014) Anthropogenic activities as a source of high prevalence of antibiotic resistant Staphylococcus aureus in the River Ganga. Appl Ecol Environ Res 12(1):33–48
Sonawane JM, Mahadevan R, Pandey A, Greener J (2022) Recent progress in microbial fuel cells using substrates from diverse sources. Heliyon 8(12)
Sutherland DL, McCauley J, Labeeuw L, Ray P, Kuzhiumparambil U, Hall C et al (2021) How microalgal biotechnology can assist with the UN Sustainable Development Goals for natural resource management. Curr Res Environ Sustain 3:100050. https://doi.org/10.1016/j.crsust.2021.100050
Tanaka Y, Kasahara K, Izawa M, Ochi K (2017) Applicability of ribosome engineering to vitamin B12 production by Propionibacterium shermanii. Biosci Biotechnol Biochem 81:1636–1641
Tao C, Li R, Xiong W, Shen Z, Liu S, Wang B, Ruan Y, Geisen S, Shen Q, Kowalchuk GA (2020) Bio-organic fertilizers stimulate indigenous soil Pseudomonas populations to enhance plant disease suppression. Microbiome 8:1–14
Tauseef SM, Abbasi T, Abbasi SA (2013) Energy recovery from wastewaters with high-rate anaerobic digesters. Renew Sust Energ Rev 19:704–741
Timmis K, Lorenzo VD, Verstraete W, Ramos JL, Danchin A, Brüssow H, Singh BK, Timmis JK (2017a) The contribution of microbial biotechnology to economic growth and employment creation. Microb Biotechnol 10(5):1137–1144
Timmis K, De Vos WM, Ramos JL, Vlaeminck SE, Prieto A, Danchin A et al (2017b) The contribution of microbial biotechnology to sustainable development goals. Microb Biotechnol 10(5):984–987
Titlyanov E, Titlyanova T (2020) Symbiotic Relationships between microalgal zooxanthellae and reef-building coral polyps in the process of autotrophic and heterotrophic nutrition. Russ J Mar Biol 46:307–318

Trombetta T, Vidussi F, Roques C, Scotti M, Mostajir B (2020) Marine microbial food web networks during phytoplankton bloom and non-bloom periods: warming favors smaller organism interactions and intensifies trophic cascade. Front Microbiol 11:502336

Tumwesige V, Fulford D, Davidson GC (2014) Biogas appliances in sub-Sahara Africa. Biomass Bioenergy 70:40–50

United Nations (2016) Sustainable Development GOALS – 17 Goals to Transform Our World. Sustain Dev goals – United Nations. United Nations Development Program (UNDP)

United Nations. Department of Economic and Social Affairs (2022) The Sustainable Development Goals: Report 2022. UN

Urata S, Kuroda K, Tonegawa Y (2023) Sustainable development disciplines for humanity. Springer Nature. https://doi.org/10.1007/978-981-19-4859-6

Vargason AM, Anselmo AC (2018) Clinical translation of microbe-based therapies: current clinical landscape and preclinical outlook. Bioeng Transl Med 3(2):124–137

Vekemans J, Hasso-Agopsowicz M, Kang G, Hausdorff WP, Fiore A, Tayler E, Klemm EJ, Laxminarayan R, Srikantiah P, Friede M et al (2021) Leveraging vaccines to reduce antibiotic use and prevent antimicrobial resistance: an action framework. Clin Infect Dis 73:e1011–e1017. https://doi.org/10.1093/cid/ciab062

Villarreal-Delgado MF, Villa-Rodríguez ED, Cira-Chávez LA, Estrada-Alvarado MI, Parra-Cota FI, Santos-Villalobos S (2018) The genus Bacillus as a biological control agent and its implications in the agricultural biosecurity. Rev Mex Fitopatol 36:95–130

Vurukonda SSKP, Vardharajula S, Shrivastava M, SkZ A (2016) Enhancement of drought stress tolerance in crops by plant growth promoting rhizobacteria. Microbiol Res 184:13–24

Wainwright BJ, Millar T, Bowen L, Semon L, Hickman K, Lee JN, Yeo ZY, Zahn G (2023) The core mangrove microbiome reveals shared taxa potentially involved in nutrient cycling and promoting host survival. Environ Microbiome 18(1):47

Walker GM, Stewart GG (2016) Saccharomyces cerevisiae in the production of fermented beverages. Beverages 2(4):30. https://doi.org/10.3390/beverages2040030

Wang H, Luo H, Fallgren PH, Jin S, Ren ZJ (2015) Bioelectrochemical system platform for sustainable environmental remediation and energy generation. Biotechnol Adv 33(3–4):317–334

Weiman S (2015) Microbes help to drive global carbon cycling and climate change. Microb Mag 10(6):233–238

Woon JM, Khoo KS, Al-Zahrani AA, Alanazi MM, Lim JW, Cheng CK et al (2023) Epitomizing biohydrogen production from microbes: critical challenges vs opportunities. Environ Res:115780. https://doi.org/10.1016/j.envres.2023.115780

World Health Organization (2017) Poverty [WWW Document]. Heal Dev. URL. http://www.who.int/topics/poverty/en/ (Accessed 06 June 2023)

Yang M, Chen L, Wang J, Msigwa G, Osman AI, Fawzy S et al (2023) Circular economy strategies for combating climate change and other environmental issues. Environ Chem Lett 21(1):55–80

Yarwood SA (2018) The role of wetland microorganisms in plant-litter decomposition and soil organic matter formation: a critical review. FEMS Microbiol Ecol 94:fiy175

Yoha K, Nida S, Dutta S, Moses J, Anandharamakrishnan C (2021) Targeted delivery of probiotics: perspectives on research and commercialization. In: Probiotics and antimicrobial proteins, pp 1–34

Zhang HL, Li WS, Xu DN, Zheng WW, Liu Y, Chen J, Qiu ZB, Dorfman RG, Zhang J, Liu J (2016) Mucosa-reparing and microbiota-balancing therapeutic effect of Bacillus subtilis alleviates dextrate sulfate sodium-induced ulcerative colitis in mice. Exp Ther Med 12:2554–2562

Zhao W, Yin C, Hua T, Meadows ME, Li Y, Liu Y et al (2022) Achieving the sustainable development goals in the post-pandemic era. Humanit Soc Sci Commun 9(1):1–7

Zhou B, Zhang T, Wang F (2023) Microbial-based heavy metal bioremediation: toxicity and eco-friendly approaches to heavy metal decontamination. Appl Sci 13(14):8439. https://doi.org/10.3390/app13148439

Zhu LD, Hiltunen E, Antila E, Zhong JJ, Yuan ZH, Wang ZM (2014) Microalgal biofuels: flexible bioenergies for sustainable development. Renew Sust Energ Rev 30:1035–1046

Zimmerman L, Labonte B (2015) Climate change and the microbial methane banquet. Climate Alert 27(1):1–6

Zong G, Fu J, Zhang P, Zhang W, Xu Y, Cao G, Zhang R (2022). Use of elicitors to enhance or activate the antibiotic production in Streptomyces. Crit Rev Biotechnol 42:1260–1283

Part II
Sustainable Agriculture and Biotechnological Solutions

Applications in Food Security, Crop Production, and Soil Health

Chapter 2
Microorganisms for Sustainable Agriculture and Food Security

Chanda Bharti, Priya Mishra, Priyanka Verma, Ankita Bhattacharya, Pranay Arora, and Naveen Kumar Arora

Abstract Ensuring food for the growing population is the major challenge in the years to come. Microbial inoculants offer a long-term sustainable way to boost output without endangering the environment or public health. Abiotic and biotic stressors have become more common due to climate change, which have decreased productivity. Microbe-based biostimulants have shown to increase the productivity of soils including degraded agro-ecosystems. Rhizobacteria, soil or endophytic bacteria, endo- or ectomycorrhizal fungi, or cyanobacteria are examples of microbes that increase plant productivity in several direct and indirect ways, such as by solubilizing minerals, producing phytohormones, alleviating plant stress tolerance, and generating antipathogenic compounds. The development of bioformulations with greater shelf life, broader applicability, and ease of use has a lot of possibilities. This chapter discusses the variety of microorganisms utilized as bioinoculants and their mechanisms for increasing plant productivity. The chapter also discusses on the ability of microbes to fight various abiotic stresses and enhancement of productivity in such situations with future prospects in this field.

Keywords Biostimulants · Agroecosystems · Cyanobacteria · Abiotic stresses · Phytohormones · Rhizobacteria

---

C. Bharti · P. Mishra · P. Verma · A. Bhattacharya · N. K. Arora (✉)
Department of Environmental Science, School of Earth and Environmental Sciences, Babasaheb Bhimrao Ambedkar University, Lucknow, UP, India

P. Arora
School of Geography and the Environment, The University of Oxford, Oxford, United Kingdom

N. K. Arora et al. (eds.), *Environmental Sustainability and Biotechnology: Opportunities and Challenges*, Microorganisms for Sustainability 46,
https://doi.org/10.1007/978-981-96-9025-1_2

2.1 Introduction

World population is increasing and is expected to reach up to 9.7 billion by 2050, which will require 70% increase in food requirements to feed the population in a sustainable manner (FAO 2011). There is a tremendous pressure on the agricultural sector to provide food security to the increasing population. The problem of hunger varies throughout the planet, and poverty along with chronic malnutrition limits human capital development. Economic growth is only sustainable if all countries have food security strategies. Food security is needed as a major tool to end hunger along with improving nutritional status, which will improve human health and also will act as key to reach sustainable economic development with the ongoing land system changes on the planet (Bonis-Profumo et al. 2019). Food security is catalogued as one of the prime challenges under the United Nations Sustainable Development Goals (SDGs) as the second pillar of "Zero Hunger" (Hambrey 2017; Subramaniam and Masron 2021). With the increasing population and soil degradation, the need for cultivable land is increasing, which is reducing forest cover impacting biodiversity and increasing carbon emissions. Hence, there is a need for high-yield cultivars requiring less area or those that can adapt and grow in stressed or degraded lands. As the competition for land resources is also increasing, there is a need to enhance the efficiency of food production from the available arable or marginalized/stressed agro-ecosystems. Usage of chemicals in agricultural fields although has helped in improving the productivity but also resulted in many environmental and health complications in humans. Researches have shown that the application of these agrochemicals can initially result in high productivity but subsequently reduces the fertility of soil, severely impacting the soil health and biodiversity leading to land deterioration over the years (Arora et al. 2018). Additionally, indiscriminate use of these chemicals causes environmental problems like eutrophication of water bodies and outbreak of resistant plant pathogens, which ultimately lead to disturbances in food chain (Wang et al. 2019). The target of an increase in crop production must be achieved sustainably, keeping in mind the issues of climate change and soil degradation (Godfray et al. 2010; Davies et al. 2020; Bahar et al. 2020).

It has been proposed that the new agricultural interventions including organic ways of farming could ensure nearly 85% of global agricultural production in the coming 10 years, reducing the dependency on chemical fertilizers by 20% (Faizan et al. 2021; Verma et al. 2022). Many scientific evidences have been produced, linking agricultural productivities with climate change. Lack of awareness and limited understanding regarding the complex interlinks between climate changes and food production have resulted in the loss in agricultural sector in the past years (Smith et al. 2020; Toromade et al. 2024). Agriculture is a complex and sensitive sector which is impacted by slight changes in rainfall and temperature patterns; thus, climate change becomes the primary determinant of agricultural productivity. Nowadays, climate change is also responsible for biotic and abiotic stresses, due to erratic rainfall and increasing temperatures (Tewari and Arora 2013; Yasmeen et al.

2019). Changes in temperature and rainfall patterns affect the life cycles of plant pathogens, making them more resistant.

Plant growth-promoting microorganisms (PGPMs), especially plant growth-promoting rhizobacteria (PGPR), arbuscular mycorrhizal fungi (AMF), and ecto-mycorrhizal fungi have the potential to be used as green substitutes to agrochemicals in the agricultural sector and are gaining popularity for being more sustainable and environmentally sound options for improving crop productivity (Dodd and Pérez-Alfocea 2012; Tewari and Arora 2016; Etesami and Beattie 2018). PGPMs belonging to the genera *Bacillus, Pseudomonas, Rhizobium, Azotobacter,* and *Azospirrilum* among bacteria; fungi including genera *Trichoderma, Aspergillus, Piriformospora, Penicillium,* and *Phoma,* and mycorrhizal species from *Glomus, Entrophospora, Acaulospora,* and *Sclerocystis,* are the most commonly available microbes used as bioinoculants and dominating the market due to their diverse mechanisms of action and mutualistic associations with plants. Plant microbiome helps their host plants multidimensionally through nutrient acquisition by nitrogen fixation, solubilization of minerals of phosphorus (P), iron (Fe), zinc (Zn), and potassium (K), and production of phytohormones and siderophores, and suppresses pathogens by the production of antibiotics, lytic enzymes, and other metabolites such as hydrogen cyanide. Moreover, they also can produce stress-reducing metabolites, for example, 1-amino cyclopropane-1-carboxylate deaminase (ACCD), which can cleave stress ethylene, or produce various antireactive oxygen species (ROS) molecules, thus ameliorating abiotic stress (Ali et al. 2014). This chapter will mainly focus on the types of microorganisms that are currently used as biostimulants of plants, and their mechanisms will be discussed that help in enhancing plant productivity and ameliorating various environmental stresses to achieve the targets of "Zero Hunger" in a sustainable manner.

2.2 Microorganisms Involved in Sustaining Agro-Ecosystems

Modern agricultural methods heavily rely on synthetic fertilizers, pesticides, intensive tillage, and excessive irrigation, which undoubtedly aided many developing nations in meeting their food needs. Along with these benefits, they have also given rise to environmental hazards such as soil degradation, overexploitation of natural resources, and increased cost of agriculture production. In light of these concerning scenarios, use of helpful microbes known as PGPMs, as microbial inoculants (biofertilizers), present a sustainable solution for enhancing crop productivity and maintaining food security (Singh et al. 2016, 2020; Arora and Fatima 2022). PGPMs have the ability to establish a substantial presence in both soil and plants. These microorganisms play a crucial role in promoting plant growth and health by employing multiple beneficial mechanisms (Fig. 2.1).

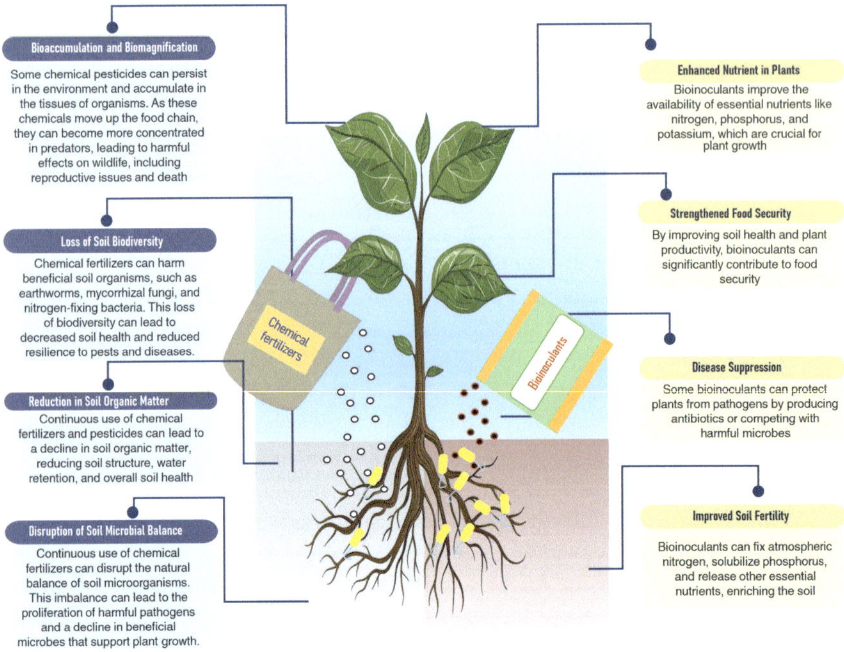

Fig. 2.1 Contrasting impacts of chemical fertilizers and bioinoculants on soil health, plant productivity, and sustainable agriculture

2.2.1 Bacteria

Microorganisms exhibit distinctive characteristics that are frequently lacking in more complex organisms, yet they play a crucial role in agricultural ecosystems. Due to their unique biological and physiochemical properties, soil bacteria are often formulated and employed as effective microorganisms (EMs) for increasing crop productivity (Qadir et al. 2024). The group of EMs known as PGPR are diverse bacteria found in the rhizosphere and on root surfaces that enhance plant growth and protect against diseases and abiotic stresses (Glick 2012; Souza et al. 2015). The role of PGPR is crucial in nitrogen fixation, iron chelation through siderophore production, ACCD production for stress management, solubilizing essential elements such as nitrogen (N), carbon (C), and phosphorus (P), and enhancing resilience from biotic and abiotic stresses. Table 2.1 illustrates various microbial inoculants which are involved in controlling phytopathogens and improving plant growth using different mechanisms. They also enhance soil characteristics, improve photosynthetic efficiency, produce growth regulators, control soilborne diseases, and aid in lignin breakdown (Hamayun et al. 2015; Zahoor et al. 2017).

The initial microorganisms used as bioinoculants were the "rhizobia," nitrogen-fixing bacteria that can live in the rhizosphere and form nodules in the roots of host plants. Family Rhizobiaceae includes genera such as *Rhizobium, Agrobactrium*

Table 2.1 Mechanisms of microorganisms involved in plant growth promotion for food security

Microbial inoculants	Genus/Species	Major mechanism	Crop	References
Bacteria (PGPRs)	*Azotobacter chroococcum* 76A	Increased photosynthetic efficiency and rosmarinic acid content	*Ocimum basilicum*	Comite et al. (2021)
	Pseudomonas thivervalensis, *Paenibacillus amylolyticus*, *Paenibacillus polymyxa*, and *Peribacillus simplex*	Increased seedling growth, enhance copper (Cu), molybdenum (Mo), manganese (Mn) and iron (Fe) concentration	Canola	Martínez-Hidalgo et al. (2021)
	Rahnella aquatilis, Enterobacter ludwiqii, Paenibacillus brasiliensis, and Bacillus subtilis spp.	Siderophore production, phosphate solubilization, N-fixation, phenazines and antibiotic production, 1-aminocyclopropane-1-carboxylate deaminase (ACCD)	Olive	Bouaichi et al. (2019)
	Paenibacillus peoriae	Produced volatile organic compounds (VOCs), increased seedling growth, above ground biomass and lateral roots	Arabidopsis thaliana	Wang et al. (2023a, b)
	Ochrobactrum sp. EB-165, *Microbacterium* sp. EB-65, *Enterobacter* sp. EB-14, and *Enterobacter cloacae* strain EB-48	Showed multiple plant growth promoting (PGP) traits, improved relative water content (RWC) and membrane stability index (MSI), provide osmotic adjustment	Sorghum	Govindasamy et al. (2020)
	Bacillus spp	Enhanced nutrient uptake, reduced malondialdehyde (MDA) content and increased superoxide dismutase (SOD), peroxidase (POD), and polyphenol oxidase (PPO) activities ascorbate peroxidase (APX), catalase (CAT) and flavonoid	*Zea mays* L.	Azeem et al. (2022)
	Azospirillum brasilense	Increased root shoot N uptake, grain yield, biomass, enhanced agronomic efficiency (AE), and apparent N-fertilizer recovery	Wheat	Galindo et al. (2022)
	Lysinibacillus fusiformis, Enterobacter sp., *Brucella intermedia, E. cloacae, Alcaligenes faecalis,* and *Enterobacter Hormaechei*	Solubilized potassium, produced siderophore, produced organic acids	Cotton	Santosh et al. (2022)
	A. brasilense, *B. subtilis*	Provided higher root growth, improved N balance index, higher photosynthetic rate, high flavonoid content	Maize	Galindo et al. (2024)
	Paeonia lactiflora	Enhance root growth, produced indole acetic acid (IAA), ACCD, enhanced photosynthetic activity, nutrient accumulation and antioxidant activity	*Arabidopsis thaliana*	Yuan et al. (2022)
	Pseudomonas hunanensis SPT26	Produced pyrolnitrin, 1- hydroxyphenazine, increased the growth parameters, antioxidant activity, antagonist against *Fusarium oxysporum*	Tomato	Verma et al. (2024a, b)
	Burkholderia gladioli	Fixed nitrogen, solubilized P and potassium (K), produced siderophore, VOCs, activated salicylic acid (SA), jasmonic acid (JA) and ethylene (ET) dependent signalling pathway,	Tobacco and table grapes	Wang et al. (2023a, b)

(continued)

Table 2.1 (continued)

Microbial inoculants	Genus/Species	Major mechanism	Crop	References
Fungus	*Chaetosphaeronema achilleae*	Produced IAA and siderophore, solubilize phosphate, secreted hydrolytic enzymes, improved chlorophyll content and biomass	Wheat	Mohamed et al. (2022)
	Colletotrichum yulongense Ai1, Colletotrichum cobbittiense Ai2, Colletotrichum alienum Ai2.1, Colletotrichum cobbittiense Ai3, and Fusarium equiseti	Showed significant production of alkaloids, phenolics, flavonoids, saponins, higher total phenolic concentration (TPC), total flavonoid concentration (TFC), enzymes such as cellulases, amylases, laccases, lipases	*Anisomeles indica* L.	Toppo et al. (2024)
	AMF with PGPR	Enhanced plant growth and physiochemical properties of soil	*Elymus nutans* Griseb	Yu et al. (2022)
	Rhizoglomus irregularis and *Funneliformis mosseae*, as well as *Trichoderma koningii*	Induced the accumulation of carotenoids, saponins, and TPC, gibberellin, auxin, and cytokinin (CTK)	*Capsicum annuum* L.	Bonini et al. (2020)
	Cladosporium tenuissimum DF11	Upregulate the expression of genes encoding the biosynthesis of tanshinone (*DXS2, CMK, HMGS2, 1, KSL1, KSL2, CYP76AH1*, and *CYP76AK3)*	*Salvia miltiorrhiza*	Chen et al. (2022)
	Trichoderma asperellum	Solubilized phosphate, provide nitrogen, improved uptake of K, sodium (Na) and Mn	Tomato	Allaga et al. (2020)
	Anthemis altissima, Matricaria parthenium, Cichorium intybus, Achillea millefolium, and A. filipendulina	Increased seed germination, IAA production	*Zea mays* L.	Hatamzadeh et al. (2023)
	Rhodotorula mucilaginous yeast	Improved plant growth and nitrogen use efficiency, upregulated NODULE *INCEPTION*-like transcription factors	*Oryza sativa L.*	Paul et al. (2020)
	Aspergillus niger MJ1 with *Pseudomonas stutzeri* DSM4166	Improved growth, soil organic matter (SOM), total nitrogen (TN), microbial communities, protein and vitamin C	Cucumber and lettuce plants	Ni et al. (2023)
	Glomus mosseae	Improving photosynthetic efficiency, antioxidant enzyme activity, soluble sugar and soluble protein content, and reduce MDA content	Grape	Jian et al. (2024)
	G. mosseae	Decreased leaf necrotic, ion leakage and MDA content but increased SPAD index, catalase and guaiacol peroxidase activities	Grapevine	Karimi and Noori (2022)

Algae	*Chlorella, Dunaliella salina, aphanothece* spp., *Athrospira maxima*	Chlorophyll content, nutrient uptake, increased shoot and root, enhanced root volume, improved osmotic adjustment	*Solanum lycopersicum* L.	Mutale-Joan et al. (2020)
	Chlorella vulgaris, Nannochloropsis salina	Photosynthetic pigment, increased rutin and gallic acid content and many bioactive compounds	*Moringa oleifera*	Al Dayel and El Sherif (2021)
	Anabaena torulosa ISB213. Nostoc calcicola ISB215, *Nostoc ellipsosporum* ISB217	Improved plant height, dry and fresh weight, leaf and flower number, leaf area, chlorophyll content, TPC and TFC, enhanced highest expression of *DXR* and *TPS2* genes, increased essential oil accumulation	*Thymus vulgaris* L.	Rasuli et al. (2024)
	C. vulgaris and Spirulina platensis	Showed highest vegetative growth, yield, chlorophyll content, improved micronutrient (Fe, Zn, Mn, and cu) uptake	Tomato	Mostafa et al. (2024)
	Asterarcys quadricellulare	Improve chlorophyll, amino acid, sugar, nitrate reductase, promoted hypocotyl expansion	Potato	Cordeiro et al. (2022)
	Chlamydomonas sp., *and Chlorella* sp.	Fixed nitrogen, showed exopolysaccharides (EPS) production, siderophore production, IAA production, enhanced seed germination	*Oryza sativa* L.	Lamb et al. (2023)
	Anabaena sp., *Aulosira* sp., *Cylindrospermum* sp., *Nostoc* sp., and *Tolypothrix* sp.	Biomass used for large scale production of food, secondary metabolites, medicines	*Oryza sativa* L.	Chittora et al. (2020)
	Anabaena variabilis SCAU30	Controlled rice sheath blight, increased yield by 14–22%, showed higher growth rate and nitrogenase activity, produced CTKs, auxins, and SA	*Oryza sativa* L.	Bao et al. (2021)
	Ulva lactuca and *Solieria* spp.	Enhanced photosynthesis, increased biomass and yield, enhanced N assimilation	*Zea mays* L.	Pérez-Álvarez et al. (2024)

Allorhizobium, Neorhizobium, Pararhizobium, Bradyrhizobium, Sinorhizobium, Azorhizobium, and *Mesorhizobium*. Among these genera, the genus *Rhizobium* has been extensively researched and holds great significance in the context of nitrogen fixation. Legume plants are closely linked to this particular bacterium, making it a crucial contributor to the production of pulses (Laranjo et al. 2014). Further, rhizobitoxine, a phytotoxin synthesized by rhizobial strains, hinders the production of ethylene (ET) by chemically targeting the enzyme ACC synthase (Nascimento et al. 2012). Mutualistic nitrogen-fixing partners such as *Mesorhizobium, Sinorhizobium, Azorhizobium, Bradyrhizobium, Rhizobium*, and *Allorhizobium* play a beneficial role in enhancing plant growth and increasing yield (Lindström and Mousavi 2010).

At present, several bacterial genera, apart from rhizobia, are employed in the development of microbial inoculants owing to their diverse metabolism. For example, various species of the genus *Bacillus* stimulate plant growth, control phytopathogens, and possess the ability to form spores, making them resistant to harsh environmental conditions. Rhizospheric *Bacillus* species are also known for their ability to synthesize various metabolites that protect plants from fungal pathogens (Prakash and Arora 2021). Plants demonstrated resistance to tobacco mosaic virus through the mechanisms of *Bacillus artrophaeus* HAB-5, which functions through the signaling pathways dependent on salicylic acid, jasmonic acid (JA), and ET. As a result of this interaction, there was an enhanced expression of genes that facilitate plant growth (Rajaofera et al. 2020). A study on eggplants revealed that *Bacillus mucilaginosus* and *Bacillus megaterium* markedly enhanced the solubilization of K and P, thereby promoting the growth characteristics of the host plant (Hamid et al. 2021). The photosynthetic efficiency of *Glycyrrhiza uralensis* improved with the application of the PGPR, *Bacillus pumilus*, which also preserved the structural integrity of chloroplasts and mitochondria in drought-stressed cells (Zhang et al. 2019).

According to Yousefi et al. (2017), *Azospirillum* spp. are recognized as bioinoculants for their ability to promote plant growth through biological nitrogen fixation (BNF), phytohormone production, synthesizing ACCD, phosphate solubilization, and siderophore synthesis. Additionally, they also exhibit biocontrol properties against phytopathogens (Prasad and Babu 2017; El-Komy et al. 2020). In vitro studies have revealed improved biomass, differentially expressed genes, Zn uptake, auxin production, and N-fixation to their host plants (Housh et al. 2021; Cook et al. 2022). The genus *Azospirillum* has been found to be associated with *Pseudomonas*, facilitating the rapid colonization of plant roots. The observation supports the notion that these genera can collaborate as bacterial inoculants to enhance plant growth (Elnahal et al. 2022). *Azotobacter* strains have also been used as biostimulants as nitrogen fixers, protecting plants from infectious diseases and synthesizing substances that promote plant growth along with the enhancement of mineral absorption (Mathivanan et al. 2015; Nosheen et al. 2019). BNF through *Azotobacter* species contributes N input to soil in between 0.3 and 15 kg per hectare per year (Saha et al. 2017). Kumar et al. (2022) demonstrated that minimizing chemical fertilizers significantly benefits yam (*Dioscorea alata*) through the effects of *Rhizophagus irregularis* and *Azotobacter*. Additionally, Singh et al. (2022) found that co-inoculating *Azotobacter chroococcum* WR5 and *Piriformospora indica* increased Zn and Fe levels in wheat plants and enhanced *ZIP* transporter gene expression.

Many *Pseudomonas* spp. have been found in controlling phytopathogens and disease management through their direct and indirect mechanisms. *Pseudomonas* spp. fix atmospheric nitrogen, produce phytohormones, solubilize K, Zn, and P, and also produce volatile organic compounds (VOCs), lytic enzymes, secondary metabolites, exopolysaccharides (EPS), ACCD, and antibiotics during stress conditions (Fatima and Arora 2021, 2023; Dimkić et al. 2022; Saranraj et al. 2022; Singh et al. 2024). Among these, hydrogen cyanide (HCN) produced by many rhizobacteria plays an important role in the biological control of pathogenic bacteria and fungi in soil. Vanitha and Ramjegathesh (2014) reported siderophore, antibiotic, and HCN synthesis by *Pseudomonas fluorescens* species that affected the proliferation of *Macrophomina phaseolina* (Tassi) Goid. Numerous studies have emphasized the significant involvement of *Pseudomonas* spp. in inhibiting a range of fungal pathogens (Win et al. 2022; Li et al. 2023), inducing systemic resistance (ISR) (Coerini et al. 2025), producing VOCs (Ling et al. 2025), and activating signals for gibberellic acid (GA) and JA (Behdad and Ganjeali 2025). Biofertilizers prepared from *Pseudomonas* spp. indirectly promoted host plant growth by producing water-soluble vitamins, siderophores, indole acetic acid (IAA), cytokinin (CTK), gibberellins, and ET, biosynthesizing cyanide, and inducing systemic resistance genes (Riaz et al. 2021). Clough et al. (2022) documented the suppressive impact of *Pseudomonas* strains on *Ralstonia solanacearum*, attributed to the synthesis of 2,4-diacetylphloroglucinol (2,4 DAPG), pyoluteorin, and orfamides, which subsequently mitigated bacterial wilt disease in tomato cultivar. Mishra et al. (2022) reported that *P. fluorescens* JM-1 produced the antifungal compound DAPG to control ear rot disease caused by *Fusarium moniliforme* and also enhanced the growth and yield of maize. Genera *Mitsuaria* and *Burkholderia* that have demonstrated efficacy in preventing soilborne pathogens in crops have also been extensively researched (Gathage et al. 2016). Some studies have shown that *Burkholderia* functions as biocontrol agents (BCAs) in tomato plants through the production of siderophores, and evidences of N-fixation were also found. Consequently, it is crucial to elucidate the mechanisms underlying their activity and their interaction with other biocontrol agents (Russo et al. 2019; Sarli et al. 2021). *Burkholderia gladioli* has been found to effectively protect tobacco and table grapes against *Botrytis cinerea* (Wang et al. 2023a, b). *Burkholderia phytofirmans* PsJN, on the other hand, was found to be effective in improving the physiology and yield of Quinoa under salinity stress (Yang et al. 2020). Further, *Burkholderia cepacia* complex (Bcc) has been used as BCA and for bioremediation of xenobiotic compounds (Jhutty 2019; Tavares et al. 2020). However, Bcc has the potential to lead to severe infections in humans, especially among individuals suffering from cystic fibrosis (Gathage et al. 2016; Gutiérrez Santana and Coria Jiménez 2024).

Furthermore, strains of *Micrococcus, Ralstonia, Enterobacter, Pantoea, Acinetobacter, Aeromonahy, Burkholderia, Microbacterium,* and *Alcaligens* have the ability to induce systemic resistance, improve tolerance to salinity, and aid in the fixation of atmospheric nitrogen (Ali et al. 2017). Moreover, there are other bacterial species that are associated with the ability to withstand various forms of biotic and abiotic stress. For instance, *Alcaligens, Serratia,* and *Bacillus* are known for their tolerance toward salinity stress and help plants under salt stress conditions (Fatima et al. 2020; Kulkova et al. 2024; Bharti et al. 2024). Some strains of *Enterobacter* can promote plant growth and increase IAA, P, and Fe concentrations as well as accumulate heavy metals (Li et al. 2020).

Azotobacter and *Bacillus* are linked to enhancing nutrient uptake efficiency (Baghel et al. 2020; dos Santos et al. 2022; Ojija and Aloo 2023). When introduced to alfalfa plants experiencing saline stress, the endophytic and rhizospheric strains of *Halomonas, Bacillus, Virgibacillus*, and *Zhihengiluella*, which were isolated from the halophytic species *Salicornia rubra, Sarcocornia utahensis,* and *Allenrolfea occidentalis* in Central Utah (with salinity levels ranging from 5% to 27%), exhibited the ability to enhance growth and improve stress resilience (Kearl et al. 2019). Further, certain bacteria found in the wheat microbiome, including *Bacillus subtilis, Bacillus amyloliquefaciens*, and members of the *Paenibacillus* genus, have been successfully utilized to address wheat diseases caused by pathogens such as *Fusarium graminearum, F. moniliforme, Gaeumannomyces tritici, and Bipolaris sorokiniana. This is achieved through the activation of systemic defence mechanisms. Therefore, bacterial bioinoculants are helpful in plant growth promotion as well as plant protection against pathogenic agents (Samain et al.* 2019; Chaparro-Encinas et al. 2022; Mashabela et al. 2023).

2.2.2 Fungi

Plant diseases pose a significant risk to the health and productivity of both natural and man-made environments. Use of synthetic fungicides to manage plant diseases has resulted in a negative impact on human health, wildlife, and the ecosystem (Steinberg and Gurr 2020). Beneficial fungi possess the capacity to outcompete pathogens in the acquisition of resources and inhabiting the same environments, including plant tissues, rhizosphere, or phyllosphere (Ghorbanpour et al. 2018). These fungi promote plant growth and development through various mechanisms, including converting insoluble nutrients into soluble forms and producing plant growth regulators, enhancing ISR, and producing VOCs. Fungi are also helpful in reducing biotic stresses such as plant pests and pathogens, as well as abiotic stresses like drought, salinity, temperature, and heavy metals (Yadav et al. 2020). Various types of fungi, known as plant-growth-promoting fungi (PGPF), including *Aspergillus fumigatus, Aspergillus niger, Trichoderma harzianum, Mucor mucedo, Phoma* spp., and *Taxomyces andreanae,* have actively engaged in contributing to sustainable farming practices and enhancing crop yields (Das et al. 2021). *Penicillium janthinellum* LK5, a type of endophytic fungus found in *Solanum lycopersicum* Mill (tomato plant), has the ability to produce bioactive gibberellins GA3, GA4, and GA7. This fungus has been observed to enhance the growth of rice seedlings and tomatoes, effectively reducing sodium toxicity in these plants (Khan et al. 2014). Tarroum et al. (2022) reported that *Penicillium olsonii,* isolated from the rhizosphere soil of *Aeluropus littoralis,* significantly enhanced the growth parameters, proline content, catalase (CAT), and superoxide dismutase (SOD) activities of tobacco plants under salt stress conditions.

Trichoderma is a beneficial fungus with the ability to control many plant pathogens, pests, and weeds by two main mechanisms: first, mycoparasitism, and second, nonmycoparasitism, in different environmental conditions (Al-Ani 2018). It reduces the reliance on chemical NPK fertilizers, enhances the absorption of micronutrients, and aids in the solubilization of phosphates (Kamal et al. 2018). According to Molla et al. (2012), the utilization of a combination of 50% nitrogen fertilizer and 50% *Trichoderma*-enriched biofertilizer resulted in a significant increase in the yield of mustard and tomato by 108% and 125% respectively, as compared to the control condition. Some *Trichoderma* species can produce siderophores (iron-chelating compounds). Through the sequestration of ferric ions in their environment, *Trichoderma* species can hinder the growth and activity of soilborne and post-harvest pathogens like *Botrytis cinerea* (Harman et al. 2004; Sood et al. 2020). They secrete extracellular enzymes, including endochitinases, β-1,3-glucanases, and proteases, which are primarily responsible for breaking down the cell walls of fungal pathogens (Tchameni et al. 2020). Some antibiotic compounds such as trichorzianins and antibiotic peptides obtained from the culture extracts of *Trichoderma* spp. showed inhibitory effects on *Phytophthora* and *Pythium ultimum* (Silva et al. 2019).

Baron et al. (2020) documented that *Purpureocillium lilacinum, Purpureocillium lavendulum,* and *Metarhizium marquandii* were able to promote soybean, bean, and maize dry matter and the availability of important nutrients such as N and P. *Muscodor* and *Nodulisporium* have been used as soil inoculants and widely recognized for inhibiting pathogenic fungi by VOCs' production (Kaddes et al. 2019). Baron et al. (2020) further illustrated the capability of *P. lilacinum, P. lavendulum,* and *M. marquandi* to synthesize IAA and solubilize P from fluoratite. The selected strains improved growth in soybean, bean, and maize by enhancing the dry matter in shoots and roots and increasing essential nutrients like P and N, attributed to endophytic colonization. Recently, during a study, *Aspergillus chevalieri* AGH-1 OP593083 and *Aspergillus egyptiacus* AGH-4 OP593080 were found as effective BCAs against *Alternaria solani.* The isolates were capable of producing siderophores, IAA, and solubilizing phosphates in the soil. Additionally, *A. chevalieri* and *A. egyptiacus* enhanced *Vicia faba* plant immunity by increasing phenol, proline, and antioxidant activities [peroxidase (POD) and polyphenol oxidase (PPO)] and reducing oxidative stress by lowering malondialdehyde (MDA) and hydrogen peroxide (H_2O_2) levels (Daigham et al. 2024). Similarly, multiple beneficial species of fungi have been described that enhanced finger millet plant growth using different mechanisms (Choudhary et al. 2020; Mahadik et al. 2024; Mahadik and Kumudini 2024). *Capsicum annuum* L. inoculated with PGPF showed 78.75% more protection against anthracnose disease, higher accumulation of lignin and callose, along with enhanced defense enzymes (Naziya et al. 2019).

2.2.3 Mycorrhiza

In response to the prevailing climate risks, most plants have a variety of defence mechanisms to ensure their survival and to sustain their growth and development. The association of AMF with multiple beneficial attributes for plants has shown to be very important, providing several benefits to their symbiotic partners (Imran and Amanullah 2022). In these relationships, non-moving substances like P, Zn, and copper (Cu) are taken up by the fungal symbiont and transferred to the plant (Ortas 2022). The enhanced uptake of different mineral nutrients, such as P, in mycorrhizal plants is credited to the expansion of the surface area available for nutrient absorption through the hyphal network (Bhattacharjee and Dey 2014; Yadav et al. 2018). Plants that have formed mycorrhizal associations show a substantial increase, up to 50 times, in the absorptive surface area of their roots compared to plants without such associations, as reported earlier (Ilyas et al. 2022). AMF maintain symbiotic association with more than 70% of terrestrial plants, supplying them essential nutrients (P, N, K, Cu, Ca, and Zn) and water in return for sugars (Wagg et al. 2019; Yu et al. 2022).

AMF mycelia have the ability to enhance soil moisture retention through the improvement of soil aggregation (Bedini et al. 2009). Plants colonized by AMF can take up a larger amount of P, almost 80% of the plant's P uptake, from lower concentrations in the soil solution in comparison to nonmycorrhizal plants (Douds et al. 2007; Rakshit and Bhadoria 2010; Yang et al. 2023a). Furthermore, introducing AMF during stress conditions could elevate the concentrations of total phenols, flavonoids, and also defence enzymes such as phenylalanine ammonia-lyase (PAL), and lipoxygenases (LOXs) in various agricultural crops (Pan et al. 2023). AMF is purportedly the genuine substitute for enhancing the growth and development of plants that are commonly found in low-fertility soil (Yang et al. 2023b).

Li et al. (2022) reported that *Piper nigrum* L. cv. Bragantina plants exhibited variations in the chemical compositions of their leaf essential oils following AMF inoculation, with elemol levels rising from 11.4% to 23.2% and α-bisabolol levels increasing from 32.3% to 48.5% (Yang et al. 2023a). Dual inoculation of PGPF and PGPR has shown the increased concentration of K^+ accompanied by an effective decrease of Na^+ in maize tissues (Moreira et al. 2020). El-Sawah et al. (2021) found stimulating effects of bioinoculants produced from *Bradyrhizobium* sp., *B. subtilis*, and AMF, individually or in combinations, resulting in enhanced biomass, leaf area index, chlorophyll content, proteins, carbohydrates, fats, and starch of guar plant. Additionally, bioinoculant treatments showed improved soil microbial activity by increasing dehydrogenase, phosphatase, protease, and invertase enzymes.

N-fixing cyanobacterium *Nostoc punctiforme* (Álvarez et al. 2020) and the mycorrhizal fungus *R. irregularis* (Wang et al. 2020) promoted the growth of rice plants by significantly improving nitrogen acquisition. The symbiotic association between the host plant and AMF also improves their physiological resilience against abiotic stressors and pathogens (Emmanuel and Babalola 2020). An enhancement in growth has been noted with the use of AMF in tarragon and lavender plants

(Golubkina et al. 2020a). Biomass increase ranging from 20% to 35% was observed following the application of AMF in *Artemisia dracunculus* and *Hyssopus officinalis*. Further, AMF inoculation increased the accumulation of Se, indium, and Zn in *A. dracunculus* while decreased the levels of heavy metals in *H. officinalis*. Additionally, the application of an AMF-based formulation resulted in the growth and yield increase of 1.4 and 1.45 times respectively, in *Allium cepa* and *Allium sativum* (Golubkina et al. 2020b). Yu et al. (2022) reported that *Elymus nutans* inoculated with AMF and PGPR showed higher growth parameters and soil nutrients while significantly decreased electrical conductivity and pH.

2.2.4 Algae

Algae encompass a wide range of mostly photosynthetic microorganisms, including both prokaryotic such as cyanobacteria (microalgae) and eukaryotic species like green algae, euglenoids, and diatoms (macroalgae) (Abideen et al. 2022). Blue-green algae (BGA), also known as cyanobacteria, can be used as biofertilizers as they can fix atmospheric nitrogen in specialized heterocyst cells and play an important role in plant growth promotion. Various microalgae have been used as biofertilizers including *Acutodesmus dimorphus*, *Spirulina platensis*, *Chlorella vulgaris*, *Scenedesmus dimorphus*, and *Anabaena* (Choudhary et al. 2024). BGA encompass a diverse range of species including *Oscillatoria, Nostoc*, *Anabaena*, *Tolypothrix*, *Aulosira*, and many others. Apart from BGA, red and brown algae have also been studied as potential biofertilizers (Paudel et al. 2012).

Algal extracts are rich in essential nutrients like Zn, K, Mg, Ca, P, S, K, Mo, Cu, Co, and Fe. The P content in the biomass ranges from 0.73% to 1.46% w/w, with the highest concentration observed in marine microalgae *Tetraselmis chuii* and freshwater oleaginous microalgae *Botryococcus braunii* (1.45–1.46% w/w) (Tibbetts et al. 2015). They also contain vitamins, and polyamines that enhance the nutritional value, stimulate vegetative growth, boost yield, and improve fruit quality in a variety of orchards and vineyards. The presence of phytohormones like cytokinins (trans-Zeatin, dihydrozeatin, isopentyladenine, and kinetin), gibberellins (GA1, GA3, GA4, GA20, and GA29), auxins, and abscisic acid (ABA) in microalgal extracts is thought to contribute their vital role in plant growth promotion (Rupawalla et al. 2022; Parmar et al. 2023). Algae can survive in a variety of environmental conditions such as high pH and salt and generally used for the remediation of saline and calcareous soil (Haggag et al. 2018). Microalgae *Coelastrella* sp. SDEC-28, *Dunaliella salina* SDEC-36, and *Spirulina subsalsa* FACHB-351 have the potential to rehabilitate salt-affected soil (Chen et al. 2024a, b). BGA are known to produce various phenolic compounds, polysaccharides, terpenoids, osmolytes, free fatty acids, chlorophylls, xanthophylls, c-phycoerythrin and c-phycocyanin, and carotenes, which synergistically act for plant growth promotion and tolerance to abiotic stresses (Santini et al. 2021). Additionally, microalgae utilize photosynthesis to capture and store CO_2 for carbon sequestration purposes, while some algae also produce EPS that enhance the quality of soil composition, phytochelatins that prevent desiccation of intracellular contents, soluble amino acids that play a crucial role in stress management by acting as osmoprotectants (glycine betaine and proline), and antioxidants such as SOD, POD, CAT, and ascorbate peroxidase (APX) (Jaimes Duarte et al. 2018; Bulgari et al. 2019). Moreover, cyanobacteria such as *S. platensis*, *Phormidium* spp., *Nostoc* spp., *Anabaena fertilissima* PUPCCC 410.5, *Synechocystis* spp., and *Galdieria sulphuraria* are known to produce significant amounts of C-phycocyanin (CPC). *S. platensis* is utilized for the production of CPC for commercial purposes, which serves as a natural blue colorant in food products and as an antioxidant in therapeutic applications (Patel et al. 2022).

Cyanobacteria are abundant in paddy fields, referred to as paddy organisms. These organisms are known for their ability to produce IAA and GA, solubilize phosphate, and are capable of fixing 20–30 kg of nitrogen per hectare in submerged rice cultivation (Sammauria et al. 2020; Afkairin et al. 2021). *Anabaena azollae* is a type of cyanobacterium that establishes a symbiotic relationship with the water fern *Azolla*. Recently, several species of *Azolla, including A. caroliniana, A. microphylla, A. filiculoides, and A. mexicana*, have been introduced in India due to their potential for substantial biomass production (Marzouk et al. 2023). Application of *Azolla*, rice straw, and reduced dose of N, P, and K exhibited the highest P (73.57 kg/ha) uptake and phosphate use efficiency (46.24%) in rice (Marzouk et al. 2024). Cyanobacterial genera, including *Nostoc, Anabaena*, and *Oscillatoria*, contribute to the enhancement of microbial communities, augment nitrogen availability, and improve soil properties. Additionally, they facilitate seed germination and encourage increased root and shoot development, dry biomass, and overall yield in rice and wheat crops (Sammauria et al. 2020). In a study, cyanobacterium *Aliinostoc* sp. YYLX235 was isolated from acidified paddy soil. The pot and field experiment demonstrated that *Aliinostoc* increased rice grain yield while reduced the cadmium accumulation in grain (Hu et al. 2022).

2.3 Microorganisms as Bioinoculants to Enhance Productivity and Control Pests

2.3.1 Bioinoculants to Enhance Crop Productivity

Microbial bioinoculants or microbial biostimulants, are a sustainable approach gradually replacing the overuse of chemical fertilizers and pesticides worldwide. A bioinoculant is a formulation containing beneficial microorganisms, such as bacteria, fungi, or algae, which enhance plant growth by improving nutrient availability, fixing atmospheric nitrogen, decomposing organic matter, or suppressing plant pathogens (Vessey 2003). These bioinoculants are not nutrients themselves but encourage the utilization of nutrients by plants for better growth and development (Castiglione et al. 2021). According to du Jardin (2015), biostimulant is a substance or microorganism that enhances plant growth, nutrient uptake, and stress tolerance by stimulating natural physiological processes, without providing direct nutrition or acting as a biopesticide. In recent years, the use of bioinoculants in agriculture has been increasing, gaining popularity for sustainable farming practices. The market is expected to grow significantly, driven by the production of organic food and improved soil health. Asia Pacific and Africa are major fertilizer consumers due the rising food demand driven by population growth. Governments in Asian countries are now promoting eco-friendly biofertilizers and organic manure to counteract the adverse effects of chemical fertilizers.

There is a diverse array of microorganisms globally used as biostimulants, and their market is expanding rapidly. Among biostimulants, nitrogen fixers hold a significant proportion. The market for microbial bioinoculants was projected to be valued at USD 3.1 billion in 2023 and will reach USD 5.2 billion by 2028, with a CAGR of 10.9%.[1] In India, a major part of the biostimulants' market comprises biofertilizers, valued at USD 99.59 million in 2021. This market is projected to expand from USD 110.07 million in 2022 to USD 243.61 million by 2029, reflecting a CAGR of 12.02%.[2] The market is anticipated to grow further, driven by the rising demand for organic food, the adoption of sustainable farming practices, and the preference for fertilizers that align with modern agricultural methods.

In India, the field application of *Rhizobium*-based biostimulants began in the early 1900s, with the first commercial development starting in 1956 (Reddy et al. 2011). Later, other microbial biofertilizers, including *Azotobacter, Azospirillum*, and blue-green algae, came into play. Currently, these biofertilizers are surging in the market. Some global market players developing biofertilizers for various crop types include Novozymes from Denmark, Mapleton Agri Biotec from Canada, Biomax from Singapore, Rashtriya Chemicals and Fertilizers Ltd. (RCF) from India, Fertoz from Australia, Ficosterra from Spain, Groundwork BioAg from Israel, Kula Bio from the USA, Earnest Earth from the USA, and GSFC Ltd. from India. These companies are leading the way in sustainable agricultural solutions by developing innovative biofertilizers that enhance soil health and crop productivity.[3]

2.3.1.1 Biopesticides

Microbial biopesticides, which include bacteria, fungi, and certain viruses, are specialized to control a range of plant pests and diseases. They support plant survival under biotic stress by suppressing pathogens, inducing plant resistance, and improving nutrient uptake (Mishra et al. 2019). Despite this, chemical pesticides are widely favored by farmers due to their effectiveness in managing a broad spectrum of pests

[1] https://www.marketsandmarkets.com/Market-Reports/compound-biofertilizers-customized-fertilizers-market-856.html

[2] https://www.fortunebusinessinsights.com/india-biofertilizers-market-106527

[3] https://datahorizzonresearch.com/blog/top-biofertilizers-companies-160

(Meissle et al. 2010; Gill and Garg 2014; Weldeslassie et al. 2018; Bashir et al. 2020). The biopesticides market is projected to be valued at USD 6.72 billion in 2025 and is expected to grow to USD 11.41 billion by 2030, with a CAGR of 11.19% during the forecast period from 2025 to 2030.[4] The biopesticides market is categorized based on the form (biofungicides, bioherbicides, and bioinsecticides), crop type (cash crops, horticultural crops, and row crops), and region (Africa, Asia-Pacific, Europe, the Middle East, North America, and South America).[5] Key players in this sector include Koppert Biological Systems from the Netherlands, Marrone Bio Innovations Inc. from the United States, Certis USA L.L.C. from the United States, and Andermatt Biocontrol AG from Switzerland. These companies are pioneers in developing sustainable pest control solutions that reduce chemical pesticide reliance and promote environmentally friendly agricultural practices.[6]

Among bacteria, *Bacillus thuringiensis* (Bt) is the most studied and commercially used bioinsecticide. Bacilli are the choice of interest for combating fungal diseases. For instance, Dobrzyński et al. (2023) highlighted the efficacy of the *B. pumilus* strain against phytopathogens such as *Arthrobotrys conoides*, *Fusarium solani*, *Fusarium oxysporum*, *Sclerotinia sclerotiorum*, *Rhizoctonia solani*, and *Fagopyrum esculentum*. Apart from Bt-based products, a large segment of the market is driven by other Bacilli-containing products. Similarly, *Brevibacillus laterosporus* is another bacterium used in integrated pest management (Ruiu et al. 2014; Marche et al. 2018).

Pseudomonas is another important bacterial genus well-documented for their biocontrol potential against various phytopathogens, demonstrating significant efficacy in the biopesticides market. Research highlights their ability to produce antifungal metabolites that protect crops from pathogenic fungi. For instance, Verma et al. (2024a, b) reported that *Pseudomonas hunanensis* SPT26 produces a range of antifungal compounds that safeguard tomato plants from *F. oxysporum*. Similarly, a study by Mishra et al. (2022) demonstrated that the metabolite 2,4-diacetylphloroglucinol (DAPG), produced by *P. fluorescens* JM-1, effectively manages ear rot disease in *Zea mays* L., caused by *Fusarium moniliforme*. Beyond these findings, *Pseudomonas* strains exhibit remarkable resilience under stress conditions and play a crucial role in suppressing fungal phytopathogens. They have proven effectiveness against several plant diseases, including damping-off disease caused by *Pythium* spp., okra root rot caused by *F. solani*, Rhizoctonia root rot linked to *R. solani*, and foliage blight disease caused by *Phytophthora nicotianae* (Mehmood et al. 2023).

Among biofungicides, *Trichoderma* spp., *Aspergillus* spp., and *Penicillium* spp. are the major players. However, *Beauveria bassiana* is a major fungus available in the market as an insecticide. The *B. bassiana* market, valued at USD 89.2 million in

[4] https://www.mordorintelligence.com/industry-reports/global-biopesticides-market-industry

[5] https://www.mordorintelligence.com/industry-reports/global-biopesticides-market-industry

[6] https://www.fortunebusinessinsights.com/blog/top-10-companies-in-the-biopesticides-market-10480

2022, is anticipated to grow to USD 195.1 million by 2032, with a CAGR of 8.4% from 2023 to 2032. This growth is fueled by the rising demand for organic pest control in agriculture, spurred by concerns over chemical residues and environmental sustainability.[7]

Chemical pesticides have long been favored by farmers due to their potential to manage a wide array of pests. However, growing awareness of the negative environmental and health consequences associated with their use is driving a shift toward more sustainable solutions like microbial biopesticides. Although there are still hurdles related to cost and feasibility, continuous research and advancements are improving their efficacy and market potential. As the trend toward organic farming gains momentum worldwide, the demand for microbial biopesticides is set to increase, paving the way for a more sustainable and healthier future for agriculture and food security.

2.3.2 *Microorganisms for Combating Abiotic Stresses*

Changing climatic conditions across the globe have adverse impacts on agro ecosystems, their health, growth, and productivity. Greenhouse gases (carbon dioxide, methane, fluorinated gases, oxides of nitrogen, etc.) are key factors behind global warming as they trap the heat radiation of sun. Global warming is responsible for rapidly increasing temperatures, imposing heat stress upon the crops, resultantly decreasing the crop yield and quality. Increasing population is becoming a serious threat for accomplishing the target of food security. Climate change has resulted in increased incidences of droughts, floods, temperature extremes, and other forms of abiotic stresses resulting in overall land degradation. Degradation of agricultural lands affects the crop production due to abiotic stresses.

Microorganisms play a crucial role in the management of crop production under abiotic-stressed environments in a sustainable manner. Abiotic stresses are significant issues in the path of achieving sustainable development goals (SDGs) as near about 9% of agricultural area is productive for crop production, while the rest of 91% is being stressed from various abiotic stresses (Minhas et al. 2017). Employing beneficial microorganisms for enhancing the soil fertility and crop production is the significant part of sustainable agriculture, which could fulfill the upcoming demands of agriculture, food supply, and food security (Dilnashin et al. 2020). Plants are entirely dependent on soil and the external environment for their development, healthy growth, and nutrient supply. These abiotic stresses affect the soil and environment, which indirectly affect the development, growth, and productivity of plants. Different stressed conditions, such as soil salinity, drought, water-logging stress, heat stress, cold stress, and heavy-metal stress, affect the morphological and physiological parameters through diverse mechanisms. Microorganisms and their bioinoculants could contribute to combating the deleterious effects of the abiotic stresses.

---

[7] https://www.alliedmarketresearch.com/beauveria-bassiana-market-A52991

2.3.2.1 Soil Salinity

Salinization of soil affects the plant growth by interfering with the water relations, which reduces the osmotic potential. Salt buildup on the top layer of soil causes water logging, which reduces water absorption in the soil and generates a shortage of water in the plant roots (Yadav et al. 2020). Microorganisms when applied to the saline soil enhance the soil quality by improving the various physical and chemical properties of soil. They also increase the microbial biomass in soil, which improves the soil texture, water-holding capacity, and osmotic potential (Bhardwaj and Kumar 2020). Diazotrophic bacteria are also found to be significant in the plant growth promotion through various mechanisms such as the regulation of phytohormones, radical scavenging activity, and osmolyte production, inducing plants tolerance under saline conditions (Verma et al. 2024a, b). Enhancing the effects of antioxidants eliminates the ROS, which decreases oxidative damage caused by the abiotic stress (Talaat 2015). Application of salt-tolerant endophytic *Bacillus altitudinis* NKA32 was found to enhance the productivity of rice growing under saline conditions by the production of ACCD along with proline, trehalose, antioxidants, and sodium-ion accumulation at various salt concentrations (Bharti et al. 2024). Agricultural output in the cultivation of rice was observed to be increased when treated with the salt-tolerant strain *Pseudomonas atacamensis* KSS-6 combined with organic manure. The treatment also stimulated the nutritional value of grains and nutritional status of soil by remediating the salt-affected agro ecosystems (Arora et al. 2024). Production of secondary metabolites by the halo-tolerant microorganisms also supports the plants' survival, growth, development and productivity. EPS-producing strain *Pseudomonas entomophila* PE3 promoted the production and growth of sunflower under salinity through the production of trisaccharide sugar melezitose, which further increased with the increasing salt level (Fatima and Arora 2023). The study conducted by Forouzi et al. (2020) found that the inoculation of endophytic fungus *Piriformospora indica* and Kj152148 *Streptomyces* sp. and K152149 *Streptomyces* sp. decreased the deleterious effects of salinity through augmenting the radical scavenging activity, phenols, flavonoids, proline content, and antioxidants in Stevia plant.

2.3.2.2 Drought

Drought conditions bring disturbance in water relation and flow between xylem and adjoining elongating cells, which participate in the growth of plants. This disturbance causes hindrance in mitosis due to which the leaf area and length of the plant are reduced and overall retard the plant growth and expansion (Yadav et al. 2020). Inoculation of drought-tolerant microorganisms in agricultural field experiencing the scarcity of water is found fruitful in fighting against such stressful conditions. A study conducted by Admassie et al. (2022) showed that treatment of pepper seeds and plant with drought-resistant bacteria (*Enterobacter hormaechei*, *Rhizobium* spp., and *P. fluorescens*) reduced the effect of drought stress and promoted the plant growth by auxin and EPS production. Consortium of PGPR (*Micrococcus yunnanensis*) and AMF (*Claroideoglomus etunicatum*) were found to stimulate the growth of *Dracocephalum moldavica* under water-deficit conditions. The study revealed that consortium significantly protected the plant from drought stress and boosted the growth through increasing nutrient uptake, photosynthesis, and production of secondary metabolites (Ghanbarzadeh et al. 2020). When observed alone or in combination with the AMF (*Glomus coronatum, Glomus constrictum, or Glomus claroideum*), *B. megaterium* proved to be the most effective bacterium during drought (in consecutive harvests). The rhizosphere and endorhizosphere zones were invaded by *B. megaterium*. Under drought conditions, their inoculation raised the water content and biomass of the shoots and roots. The efficiency of bacterial IAA in boosting plant growth and shoot water content, which increases the plant resistance to drought, may be explained by their production under stressful situations (Marulanda et al. 2009). The tiny millet leaf apoplast contains *B. amyloliquefaciens* LAS10, which can withstand a lower osmotic potential of −3.6 MPa. Under adverse conditions, the strain generates bioactive compounds that help plants withstand drought stress. *B. amyloliquefaciens* is reported to increase crop yield in millets under drought conditions (Rajkumar et al. 2024). Drought-tolerant PGPR, *Pseudomonas aeruginosa* strains MK513745, MK513746, MK513747, MK513748, and MK513749, were employed for the growth promotion of *Vigna radiata* grown under drought stress. These strains possessed several characteristics, including the ability to produce IAA, fix nitrogen, solubilize P, and produce siderophores. The findings indicated that inoculations containing these strains that produce IAA would be appropriate for promoting plant growth in regions experiencing water scarcity (Uzma et al. 2022). The study was performed to observe the response of wheat plants, inoculated with drought-tolerant *B. megaterium* (MU2) and *Bacillus licheniformis* (MU8) under drought stress. Outcomes from the study revealed that isolates enhanced plant biomass, photosynthetic pigments, osmolytes, and relative water content after effectively colonizing the wheat roots. *B. megaterium*-inoculated wheat plants showed enhanced plant tolerance to drought stress. Therefore, these strains may be useful bio-enhancers and biofertilizers for growing wheat in dry and semi-arid areas (Rashid et al. 2022).

2.3.2.3 Water Logging and Floods

Water-logging stress (WS) is also one of the major barriers in sustainable agriculture and food security. Farmers face huge challenges in achieving adequate and quality production in the areas suffering from heavy rainfall and floods. Heavy rainfall generates the water-logged conditions, which affect the crops by altering the physiochemical properties of soil and oxygen level (Ashraf 2012). This stress impacts approximately 10–15 million hectares of crops such as wheat annually, leading to yield reductions by up to 20–50% (Pais et al. 2022). WS triggers the production of ET and polyamines, including spermine, putrescine, and spermidine, in plants, while microbial biostimulants mitigate WS by enhancing antioxidant activities and reducing ROS levels (Rauf et al. 2021). Several biostimulants, like ExelGrow, use algae extracts, particularly from *Ascophyllum nodosum*, a brown seaweed, rich in phytohormones (ABA, CTK, auxins), organic osmolytes, essential nutrients, vitamins, amino acids, and bioactive compounds, including polysaccharides, antioxidants, peptides, betaines, and secondary metabolites (De Saeger et al. 2020; Buga and Petek 2023). Using algal extracts along with other biostimulants is reported to have improved the cellular respiration, ATP synthesis, and yield of soybeans under hypoxia conditions (de Andrade Silva et al. 2023). Wu et al. (2024) studied that application of microbe-based fertilizers and conventional organic fertilizers in rapeseed enriched the rhizospheric microbiota with phylum *Actinobacteriota* and *Verrucomicrobiota,* respectively. Promotion in the population of these bacteria stimulated the plant growth, survival, productivity and tolerance against flood stress. Inoculation of *Paraglomus occultum* (AMF) in cucumber elevated the tolerance against water-logged conditions by the expression of stress-responsive gene, i.e., 11 out of 12 heat shock protein 70 (*CsHsp70*) genes and 12 out of 14 plasma membrane intrinsic protein (*CsPIP*) genes in the roots of the plant under stress. Osmolyte level was observed to be increased, which accelerated the plant growth and biomass (Xiang et al. 2024).

2.3.2.4 Temperature

The world is focusing on the increase of food productivity by adapting sustainable agricultural practices for ensuring the food security. Global warming situation is resulting into increasing air temperature, which is further responsible for the rise in soil temperature, causing injuries and damages to the major crops (Islam et al. 2019). Cold stress and frost conditions reduce the yield of important crops by disrupting the vital development and growth process in these plants (Soualiou et al. 2022). Plants adapt to the extreme temperature conditions (heat and cold stress) by morphological and biochemical changes during their various developmental stages, while they are exposed to those conditions for a longer time period. The process of finding stress-tolerant varieties of crops through gene transformation technology is time-consuming and exorbitant. In order to finding out the sustainable and cost-effective alternative, heat and cold stress-tolerant microorganisms are emerging out as a constructive solution. Two bacterial consortia having combination of pychrotolerant strains *Pseudomonas* sp. CcR1d + *Pseudomonas* sp. NUR4a and *Pseudomonas* sp. TmR5a + *Curtobacterium* sp. BmP22c augmented the growth of tomato plant grown below 25 °C. Auxin production, phosphate-solubilizing activities performed by bacterial consortia, along with the expression of *nifH* and *acdS* genes enhanced the development and yield of tomato plant under cold stress (Vega-Celedón et al. 2021). Research was conducted to study the effect of thermotolerant *Bacillus cereus* SA1 on the growth enhancement of soybean plant growing under high temperatures. The study confirmed that *B. cereus* was competent in combating the negative effect of heat stress by balancing the abscisic acid and salicylic acid. SA1 also showed to significantly increase the antioxidant activity, decrease the amino acid content, and overexpress the stress-responsive genes GmLAX3 and GmAKT2, which reduced the induction of ROS causing oxidative damage to plants under heat stress (Khan et al. 2020). Research has shown that AMF significantly bolster cold stress by reducing the MDA content, maintaining membrane integrity, enhancing root hydraulic conductivity, and augmenting antioxidant capacity (Ding et al. 2020). Reva et al. (2021) illustrated that the inoculation of AMF enhanced the resilience, yield, and fruit quality of tomatoes, peppers, and cucumbers when subjected to heat stress, thereby promoting sustainable agricultural methods. Liu et al. (2023a, b) noted that the consortium of AMF strains enhances the growth and efficiency of photosystem II (PSII) (Fv/Fm), K^+: Na^+ ratio, and relative water content in peanuts, while decreasing the ROS accumulation under stressful conditions. Umar et al. (2024) demonstrated that AMF facilitates the absorption of nutrients to support plant growth under less-than-ideal temperature conditions. AMF also enhances crop resilience to heat stress, which is increasingly important due to climate change. Wei et al. (2023) reported that the combined application of melatonin (MT) with AMF increased the MT, GA, and CTK biosynthetic genes in ryegrass. Additionally, the treatment reduced electrolyte leakage, H_2O_2, and MDA levels, respectively, resulting in improved plant growth and a decrease in leaf senescence. Examples of associations between plants and microorganisms and mechanisms involved that provide resilience against abiotic stresses are mentioned in Table 2.2.

Incorporation of microbe-based bioinoculants in such stress conditions could help in the amplification of agricultural output. Hence, it can be concluded that the application of microbe-based bioformulations and other microbial technologies could be favorable in the promotion of agricultural produce by managing the crop production under different abiotic stressed conditions occurring due to changing climatic conditions and global warming.

2.4 Future Prospects

Microorganisms are going to play a very important role in achieving food security and targets of SDGs, particularly SDG2. They are not only important for maintaining soil health and sustainability but are being used to produce products that can help improve soil health in agro-ecosystems (Banerjee and Heijden 2022). Bioinoculants (biostimulants, biofertilizers, and biopesticides) made using the consortium of microbes utilize diverse features of PGPMs to enhance crop productivity and control pests even under stress environments like droughts, floods, high salinity, and temperature extremes, which are expected to increase under the changing climate change scenario (Naamala and Smith 2020). Although the interest in the use of microbe-based products in agriculture gained momentum in 1980s due to their eco-friendly nature and role in agricultural sustainability, their use remains limited in the real-world scenarios. Even though the production of biofertilizers and biopesticides has steadily increased at an average rate of 10% and 12% CAGR, from 2015 to 2023, respectively, at present, biofertilizers and biopesticides only make up 4–5%

Table 2.2 Mechanisms of microbial biostimulants for combating abiotic stresses

Microorganisms			Abiotic stress	Mechanism	Plant	References
Bacteria (PGPRs)	*Pseudomonas* sp.	*Pseudomonas putida* strain AHK_SHA007	Salinity	Induction of systemic tolerance in plants	*Sorghum bicolor*	Azeem et al. (2025)
		Pseudomonas fluorescens and *Enterococcus hirae*	Salinity	ACC deaminase activity and *nif*H, *acds*, *pqq*, and *ipdc* genes	*Vigna radiata*	Kumawat et al. (2024)
		Pseudomonas aeruginosa HG28–5	Salinity	Up-regulation of Na^+/K^+ homeostasis and ABA signaling pathway	Tomato	Dong et al. (2024)
		Pseudomonas palleroniana GZNU148	Drought	Increasing the expression levels of drought tolerance related genes (*sucrose synthase 4*, *glucose-6-phosphate isomerase*, etc.) in roots, and increasing secondary metabolites in rhizosphere soil	*Themeda japonica*	Chen et al. (2024a, 2024b)
		P. fluorescens FY32	Drought	Increased stress tolerance by higher activities of antioxidant enzymes	*Brassica Napus L.*	Kazemi Oskuei et al. (2024)
		Pseudomonas chlororaphis RO1 and *P. fluorescens* RO7	Drought	Higher superoxide dismutase (SOD) and catalase enzyme activity	*Solanum tuberosum* L.	Rafiei et al. (2024)
		P. fluorescens and *methyl Jasmonate*	Drought	Upregulation of antioxidative enzymes, including superoxide dismutase, catalase, and ascorbate peroxidase and glutathione reductase	*Brassica juncea*	Khan et al. (2024)
		Pseudomonas koreensis S4T10	Salinity and drought	Activated the antioxidant defence by modulating glutathione (GSH), superoxide dismutase (SOD), peroxidase (POD), and polyphenol oxidase (PPO) activities	*Arabidopsis thaliana*	Kalleku et al. (2024)
		P. putida AKMP7	Drought	Downregulation of SA levels	*Arabidopsis thaliana*	Gopalan et al. (2024)
		Pseudomonas turukhanskensis with *rhizobium leguminosarum 128C53*	Heat stress	ACC deaminase activity and enhanced biofilm formation	*Pisum sativum*	Ben Gaied et al. (2024)
		P. fluorescens	Heavy-metal stress	Induction of AtPCR2 gene expression	*Arabidopsis thaliana*	Reddy et al. (2023)
		P. putida KNP9	Heavy-metal stress	Tolerance and metal bioaccumulation for Pb and cd	Soybean	Tripathi et al. (2023)
		Pseudomonas geniculata	Heavy-metal stress	Increased level of antioxidant enzymes and decreased oxidative stress indicators	*Vigna radiata* L.	Madhogaria et al. (2024)

	Streptomyces sp.	*Streptomyces* sp. *(AB-11)*	Heat stress	Production of siderophores, indole acetic acid, phosphate solubilization, calcite solubilization, nitrogen fixation, and biocontrol against *Pseudomonas syringae* and *Xanthomonas campestris*	*Cicer arietinum*	Thakur and Yadav (2023)
		Streptomyces mutabilis	Salinity	Nitrogen fixation, phosphate-solubilizing activities, production of indole-3-acetic acid and ammonia, and salt tolerance	*Triticum aestivum* L.	Hesham et al. (2025)
		Streptomyces sp. *NRC21696*	Heavy-metal stress	Heavy-metal-resistant and removal against Ni^{2+}, Zn^{2+}, Fe^{3+}, Cu^{2+}, Cd^{2+}, Ag^{+}, Al^{3+}, Hg^{2+}, Mn^{2+}, As^{3+}, As^{5+}, Cr^{6+}, Co^{2+}, and Pb^{2+}	*Zea mays L.*	AL-Huqail and El-Bondkly (2022)
		Streptomyces chartreuses WZS021	Drought	Enhancing phenylalanine biosynthesis	Sugarcane	Pang et al. (2024)
	Bacillus sp.	*Bacillus subtilis*	Salinity	Induction of systemic tolerance	*Glycine max* L.	Yasmin et al. (2020)
		Bacillus halotolerans, Pseudomonas sihuiensis, and *Bacillus atrophaeus* with folic acid	Drought	Production of metabolites and antioxidants	Wheat	Tanveer et al. (2024)
		Bacillus mycoides	Heavy-metal stress	Reducing the toxic effects of Ni and Zn which suppressed heavy-metal toxicity	*Triticum aestivum*	Shahzad et al. (2024)
		Bacillus glycinifermentans	Heavy-metal stress	Improving barley seedling development	*Hordeum vulgare* L.	Belhassan et al. (2024)
		Bacillus mojavensis I4	Salinity	Controlled membrane damage by lowering the electrolyte leakage and malondialdehyde content	Durum wheat	Ghazala et al. (2023)
		Bacillus safensis EH143	Salinity and cadmium stress	Improving P and Ca^{2+} uptake and the K^{+}/Na^{+} ratio and reducing cd uptake (49% under Cd stress) in shoots	Soybean	Kwon et al. (2024)
		Bacillus CUM6, Bacillus DUM7, and Bacillus PIS7	Heavy-metal stress	Heavy-metal uptake (Cd, Pb, and Cr) decreased linearly with increasing phosphorus levels	Maize	Nawaz et al. (2023)
		Bacillus safensis strain SCAL1	Heat stress	Production of plant growth regulators, antioxidant enzyme activities, and chlorophyll content	*Solanum lycopersicum* L.	Mukhtar et al. (2023)
		Bacillus amyloliquefaciens	Drought and salinity stress	Altered expression of stress-related and transcription factor genes (*USP, DEF, CYP450, GST, MYB*, and *bZIP*) and stress-protective mechanisms	*Oryza sativa* L.	Bisht et al. (2025)

(continued)

Table 2.2 (continued)

Microorganisms			Abiotic stress	Mechanism	Plant	References
Fungi (PGPF)	*Trichoderma* sp.	*Trichoderma asperellum*	Drought	Increased photosynthesis rate, stomatal conductance, and water use efficiency, increasing the superoxide dismutase and peroxidase enzyme activities	Sugarcane	Scudeletti et al. (2021)
		Trichoderma koningii and Trichoderma harzianum	Salinity	Stimulating photosynthetic pigment, salicylic acid, and antioxidants which restricted oxidative and nitrosative stresses and membrane damage	*Phaseolus vulgaris*	Abdelrhim et al. (2024)
		T. harzianum strain, IIPRTh-3	Salinity	Defense enzyme induction	Pigeon pea	Mishra et al. (2024)
		T. harzianum	Drought	Modification of plant root architecture and redistribution of photosynthates	*Glycine max* and *Phaseolus vulgaris* L.	Battaglia et al. (2024)
		Trichoderma sp. T-4 and *P. fluorescence* PGPR-7	Heavy metal stress	Decreased the proline, MDA content, POD, and CAT activities	*Cicer arietinum* L.	Syed et al. (2023)
		Trichoderma longibrachiatum	Drought	Regulating physiological responses and soil microbial community	*Pinus massoniana*	Yu et al. (2023)
	Other strains	*Piriformospora indica* and *P. fluorescens*	Salinity	Induction of systemic tolerance by the promotion of antioxidant enzyme activities	*Salvia officinalis*	Aslani et al. (2024)
		Piriformospora indica	Salinity	Lipid desaturation, activation of antioxidant enzymes, and phytohormone production	*Solanum melongena* L.	Seshagiri et al. (2025)
		Serendipita indica (with mixture of *Pseudomonas areuginosa*, *P. putida*, and *P. fluorescens*)	Salinity	Increasing the absorption of water and nutrients, preserving photosynthetic pigments and reducing the accumulation of toxic ions	*Ocimum basilicum* L.	Taheri Boukani et al. (2024)
		Glomus mosseae and *P. fluorescens*	Salinity	Promotion in salt tolerance and plant defence mechanisms	*Punica granatum*	Ahmadi et al. (2024)

Arbuscular mycorrhizal fungi (AMF)	*Glomus* sp.	*Glomus lamellosum* and *Glomus etunicatum*	Drought	Lower the accumulation of malondialdehyde (MDA) and increased the accumulation of enzymes and osmotic substances	*Cinnamomum migao*	Liao et al. (2021)
		Glomus sp. and *Bacillus* sp.	Drought	Stabilization of cell redox status, downregulating pectin, affecting their methylation degree, and increasing hemicellulose content	*Zea mays* L.	Wilmowicz et al. (2022)
		Glomus intraradices	Salinity	Enhancing plant stress tolerance	*Salvia officinalis* L.	Tunçtürk et al. (2024)
		Glomus mosseae	Heavy-metal and heat stress	*Cu/Zn*-superoxide dismutase gene expression, increased ascorbate peroxidase activity, peroxidase (POD) gene expression, and soluble proteins content	*Medicago sativa* L.	Gao et al. (2023)
		Glomus species	Heat stress	Increase in photosynthetic parameters like quantum yield of photosystem (PS) II, PSI, electron transport, and fractions of open reaction centers	Maize	Mathur et al. (2021)
		Claroideoglomus lamellosum and *Claroideoglomus etunicatum*	Drought	Improved photosynthesis, activated antioxidant enzymes, regulated cell osmotic state, and enhanced the drought tolerance	*Cinnamomum migao*	Yan et al. (2022)
	Other strains	*Septoglomus constrictum* and *Funneliformis mosseae*	Drought and heat stress	Maximal PSII photochemical efficiency, proline accumulation, antioxidant enzymes, and lowered ROS markers	Tomato	Duc et al. (2023)
		Diversispora versiformis	Heat stress	Up-regulated the expression of *CsPIP1;6*, *CsPIP2;1*, and *CsPIP2;6* and *CsHsp70s* downregulated	Cucumber	Liu et al. (2023a, b)
		Rhizophagus irregularis and melatonin	Heat stress	Up-regulated genes for CTK, GA and MT signaling and responses, downregulated genes for ABA signaling and responses	*Lolium perenne* L.	Wei et al. (2023)

of the global fertilizer and pesticide market (Šunjka and Mechora 2022). The limited utilization of bioinoculants can be attributed to a combination of factors, such as variable efficacy, depending on external environmental conditions like soil type, crop varieties, storage conditions, lack of regulations and quality control, and paucity of awareness among farmers about their use and benefits (Saritha and Tollamadugu 2019).

Production of bioinoculants should involve an assessment of the type of agro-ecosystems they are supposed to be used in. This involves studying the soil microbial structure and soil properties like pH, temperature, water activity, nutrient content, etc. Modern biotechnological techniques like metagenomics and metatranscriptomics can help with the identification of key viable but nonculturable microbes (VNCM) within the soil microbiome that might be useful for bioinoculant production and efficacy, allowing their use for the production of impactful inoculums (Li et al. 2020). Furthermore, gene modification techniques can be utilized to engineer microbial consortia as inoculums for bioinoculants. Such consortium can display multiple characteristics that allow them to improve crop productivity within different agro-environments, thus reducing the need for producing several inoculants for different agro-environments. Bioinformatics can play an important role in this by working as a directory of genetic data and allowing scientists to pick the traits required for different agro-ecosystems in a single consortium (Sahu et al. 2023). Designing such consortia-based bioinoculants is being looked at as a promising solution for improving crop productivity (Nunes et al. 2023). However, the technology is still in infancy, and more on-ground research is required before large-scale implementation. Moreover, gene editing and use of genetically engineered microbes is altogether a different challenge, with various hurdles related to their acceptability and regulations, varying according to regions and countries.

Apart from improving bioinoculant efficiency, it is also critical to maintain their quality. Absence of regulatory parameters for bioinoculants has led to the production and use of low-quality biofertilizers and biopesticides, resulting in reduced benefits and lack of trust among farmers (Rai et al. 2023). Considering the fact that the global bioinoculant market value is expected to reach USD 3.1 billion by the end of 2024, from USD 1 billion in 2016, setting up international standards, for example, ISO standards, for bioinoculant production and following strict quality control protocols can help ensure high-quality products, which in turn will ensure the confidence of farmers (Joshi and Gauraha 2022). Increasing farmer awareness about the long-term benefits of biofertilizer and biopesticide use is also critical for their success. This can be achieved by educating farmers about the collective benefits gained from using bioinoculants, providing subsidies on bioinoculants to make them more accessible, and introducing schemes to promote their use (Ataei et al. 2022; Li et al. 2023). Accomplishing this would require a collective effort from governments and various organizations around the world.

Conservation of the soil microbiome is as important to achieving food security sustainably as using microbial-based products. The soil microbiome is a highly diverse and critical part of the environment for achieving agricultural sustainability due to the invaluable ecosystem services offered by it, including nutrient cycling, pathogen control, and plant–microbe symbiotic relations, which ensure health and vigor of the plant (Arora 2023; Suman et al. 2022). Utilizing the services offered by the soil microbiome can help reduce the dependency on agri-chemicals. Soil microbiome engineering, i.e., engineering crops to produce root exudates that select beneficial microbes in the rhizosphere, is being researched as a technique to improve crop productivity (Jansson et al. 2023). On the other hand, quorum-sensing and quorum-quenching biomolecules can be added in bioinoculants to promote the growth of beneficial microbes and reduce pathogens in the rhizosphere (Majdura et al. 2022). Additionally, CRISPR technology is also being tested to be used for in situ engineering of soil microbes that can help introduce traits to produce resistance to abiotic stresses among the soil microbial community (Nethery et al. 2022). However, these techniques are still in early stages of development, and therefore, more robust research is required to bring them into service. It has become increasingly vital to include the soil microbiome in the SDG agenda, either by including them in a current SDG target or by introducing a new SDG, specifically aimed at their conservation (Arora 2023). This will help in raising awareness about the importance of the soil microbiome and force nations globally to act on its conservation.

According to the State of the Food Security and Nutrition in the World report (2023), between 691 and 783 million faced hunger in 2022, while 2.4 billion people suffered from food insecurity globally. Considering the fact that the human population is expected to rise and reach 9.8 billion by 2050, people facing food insecurity will also increase in the future (UN 2023). Tackling this will require an increase in the productivity of agricultural lands globally in a sustainable way. Microorganisms will play a key role in achieving this by harnessing the ecosystem services offered by them and utilizing these to revolutionize the agricultural sector. Microbes will be critical for not only increasing crop productivity but also to tackle the effects of a changing climate on agricultural production (Malhi et al. 2020). The need of the hour is therefore to invest in microbial-based solutions to achieve a food-secure planet.

2.5 Conclusions

The most fundamental responsibility of a nation is to guarantee food security for its citizens. There is a lot of pressure on the agricultural and associated industries for the same. Chemical fertilizers have been useful in increasing the crop output, but over time, their overuse has caused plants to become less productive and has also affected soil quality, leading to a rise in arid, less fertile areas. On the other side, chemical pesticides are absorbed into the food chain, affecting many life forms and causing ecosystem imbalances. These xenobiotics also have an effect on human health. This issue is made worse by the climate change scenario. Crops are extremely vulnerable to changes in precipitation and temperature patterns. Research over the last few decades has concentrated on creating more sustainable agricultural practices that can be employed as a more comprehensive strategy to provide food for everyone. One of the most promising, economical, and environmentally friendly approaches to sustainable agriculture is the use of microorganisms. Numerous studies in this field have confirmed the importance of microorganisms in resolving important environmental issues. Along by preventing the negative effects of abiotic stressors and restoring soil health, PGPR and other microbial biofertilizers and biopesticides aid in the holistic growth of plants. Although the market for biopesticides and biofertilizers has grown rapidly in recent years, there is still a lot of uncharted territory in this area. Researchers in the field must address a number of limitations in the development of bioformulation, such as improving the products' shelf life, creating convenient, readily applicable products, and expanding their applicability. This is going to be important to ensure food security in a sustainable manner.

References

Abdelrhim AS, Hemeda NF, Mwaheb MA, Omar MO, Dawood MF (2024) The role of *Trichoderma koningii* and *Trichoderma harzianum* in mitigating the combined stresses motivated by *Sclerotiniasclerotiorum* and salinity in common bean (*Phaseolus vulgaris*). Plant Stress 11:100370

Abideen Z, Hanif M, Munir N, Nielsen BL (2022) Impact of nanomaterials on the regulation of gene expression and metabolomics of plants under salt stress. Plan Theory 11(5):691

Admassie M, Woldehawariat Y, Alemu T, Gonzalez E, Jimenez JF (2022) The role of plant growth-promoting bacteria in alleviating drought stress on pepper plants. Agric Water Manag 272:107831

Afkairin A, Ippolito JA, Stromberger M, Davis JG (2021) Solubilization of organic phosphorus sources by cyanobacteria and a commercially available bacterial consortium. Appl Soil Ecol 162:103900

Ahmadi SRZ, Seifi E, Varasteh F, Akbarpour V (2024) Mitigating salinity stress in pomegranate: effects of *Pseudomonas fluorescens* and *Glomus mosseae* on stress responses of red angel and wonderful cultivars. Sci Hortic 330:113036

Al Dayel MF, El Sherif F (2021) Evaluation of the effects of *Chlorella vulgaris, Nannochloropsis salina*, and *Enterobacter* cloacae on growth, yield and active compound compositions of Moringa oleifera under salinity stress. Saudi J Biol Sci 28(3):1687–1696

Al-Ani LKT (2018) *Trichoderma* from extreme environments: physiology, diversity, and antagonistic activity. In: Egamberdieva D, Birkeland NK, Panosyan H, Li WJ (eds) Extremophiles in Eurasian ecosystems: ecology, diversity, and applications. Microorganisms for sustainability, vol 8. Springer, Singapore, pp 389–403

AL-Huqail A, El-Bondkly MA (2022) Improvement of *Zea mays* L. growth parameters under chromium and arsenic stress by the heavy metal-resistant *Streptomyces* sp. NRC21696. Int J Environ Sci Technol 19:5301–5322

Ali S, Charles TC, Glick BR (2014) Amelioration of high salinity stress damage by plant growth-promoting bacterial endophytes that contain ACC deaminase. Plant Physiol Biochem 80:160–167

Ali S, Charles TC, Glick BR (2017) Endophytic phytohormones and their role in plant growth promotion. In: Doty S (eds) Functional importance of the plant microbiome. Springer, Cham, pp 89–105

Allaga H, Bóka B, Poór P, Nagy VD, Szűcs A, Stankovics I et al (2020) A composite bioinoculant based on the combined application of beneficial bacteria and fungi. Agronomy 10(2):220

Álvarez C, Navarro JA, Molina-Heredia FP, Mariscal V (2020) Endophytic colonization of rice (*Oryza sativa* L.) by the symbiotic strain *Nostoc punctiforme* PCC 73102. MPMI 33(8):1040–1045

Arora P (2023) Life below land: the need for a new sustainable development goal. Front. Environ Sci 11

Arora NK, Fatima T (2022) Role of plant growth promoting microbes in managing sustainability of stressed agroecosystems. ESG 5(1):1–3

Arora NK, Fatima T, Mishra I, Verma M, Mishra J, Mishra V (2018) Environmental sustainability: challenges and viable solutions. ESG 1:309–340

Arora NK, Mishra J, Singh P, Fatima T (2024) Salt-tolerant plant growth-promoting *Pseudomonas atacamensis* KSS-6 in combination with organic manure enhances rice yield, improves nutrient content and soil properties under salinity stress. J Basic Microbiol 64(6):2300767

Ashraf MA (2012) Waterlogging stress in plants: a review. Afr J Agric Res 7(13):1976–1981

Aslani Z, Hassani A, Mandoulakani BA, Barin M, Maleki R (2024) The symbiotic association with *Piriformospora indica* and *Pseudomonas fluorescens* improves salt tolerance in sage (*Salvia officinalis*) plants. Plant Soil 495(1):391–410

Ataei P, Karimi H, Klöckner CA, Es'haghi SR, Zarei R (2022) The promotion of biofertilizer application on farms: farmers' intentional processes. Environ Technol Innov 28(1):102722

Azeem M, Haider MZ, Javed S, Saleem MH, Alatawi A (2022) Drought stress amelioration in maize (*Zea mays* L.) by inoculation of *Bacillus* spp. strains under sterile soil conditions. Agriculture 12(1):50

Azeem M, Sultana R, Ahmed N, Abbasi MW, Hasan KA, Dong R, Alamri S, Alfagham AT (2025) Salinity stress resilience in *Sorghum bicolor* through *Pseudomonas*-mediated modulation of growth, antioxidant system, and eco-physiological adaptations. ACS Omega 10(1):940–954

Baghel V, Thakur JK, Yadav SS, Manna MC, Mandal A, Shirale AO, Sharma P, Sinha NK, Mohanty M, Singh AB, Patra AK (2020) Phosphorus and potassium solubilization from rock minerals by endophytic *Burkholderia* sp. strain FDN2-1 in soil and shift in diversity of bacterial endophytes of corn root tissue with crop growth stage. Geomicrobiol J 37(6):550–563

Bahar NH, Lo M, Sanjaya M, Van Vianen J, Alexander P, Ickowitz A, Sunderland T (2020) Meeting the food security challenge for nine billion people in 2050: what impact on forests? Glob Environ Chang 62:102056

Banerjee S, Heijden MGA (2022) Soil microbiomes and one health. Nat Rev Microbiol 21(1):6–20

Bao J, Zhuo C, Zhang D, Li Y, Hu F, Li H, Su Z, Liang Y, He H (2021) Potential applicability of a cyanobacterium as a biofertilizer and biopesticide in rice fields. Plant Soil 463:97–112

Baron NC, de Souza PA, Rigobelo EC (2020) *Purpureocillium lilacinum* and *Metarhizium marquandii* as plant growth-promoting fungi. Peer J 8:e9005. https://doi.org/10.1007/s11104-021-04899-9

Bashir MA, Atta S, Nisar MS, Khan AK, Batool M, Khan KA et al (2020) Management of sucking insects pest complex of cotton through foliar spray of insecticides. Fresenius Environ Bull 29(7A):5777–5785

Battaglia ME, Martinez SI, Covacevich F, Consolo VF (2024) *Trichoderma harzianum* enhances root biomass production and promotes lateral root growth of soybean and common bean under drought stress. Ann Appl Biol 185(1):36–48

Bedini S, Pellegrino E, Avio L, Pellegrini S, Bazzoffi P, Argese E, Giovannetti M (2009) Changes in soil aggregation and glomalin-related soil protein content as affected by the arbuscular mycorrhizal fungal species *Glomus mosseae* and *Glomus intraradices*. Soil Biol Biochem 41(7):1491–1496

Behdad A, Ganjeali A (2025) Phytohormones and microbial elicitation on glycyrrhizin production and gene expression in the hairy root of *Glycyrrhiza glabra* L. PCTOC 160(1):4

Belhassan M, Farhat A, Abed HE et al (2024) Isolation and identification of a new *Bacillus glycinifermentans* strain from date palm rhizosphere and its effect on barley seeds under heavy metal stress. Braz J Microbiol 55(1):843–854

Ben Gaied R, Sbissi I, Tarhouni M, Brígido C (2024) Enhancing *Pisum sativum* growth and symbiosis under heat stress: the synergistic impact of co-inoculated bacterial consortia and ACC deaminase-lacking *Rhizobium*. Arch Microbiol 206(5):203

Bhardwaj S, Kumar P (2020) Salinity stress, its physiological response and mitigating effects of microbial bio inoculants and organic compounds. J Pharmacogn Phytochem 9(4):1297–1303

Bharti C, Fatima T, Mishra P, Verma P, Bhattacharya A, Alaylar B, Arora NK (2024) Salt-tolerant endophytic *Bacillus altitudinis* NKA32 with ACC deaminase activity modulates physiochemical mechanisms in rice for adaptation in saline ecosystem. ESG 7:231–249

Bhattacharjee R, Dey U (2014) Biofertilizer, a way towards organic agriculture: a review. Afr J Microbiol Res 8(24):2332–2343

Bisht N, Singh T, Ansari M, Joshi H, Mishra SK, Chauhan PS (2025) Plant growth-promoting *Bacillus amyloliquefaciens* orchestrate homeostasis under nutrient deficiency exacerbated drought and salinity stress in *Oryza sativa* L. seedlings. Planta 261(1):1–19

Bonini P, Rouphael Y, Miras-Moreno B, Lee B, Cardarelli M, Erice G et al (2020) A microbial-based biostimulant enhances sweet pepper performance by metabolic reprogramming of phytohormone profile and secondary metabolism. Front Plant Sci 11:567388

Bonis-Profumo G, McLaren R, Fanzo J (2019) Ravaged landscapes and climate vulnerability: the challenge in achieving food security and nutrition in post-conflict Timor-Leste. In: Advances in food security and sustainability. Elsevier, pp 97–132

Bouaichi A, Benkirane R, El-kinany S, Habbadi K, Lougraimzi H, Sadik S et al (2019) Potential effect of antagonistic bacteria in the management of olive knot disease caused by *Pseudomonas savastanoi* pv. savastanoi. J Microbiol Biotechnol Food Sci 8(4):1035–1040

Buga N, Petek M (2023) Use of biostimulants to alleviate anoxic stress in waterlogged cabbage (Brassica oleracea var. capitata)—a review. Agriculture 13(12):2223

Bulgari R, Franzoni G, Ferrante A (2019) Biostimulants application in horticultural crops under abiotic stress conditions. Agronomy 9(6):306

Castiglione AM, Mannino G, Contartese V, Bertea CM, Ertani A (2021) Microbial biostimulants as response to modern agriculture needs: composition, role and application of these innovative products. Plan Theory 10(8):1533

Chaparro-Encinas LA, Parra-Cota FI, Cruz-Mendívil A, Santoyo G, Peña-Cabriales JJ, Castro-Espinoza L, de Los Santos-Villalobos S (2022) Transcriptional regulation of cell growth and reprogramming of systemic response in wheat (*Triticum turgidum* subsp. durum) seedlings by *Bacillus paralicheniformis* TRQ65. Planta 255(3):56

Chen H, Chen J, Qi Y, Chu S, Ma Y, Xu L et al (2022) Endophytic fungus *Cladosporium tenuissimum* DF11, an efficient inducer of tanshinone biosynthesis in *Salvia miltiorrhiza* roots. Phytochemistry 194:113021

Chen H, Yu S, Yu Z, Ma M, Liu M, Pei H (2024a) Phycoremediation potential of salt-tolerant microalgal species: motion, metabolic characteristics, and their application for saline–alkali soil improvement in eco-farms. Microorganisms 12(4):676

Chen L, Gong J, Jin J, Wang L, Chen X, Wang C et al (2024b) Intrinsic and extrinsic regulatory mechanisms of *Pseudomonas palleroniana* GZNU148 for enhancing *Themeda japonica* tolerance to drought stress. Plant Soil 504:1–24

Chittora D, Meena M, Barupal T, Swapnil P, Sharma K (2020) Cyanobacteria as a source of biofertilizers for sustainable agriculture. Biochem Biophys Rep 22:100737

Choudhary R, Rawat G, Kumar V, Kumar V (2020) Diversity and function of microbes associated with rhizosphere of finger millet (*Eleusine coracana*). In: Sharma SK, Singh UB, Sahu PK, Singh HV, Sharma PK (eds) Rhizosphere microbes. Microorganisms for Sustainability. Springer, Singapore, pp 431–451

Choudhary N, Tripathi A, Singh PK, Parikh HS, Tiwari A (2024) Application of algae for enhanced plant growth and food productivity. SMAB 4(2):564–574

Clough SE, Jousset A, Elphinstone JG, Friman VP (2022) Combining in vitro and in vivo screening to identify efficient *Pseudomonas* biocontrol strains against the phytopathogenic bacterium *Ralstonia solanacearum*. Microbiology 11(2):e1283

Coerini LF, Mulato ATN, Martins-Junior J, Persinoti GF, de Castro Oliveira JV (2025) Inhibition of *Xanthomonas* growth by bioactive volatiles from *Pseudomonas* sp. triggers remarkable changes in the phytopathogen transcriptome. Microbiol Res 290:127971

Comite E, El-Nakhel C, Rouphael Y, Ventorino V, Pepe O, Borzacchiello A et al (2021) Bioformulations with beneficial microbial consortia, a bioactive compound and plant biopolymers modulate sweet basil productivity, photosynthetic activity and metabolites. Pathogens 10(7):870

Cook J, Degon Z, Ruiz D, Pope J, Rahmatallah Y, Mukherjee A (2022) The plant growth-promoting bacteria, *Azospirillum brasilense*, induce a diverse array of genes in rice shoots and promote their growth. Plant Growth Regul 97(1):143–155

Cordeiro ECN, Mógor ÁF, de Oliveira AJ, Mógor G, de Lara GB, Marques HMC (2022) Microalga biofertilizer triggers metabolic changes improving onion growth and yield. Horticulture 8(3):223

Daigham GE, Mahfouz AY, Abdelaziz AM, Nofel MM, Attia MS (2024) Protective role of plant growth-promoting fungi *Aspergillus chevalieri* OP593083 and *Aspergillus egyptiacus* OP593080 as biocontrol approach against Alternaria leaf spot disease of *Vicia faba* plant. Biomass Convers Biorefinery 14(18):23073–23089

Das S, Singh S, Verma JP, Mukherjee A (2021) Fungi: a potential candidate for sustainable agriculture and agroecosystem. In: Verma JP, Catriona A, Macdonald, Gupta VK, Podile AR (eds) New and future developments in microbial biotechnology and bioengineering. Elsevier, pp 159–164

Davies WJ, Ward SE, Wilson A (2020) Can crop science really help us to produce more better-quality food while reducing the world-wide environmental footprint of agriculture? Front Agric Sci Eng 7(1):28–44

de Andrade Silva R, Lessa Silva W, Farias Damasceno L, Luís Oliveira Cunha M, Angelica Carvalho Mendes N, Aparecido Manzani Lisboa L (2023) Physiological and productive role of biostimulants in alleviating hypoxia stress in soybean grown under field conditions. Gesunde Pflanzen 75(6):2713–2721

De Saeger J, Van Praet S, Vereecke D, Park J, Jacques S, Han T, Depuydt S (2020) Toward the molecular understanding of the action mechanism of Ascophyllum nodosum extracts on plants. J Appl Phycol 32(1):573–597

Dilnashin H, Birla H, Hoat TX, Singh HB, Singh SP, Keswani C (2020) Applications of agriculturally important microorganisms for sustainable crop production. In: Molecular aspects of plant beneficial microbes in agriculture. Academic, pp 403–415

Dimkić I, Janakiev T, Petrović M, Degrassi G, Fira D (2022) Plant-associated *Bacillus* and *Pseudomonas* antimicrobial activities in plant disease suppression via biological control mechanisms-a review. Physiol Mol Plant Pathol 117:101754

Ding Y, Shi Y, Yang S (2020) Molecular regulation of plant responses to environmental temperatures. Mol Plant 13(4):544–564

Dobrzyński J, Wróbel B, Górska EB (2023) Taxonomy, ecology, and cellulolytic properties of the genus *Bacillus* and related genera. Agriculture 13(10):1979

Dodd IC, Pérez-Alfocea F (2012) Microbial amelioration of crop salinity stress. J Exp Bot 63(9):3415–3428

Dong H, Wang Y, Di Y, Qiu Y, Ji Z, Zhou T et al (2024) Plant growth-promoting rhizobacteria *Pseudomonas aeruginosa* HG28-5 improves salt tolerance by regulating Na+/K+ homeostasis and ABA signaling pathway in tomato. Microbiol Res 283:127707

dos Santos IB, Pereira APDA, de Souza AJ, Cardoso EJBN, da Silva FG, Oliveira JTC et al (2022) Selection and characterization of *Burkholderia* spp. for their plant-growth promoting effects and influence on maize seed germination. Front. Soil Sci 1:805094

Douds Jr DD, Nagahashi G, Reider C, Hepperly PR (2007) Inoculation with arbuscular mycorrhizal fungi increases the yield of potatoes in a high P soil. Biol Agric Hortic 25(1):67–78

du Jardin P (2015) Plant biostimulants: definition, concept, main categories and regulation. Scientia Hortic 196:3–14

Duc NH, Szentpéteri V, Mayer Z, Posta K (2023) Distinct impact of arbuscular mycorrhizal isolates on tomato plant tolerance to drought combined with chronic and acute heat stress. Plant Physiol Biochem 201:107892

El-Komy MH, Hassouna MG, Abou-Taleb EM, Al-Sarar AS, Abobakr Y (2020) A mixture of *Azotobacter*, *Azospirillum*, and *Klebsiella* strains improves root-rot disease complex management and promotes growth in sunflowers in calcareous soil. Eur J Plant Pathol 156(3):713–726

Elnahal AS, El-Saadony MT, Saad AM, Desoky ESM, El-Tahan AM, Rady MM et al (2022) The use of microbial inoculants for biological control, plant growth promotion, and sustainable agriculture: a review. Eur J Plant Pathol 162(4):759–792

El-Sawah AM, El-Keblawy A, Ali DFI, Ibrahim HM, El-Sheikh MA, Sharma A et al (2021) Arbuscular mycorrhizal fungi and plant growth-promoting rhizobacteria enhance soil key enzymes, plant growth, seed yield, and qualitative attributes of guar. Agriculture 11(3):194

Emmanuel OC, Babalola OO (2020) Productivity and quality of horticultural crops through co-inoculation of arbuscular mycorrhizal fungi and plant growth promoting bacteria. Microbiol Res 239:126569

Etesami H, Beattie GA (2018) Mining halophytes for plant growth-promoting halotolerant bacteria to enhance the salinity tolerance of non-halophytic crops. Front Microbiol 9:148

Faizan M, Rajput VD, Al-Khuraif AA, Arshad M, Minkina T, Sushkova S, Yu F (2021) Effect of foliar fertigation of chitosan nanoparticles on cadmium accumulation and toxicity in *Solanum lycopersicum*. Biol 10(7):666

FAO (2011) Linking sustainability and climate financing: implications for agriculture. FAO, Rome

Fatima T, Arora NK (2021) *Pseudomonas entomophila* PE3 and its exopolysaccharides as biostimulants for enhancing growth, yield and tolerance responses of sunflower under saline conditions. Microbiol Res 244:126671

Fatima T, Arora NK (2023) Structural, compositional characterization of novel exopolysaccharide from halotolerant strain of *Pseudomonas entomophila* PE3 and their role in growth promotion of sunflower under saline condition. Environ Exp Bot 213:105418

Fatima T, Mishra I, Verma R, Arora NK (2020) Mechanisms of halotolerant plant growth promoting Alcaligenes sp involved in salt tolerance and enhancement of the growth of rice under salinity stress. 3 Biotech 10:1–12

Forouzi A, Ghasemnezhad A, Nasrabad RG (2020) Phytochemical response of Stevia plant to growth promoting microorganisms under salinity stress. S Afr J Bot 134:109–118

Galindo FS, Pagliari PH, Fernandes GC, Rodrigues WL, Boleta EHM, Jalal A et al (2022) Improving sustainable field-grown wheat production with *Azospirillum brasilense* under tropical conditions: a potential tool for improving nitrogen management. Front Environ Sci 10:821628

Galindo FS, Thiengo CC, Pagliari PH, Bernardes JVS, dos Santos GD, Longato PAF (2024) Interactive effects of bacterial consortia and basal nitrogen fertilization on initial maize growth: an investigation based on physiological parameters and 15N isotopic analysis. Plant Growth Regul:1–17

Gao YF, Jia X, Zhao YH, Ding XY, Zhang CY, Feng XJ (2023) *Glomus mosseae* improved the adaptability of alfalfa (*Medicago sativa* L.) to the coexistence of cadmium-polluted soils and elevated air temperature. Front Plant Sci 14:1064732

Gathage JW, Lagat ZO, Fiaboe KKM, Akutse KS, Ekesi S, Maniania NK (2016) Prospects of fungal endophytes in the control of *Liriomyza leafminer* flies in common bean Phaseolus vulgaris under field conditions. BioControl 61:741–753

Ghanbarzadeh Z, Mohsenzadeh S, Rowshan V, Zarei M (2020) Mitigation of water deficit stress in *Dracocephalum moldavica* by symbiotic association with soil microorganisms. Sci Hortic 272:109549

Ghazala I, Chiab N, Saidi MN, Gargouri-Bouzid R (2023) The plant growth-promoting bacteria strain Bacillus mojavensis I4 enhanced salt stress tolerance in durum wheat. Curr Microbiol 80(5):178

Ghorbanpour M, Omidvari M, Abbaszadeh-Dahaji P, Omidvar R, Kariman K (2018) Mechanisms underlying the protective effects of beneficial fungi against plant diseases. BioControl 117:147–157

Gill HK, Garg H (2014) Pesticide: environmental impacts and management strategies. Pesticides-Toxic Aspects 8(187):10–5772

Glick BR (2012) Plant growth-promoting bacteria: mechanisms and applications. Scientifica 2012(1):963401

Godfray HCJ, Beddington JR, Crute IR, Haddad L, Lawrence D, Muir JF et al (2010) Food security: the challenge of feeding 9 billion people. Science 327(5967):812–818

Golubkina N, Krivenkov L, Sekara A, Vasileva V, Tallarita A, Caruso G (2020a) Prospects of arbuscular mycorrhizal fungi utilization in production of Allium plants. Plan Theory 9(2):279

Golubkina N, Logvinenko L, Novitsky M, Zamana S, Sokolov S, Molchanova A et al (2020b) Yield, essential oil and quality performances of *Artemisia dracunculus*, *Hyssopus officinalis* and *Lavandula angustifolia* as affected by arbuscular mycorrhizal fungi under organic management. Plan Theory 9(3):375

Gopalan NR, Priya SY, Mohapatra S (2024) The rhizobacterial strain, *Pseudomonas putida* AKMP7, causes conditional pathogenesis in *Arabidopsis thaliana* via negative regulation of salicylic acid signaling, under water stress. Plant Physiol Biochem 206:108262

Govindasamy V, George P, Kumar M, Aher L, Raina SK, Rane J et al (2020) Multi-trait PGP rhizobacterial endophytes alleviate drought stress in a senescent genotype of sorghum [*Sorghum bicolor* (L.) Moench]. 3 Biotech 10:1–14

Gutiérrez Santana JC, Coria Jiménez VR (2024) *Burkholderia cepacia* complex in cystic fibrosis: critical gaps in diagnosis and therapy. Ann Med 56(1):2307503

Haggag WM, Hoballah MME, Ali RR (2018) Applications of nano biotechnological microalgae product for improve wheat productivity in semi-arid areas. Int J Agric Technol 14(5):672–692

Hamayun M, Hussain A, Khan SA, Irshad M, Khan AL, Waqas M et al (2015) Kinetin modulates physio-hormonal attributes and isoflavone contents of soybean grown under salinity stress. Front Plant Sci 6:377

Hambrey J (2017) The 2030 agenda and the sustainable development goals: the challenge for aquaculture development and management. FAO fisheries and aquaculture circular C1141

Hamid B, Zaman M, Farooq S, Fatima S, Sayyed RZ, Baba ZA et al (2021) Bacterial plant biostimulants: a sustainable way towards improving growth, productivity, and health of crops. Sustain For 13(5):2856

Harman GE, Howell CR, Viterbo A, Chet I, Lorito M (2004) Trichoderma species—opportunistic, avirulent plant symbionts. Nat Rev Microbiol 2(1):43–56

Hatamzadeh S, Rahnama K, White JF, Oghaz NA, Nasrollahnejad S, Hemmati K (2023) Investigation of some endophytic fungi from five medicinal plants with growth promoting ability on maize (*Zea mays* L.). J Appl Microbiol 134(1):lxac015

Hesham AEL, Mahmoud E, Hozzein WN, Khaled K (2025) Isolation and identification of endophytic bacteria from *Mentha longifolia* and their application for the enhancement of wheat growth under salt conditions. Egypt J Bot 65(1):146–158

Housh AB, Benoit M, Wilder SL, Scott S, Powell G, Schueller MJ, Ferrieri RA (2021) Plant-growth-promoting bacteria can impact zinc uptake in *Zea mays*: an examination of the mechanisms of action using functional mutants of *Azospirillum brasilense*. Microorganisms 9(5):1002

Hu T, Chen A, Jiang Y, Sun C, Luo S, Shao J (2022) Application of a newly recorded diazotrophic cyanobacterium in acidified and cd contaminated paddy soil: promotes rice yield and decreases Cd accumulation. Sci Total Environ 814:152630

Ilyas F, Ali MA, Modhish A, Ahmed N, Hussain S, Bilal M et al (2022) Synchronisation of zinc application rates with arbuscular mycorrhizal fungi and phosphorus to maximise wheat growth and yield in zinc-deficient soil. Crop Pasture Sci 74(3):157–172

Imran, Amanullah (2022) Phosphorus biofortification and uptake in maize enhanced with integrated phosphorus management. Phosphorus Sulfur Silicon Relat Elem 197(7):766–776

Islam MZ, Islam MR, Mamun MAA, Haque MM, Ahmed JU, Akter N, Karim MA (2019) Genetic variability of submergence tolerance in rice related to yield and yield contributing traits. J Plant Stress Physiol 5:22–27

Jaimes Duarte I, Humberto Álvarez Hernández S, Laura Ibañez A, Rodríguez Canto A (2018) Macroalgae as soil conditioners or growth promoters of *Pisum sativum* (L). ARRB 27(6):1–8

Jansson JK, McClure R, Egbert RG (2023) Soil microbiome engineering for sustainability in a changing environment. Nat Biotechnol 41(12):1716–1728

Jhutty A (2019) Molecular detection and characteristics of *Burkholderia Cepacia* complex strains isolated as industrial contaminants. The University of Liverpool (United Kingdom)

Jian P, Zhang H, Zha Q, Xi X, Yin X, Sun P, Zhang D (2024) Impact of AMF on grape growth and stress tolerance under high temperature conditions. Available at SSRN 5076272

Joshi SK, Gauraha AK (2022) Global biofertilizer market: emerging trends and opportunities. In: Trends of applied microbiology for sustainable economy. Academic, London, pp 689–697

Kaddes A, Fauconnier ML, Sassi K, Nasraoui B, Jijakli MH (2019) Endophytic fungal volatile compounds as solution for sustainable agriculture. Molecules 24(6):1065

Kalleku JN, Ihsan S, Al-Azzawi TNI, Khan M, Hussain A, Chebitok F et al (2024) Halotolerant *Pseudomonas koreensis* S4T10 mitigate salt and drought stress in *Arabidopsis thaliana*. Physiol Plant 176(2):e14258

Kamal MK, Gautam MK, Uday AM, Singh P, Singh RK (2018) To evaluate the effect of INM on content and uptake of nutrients on wheat. J Pharmacognosy Phytochem 7(3):1425–1429

Karimi R, Noori A (2022) *Streptomyces rimosus* rhizobacteria and *Glomus mosseae* mycorrhizal fungus inoculation alleviate salinity stress in grapevine through morphophysiological changes and nutritional balance. Sci Hortic 305:111433

Kazemi Oskuei B, Bandehagh A, Farajzadeh D, Asgari Lajayer B, Shu W, Astatkie T (2024) Effects of *Pseudomonas Fluorescens* FY32 on Canola (*Brassica Napus* L.) cultivars under drought stress induced by polyethylene glycol. J Crop Health 76(1):251–260

Kearl J, McNary C, Lowman JS, Mei C, Aanderud ZT, Smith ST (2019) Salt-tolerant halophyte rhizosphere bacteria stimulate growth of alfalfa in salty soil. Front Microbiol 10:1849

Khan MR, Ashraf S, Rasool F, Salati K, Mohiddin FA, Haque Z (2014) Field performance of *Trichoderma* species against wilt disease complex of chickpea caused by *Fusarium oxysporum* f. sp. *ciceri* and *Rhizoctonia solani*. Turk J Agric For 38(4):447–454

Khan MA, Asaf S, Khan AL, Jan R, Kang SM, Kim KM, Lee IJ (2020) Thermotolerance effect of plant growth-promoting *Bacillus cereus* SA1 on soybean during heat stress. BMC Microbiol 20:1–14

Khan V, Mubashshir M, Umar S, Iqbal N (2024) Methyl Jasmonate and *Pseudomonas fluorescens* synergistically boost Antioxidative defense, secondary metabolites, and Osmolyte production to enhance drought resilience in mustard. J Plant Growth Regul:1–19

Kulkova I, Wróbel B, Dobrzyński J (2024) *Serratia* spp. as plant growth-promoting bacteria alleviating salinity, drought, and nutrient imbalance stresses. Front Microbiol 15:1342331

Kumar A, Naqvi SDY, Kaushik P, Khojah E, Amir M, Alam P, Samra BN (2022) *Rhizophagus irregularis* and nitrogen fixing azotobacter enhances greater yam (*Dioscorea alata*) biochemical profile and upholds yield under reduced fertilization. Saudi J Biol Sci 29(5):3694–3703

Kumawat KC, Sharma P, Sirari A, Sharma B, Kumawat G, Nair RM, Bindumadhava H (2024) Co-existence of halo-tolerant *Pseudomonas fluorescens* and *Enterococcus hirae* with multifunctional growth promoting traits to ameliorate salinity stress in *Vigna radiata*. Chemosphere 349:140953

Kwon EH, Adhikari A, Imran M, Hussain A, Gam HJ, Woo JI et al (2024) Novel melatonin-producing *Bacillus safensis* EH143 mitigates salt and cadmium stress in soybean. J Pineal Res 76(4):e12957

Lamb TI, Berghahn E, Pita FM, de Oliveira NL, dos Reis Blasi ÉA, Hofstetter JS (2023) Isolation and selection of microalgae capable of stimulating rice plant development and seed production. Algal Res 74:103203

Laranjo M, Alexandre A, Oliveira S (2014) Legume growth-promoting rhizobia: an overview on the *Mesorhizobium* genus. Microbiol Res 169(1):2–17

Li R, Pang Z, Zhou Y, Fallah N, Hu C, Lin W, Yuan Z (2020) Metagenomic analysis exploring taxonomic and functional diversity of soil microbial communities in sugarcane fields applied with organic fertilizer. Biomed Res Int 1:9381506

Li J, Charles LS, Yang Z, Du G, Fu S (2022) Differential mechanisms drive species loss under artificial shade and fertilization in the alpine meadow of the Tibetan Plateau. Front Plant Sci 13:832473

Li B, Qiao Y, Yao R (2023) What promote farmers to adopt green agricultural fertilizers? Evidence from 8 provinces in China. J Clean Prod 426:139123

Liao X, Chen J, Guan R, Liu J, Sun Q (2021) Two arbuscular mycorrhizal fungi alleviates drought stress and improves plant growth in *Cinnamomum migao* seedlings. Mycobiology 49(4):396–405

Lindström K, Mousavi SA (2010) *Rhizobium* and other N-fixing symbioses. eLS. https://doi.org/10.1002/9780470015902.a0021157

Ling L, Yue R, Wang Y, Feng L, Yang L, Li Y et al (2025) Volatile organic compounds from *Stenotrophomonas geniculata* J-0 as potential biofumigants manage bulb rot caused by *Fusarium oxysporum* in postharvest Lanzhou lily. World J Microbiol Biotechnol 41(1):1–17. https://doi.org/10.1007/s11274-024-04228-z

Liu XR, Rong ZY, Tian X, Hashem A, Abd-Allah EF, Zou YN, Wu QS (2023a) Mycorrhizal fungal effects on plant growth, osmolytes, and CsHsp70s and CsPIPs expression in leaves of cucumber under a short-term heat stress. Plan Theory 12(16):2917

Liu Y, Lu J, Cui L, Tang Z, Ci D, Zou X et al (2023b) The multifaceted roles of Arbuscular Mycorrhizal fungi in peanut responses to salt, drought, and cold stress. BMC Plant Biol 23(1):36

Madhogaria B, Banerjee S, Chakraborty S, Dhak P, Kundu A (2024) Alleviation of heavy metals chromium, cadmium and lead and plant growth promotion in *Vigna radiata* L. plant using isolated *Pseudomonas geniculata*. Int Microbiol:1–17

Mahadik SP, Kumudini BS (2024) Halotolerant fungi secreting phytohormones and volatile organic compounds enhance growth and mineral content in finger millet under salinity stress. Plant Stress 11:100426

Mahadik SP, Patil SV, Kumudini BS (2024) Bioprospecting rhizosphere fungi endowed with multifarious plant growth-promoting potential to enhance finger millet growth under salinity stress. Plant Growth Regul 104:1483–1505

Majdura J, Jankiewicz U, Gałązka A, Orzechowski S (2022) The role of quorum sensing molecules in bacterial–plant interactions. Metabolites 13(1):114

Malhi GS, Kaur M, Kaushik P (2020) Impact of climate change on agriculture and its mitigation strategies: a review. Sustain For 13(3):1318

Marche G, Camiolo S, Porceddu A, Ruiu L (2018) Survey of *Brevibacillus laterosporus* insecticidal protein genes and virulence factors. J Invertebr Pathol 155:38–43

Martínez-Hidalgo P, Flores-Félix JD, Sánchez-Juanes F, Rivas R, Mateos PF, Regina S et al (2021) Identification of canola roots endophytic bacteria and analysis of their potential as biofertilizers for canola crops with special emphasis on sporulating bacteria. Agronomy 11(9):1796

Marulanda A, Barea JM, Azcón R (2009) Stimulation of plant growth and drought tolerance by native microorganisms (AM fungi and bacteria) from dry environments: mechanisms related to bacterial effectiveness. J Plant Growth Regul 28:115–124

Marzouk SH, Tindwa HJ, Amuri NA, Semoka JM (2023) An overview of underutilized benefits derived from *Azolla* as a promising biofertilizer in lowland rice production. Heliyon 9(1):e13040

Marzouk SH, Tindwa HJ, Amuri NA, Chande HH, Semoka JM (2024) Enhancing phosphorus use efficiency and soil quality indicators in lowland paddy ecosystem through *Azolla*, rice straw, and NPKS fertilizers. Front Agron 6:1376110

Mashabela MD, Tugizimana F, Steenkamp PA, Piater LA, Dubery IA, Terefe T, Mhlongo MI (2023) Metabolomic evaluation of PGPR defence priming in wheat (*Triticum aestivum* L.) cultivars infected with *Puccinia striiformis* f. sp. tritici (stripe rust). Front Plant Sci 14:1103413

Mathivanan R, Umavathi S, Ramasamy PK, Thangam Y (2015) Influence of vermicompost on the activity of the plant growth regulators in the leaves of the Indian butter bean plant, *Dolichos lab* L. Int J Adv Res Biol Sci 2(7):84–89

Mathur S, Agnihotri R, Sharma MP, Reddy VR, Jajoo A (2021) Effect of high-temperature stress on plant physiological traits and mycorrhizal symbiosis in maize plants. J Fungi 7(10):867

Mehmood N, Saeed M, Zafarullah S, Hyder S, Rizvi ZF, Gondal AS et al (2023) Multifaceted impacts of plant-beneficial *Pseudomonas* spp. in managing various plant diseases and crop yield improvement. ACS Omega 8(25):22296–22315

Meissle M, Mouron P, Musa T, Bigler F, Pons X, Vasileiadis VP et al (2010) Pests, pesticide use and alternative options in European maize production: current status and future prospects. J Appl Entomol 134(5):357–375

Minhas PS, Rane J, Pasala RK (2017) Abiotic stresses in agriculture: an overview. In: Minhas P, Rane J, Pasala R (eds) Abiotic stress management for resilient agriculture. Springer, Singapore, pp 3–8

Mishra A, Arshi A, Mishra SP, Bala M (2019) Microbe-based biopesticide formulation: a tool for crop protection and sustainable agriculture development. In: Arora P (ed) Microbial technology for the welfare of society. Microorganisms for sustainability, vol 17. Springer, Singapore, pp 125–145

Mishra J, Mishra I, Arora NK (2022) 2, 4-Diacetylphloroglucinol producing *Pseudomonas fluorescens* JM-1 for management of ear rot disease caused by *Fusarium moniliforme* in *Zea mays* L. 3 Biotech 12(6):138

Mishra RK, Pandey S, Mishra M, Singh Rathore U, Singh B (2024) Identification and characterization of high salinity tolerant *Trichoderma harzianum* strain, IIPRTh-3 for mitigating salt stress, wilt incidence and promoting plant growth in pigeon pea. Arch Phytopathol Pflanzenschutz 57(9):691–704

Mohamed AH, Abd El-Megeed FH, Hassanein NM, Youseif SH, Farag PF, Saleh SA et al (2022) Native rhizospheric and endophytic fungi as sustainable sources of plant growth promoting traits to improve wheat growth under low nitrogen input. J Fungi 8(2):94

Molla AH, Manjurul Haque M, Amdadul Haque M, Ilias GNM (2012) *Trichoderma*-enriched biofertilizer enhances production and nutritional quality of tomato (*Lycopersicon esculentum* Mill.) and minimizes NPK fertilizer use. Agric Res 1:265–272

Moreira H, Pereira SI, Vega A, Castro PM, Marques AP (2020) Synergistic effects of arbuscular mycorrhizal fungi and plant growth-promoting bacteria benefit maize growth under increasing soil salinity. J Environ Manag 257:109982

Mostafa MM, Hammad DM, Reda MM, El-Sayed AEKB (2024) Water extracts of *Spirulina platensis* and *Chlorella vulgaris* enhance tomato (*Solanum lycopersicum* L.) tolerance against saline water irrigation. Biomass Convers Biorefinery 14(17):21181–21191

Mukhtar T, Ali F, Rafique M, Ali J, Afridi MS, Smith D et al (2023) Biochemical characterization and potential of *Bacillus safensis* strain SCAL1 to mitigate heat stress in *Solanum lycopersicum* L. Plant Growth Regul 42(1):523–538

Mutale-Joan C, Redouane B, Najib E, Yassine K, Lyamlouli K, Laila S et al (2020) Screening of microalgae liquid extracts for their bio stimulant properties on plant growth, nutrient uptake and metabolite profile of *Solanum lycopersicum L.* Sci Rep 10(1):2820

Naamala J, Smith DL (2020) Relevance of plant growth promoting microorganisms and their derived compounds, in the face of climate change. Agronomy 10(8):1179

Nascimento FX, Brígido C, Glick BR, Oliveira S (2012) ACC deaminase genes are conserved among *Mesorhizobium* species able to nodulate the same host plant. FEMS Microbiol Lett 336(1):26–37

Nawaz A, Nawaz H, Khan K, Haq MU, Khan H, Manan U, Tariq M (2023) Integrated effect of heavy metal-tolerant rhizobacteria and phosphorus on maize growth and phosphorus bioavailability in contaminated soil. JSPAE 2(1):21–52

Naziya B, Murali M, Amruthesh KN (2019) Plant growth-promoting fungi (PGPF) instigate plant growth and induce disease resistance in *Capsicum annuum* L. upon infection with *Colletotrichum capsici* (Syd.) Butler & Bisby. Biomol Ther 10(1):41

Nethery MA, Roberts A, Barrangou R (2022) CRISPR-based engineering of phages for in situ bacterial base editing. PNAS 119(46):e2206744119

Ni H, Wu Y, Zong R, Ren S, Pan D, Yu L et al (2023) Combination of *Aspergillus Niger* MJ1 with *Pseudomonas stutzeri* DSM4166 or mutant *Pseudomonas fluorescens* CHA0-nif improved crop quality, soil properties, and microbial communities in barrier soil. Front Microbiol 14:1064358

Nosheen A, Bano A, Naz R, Yasmin H, Hussain I, Ullah F et al (2019) Nutritional value of *Sesamum indicum* L. was improved by *Azospirillum* and *Azotobacter* under low input of NP fertilizers. BMC Plant Biol 466(19):1–12

Nunes PS, Lacerda-Junior GV, Mascarin GM et al (2023) Microbial consortia of biological products: do they have a future? Biological Control 188:105439

Ojija F, Aloo BN (2023) The role of rhizobia toward food production, food and soil security through microbial agro-input utilization in developing countries. Case Stud Chem Environ Eng 8:100404

Ortas I (2022) The role of mycorrhiza in food security and the challenge of climate change. Int J Agric Appl Sci 3(1):1–11

Pais IP, Moreira R, Semedo JN, Ramalho JC, Lidon FC, Coutinho J et al (2022) Wheat crop under waterlogging: potential soil and plant effects. Plan Theory 12(1):149

Pan C, Yang K, Erhunmwunsee F, Li YX, Liu M, Pan S et al (2023) Inhibitory effect of cinnamaldehyde on *Fusarium solani* and its application in postharvest preservation of sweet potato. Food Chem 408:135213

Pang F, Solanki MK, Xing YX, Dong DF, Wang Z (2024) *Streptomyces* improves sugarcane drought tolerance by enhancing phenylalanine biosynthesis and optimizing the rhizosphere environment. Plant Physiol Biochem 217:109236

Parmar P, Kumar R, Neha Y, Srivatsan V (2023) Microalgae as next generation plant growth additives: functions, applications, challenges and circular bioeconomy-based solutions. Front Plant Sci 14:1073546

Patel AK, Albarico FPJB, Perumal PK, Vadrale AP, Nian CT, Chau HTB et al (2022) Algae as an emerging source of bioactive pigments. Bioresour Technol 351:126910

Paudel Y, Pradhan S, Pant B, Prasad B (2012) Role of blue green algae in rice productivity. Agric Biol J North America 3(8):332–335

Paul K, Saha C, Nag M, Mandal D, Naiya H, Sen D et al (2020) A tripartite interaction among the basidiomycete *Rhodotorula mucilaginosa*, N2-fixing endobacteria, and rice improves plant nitrogen nutrition. Plant Cell 32(2):486–507

Pérez-Álvarez S, Ochoa-Chaparro EH, Anchondo-Páez JC, Escobedo-Bonilla CM, Rascón-Solano J, Magallanes-Tapia MA et al (2024) Nitrogen assimilation, biomass, and yield in response to application of algal extracts, *rhizobium* sp., and *Trichoderma asperellum* as biofertilizers in hybrid maize. Nitrogen 5(4):1031–1047

Prakash J, Arora NK (2021) Novel metabolites from *Bacillus safensis* and their antifungal property against *Alternaria alternata*. Antonie Van Leeuwenhoek 114(8):1245–1258

Prasad AA, Babu S (2017) Compatibility of *Azospirillum brasilense* and *Pseudomonas fluorescens* in growth promotion of groundnut (*Arachis hypogea* L.). An Acad Bras Ciênc 89(02):1027–1040

Qadir M, Hussain A, Iqbal A, Shah F, Wu W, Cai H (2024) Microbial utilization to nurture robust agroecosystems for food security. Agronomy 14(9):1891

Rafiei M, Rahimi A, Azari A, Sahhafi SR, Aein A, Khodaygan P (2024) Effect of different strains of *Pseudomonas* bacteria on physiological traits and yield components of potato (*Solanum tuberosum* L.) under drought stress conditions. J Plant Process Func 13(60):265–288

Rai PK, Rai A, Sharma NK, Singh T, Kumar Y (2023) Limitations of biofertilizers and their revitalization through nanotechnology. J Clean Prod 138194:138194

Rajaofera MJN, Wang Y, Jatoi ZA, Jin P, Cui H, Lin C, Miao W (2020) *Bacillus atrophaeus* HAB-5 secretion metabolites preventing occurrence of systemic diseases in tobacco plant. E J Plant Pathol 156:159–172

Rajkumar M, Narayanasamy S, Uthandi S (2024) Apoplast-associated *Bacillus amyloliquefaciens* LAS10 for plant growth promotion and drought stress tolerance in little millet (*Panicum sumatransae*). Plant Stress 11:100340

Rakshit A, Bhadoria PS (2010) Role of VAM on growth and phosphorus nutrition of maize with low soluble phosphate fertilization. Acta Agronómica 59(1):111–118

Rashid U, Yasmin H, Hassan MN, Naz R, Nosheen A, Sajjad M et al (2022) Drought-tolerant *Bacillus megaterium* isolated from semi-arid conditions induces systemic tolerance of wheat under drought conditions. Plant Cell Rep 41:549–569

Rasuli N, Riahi H, Shariatmadari Z, Nohooji MG, Dehestani A, Mehraban Joubani P (2024) Growth enhancement, metabolic profile improvement, and DXR and TPS2 gene expression changes in *Thymus vulgaris* L. by cyanobacterial inoculation. J Sci Food Agric 104(10):5751–5763
Rauf M, Awais M, Ud-Din A, Ali K, Gul H, Rahman MM et al (2021) Molecular mechanisms of the 1-aminocyclopropane-1-carboxylic acid (ACC) deaminase producing *Trichoderma asperellum* MAP 1 in enhancing wheat tolerance to waterlogging stress. Front Plant Sci 11:614971
Reddy KG, Madhavi GB, Reddy ASR, Yellareddygari SK, Reddy MS (2011) Current status of biofertilizers development, farmers acceptance, utilization and future perspective in Andhra Pradesh, India
Reddy CS, Cho M, Kaul T, Joeng JT, Kim KM (2023) *Pseudomonas fluorescens* imparts cadmium stress tolerance in *Arabidopsis thaliana* via induction of AtPCR2 gene expression. J Genet Eng Biotechnol 21(1):8
Reva M, Cano C, Herrera MA, Bago A (2021) Arbuscular mycorrhizal inoculation enhances endurance to severe heat stress in three horticultural crops. Hortc Sci 56(4):396–406
Riaz U, Murtaza G, Anum W, Samreen T, Sarfraz M, Nazir MZ (2021) Plant growth-promoting rhizobacteria (PGPR) as biofertilizers and biopesticides. In: Hakeem KR, Dar GH, Mehmood MA, Bhat RA (eds) Microbiota and biofertilizers. Springer, Cham, pp 181–196
Ruiu L, Satta A, Floris I (2014) Administration of *Brevibacillus laterosporus* spores as a poultry feed additive to inhibit house fly development in feces: a new eco-sustainable concept. Poultry Sci 93:519–526
Rupawalla Z, Shaw L, Ross IL, Schmidt S, Hankamer B, Wolf J (2022) Germination screen for microalgae-generated plant growth biostimulants. Algal Res 66:102784
Russo ML, Scorsetti AC, Vianna MF, Allegrucci N, Ferreri NA, Cabello MN, Pelizza SA (2019) Effects of endophytic *Beauveria bassiana* (Ascomycota: Hypocreales) on biological, reproductive parameters and food preference of the soybean pest *Helicoverpa gelotopoeon*. J King Saud Univ Sci 31(4):1077–1082
Saha B et al (2017) Biological nitrogen fixation for sustainable agriculture. In: Meena V, Mishra P, Bisht J, Pattanayak A (eds) Agriculturally important microbes for sustainable agriculture. Springer, Singapore, pp 81–128
Sahu SK, Waseem M, Aslam MM (2023) Editorial: bioinformatics, big data and agriculture: a challenge for the future. Front Plant Sci 14:1271305
Samain E, Aussenac T, Selim S (2019) The effect of plant genotype, growth stage, and *Mycosphaerella graminicola* strains on the efficiency and durability of wheat-induced resistance by *Paenibacillus* sp. strain B2. Front Plant Sci 10:587
Sammauria R, Kumawat S, Kumawat P, Singh J, Jatwa TK (2020) Microbial inoculants: potential tool for sustainability of agricultural production systems. Arch Microbiol 202(4):677–693
Santini G, Biondi N, Rodolfi L, Tredici MR (2021) Plant biostimulants from cyanobacteria: An emerging strategy to improve yields and sustainability in agriculture. Plan Theory 10(4):643
Santosh S, Velmourougane K, Idapuganti RG, Manikandan A, Blaise D (2022) Potassium solubilizing potential of native bacterial isolates from cotton rhizosphere of rainfed vertisols. Natl Acad Sci Lett 45(3):209–212
Saranraj P et al (2022) Plant growth-promoting and biocontrol metabolites produced by Endophytic *Pseudomonas fluorescence*. In: Sayyed RZ, Uarrota VG (eds) Secondary metabolites and volatiles of PGPR in plant-growth promotion. Springer, Cham, pp 349–381
Saritha M, Tollamadugu NP (2019) The status of research and application of biofertilizers and biopesticides: global scenario. In: Buddolla V (ed) Recent developments in applied microbiology and biochemistry. Academic, pp 195–207
Sarli DA, Sánchez LA, Delgado OD (2021) *Burkholderia gladioli* MB39 an antarctic strain as a biocontrol agent. Curr Microbiol 78(6):2332–2344
Scudeletti D, Crusciol CAC, Bossolani JW, Moretti LG, Momesso L, Servaz Tubana B (2021) *Trichoderma asperellum* inoculation as a tool for attenuating drought stress in sugarcane. Front Plant Sci 12:645542
Seshagiri S, Varma A, Tallapragada P (2025) *Piriformospora indica* mediated amelioration of salt stress in *Solanum melongena* L. plants under greenhouse- a study on stomatal opening/closure. Geomicrobiol J:1–11

Shahzad A, Aslam U, Ferdous S, Qin M, Siddique A, Billah M et al (2024) Combined effect of endophytic *Bacillus mycoides* and rock phosphate on the amelioration of heavy metal stress in wheat plants. BMC Plant Biol 24(1):125

Silva RN, Monteiro VN, Steindorff AS, Gomes EV, Noronha EF, Ulhoa CJ (2019) Trichoderma/pathogen/plant interaction in pre-harvest food security. Fungal Biol 123(8):565–583

Singh JS, Kumar A, Rai AN, Singh DP (2016) Cyanobacteria: a precious bio-resource in agriculture, ecosystem, and environmental sustainability. Front Microbiol 7:186282

Singh A, Kumari R, Yadav AN, Mishra S, Sachan A, Sachan SG (2020) Tiny microbes, big yields: microorganisms for enhancing food crop production for sustainable development. In: Rastegari AA, Yadav AN, Yadav N (eds) New and future developments in microbial biotechnology and bioengineering. Elsevier, pp 1–15

Singh S, Annapurna K, Shrivastava N, Varma A (2022) Symbiotic interplay of *Piriformospora indica* and *Azotobacter chroococcum* augments crop productivity and biofortification of zinc and iron. Microbiol Res 262:127075

Singh P, Singh RK, Li HB, Guo DJ, Sharma A, Verma KK et al (2024) Nitrogen fixation and phytohormone stimulation of sugarcane plant through plant growth promoting diazotrophic *Pseudomonas*. Biotechnol Genet Eng Rev 40(1):15–35

Smith P, Calvin K, Nkem J, Campbell D, Cherubini F, Grassi G, Korotkov V, Le Hoang A, Lwasa S, McElwee P, Nkonya E (2020) Which practices co-deliver food security, climate change mitigation and adaptation, and combat land degradation and desertification? Glob Chang Biol 26(3):1532–1575

Sood M, Kapoor D, Kumar V, Sheteiwy MS, Ramakrishnan M, Landi M et al (2020) Trichoderma: the "secrets" of a multitalented biocontrol agent. Plan Theory 9(6):762

Soualiou S, Duan F, Li X, Zhou W (2022) Crop production under cold stress: an understanding of plant responses, acclimation processes, and management strategies. Plant Physiol Biochem 190:47–61

Souza RD, Ambrosini A, Passaglia LM (2015) Plant growth-promoting bacteria as inoculants in agricultural soils. Genet Mol Biol 38:401–419

Steinberg G, Gurr SJ (2020) Fungi, fungicide discovery and global food security. Fungal Genet Biol 144:103476

Subramaniam Y, Masron TA (2021) The impact of economic globalization on biofuel in developing countries. In: Energy conversion and management: X, vol 10. Elsevier, Amsterdam, p 100064

Suman J, Rakshit A, Ogireddy SD, Singh S, Gupta C, Chandrakala J (2022) Microbiome as a key player in sustainable agriculture and human health. Front Soil Sci 2:821589

Šunjka D, Mechora Š (2022) An alternative source of biopesticides and improvement in their formulation—recent advances. Plan Theory 11(22):3172

Syed A, Elgorban AM, Bahkali AH, Eswaramoorthy R, Iqbal RK, Danish S (2023) Metal-tolerant and siderophore producing *Pseudomonas fluorescence* and *Trichoderma* spp. improved the growth, biochemical features and yield attributes of chickpea by lowering cd uptake. Sci Rep 13(1):4471

Taheri Boukani K, Hassani A, Rasouli-Sadaghiani M, Barin M, Maleki R (2024) Effect of inoculation with plant growth-promoting microorganisms on some morphophysiological characteristics and essential oil and nutrients content of basil (*Ocimum basilicum* L.) under NaCl salinity stress. IJMAPR 40(1):104–129

Talaat NB (2015) Effective microorganisms improve growth performance and modulate the ROS-scavenging system in common bean (*Phaseolus vulgaris* L.) plants exposed to salinity stress. J Plant Growth Regul 34:35–46

Tanveer Y, Yasmin H, Nosheen A, Farah MA, Altaf MA (2024) Synergizing *Bacillus halotolerans*, *Pseudomonas sihuiensis* and *Bacillus atrophaeus* with folic acid for enhanced drought resistance in wheat by metabolites and antioxidants. BMC Plant Biol 24(1):1003

Tarroum M, Romdhane WB, Al-Qurainy F, Ali AAM, Al-Doss A, Fki L, Hassairi A (2022) A novel PGPF *Penicillium olsonii* isolated from the rhizosphere of *Aeluropus littoralis* promotes plant growth, enhances salt stress tolerance, and reduces chemical fertilizers inputs in hydroponic system. Front Microbiol 13:996054

Tavares M, Kozak M, Balola A, Sá-Correia I (2020) *Burkholderia cepacia* complex bacteria: a feared contamination risk in water-based pharmaceutical products. CMR 33(3):10–1128
Tchameni SN, Cotârleț M, Ghinea IO, Bedine MAB, Sameza ML, Borda D et al (2020) Involvement of lytic enzymes and secondary metabolites produced by *Trichoderma* spp. in the biological control of *Pythium myriotylum*. Int Microbiol 23:179–188
Tewari S, Arora NK (2013) Plant growth promoting rhizobacteria for ameliorating abiotic stresses triggered due to climatic variability. CCES 1(2):95–103
Tewari S, Arora NK (2016) Soybean production under flooding stress and its mitigation using plant growth-promoting microbes. In: Miransari M (ed) Environmental stresses in soybean production. Academic, pp 23–40
Thakur R, Yadav S (2023) Thermotolerant and halotolerant *Streptomyces* sp. isolated from *Ajuga parviflora* having biocontrol activity against *Pseudomonas syringae* and *Xanthomonas campestris* acts as a sustainable bioadditive in growth promotion of *Cicer arietinum*. Physiol Mol Plant Pathol 127:102059
Tibbetts SM, Milley JE, Lall SP (2015) Chemical composition and nutritional properties of freshwater and marine microalgal biomass cultured in photobioreactors. J Appl Phycol 27:1109–1119
Toppo P, Jangir P, Mehra N, Kapoor R, Mathur P (2024) Bioprospecting of endophytic fungi from medicinal plant *Anisomeles indica* L. for their diverse role in agricultural and industrial sectors. Sci Rep 14(1):588
Toromade AS, Soyombo DA, Kupa E, Ijomah TI (2024) Reviewing the impact of climate change on global food security: challenges and solutions. Int J Appl Res Soc Sci 6(7):1403–1416
Tripathi M, Kumar S, Makarana G, Goel R (2023) Metal-tolerant bioinoculant *Pseudomonas putida* KNP9 mediated enhancement of soybean growth under heavy metal stress suitable for biofuel production at the metal-contaminated site. Energies 16(11):4508
Tunçtürk R, Tunçtürk M, Rezaee Danesh Y, Najafi S, Toprak T (2024) The effects of *Glomus intraradices* on seedling growth parameters of Sage (*Salvia officinalis* L.) under salinity stress conditions. Int J Agric Technol 20(6):2579–2588
Umar A, Mwaheb MA, Ameen F, Almomani F, Dufossé L, Gancarz M (2024) Role of ectomycorrhizal colonization in enhancement of nutrients for survival of plants collected from mountainous cold stress areas. BMC Microbiol 24(1):304
UN (2023) World population projected to reach 9.8 billion in 2050, and 11.2 billion by 2100. Available at: https://www.un.org/tr/desa/world-population-projected-reach-98-billion-2050-and-112-billion-2100
Uzma M, Iqbal A, Hasnain S (2022) Drought tolerance induction and growth promotion by indole acetic acid producing *Pseudomonas aeruginosa* in *Vigna radiata*. PLoS One 17(2):e0262932
Vanitha S, Ramjegathesh R (2014) Bio control potential of *Pseudomonas fluorescens* against coleus root rot disease. J Plant Pathol Microb 5(1):216
Vega-Celedón P, Bravo G, Velásquez A, Cid FP, Valenzuela M, Ramírez I et al (2021) Microbial diversity of psychrotolerant bacteria isolated from wild flora of Andes mountains and Patagonia of Chile towards the selection of plant growth-promoting bacterial consortia to alleviate cold stress in plants. Microorganisms 9(3):538
Verma KK, Song XP, Joshi A, Tian DD, Rajput VD, Singh M et al (2022) Recent trends in nanofertilizers for sustainable agriculture under climate change for global food security. Nanomater 12(1):173
Verma P, Bhattacharya A, Bharti C, Arora NK (2024a) Antifungal metabolites produced by *Pseudomonas hunanensis* SPT26 effective in biocontrol of fusarium wilt of *Lycopersicum esculentum* under saline conditions. World J Microbiol Biotechnol 40(10):305
Verma P, Bhattacharya A, Bharti C, Arora NK (2024b) Diazotrophs for enhancement of productivity of major cereal crops under abiotic stresses. In: Arora NK, Bouizgarne B (eds) Microbial biotechnology for sustainable agriculture Volume 2. Microorganisms for Sustainability, vol 34. Springer, Singapore, pp 131–158
Vessey JK (2003) Plant growth promoting Rhizobacteria as biofertilizers. Plant Soil 255:571–586
Wagg C, Schlaeppi K, Banerjee S, Kuramae EE, van der Heijden MG (2019) Fungal-bacterial diversity and microbiome complexity predict ecosystem functioning. Nat Commun 10(1):4841

Wang J, Fu Z, Qiao H, Liu F (2019) Assessment of eutrophication and water quality in the estuarine area of Lake Wuli, Lake Taihu, China. Sci Total Environ 650:1392–1402

Wang J, Li R, Zhang H, Wei G, Li Z (2020) Beneficial bacteria activate nutrients and promote wheat growth under conditions of reduced fertilizer application. BMC Microbiol 20(38):1–12

Wang D, Luo WZ, Zhang DD, Li R, Kong ZQ, Song J et al (2023a) Insights into the biocontrol function of a *Burkholderia gladioli* strain against *Botrytis cinerea*. Microbiol Spect 11(2):e04805–e04822

Wang K, Lin Z, Dou J, Jiang M, Shen N, Feng J (2023b) Identification and surveys of promoting plant growth VOCs from biocontrol bacteria *Paenibacillus peoriae* GXUN15128. Microbiol Spect 11(3):e04346–e04322

Wei H, He W, Kuang Y, Wang Z, Wang Y, Hu W et al (2023) Arbuscular mycorrhizal symbiosis and melatonin synergistically suppress heat-induced leaf senescence involves in abscisic acid, gibberellin, and cytokinin-mediated pathways in perennial ryegrass. Environ Exp Bot 213:105436

Weldeslassie T, Naz H, Singh B, Oves M (2018) Chemical contaminants for soil, air and aquatic ecosystem. In: Oves M, Zain Khan M, Ismail IMI (eds) Modern age environmental problems and their remediation. Springer, Cham, pp 1–22

Wilmowicz E, Kućko A, Bogati K, Wolska M, Świdziński M, Burkowska-But A, Walczak M (2022) *Glomus* sp. and *Bacillus* sp. strains mitigate the adverse effects of drought on maize (*Zea mays* L.). Front Plant Sci 13:958004

Win KT, Kobayashi M, Tanaka F, Takeuchi K, Oo AZ, Jiang CJ (2022) Identification of *Pseudomonas* strains for the biological control of soybean red crown root rot. Sci Rep 12(1):14510

Wu X, Wu J, Zhou B, Hong B, Zhao D, Guan M (2024) Effects of fertilization patterns on the growth of rapeseed seedlings and rhizosphere microorganisms under flooding stress. Agron 14(3):525

Xiang N, Liu Z, Tian X, Wang D, Hashem A, Abd-Allah EF et al (2024) Improved waterlogging tolerance in roots of cucumber plants after inoculation with Arbuscular Mycorrhizal fungi. Horticulture 10(5):478

Yadav SK, Soni R, Rajput AS (2018) Role of microbes in organic farming for sustainable agroecosystem. In: Panpatte D, Jhala Y, Shelat H, Vyas R (eds) Microorganisms for green revolution, Microorganisms for sustainability, vol 7. Springer, Singapore, pp 241–252

Yadav S, Modi P, Dave A, Vijapura A, Patel D, Patel M (2020) Effect of abiotic stress on crops. SCP 3(17):5–16

Yan Q, Li X, Xiao X, Chen J, Liu J, Lin C et al (2022) Arbuscular mycorrhizal fungi improve the growth and drought tolerance of *Cinnamomum migao* by enhancing physio-biochemical responses. Ecol Evol 12(7):e9091

Yang A, Akhtar SS, Fu Q, Naveed M, Iqbal S, Roitsch T, Jacobsen SE (2020) *Burkholderia phytofirmans* PsJN stimulate growth and yield of quinoa under salinity stress. Plan Theory 9(6):672

Yang Y, Dou Y, Wang B, Xue Z, Wang Y, An S, Chang SX (2023a) Deciphering factors driving soil microbial life-history strategies in restored grasslands. iMeta 2(1):e66

Yang S, Imran, Ortas I (2023b) Impact of mycorrhiza on plant nutrition and food security. J Plant Nutr 46(13):3247–3272

Yasmeen T, Tariq M, Iqbal S, Arif MS, Riaz M, Shahzad SM et al (2019) Ameliorative capability of plant growth promoting rhizobacteria (PGPR) and arbuscular mycorrhizal fungi (AMF) against salt stress. In: Hasanuzzaman M, Hakeem K, Nahar K, Alharby H (eds) Plant abiotic stress tolerance. Springer, Cham, pp 409–448

Yasmin H, Naeem S, Bakhtawar M, Jabeen Z, Nosheen A, Naz R et al (2020) Halotolerant rhizobacteria *pseudomonas pseudoalcaligenes* and *Bacillus subtilis* mediate systemic tolerance in hydroponically grown soybean (*Glycine max* L.) against salinity stress. PLoS One 15(4):e0231348

Yousefi S, Kartoolinejad D, Bahmani M, Naghdi R (2017) Effect of *Azospirillum lipoferum* and *Azotobacter chroococcum* on germination and early growth of hopbush shrub (*Dodonaea viscosa* L.) under salinity stress. J Sustain For 36(2):107–120

Yu L, Zhang H, Zhang W, Liu K, Liu M, Shao X (2022) Cooperation between arbuscular mycorrhizal fungi and plant growth-promoting bacteria and their effects on plant growth and soil quality. PeerJ 10:e13080

Yu C, Jiang X, Xu H, Ding G (2023) *Trichoderma longibrachiatum* inoculation improves drought resistance and growth of *Pinus massoniana* seedlings through regulating physiological responses and soil microbial community. J Fungi 9(7):694

Yuan Y, Zu M, Sun L, Zuo J, Tao J (2022) Isolation and screening of 1-aminocyclopropane-1-carboxylic acid (ACC) deaminase producing PGPR from *Paeonia lactiflora* rhizosphere and enhancement of plant growth. Sci Hortic 297:110956

Zahoor M, Irshad M, Rahman H, Qasim M, Afridi SG, Qadir M et al (2017) Alleviation of heavy metal toxicity and phytostimulation of *Brassica campestris* L. by endophytic *Mucor* sp. MHR-7. Ecotoxicol Environ Saf 142:139–149

Zhang W, Xie Z, Zhang X, Lang D, Zhang X (2019) Growth-promoting bacteria alleviates drought stress of *G. uralensis* through improving photosynthesis characteristics and water status. J Plant Interact 14(1):580–589

Chapter 3
Biochar-Mediated Soil Amendment for Sustainable Agriculture

Seuj Priya Borah, Padmasri Ghosh, Chira Ranjan Bhattacharjee, Arun Jyoti Nath, and Shuvasish Choudhury

Abstract Soil, a complex ecosystem of interdependent life-support systems comprising air, water, minerals, and organic matter, including closely interacting macro- and microorganisms, plays a fundamental role in food and nutritional security. Soil degradation due to pollution and other associated factors threatens the quality and quantity of food produced, directly impacting human life. Thus, improving soil quality is a major challenge to sustain productive agriculture. Biochar, a carbon-rich material mainly produced from waste biomass through pyrolysis, has received considerable attention recently as a potentially valuable soil amendment to improve soil quality in various ways. Despite the accelerated global initiatives in research involving biochar and its multifarious applications, several critical aspects still remain unaddressed. Soil amendment through biochar has proved to be a cost-effective and sustainable option for improving soil health, thus ensuring agricultural productivity vis-à-vis food and nutritional security. This chapter aims to provide insights into the diverse aspects of biochar, such as the techniques involved in biochar production, its physical and chemical attributes, and value addition in terms of engineered biochar. The role of biochar in influencing soil microflora, carbon sequestration, and nutrient cycling has been discussed. The emerging trends of biochar and engineered biochar such as biochar–nanocomposites, biochar–microbial composites, biochar–clay (mineral) composites, and biochar–enzyme composites have been discussed for their functional role in the sustainable enhancement of soil health and increase in agricultural productivity.

S. P. Borah · P. Ghosh · S. Choudhury (✉)
Department of Life Science and Bioinformatics, Assam University, Silchar, India
e-mail: shuvasish.choudhury@aus.ac.in

C. R. Bhattacharjee
Department of Chemistry, Assam University, Silchar, India

A. J. Nath
Department of Ecology and Environmental Science, Assam University, Silchar, India

N. K. Arora et al. (eds.), *Environmental Sustainability and Biotechnology: Opportunities and Challenges*, Microorganisms for Sustainability 46,
https://doi.org/10.1007/978-981-96-9025-1_3

Keywords Soil Remediation · Sustainable Agriculture · Engineered Biochar · Biochar Composites

3.1 Introduction

In the era of climate change, food security and malnutrition are of utmost concern for humanity. The 2023 global hunger index of 18.3 is considered reasonable compared to 19.1 in 2015. Still, it indicates that much is yet to be achieved to provide humanity with a balanced and nutritious diet. One of the most significant challenges to food security is the degradation of soil quality, which affects crop quality and productivity in myriad ways. The need to enhance soil quality to boost crop productivity and quality continues to be a complicated challenge for plant breeders to ensure food security for all. In the last few decades, it has been strongly perceived that there is a need to develop strategies to improve agricultural productivity sustainably. Today, significant developments have been made regarding sustainable agriculture practices reducing reliance on chemical or inorganic soil amendments. The sustainable development goals (SDGs) laid down by the United Nations firmly emphasize zero hunger and other achievable targets by 2030 (Vagsholm et al. 2020; Arora and Mishra 2022).

Agriculture is the basis of the nourishment of human life on this planet. Without food, no one survives. The term '*sustainable agriculture*' encompasses the goal of providing humankind with hygienic food by monitoring all aspects of agriculture in relation to the environment. The indiscriminate use of chemical and inorganic amendments, including pesticides, weedicides, etc., degrades soil quality and alters biotic and abiotic components. For example, heavy metals, metalloids, microplastics, pesticide residues, pharmaceutical waste, etc. have become major pollutants in agricultural soil, resulting in the loss of organic matter, intense acidification or alkalization, and soil erosion (Jainpeng 1999; Tang et al. 2013; Egamberdieva et al. 2021; Tian et al. 2022; Zhao et al. 2022; Angon et al. 2024). Thus, the central dogma of sustainable agriculture hovers around to improve and maintain agricultural soil quality. With the large-scale availability of rich bioresources, mainly in rural areas, sustainable agricultural practices could boost organic agriculture and support the rural economy (Ding et al. 2016; Herman 2024). Biochar, derived mainly from different bioresources, including biowastes, has emerged as an attractive option for sustaining agricultural soil quality. Biochar, rich in carbon, is primarily obtained from biomass through pyrolysis. The use of biochar for soil nutrient retention and quality improvement dates back to 2000 years ago when the native Amazonians in Agrarian civilization amended the nutrient-poor tropical soil with biochar and organic matter (Glaser et al. 2001; Glaser and Birkk 2012). Originally referred to as Terra Preta or Amazonian, dark earth biochar was produced by igniting biomass, followed by burying and smouldering under a low-oxygen condition (Glaser and

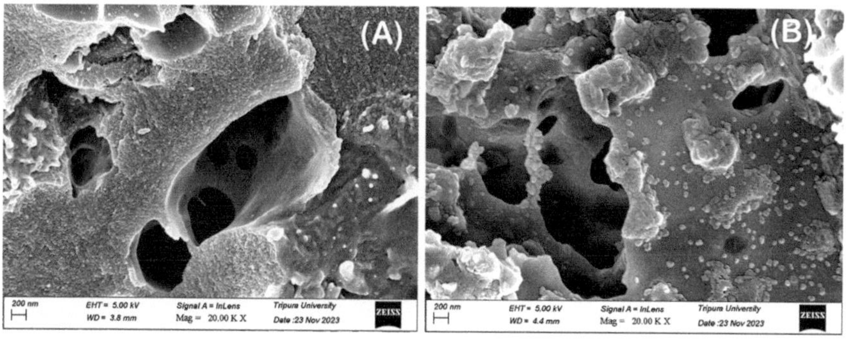

Fig. 3.1 Scanning electron microscopy (SEM) images showing porous structures of biochar obtained from tea processing (**a**) and tea pruning (**b**) waste by pyrolysis

Birkk 2012). This traditional slash-and-char agriculture method presumably affected about 50% carbon sequestration compared to the slash-and-burn technique with only ~3% carbon sequestration (Woods et al. 2009; Glaser and Birkk 2012; Nair et al. 2023). Chemically, most biochar contains about 65% carbon with rather large amounts of inorganic elements and functional groups with a porous architecture (Gaskin et al. 2008; Kim et al. 2012; Wijitkosum and Jiwnok 2019, Yadav et al. 2023). The chemical characteristics primarily depend on the biomass source and pyrolysis method (Tang et al. 2013; Wijitkosum and Jiwnok 2019). It has a carbonaceous skeleton with diverse functional groups and porous structure (Lin et al. 2012; Stylianou et al. 2020) (Fig. 3.1).

Biochar can sequester soil contaminants and improve the soil physicochemical and biological properties (Lehmann et al. 2006; Tang et al. 2013; Yadav et al. 2023; Nepal et al. 2023; Pandian et al. 2024). Today, the property of biochar in the remediation of agrochemicals, polyaromatic hydrocarbons, heavy metals, polychlorinated biphenyls, aromatic dyes, volatile organic compounds, and diverse groups of toxic contaminants from the soil is well recognized (Beesley et al. 2010; Zheng et al. 2010; Xu et al. 2012; Ahmad et al. 2014; Jung et al. 2015; Oliveria et al. 2017; Yadav et al. 2023; Nguyen et al. 2023; Itam et al. 2024) (Fig. 3.2). During the past decade, the interest in using biochar as a potential soil remediation and reclamation agent has significantly increased. Besides, biochar application and managed irrigation practices also help to alleviate methane (CH_4) emissions and improve the soil's organic carbon content (Sriphiron et al. 2020; Nan et al. 2021; Harrison et al. 2024). With immense potential for soil remediation and improving soil physicochemical and biological characteristics, biochar is evolving as one of the most viable options toward organic and sustainable agriculture. In this review, we discuss the different aspects of biochar comprising techniques in biochar production, its application in improving soil quality, and the significance of engineered biochar.

Fig. 3.2 Role of biochar in improving soil physicochemical and biological properties

3.2 Techniques of Biochar Production

Biochars usually have a high surface area, porous structure, high cation-exchange properties, and various functional groups, making them a potential candidate for environmental remediation and management (Sharma et al. 2023). Various types of biomasses, including biowastes, can be used to produce biochar. Traditionally, thermochemical methods are employed to prepare biochar, among which pyrolysis is the most preferred technique (Ghodke et al. 2021). Besides the pyrolytic method for biochar production, torrefaction, hydrothermal carbonization, and gasification processes are also used (Sharma et al. 2023). The pyrolysis method for biochar production usually varies depending on the type of biomass used. For example, conventional pyrolysis is used to prepare biochar from agricultural wastes and related biomasses. Such pyrolytic processes can be either slow or fast. The slow pyrolysis is achieved at a relatively low temperature, usually between 350 and 500 °C. This process affects the biochar's carbon content, pore size, and surface area. In the fast pyrolysis method, the heating rate spontaneously increased at 2–200 °C per second, which converts the bulk of biomass to bio-oil with a low biochar output (Li et al. 2023; Sharma et al. 2023). Slow pyrolysis is preferred for biochar production, rather than fast pyrolysis, since fast pyrolysis yields only 10–20% biochar from the biomass, and the rest is bio-oil (Li et al. 2020a; Vuppaladadiyam et al. 2023).

Biochar production is also achieved through vacuum pyrolysis, in which nitrogen purges remove oxygen at temperatures ranging from 400 to 600 °C (Mohan et al. 2014; Sharma et al. 2023; Mishra et al. 2023). Vacuum pyrolysis is often considered a better option for biochar production as it facilitates high-quality biochar with better pore structure and surface area than those obtained from conventional pyrolytic methods (Foong et al. 2020; Sharma et al. 2023). Another pyrolytic method for biochar production is microwave pyrolysis, which is believed to be more effective for biochar production than other pyrolytic methods (Sharma et al. 2023). This process is extremely fast, though the production cost is comparatively high (Haeldermans et al. 2020). The biochar obtained by microwave pyrolysis is of superior quality regarding surface area and porosity (Pandey et al. 2022a, b). Recently, the green pyrolysis technique has acquired popularity for biochar production. Herein, solar radiation is used for pyrolysis. However, the method needs a complex experimental setup, and the biochar yield is essentially low as compared to the frequently used conventional methods. However, solar pyrolysis is likely a viable future option as it will minimize the dependence on electricity and other fuels to achieve pyrolysis (Chintala et al. 2017). In addition to these methods, gasification, a thermochemical method, is also used to convert biomass to syngas and biochar at exceedingly high temperatures in the absence of air (Shenbagaraj et al. 2020). However, the gasification process for biochar production is costly, though the method yields high-quality gaseous hydrocarbons and biochar (Yao et al. 2018; Marousek and Gavurova 2022). Comparatively, biochar production through conventional pyrolysis methods is essentially cost-effective with moderate to high biochar yield. The drum method is another widely used conventional biochar production method. This is a very cost-effective method, affording a moderately high biochar yield. This method uses a metallic drum as a charring kiln to load and ignite the biomass to prepare the biochar. Once the flame inside the kiln turns blue and the smoke starts to disappear, the top and bottom of the drum are sealed with clay and allowed to cool down, after which the desired biochar is obtained. The drum method is economically viable for large-scale biochar production as it involves very low-cost pyrolysis setup and transportation, as biochar can be prepared near the biomass source. The comparison of various methods of biochar production is shown in Fig. 3.3.

3.3 Physical and Chemical Properties of Biochar

Biochar being a carbon-rich material, its physical and chemical properties are influenced mainly by the type of biomass, production temperature, and other related factors. Biochar is stable mainly due to its networked carbon structure and differs from organic carbon due to higher amounts of aromatic carbon (Schmidt and Noack 2000; Tomczyk et al. 2020). The physicochemical properties of biochar are mainly influenced by the pyrolytic temperature to which the biomass is exposed. Since different biomasses have different chemical properties, the physicochemical properties of biochar upon pyrolysis also differ considerably (Tomczyk et al. 2020; Li and Skelly 2023). Other factors that also influence the physicochemical properties of biochar include pressure and dwelling time during biochar production (Tomczyk et al. 2020).

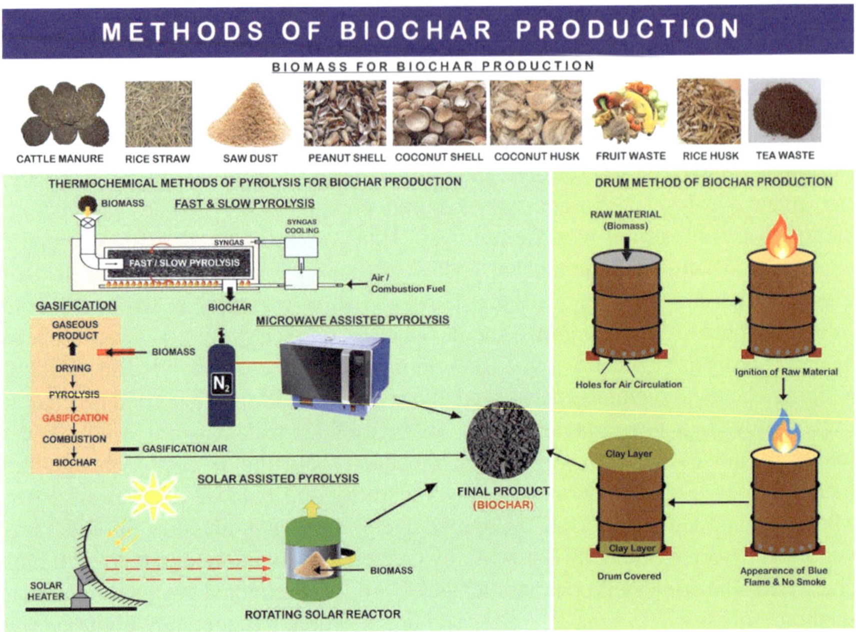

Fig. 3.3 Different thermochemical and conventional techniques for biochar production from various biomasses

At the initial stages of biochar production, the temperature at which the biomass is exposed facilitates the loss of moisture content and the volatilization process. This is called the pre-pyrolysis stage (Lee et al. 2017). As the moisture evaporates, carboxylic acid (-COOH) and carbonyl (-CO) groups are formed (Lee et al. 2017). As the temperature rises, the hemicellulose and cellulose material are degraded, followed by lignin and other organic matter degradation (Cardenas-Aguair et al. 2017; Devi et al. 2021). The variation in pyrolytic temperature affects the degradation process of organic constituents, which affects the chemical composition, surface area, porosity, and functional groups (Ding et al. 2014; Tomczyk et al. 2020). Studies have shown that a drastic increase in pyrolytic temperature increases the surface area and porosity of the biochar considerably. However, at low temperatures (~ <500 °C), biochar is strongly hydrophilic as lignin degradation slows down and is not converted to polycyclic aromatic hydrocarbon (Tomczyk et al. 2020). In contrast, the biochar obtained at higher temperatures is thermostable and strongly hydrophobic (Ghani et al. 2013; Tomczyk et al. 2020). Further, at high pyrolytic temperatures, the rate of degradation of cellulose material is also high, resulting in the formation of a large number of micropores due to the amorphous carbon structures (Vamvuka and Sfakiotakis 2011; Zhao et al. 2017; Tomczyk et al. 2020).

Fast chemical transformation occurs when biomass is exposed to very high pyrolytic temperatures. Spectroscopic studies using infrared (FT-IR) spectrometry revealed that new functional groups, typically oxygenated hydrocarbons, are formed

(Ghani et al. 2013; Mia et al. 2017; Tomczyk et al. 2020). Some studies also demonstrated that biochar prepared at temperatures ranging from 600 to 700 °C is strongly hydrophobic and possesses well-structured carbon layers with very low H and O contents (Tomczyk et al. 2020). When prepared at 300 to 400 °C, biochar is more organic and is diverse in structure (Tomczyk et al. 2020). Biochar obtained through pyrolysis at high temperatures has inferior cation-exchange capacity due to the loss of surface functional groups. It was later demonstrated that the cation-exchange capacity of biochar depends upon the presence of oxygen-rich functional groups on its surface (Banik et al. 2018; Tomczyk et al. 2020). Another important feature that affects biochar's physicochemical characteristics is the volatilization process. Biochar derived at high temperatures has a low volatile matter content, which is lost during the process of low-molecular-weight liquids and gases (Zhao et al. 2017).

In contrast, the biochar obtained at low temperatures (~400 °C) has diverse functional groups such as hydroxyl (-OH), carbonyl (-CO), carboxyl (-COOH), and hemiacetal (Tomczyk et al. 2020). The diversity of functional groups varies considerably at different stages of pyrolysis, and the functional groups change in due course of the pyrolysis process. Another critical factor is the carbon content of biochar. It was observed that the biochar obtained at high temperatures has a more extensive ash content (up to almost 19%), which in turn increases the concentration of essential elements such as magnesium (Mg), calcium (Ca), phosphorus, (P) and potassium (K) up to more than 30% (Zama et al. 2017; Tomczyk et al. 2020). The percentage of carbon content in biochar varies considerably, mainly due to the origin of the biomass. Biochar obtained from different biomass sources has different carbon contents but is also influenced by the pyrolytic temperature at which biochar is produced. Studies have shown that biochar derived from poultry litter, orange peel, apple stem, rice straw, etc. has different carbon contents because of different origins, besides the temperature at which they are exposed (Tomczyk et al. 2020). For example, biochar from poultry litter at very high pyrolytic temperatures has a low carbon content (~36%) (Tomczyk et al. 2020). Thus, the temperature critically affects biochar's physicochemical characteristics by influencing its morphology and functional groups.

3.4 Biochar and Carbon Sequestration

Soil as a substrate provides all the required elements for the plant's growth and development. In the race for carbon neutrality, devising appropriate strategies to develop carbon sinks to reduce atmospheric CO_2 is the global agenda (Nath et al. 2021). Compared to engineering techniques, natural soils are essential carbon sinks through carbon capture and storage as they contain more carbon than that stored in terrestrial vegetation and the atmosphere combined. Biochar's application to soil can sequester long-term carbon because of its high stability and easy and large-scale availability. Additionally, biochar influences soil carbon content, nutrient cycling and microbial diversity, soil texture, pH, moisture content, etc. (Nannipieri et al. 2017; Tan et al. 2017; Nath et al. 2018; Mandal et al. 2021). Today, there is a pressing need to reduce greenhouse gas emissions via carbon sequestration. Biochar can mediate the process of carbon sequestration in soil in various ways. Primarily, biochar amends carbon in the soil and increases the soil's organic carbon (SOC) pool, leading to a higher carbon stock than soils without biochar amendments (Li and Tasnady 2023).

3.5 Biochar and Soil Microflora

Using biochar to improve soil quality has attracted interest in investigating its properties in relation to soil microbial composition. Biochar considerably affects the soil microbial activity by restructuring the soil microflora (Nielsen et al. 2014; Ahmad et al. 2016). Zhu et al. (2017) Several reviews that dealt with biochar application in soil and its influence on soil microorganisms appeared. Further, biochar can shelter the microbes in its porous structures, providing nutrients essential for microbial growth viz-a-viz influencing the soil's physicochemical attributes to enhance microbial growth. Besides these effects, biochar reduces the bioavailability of soil contaminants that are detrimental to microbial growth (Stefaniuk and Olezzcuzuk 2016). Various studies have demonstrated that soils amended with biochar facilitate microbial growth by providing essential nutrients such as nitrogen (N), phosphorus (P), potassium (K), magnesium (Mg), etc. (Rodriguez-Vila et al. 2016; Zhu et al. 2017).

It is well established that biochar amendment to soil strongly influences its physicochemical properties that are well linked with microbial growth and diversity (Spokas et al. 2011; Sun et al. 2016; Gorovtsov et al. 2020). Microbial colonization of the biochar is mainly related to the chemical composition of the biochar, which varies depending on the feedstock material used and the pyrolysis temperature at which biochar is prepared. The soil physicochemical properties, organic composition, and microbial diversity also matter substantially (Gorovtsov et al. 2020). Further, biochar aging has also been attributed to its microbial colonization (Gorovtsov et al. 2020). It has been seen that microbial colonization is favored in aged biochar compared to freshly prepared biochar since the latter may impart undesirable consequence due to unwanted pyrolysis products (Mukherjee et al. 2014; Gorovtsov et al. 2020). Overall, the biochar amendment process for improving soil microbial diversity has been well demonstrated in several studies. For example, the formation of macroaggregates of *Actinobacteria* around biochar has been well documented (Zheng et al. 2018; Gorovtsov et al. 2020).

The interaction of biochar with microorganisms in the soil is a valuable way to detoxify inorganic and organic pollutants. Many studies have demonstrated that when applied with heavy metal-resistant bacterium such as *Bacillus subtilis*, biochar improves the absorption capacity of heavy metals such as lead (Pb) and mercury (Hg) (Gorovtsov et al. 2020). Biochar amendment along with *Bacillus cereus* and *Pseudomonas japonica* to soil contaminated with hexavalent chromium (Cr^{+6})-contaminated soil resulted in a significant reduction of Cr–induced phytotoxicity as compared to those with biochar amendment but in the absence of the bacterial inoculation (Arshad et al. 2017). Thus, biochar–microbe interaction improves soil characteristics substantially and enhances the remediation capacity of biochar. In another study, it was also demonstrated that biochar-amended soil, along with bacteria, limits the phyto-availability and toxicity of hexavalent Cr^{6+} in wheat seedlings by reducing it to trivalent (Cr^{3+}) forms, and likewise, biochar supplementation with *Neorhizobium hautlense* also reported mitigating Pb and cadmium (Cd)-induced phytotoxicity (Wang et al. 2016). Besides detoxifying heavy metals, biochar and microbial interaction also mitigate organic pollutants in soil. Biochar ladened with immobilized *Pseudomonas putida* effectively degraded polyaromatic hydrocarbons (Gorovtsov et al. 2020). Degradation of other organic pollutants such as phenanthrene, fluoranthene, and pyrene could be achieved by the amendment of biochar with immobilized *Mycobacterium gilvum* (Xiong et al. 2017; Gorovtsov et al. 2020).

3.6 Biochar and Nutrient Cycling

Fertile soils have high agricultural yield and good crop quality. Soil fertility could be managed effectively by regulating the nutrient cycling in agroecosystems. Today, nutrient cycling in agroecosystems is managed by employing inorganic or chemical fertilizers or related amendments. However, its long-term use has several adverse implications for the environment as well as human health. Further, in many developing and underdeveloped nations, the accessibility of subsidized fertilizers to marginal farmers is also a major constraint in losing the soil nutrient pool. In addition to these aspects, factors such as unregulated irrigation practices, poor soil conservation strategies, changing precipitation patterns, recurrent droughts, leaching, soil acidity, soil alkalinity, etc., also contribute toward degrading the nutrient availability in agroecosystems. Improving the nutrient pool and increasing its availability is a challenge. Using sustainable resources to improve soil nutrient cycling seems to be a viable option. One such way is using "biochar" (Chenu et al. 2019). Today, biochar is considered to improve the soil physicochemical properties and its attributes with the potential to integrate agricultural practice with the benefits of climate change mitigation (Burgeon et al. 2022). Besides, the use of biochar is largely encouraged for soil nutrient cycling under the United Nations (UN) SDGs by 2030 (Patro et al. 2024).

The capacity of biochar to influence the nutrient cycling in the soil also depends upon the type of biochar, its structural features, the feedstock material from which it is derived, pyrolysis temperature, functional groups, etc. (Yun et al. 2019; Jindo et al. 2020). The biochar characteristics that determine its efficiency toward nutrient cycling in the soil include the elemental composition, surface area, pore size, pH, cation-exchange capacity, and calcium carbonate equivalent (Ippolito et al. 2014). For example, nitrogen (N) availability in the form of nitrate in biochar is usually significantly low. However, studies have also shown that biochar influences N availability in the soil through various other processes (Ippolito et al. 2014; Jindo et al. 2020). It has been concluded that the initial nutrient concentration of the feedstock material and pyrolysis temperature are two important factors that strongly influence the nutrient cycling capacity of biochar (Gaskin et al. 2008; Cantrell et al. 2012; Kloss et al. 2012; Ippolito et al. 2014). Biochar has also been reported to promote biological nitrogen fixation with almost 63% enhancement, mainly in acidic soils (Jindo et al. 2020). Biochar obtained from sugarcane bagasse at a pyrolysis temperature of 350 °C improves N fixation by 83% and increases the nodulation to almost 90% in mash beans compared to those without any biochar amendment (Azeem et al. 2019). This feature of biochar in nutrient cycling is closely related to soil's physicochemical characteristics and microbial diversity (Schmalenberger and Fox 2016). Studies in mung bean plants have also shown that the amendment of biochar derived from *Eucalyptus* improves nitrogen fixation by 73% in soil enriched in boron (B) and molybdenum (Mo) (Rodon et al. 2007).

Compared to other elements, biochar's K concentration is relatively high due to its higher volatilization temperature (DeLucaa et al. 2012). The concentration of K in biochar depends mainly on the type of feedstock materials being used. Biochar derived from feedstock materials such as animal waste, rice straw, and seaweeds has a high K concentration since such biomasses have a larger K content (Peng et al. 2011; Eibisch et al. 2015; Subedi et al. 2016). It was also demonstrated that pyrolysis temperature increases the water-soluble K by 47% (Zheng et al. 2013). Biochar-mediated K cycling in soil depends mainly on soil texture, pH, ion-exchange capacity, and moisture content (Jindo et al. 2020). Applying K-rich biochar as soil amendment significantly improves the soil K content, thus promoting plant growth and yield (Biederman and Stanley 2013; Purakayastha et al. 2015; Wu et al. 2019). Another essential element needed for plant growth is phosphorus P. Phosphorus is primarily amended in the soil through inorganic phosphate fertilizers, and its availability to the plants largely depends upon the physicochemical attributes of the soil. Biochar has been reported to influence P cycling in the soil by increasing its concentration (Jindo et al. 2020). The vital aspect that affects P cycling through biochar is the pyrolysis temperature. The biochars obtained at temperatures ranging from 360 to 600 °C transform organic P into inorganic P, and its speciation changes with an increase in the pyrolysis temperature (Uchiyama et al. 2015; Schneider and Haderlein 2016; Brunn et al. 2017). Phosphate-rich biochar can increase soil pH and form a complex with aluminum (Al) and iron (Fe), thus regulating high soil acidic conditions (Jindo et al. 2020). However, in alkaline soils, biochar has no major role in P cycling (Jindo et al. 2020). Besides exerting a significant influence in N, P, and K cycling in the soil, biochar also neutralizes high soil acidity when applied in combination with Mg and calcium (Ca) (Gezahegn et al. 2019).

3.7 Engineered Biochar for Soil Remediation

In recent years, engineered biochar has been one of the emerging trends in biochar production and its value addition. Such type of biochar includes biochar–nanocomposites, biochar–microbe composites, biochar–clay (mineral) composites, and biochar–enzyme composites that are designed to have broad-spectrum functional roles (Fig. 3.4). Since the biochar has different functional roles, the prime objective behind engineering biochar in various ways is to improve its role by influencing its physicochemical and structural properties. Biochar obtained from various sources can act as an absorbent material for toxic organic and inorganic contaminants. Adding value attributes to it may be a prospective approach to increase its functional efficiency in myriad ways. Thus, the concept of engineered biochar composites is intended to improve the biochar quality.

3.7.1 *Biochar–Nanometal Oxide Composites*

The preparation of biochar–nanocomposites is one such approach for value addition to biochar, primarily aimed at improving the functional attributes of biochar. Biochar–nanocomposites can be designed with nanometal oxides (Zhao et al. 2021).

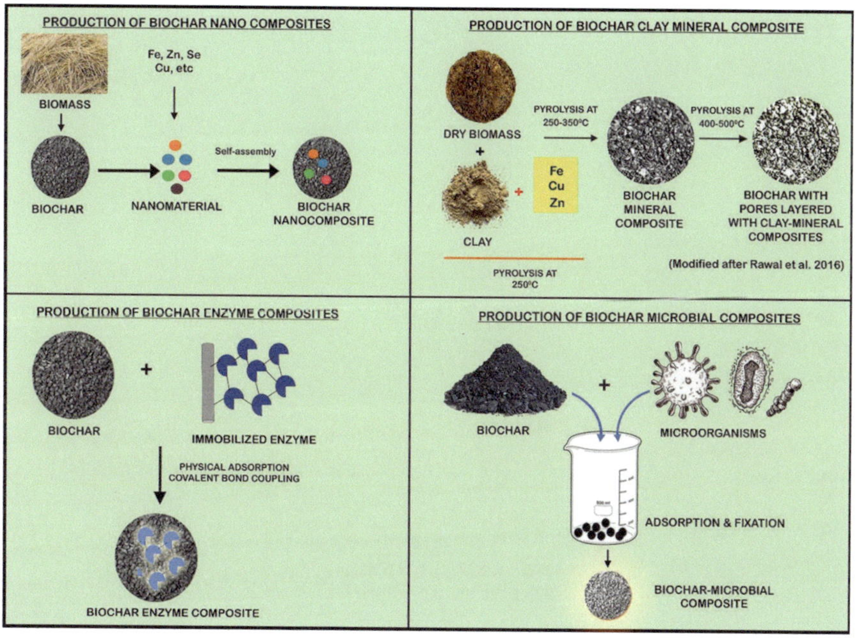

Fig. 3.4 Engineered biochar composites with nanomaterials, clay–mineral, enzymes, and microorganisms

The preparation of nanometal oxide biochar could be attained by various chemical processes such as impregnation, coprecipitation, pyrolysis, etc. (Zhao et al. 2021) (Fig. 3.5). In the process of impregnation, the biochar is immersed in a solution of active nanometal oxides, resulting in the formation of biochar nanocomposites having strong cation-exchange properties, high functional group diversity, and improved porosity (Wang et al. 2017; Zhang et al. 2020; Zhao et al. 2021). For example, Zhang et al. (2020) showed that biochar impregnated with Fe-metal oxide increases the efficiency of biochar materials in degrading sulfamethoxazole. Other than Fe, materials manganese ($KMnO_4$) and magnesium ($MgCl_2$) have been demonstrated to extract arsenic (As) from the aqueous and soil media (Yu et al. 2015; Lin et al. 2019). Studies have shown that at acidic pH (5.0), Mn oxide biochar nanocomposites can absorb toxic heavy metals such as Pb and Cr, and the efficacy of such an extract is almost five times higher than those compared to the original biochar (Lin

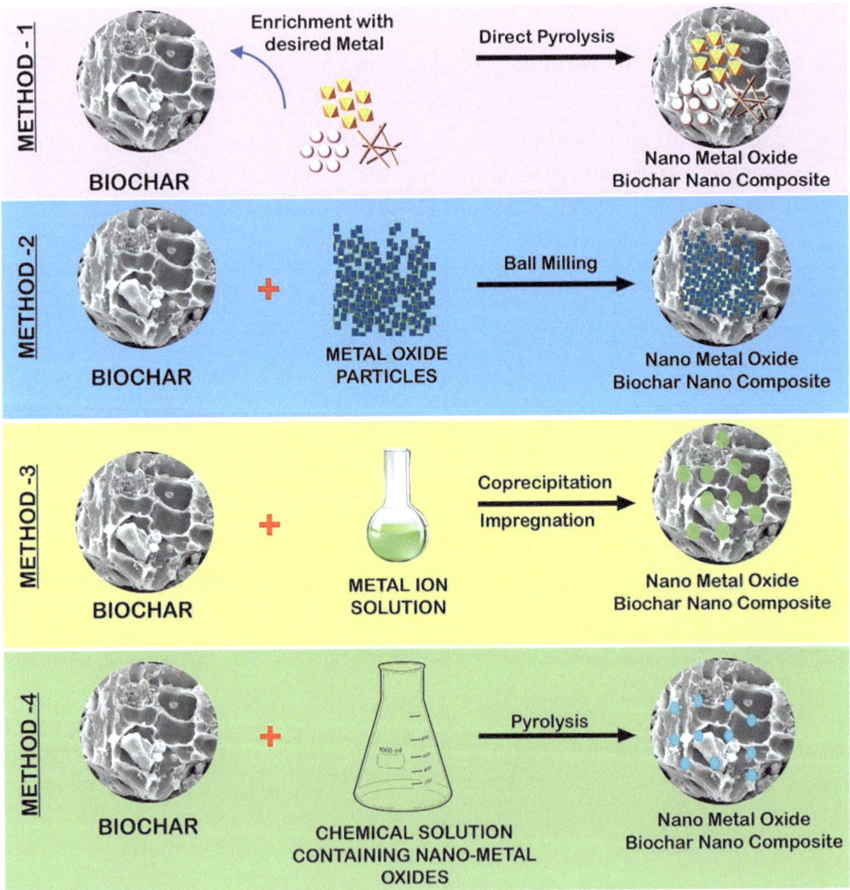

Fig. 3.5 Various methods of preparation of nanometal oxide biochar composites. (Modified from Zhao et al. (2021))

et al. 2019). Besides this, the extraction efficiency is almost close to 98% (Lin et al. 2019). It was also demonstrated that the extraction of Cr is enhanced due to chemical interaction with Mn doped to biochar's surface (Lin et al. 2019; Zhao et al. 2021). At the high and low acidic conditions of pH 3–7, nanometal oxide biochar composites can extract exceeding high levels of As from soil, which has been attributed to the presence of oxygen-containing functional groups.

Another method of preparation of nanometal oxide biochar composites is by the process of coprecipitation. The coprecipitation process involves various chemical modifications of the solution, such as pH adjustment, chemical reduction, etc. (Zhao et al. 2021). Compared to other methods, the process of chemical coprecipitation is highly cost-effective and yields nanomaterials with improved purity and homogeneity compared to nanomaterials synthesized through other methods (Zhao et al. 2021). Though the coprecipitation method of nanomaterial synthesis has advantages, it also has some disadvantages in terms of causing chemical pollution (Zhao et al. 2021). For example, Mn oxide nanomaterials synthesized through the coprecipitation method were found to absorb a higher level of antibiotic residues such as di-*n*-butyl phthalate and oxytetracycline from the aqueous medium as compared to biochar without any nano-doping (Gao et al. 2018). The higher efficacy of other metal oxide nanocomposites, such as nickel oxide (NiO) prepared through chemical coprecipitation, was found to extract high levels of Pb as compared to biochar without any nano-doping (Saravanakumar et al. 2019).

Another method of preparation of nanometal oxide biochar composites is direct pyrolysis, where the biomass is enriched with metal oxides and subjected to pyrolysis (Zhao et al. 2021). In this process, nanocomposites are easily derived as biomass remains enriched with the desired metal oxide, which enables smooth preparation of the composites; however, it is difficult to regulate the proportion of metal oxides in the nanocomposites (Yao et al. 2013). Studies have demonstrated that metal oxide biochar nanocomposites prepared by direct pyrolysis from a metal-enriched bioresource, such as Mg-enriched tomato seedlings, have a strong capacity to remove phosphate (Yao et al. 2013). Likewise, ZnO–nano-biochar composites prepared through direct pyrolysis from zinc (Zn)-infested corn roughage were found to be more effective in the removal of contaminants such as Pb, Cr, and copper (Cu) from the soil as compared to biochar alone (Li et al. 2018). Alternative methods for preparing nanometal oxide biochar composites include a sol–gel method in which biochar is mixed in a chemical solution containing metal oxides and subjected to pyrolysis. This method has successfully obtained biochar composites for zirconium metal oxide (ZrO), ZnO, and FeO (Irawan et al. 2023). The ball milling method is a low-cost way to prepare biochar nanocomposites. This method facilitates reduction in the particle size of the metal oxide, with no hazardous chemical reactions involved (Wang et al. 2019a, b; Qin et al. 2020; Zhao et al. 2021).

The ability of nanometal oxide–biochar composites to absorb and extract organic and inorganic pollutants from the soil and aqueous medium depends on their physicochemical properties. These biochar composites are widely used for the remediation of soil and aquatic environments. MnO-doped biochar derived from biomasses from potato stems and leaves, pine needles, corn roughage, and vinasse was found to adsorb organic pollutants such as antibiotics (Li et al. 2018; Xiang et al. 2020; Xu et al. 2021), di-*n*-butyl phthalate (Li et al. 2018), and methylene blue, effectively (Sumalinog et al. 2018). Likewise, biochar derived from bamboo, corn roughage, peanut shell, and sewage sludge doped with the nanometal oxides of Zn, Cu, cobalt (Co), and titanium (Ti) have improved capacity of adsorbing organic pollutants and antibiotics (Lai et al. 2019; Gholami et al. 2019; Heo et al. 2019; Zhai et al. 2019; Guan et al. 2020). The nanometal oxide biochar composites are highly efficient in the adsorption and extraction of inorganic pollutants. These biochar composites act on inorganic pollutants such as heavy metals by adsorption (Bai et al. 2020). Biochar obtained from various biomasses and doped with the metal oxides of Fe, Zn, Mn, and Ni have been successfully used to adsorb toxic elements such as Cd, Cr, Pb, and As (Zhao et al. 2021). Fe–Mn oxide-doped biochar prepared from feedstocks such as corn straw, rice husk, and peanut shell has been demonstrated to adsorb heavy metals such as Cd and Cu and metalloids like As with enormous adsorption capacity as compared to simple biochar from aqueous medium (Zhou et al. 2018; Guo et al. 2019; Wu et al. 2020). Biochar doped with MnO nanomaterial has been demonstrated to extract Cu ions from aqueous mediums efficiently and both Cu and Cd ions from sewage (Zhou et al. 2017). The extraction of toxic heavy metals using nanometal oxide–biochars is also influenced by physicochemical factors such as pH and the humic acid content of the soil. At an acidic pH of 4.0 to 6.0 and high humic acid content, the adsorption of Cu and Cd by the nano-MnO–biochar composite is largely enhanced. The adsorbed Cu gets converted to CuO, $CuCO_3$, and $Cu(OH)_2$, while Cd gets converted to CdO and $Cd(OH)_2$ (Zhou et al. 2018). Other metal oxides, such as FeO-doped biochar, can effectively remove Cr(VI) up to almost 90% and some even up to 99% from the medium (Liu et al. 2018; Wang et al. 2019a, b). Metal oxide biochar composites prepared from the combination of FeO and MnO enormously improve the ability of biochar composites to extract the toxic elements effectively. For example, a biochar composite containing nano-FeO and MnO can adsorb trivalent As and oxidize it to its pentavalent form on the surface of metal oxide nanoparticles (Lin et al. 2019). In addition to the extraction capacity of such biochar composites for toxic elements, they can also adsorb nutrients such as P and N. Several nanometal oxides such as MgO, MnO, CeO_2, CaO, and Fe_2O_3 doped with biochar obtained from a wide range of feedstock material effectively adsorb P and N from the soil and aqueous medium (Feng et al. 2017; Xu et al. 2018; Li et al. 2020b).

The doping of biochar derived from diverse bioresources with nanometal oxides is a promising approach for mitigating environmental toxicants. The essential advantage of using metal oxide nano-doped biochars is its high dispersibility. This makes nano-doped biochars a promising and emerging substrate for environmental remediation. However, there are numerous constraints in using these biochar composites, such as their low extractability from the substrate and sometimes high-cost involvement.

3.7.2 Biochar Clay Mineral and Biochar: Enzyme Composites

Clay is a plentiful material without which life sustenance is not possible. Due to its diverse physicochemical properties and structural forms, clay has various uses. Depending upon their physical and chemical attributes, different types of clay have various adsorbent capacities as well as cation-exchange properties and mechanical stability (Djomgoue and Njopwouo 2013; Viglasova et al. 2018; Rallet et al. 2022). The physicochemical attributes of clay also influence the prospect of its value addition, such as the preparation of clay-based composites. One such is the clay or clay–mineral–biochar composite, which includes montmorillonite, attapulgite, bentonite, and illite that are primarily used for the preparation of biochar composites due to their structural attributes with high ion-exchange capacity (Adeyemo et al. 2017). Besides, such biochar composites are nonpolluting and eco-friendly (Lin et al. 2023). Studies have shown that clay–mineral biochars are usually thermostable and possess a firm carbon skeleton, contributing to the high adsorption capacity (Lin et al. 2022, 2023). The biochar clay mineral composites are primarily prepared by thermal polymerization or by the process where biochar and clay are soaked together for some time in suitable solvents (Zhang et al. 2018; Ashiq et al. 2019).

Biochar–enzyme composites are an exciting addition to improve the functional characteristics of biochar. These composites facilitate the extraction of toxicants from the environment by promoting enzymatic alteration and are considered highly eco-friendly. The composites are usually prepared through thermal polymerization of the feedstock material, and such materials could be used to load immobilized enzymes through physical adsorption and covalent bond coupling (Fernandez-Fernendez et al. 2013; Pandey et al. 2022a, b; Lin et al. 2023). Both these biochar enzyme composite preparation methods are different, as physical adsorption is a simple and low-cost process (Lin et al. 2023). The covalent bond coupling method, in comparison to the physical adsorption method, generates high-quality composites having commercial value, but at the same time also has certain disadvantages, such as loss of structural integrity and enzyme denaturation during preparation (Lin et al. 2023). The use and application of both biochar clay–mineral and enzyme composites are emerging. Using biochar composites in soil and aquatic environment remediation needs significant research inputs.

3.7.3 Biochar: Microbe Composites

One of the significant ways to improve the functional properties of biochar is the preparation of biochar–microbe composites. In such composites, biochar is a medium to carry the immobilized microorganisms. When applied to a polluted medium, such biochar composites allow the microbes to percolate, thus improving the adsorption of toxicants and pollutants. Various methods, such as biochar composites, are prepared, including adsorption, cross-linking, and encapsulation

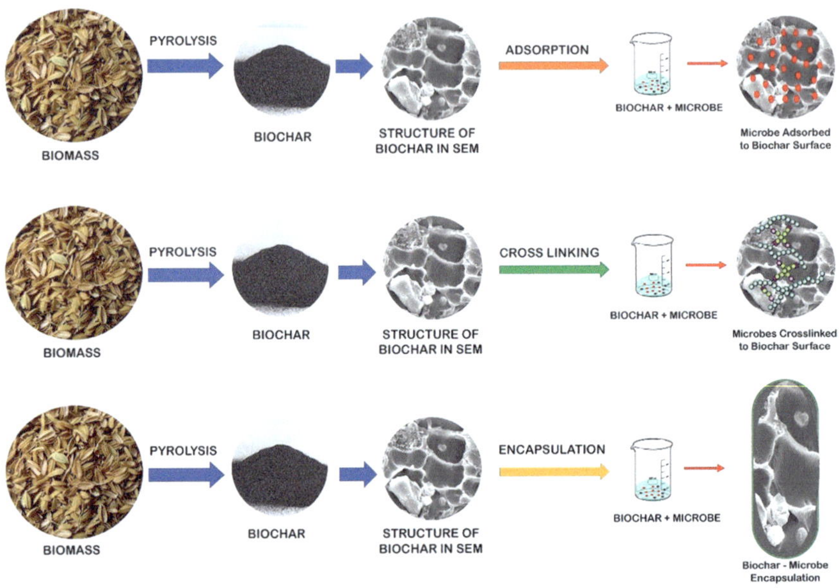

Fig. 3.6 Different techniques of preparation of biochar–microbe composites

(Fig. 3.6) (Jiang et al. 2022; Lin et al. 2023). Studies have shown that Pb immobilization could be achieved effectively by applying biochar coupled with phosphate-solubilizing bacteria (Chen et al. 2019). Applying biochar coupled with microbial inoculants also efficiently suppresses plant disease (Liu et al. 2023). The application of biochar–microbe composites is emerging and requires immense research input. Evidently, its use considerably affects the soil in diverse ways, such as immobilizing toxicants, improving fertility, etc.

3.8 Conclusions

This chapter presents a comprehensive and systematic evaluation of multifarious aspects of biochar, including the production techniques, its impact on biochar quality, and its prospective role in improving soil quality. Conventional ways to prepare biochar are effective in terms of its low-cost production and medium to high yields, depending upon the biomass. Besides this, biochar obtained through conventional methods has diverse functional groups that add to its functional perspective. Though other high-throughput methods are emerging and capable of producing very high-quality biochar, its large-scale production may question its cost-effectiveness. In years, green pyrolysis techniques have emerged by utilizing solar radiation. This approach may help reduce environmental pollution; however, it needs a complex technical setup and a very low yield. Biochar, rich in carbon and other elements, is anticipated to provide an effective solution to deteriorating soil health. Biochar can be engineered to improve its efficacy through value addition. Engineering the biochar to form biochar–nanocomposites, microbial composites, enzyme composites, and clay composites is an effective way for value addition. The nanometal oxide–biochar composites are acclaimed to be highly efficient in the adsorption and extraction of inorganic pollutants from soil. Besides its increasing use, other composites are also gaining significant attention regarding their use in soil quality improvement. Biochar microbial composites help regulate the functionality of soil microflora by providing a substrate and modulating the physicochemical attributes. In addition, biochar–enzyme and clay composites have been developed to increase the efficacy of biochar. Therefore, biochar and its composites may find wide applications in soil and aquatic environment remediation. Future research on biochar should also focus on conducting long-term field experiments to evaluate its climate change mitigation potential. The synthesis protocol for biochar and its composites and application process needs to be upgraded through chemical and agricultural engineering to industry-scale harvesting using biological waste feedstock to achieve environmental sustainability goals.

References

Adeyemo AA, Adeoye IO, Bello OS (2017) Adsorption of dyes using different types of clay: a review. Appl Water Sci 7:543–568

Ahmad M, Rajapksa AU, Lim JE, Zhang M, Bolan N, Mohan D, Vithanage M, Lee SS, Ok YS (2014) Biochar as a sorbent for contaminant management in soil and water: A review. Chemosphere 99:19–33

Ahmad M, Ok YS, Kim BY, Ahn JH, Lee YH, Zhang M, Moon DH, Al-Wabel MI, Lee SS (2016) Impact of soybean stover – and pine needle derived biochars on Pb and As mobility, microbial community and carbon stability in a contaminated agricultural soil. J Environ Manag 166:131–139

Angon PB, Islam MS, Shreejana KC, Das A, Anjum N, Poudel A, Suchi SA (2024) Sources, effects and present perspectives of heavy metals contamination: soil plant and human food chain. Heliyon 10(7):e28357

Arora NK, Mishra I (2022) Current scenario and future directions for sustainable development goal 2: a roadmap to zero hunger. Environ Sustain **5**:129–133. https://doi.org/10.1007/s42398-022-00235-8

Arshad M, Ali Khan AH, Hussian I, uz-Zaman B, Anees M, Iqbal M, Soja G, Linde C, Yousaf S (2017) The reduction of chromium (VI) phytotoxicity and phyto-availability to wheat (*Triticum aestivum* L.) using biochar and bacteria. Appl Soil Ecol 114:90–98

Ashiq A, Sarkar B, Adassooriya N, Walpita J, Rajapaksha AU, Ok YS, Vithanage M (2019) Sorption process of municipal solid waste biochar-montmorillonite composite from ciprofloxacin removal from aqueous media. Chemosphere 236:124384

Azeem M, Hayat R, Hussian Q, Ahmed M, Pan G, Ibhrahim TM (2019) Biochar improves soil quality and N_2 fixation and reduces net ecosystem CO_2 exchange in a dryland legume cropping system. Soil Tilage Res 186:172–182

Bai S, Wang L, Ma F, Zhu S, Xiao T, Yu T, Wang Y (2020) Self-assembly biochar colloids mycelial pellet for heavy metal removal from aqueous solution. Chemosphere 242:125182

Banik C, Lawrinenko M, Bakshi S, Laird DA (2018) Impact of pyrolysis temperature and feedstock on surface charge and functional group chemistry of biochars. J Environ Qual 47:452–461
Beesley L, Jimenez EM, Eyles JLG (2010) Effect of biochar and green waste compost amendments on mobility, bioavailability and toxicity of inorganic and organic contaminants in multi-element polluted soil. Environ Pollut 158:2282–2287
Biederman LA, Stanley HW (2013) Biochar and its effect on plant productivity and nutrient cycling: a meta-analysis. GCB Bioenergy 11(2):364–380
Bruun S, Harmer SL, Bekiaris G, Christel W, Zuin L, Hu Y (2017) Effect of different pyrolysis temperatures on the speciation and availability in soil of P in biochar produced from the soil fraction of manure. Chemosphere 169:377–386
Burgeon V, Fouche J, Garre S, Dehkordi RH, Colinet G, Cornelis J-T (2022) Young and century old biochar strongly affect nutrient cycling in a temperate agroecosystem. Agril Ecosyst Environ 328:107847
Cantrell KB, Hunt PG, Uchimiya M, Novak JM, Ro KS (2012) Impact of pyrolysis temperature and manure source on physiochemical characteristics of biochar. Bioresour Technol 107:419–428
Cardenas-Aguair E, Gasco G, Paz-Ferrerio J, Mendez A (2017) The effect of biochar and compost from urban organic waste on plant biomass and properties of an artificially copper polluted soil. Int Biodeterio Biodegrad 124:223–232
Chen H, Zhang J, Tang L, Su M, Tian D, Zhang L, Li Z, Hu S (2019) Enhanced Pb immobilization via the combination of biochar and phosphate solubilizing bacteria. Environ Int 127:395–401
Chenu C, Angres DA, Barre P, Derrien D, Arrouyas D, Balesdent J (2019) Increasing organic stocks in agricultural soils: knowledge gaps and potential innovations. Soil Tillage Res 188:41–52
Chintala V, Kumar S, Pandey JK, Sharma AK, Kumar S (2017) Solar thermal pyrolysis of non-edible seeds to biofuels and their feasibility assessment. Energy Convers Manag 153:482–492
DeLucaa TH, MacKenzie MD, Gundale MJ (2012) Biochar effects on soil nutrient transformation. Biochar Environ Manag Sci Technol 10:251–270
Devi M, Rawat S, Sharma S (2021) A comprehensive review of the pyrolysis process: from carbon nanomaterials synthesis to waste treatment. Oxf Open Mat Sci 1(1):itab014
Ding W, Dong X, Ime IM, Gao B, Ma LQ (2014) Pyrolytic temperature impact lead sorption mechanism by bagasse biochars. Chemosphere 105:68–74
Ding Y, Liu Y, Liu S, Li Z, Tan X, Huang X, Zeng G, Zhou L, Zheng B (2016) Biochar to improve soil fertility. A review. Agron Sustain Dev 36:36
Djomgoue P, Njopwouo D (2013) FT-IR spectroscopy applied for surface clay characterization. J Surf Eng Mater Adv Technol 3:275–282
Egamberdieva D, Jabbarov Z, Arora NK, Wirth S, Bellingrath-Kimura SD (2021) Biochar mitigates effects of pesticides on soil biological activities. Environ Sustain **4**:335–342. https://doi.org/10.1007/s42398-021-00190-w
Eibisch N, Durner W, Bechtold M, Fub R, Mikutta MA, Faz A (2015) Does water repellency of pyrochars and hydrochars counter their positive effects on soil hydraulic properties? Geoderma 245–246:31–39
Farnendez-Farnendez M, Sanroman MA, Moldes D (2013) Recent development and application of immobilized laccase. Biotechnol Adv 31:1808–1825
Feng Y, Lu H, Li Y, Xue L, Dionysiou D, Yang L, Xing B (2017) Nano cerium oxide functionalized biochar for phosphate retention: preparation, optimization and rice paddy application. Chemosphere 185:816–825
Foong SY, Latiff NSA, Liew RK, Yek PNY, Lam SS (2020) Production of biochar for potential catalytic and energy applications via microwave vacuum pyrolysis conversion of cassava stem. Mat Sci Energy Technol 3:728–733
Gao M, Zhang Y, Gong X, Song Z, Guo Z (2018) Removal mechanism of di-*n*-butyl phthalate and oxytetracycline from aqueous solutions by nano manganese dioxide modified biochar. Environ Sci Pollut Res 25:7796–7807
Gaskin JW, Steiner C, Harris K, Das KC, Bibens B (2008) Effect of low temperature pyrolysis for agricultural use. Trans Am Soc Agric Biol Eng 51:2061–2069

Gezahegn S, Sain M, Thomas S (2019) Variation in feedstock wood chemistry strongly influences biochar liming potential. Soil Syst 3(2):26

Ghani WAWAK, Mohd A, de Silva G, Bachmann RT, Taufiq-Yap YH, Rashid U, Al-Muhtaseb AH (2013) Biochar production from waste rubber – wood – sawdust and its potential use in C sequestration: chemical and physical characterization. Ind Crop Prod 44:18–24

Ghodke PK, Sharma AK, Pandey JK, Chen WH, Patel A, Ashokkumar V (2021) Pyrolysis of sewage-sludge for sustainable biofuels and value-added biochar production. J Environ Manag 298:113450

Gholami P, Dinpazhoh L, Khataee A, Orooji Y (2019) Sonocatalytic activity of biochar supported ZnO nanorod degradation of Gemifloxacin: synergy study, effects on parameters and phytotoxicity evaluation. Ultrason Sonochem 55:44–56

Glaser B, Birkk JJ (2012) State of the scientific knowledge on properties and genesis of Anthropogenic Dark Earths in Central Amazonia (terra preta de Indio). Geochim Cosmochim Acta 82:39–51

Glaser B, Haumaier L, Guggenberger ZW (2001) The "Terra Preta" phenomenon: a model for sustainable agriculture in the humid tropics. Die Naturwiss 88:37–41

Gorovtsov AV, Minkina TM, Mandzhieva SS, Perelomov LV, Soja G, Zamilina IV, Rajput VD, Sushkova SN, Mohan D, Yao J (2020) The mechanism of biochar interactions with microorganisms in soil. Environ Geochem Health 42:2495–2518

Guan K, Zhao P, Zhang J, Zhu L (2020) Synthesis and characterization of ZnO@RSDBC composites and their photo-oxidative degradation of acid orange 7 in water. J Mol Struct 1203:127425

Guo J, Yan C, Luo Z, Fang H, Hu S, Cao Y (2019) Synthesis of novel tannery HA Fe/Mn oxides -loaded biochar composites and its application in cadmium (II) and arsenic (V) adsorption. J Environ Sci 85:168–176

Haeldermans T, Campion L, Kuppens T, Vanreppelen K, Cuypers A, Scherurs S (2020) A comparative techno-economic assessment of biochar production from different residue streams using conventional and microwave pyrolysis. Bioresour Technol 318:124083

Harrison BP, Moo Z, Perez-Agredano E, Gao S, Zhang X, Ryals R (2024) Biochar-composting substantially reduces methane and air pollutant emission from dairy manure. Environ Res Lett 19:014081

Heo J, Yoon Y, Lee G, Kim Y, Han J, Park CM (2019) Enhanced adsorption of bisphenol A and sulfamethoxazole by novel magnetic $CuZnFe_2O_4$ – biochar – composite. Bioresour Technol 281:179–187

Herman E (2024) Sustainable agriculture and its impact on the rural development in EU countries: a multivariate analysis. Land 13:947

Ippolito JA, Spokas KA, Novak JM, Lentz RD, Cantrell KB (2014) Biochar elemental composition and factors influencing nutrient retention. Biochar Environ Manag Sci Technol:137–162

Irawan C, Refki MF, Hidayat R, Mu'minah R, Nata IF, Putra MD, Triantoro A (2023) Synthesis of magnetic nanoparticles coated zirconia using one-pot solvothermal processes as adsorbent for Pb(II) and Cd(II) removal. South African J Chem Engr 45: 247–255

Itam DH, Horsfall IT, Ekiyor TH (2024) Application of biochar in soil remediation: a decade of scientometrics and systematic review from 2014 to 2024. Result Eng 23:102757

Jainpeng Z (1999) Soil erosion in Guizhou province of China: a case study of Bijie prefecture. Soil Use Manag 15:68–70

Jiang Y, Yang F, Dai M, Ali I, Shen X, Hou X, Alhewairini SS, Peng C, Naz I (2022) Application of microbial immobilization technology for remediation of Cr (VI) contamination: a review. Chemosphere 286:13172

Jindo K, Audette Y, Higashikawa FS, Silva CA, Akashi K, Mastrolonardo G, Sanchez-Monedero MA, Mondini C (2020) Role of biochar in promoting circular economy in the agriculture sector: part 1: a review of the biochar roles in soil N, P and K cycles. Chem Biol Technol Agric 7:15

Jung KW, Hwang MJ, Jeong TU, Ahn KH (2015) A novel approach for preparation of modified-biochar derived from marine macroalgae: Dual purpose electro-modification for improvement of surface area and metal impregnation. Bioresour Technol 191:342–345

Kim KH, Kim JY, Cho TS, Choi JW (2012) Influence of pyrolysis temperature on physicochemical properties of biochar obtained from the fast pyrolysis of pitch pine (Pinus rigida). Bioresour Technol 118:158–162

Kloss S, Zehetner F, Dellantonio A, Hamid R, Ottner F, Liedtke V, Schwanninger M, Gerzabek MH, Soja G (2012) Characterization of slow pyrolysis biochar: effects of feedstocks and pyrolysis temperature on biochar properties. J Environ Qual 41:990–1000

Lai C, Huang F, Zeng G, Huang D, Qin L, Cheng L, Zhang C, Li B, Yi H, Liu S, Li L, Chen L (2019) Fabrication of novel magnetic $MnFe_2O_4$ / bio-char composite and heterogenous photo Fenton degradation of tetracycline in near neutral pH. Chemosphere 224:910–921

Lee XJ, Lee LY, Gan S, Thangalazhy-Gopakumar S, Ng HK (2017) Biochar potential evaluation of palm oil wastes through pyrolysis: thermolytic characterization and pyrolytic kinetic studies. Bioresour Technol 236:155–163

Lehmann J, Gaunt J, Rondon M (2006) Bio-char sequestration in terrestrial ecosystem – a review. Mitig Adapt Global 11:403–427

Li S, Skelly S (2023) Physicochemical properties and applications of biochars derived from municipal solid waste: a review. Environ Adv 13:100395

Li S, Tasnady D (2023) Biochar for soil carbon sequestration: current knowledge, mechanisms and future perspective. C (J Carbon Res) 9:67

Li C, Zhang L, Gao Y, Li A (2018) Facile synthesis of nanoZnO/ZnS modified biochar by directly pyrolyzing of zinc contaminated corn stover for Pb(II), Cu(II) and Cr(VI) removal. Waste Manag 79:625–637

Li K, Wang Z, Zhang G, Cui M, Lu Q, Yang Y (2020a) Selective production of monocyclic aromatic hydrocarbons from ex-situ catalytic fast pyrolysis of pine over the HZSM-5 catalyst with calcium formate as hydrogen source. Sustain Energy Fuels 4:538–548

Li X, Xie Y, Jiang F, Wang B, Hu Q, Tang Y, Luo T, Wu T (2020b) Enhanced phosphate removal from aqueous solution using resourceable nano-CaO_2/BC composite: behaviors and mechanisms. Sci Total Environ 709:136123

Li Y, Gupta R, Zhang Q, You S (2023) Review of biochar production via crop residue pyrolysis: development and perspective. Bioresour Technol 369:128423

Lin Y, Munroe P, Joseph S, Henderson R, Ziolkowski A (2012) Water extractable organic carbon in untreated and chemical treated biochar. Chemosphere 87:151–157

Lin L, Song Z, Huang Y, Khan ZH, Qui W (2019) Removal and oxidation of arsenic from aqueous solution by biochar impregnated with Fe-Mn oxide. Water Air Soil Pollut 230:105

Lin M, Li F, Cheng W, Rong X, Wang W (2022) Facile preparation of novel modified biochar-based supramolecular self-assembled g-C_3N_4 for enhanced visible light photocatalytic degradation of phenanthrene. Chemosphere 288:132620

Lin M, Li F, Li X, Rong X, Oh K (2023) Biochar-clay, biochar-microorganism and biochar-enzyme composites for environmental remediation: a review. Environ Chem Lett 21:1837–1862

Liu Y, Li Y, Huang J, Zhang Y, Ruan Z, Hu T, Wang JJ, Li WY, Hu HJ, Jiang GB (2018) An advanced sol gel strategy for enhancing interfacial reactivity of iron oxide nanoparticles on rosin biochar substrate to remove Cr(VI). Sci Total Environ 690:438–436

Liu Z, Zhou W, Sun Y, Peng Y, Niu J, Tan J, Wei M (2023) Biochar and its coupling with microbial inoculants for suppressing plant diseases: a review. Appl Soil Ecol 190:105025

Mandal S, Pu S, Adhikari S, Ma H, Kim D-H, Bai Y, Hou D (2021) Progress and future of biochar composites: application and reflection in the soil environment. Crit Rev Environ Sci Technol 51:219–271

Marousek J, Gavurova B (2022) Recovering phosphorous from biogas fragmentation residues indicates promising economic results. Chemosphere 291:133008

Mia S, Singh B, Djikstra FA (2017) Aged biochar affects gross nitrogen mineralization and recovery: a 15N study in two contrasting soils. Glob Chang Biol Bioenergy 9:1196–1206

Mishra RK, Jaya PKD, Narula A, MinnatChistie S, Naik US (2023) Production and beneficial impact of biochar for environmental application: a review on types of feedstocks, chemical compositions, operating parameters, techno-economic study and life cycle assessment. Fuel 343:125451

Mohan D, Sarswat A, Ok YS, Pittman CU (2014) Organic and inorganic contaminants removal from water with biochar, a renewable, low cost and sustainable adsorbent – a critical review. Bioresour Technol 160:191–202

Mukherjee A, Zimmerman AR, Hamdan R, Copper WT (2014) Physiochemical changes in pryogenic organic matter (biochar) after 15 months of field aging. Solid Earth 5(2):693–704

Nair RR, Schaate A, Klepzig LF, Turcios AE, Lecinski J, Shamsuyeva M, Endres H-J, Papenbrock H-J, Behrens P, Weichgrebe D (2023) Physico-chemical characterization of walnut shell biochar from uncontrolled pyrolysis in a garden oven and surface modification by ex-situ chemical magnetization. Clean Technol Environ Policy 25:2727–2746

Nan Q, Xin L, Qin Y, Waqas M, Wu W (2021) Exploring long term effects of biochar on mitigating methane emission from paddy soil: a review. Biochar 3:125–134

Nannipieri P, Ascher J, Ceccherini MT, Landi L, Pietramellara G, Renella G (2017) Microbial diversity and soil function. Eur J Soil Sci 68(1):12–26

Nath AJ, Lal R, Sileshi GW, Das AK (2018) Managing India's small landholder farms for food security and achieving the "4 per Thousand" target. Sci Total Environ 634:1024–1033. https://doi.org/10.1016/j.scitotenv.2018.03.382

Nath AJ, Sileshi GW, Laskar SY, Pathak K, Reang D, Nath A, Das AK (2021) Quantifying carbon stocks and sequestration potential in agroforestry systems under divergent management scenarios relevant to India's Nationally Determined Contribution. J Clean Prod 281:124831. https://doi.org/10.1016/j.jclepro.2020.124831

Nepal J, Ahmad W, Munsif F, Khan A, Zou Z (2023) Advances and prospects of biochar in improving soil fertility, biochemical quality and environmental applications. Front Environ Sci 11:1114752

Nguyen TB, Sherpa K, Bui XT, Nguyen VT, Vo TDH, Ho HTT, Chen CW, Dong CD (2023) Biochar for soil remediation: a comprehensive review of current research on pollutant removal. Environ Pollut 337:122571

Nielsen S, Minchin T, Kimber S, van Zwieten L, Gilber J, Munroe P, Joseph S, Thomas T (2014) Comparative analysis of microbial communities in agricultural soils amended with enhanced biochars or traditional fertilizers. Agric Ecosyst Environ 191:73–72

Oliveria FR, Patel AK, Jaisi DP, Adhakari S, Lu H, Khanal SK (2017) Environmental application of biochar: current status and perspective. Bioresour Technol 246:110–122

Pandey D, Daverey A, Dutta K, Arunachalam K (2022a) Bioremoval of toxic malachite green from water through simultaneous decolorization and degradation using laccase immobilized biochar. Chemosphere 297:134126

Pandey D, Daverey A, Dutta K, Yata VK, Arunachalam K (2022b) Valorization of waste pine needle biomass into biosorbent for removal of methylene blue dye from water: kinetics, equilibrium and thermodynamics study. Environ Technol Innov 25:102200

Pandian K, Vijaykumar S, Mustaffa MRAF, Subramaniam P, Chitraputhirapillai S (2024) Biochar – a sustainable soil conditioner for improving soil health, crop production and environment under changing climate – a review. Front Soil Sci 4:1376159

Patro A, Dwivedi S, Thakur A, Sahoo PK, Biswas JK (2024) Recent approaches and advancement in biochar based environmental sustainability: is biochar fulfilling the sustainable development goals? IScience 27(9):110812

Peng X, Ye LL, Wang CH, Zhou H, Sun B (2011) Temperature – and duration – dependent rice straw derived biochar: characteristics and its effects on soil properties of an Ultisol in southern China. Soil Tillage Res 112(2):159–166

Purakayastha TJ, Kumari S, Pathak H (2015) Characterization, stability and microbial effects of fourbiochars produced from crop residue. Geoderma 239–240:293–303

Qin Y, Zhu X, Su Q, Anumah A, Gao B, Lyu W, Zhou X, Xing Y, Wang B (2020) Enhanced removal of ammonium from water by ball milled biochar. Environ Geochem Health 42(6):1579–1587

Rallet D, Paltahe A, Tsamo C, Loura B (2022) Synthesis of clay-biochar composite for glyphosate removal from aqueous solution. Heliyon 8(3):e09112

Rodon MA, Lehmann J, Ramirez J, Hurtado M (2007) Biological nitrogen fixation by common bean (*Phaseolus vulgaris* L.) increase with biochar additions. Biol Fertil Soils 43(6):699–708

Rodriguez-Vila A, Forjan R, Guedes RS, Covelo EF (2016) Changes on phytoavailability of nutrients in a mine soil reclaimed with compost and biochar. Water Air Soil Pollut 227:453

Saravanakumar R, Muthukumaran K, Seivaraju N (2019) Enhanced Pb(II) ions removal by using magnetic NiO/biochar composite. Mater Res Express 6:105504

Schmalenberger A, Fox A (2016) Bacterial mobilization of nutrients from biochar amendment soil. Advances in applied microbiology, vol 94. Elsevier, Amsterdam, pp 109–159

Schmidt MWI, Noack AG (2000) Black carbon is soils and sediments: analysis, distribution, implications, and current challenges. Global BiogeochemCy 14:777–793

Schnieder F, Haderlein SB (2016) Potential effect of biochar on the availability of phosphorous – mechanistic insight. Geoderma 277:83–90

Sharma AK, Ghodke PK, Goyal N, Bobde P, Kwon EE, Lin KYA, Chen WH (2023) A critical review on biochar production from pine wastes, upgradation techniques, environmentally sustainability, and challenges. Bioresour Technol 386:129632

Shenbagaraj S, Sharma PK, Sharma AK, Raghav G, Kota KB, Ashokkumar V (2020) Performance analysis of medium scale downdraft gasifier using *Lantana camara* biomass as feeding material. Energy Sour A Recovery Util Environ Eff 00:1–15

Spokas KA, Novak JM, Stewart CE, Cantrell KB, Uchiyama M, DuSaire MG, Ro KS (2011) Qualitative analysis of volatile organic compounds of biochar. Chemosphere 85(5):869–882

Sriphiron P, Chidthaisong A, Yagi K, Tripetchkul S, Towprayoon S (2020) Evaluation of biochar application combined with alternate wetting and drying (AWD) water management in rice field as methane mitigation option for farmer's adoption. Soil Sci Plant Nutr 66(1):235–246

Stefaniuk M, Olezzcuzuk P (2016) Addition of biochar to sewage sludge decreases freely dissolved PAHs content and toxicity of sewage sludge amendment soil. Environ Pollut 218:242–251

Stylianou M, Christou A, Dalias P, Polycarpou P, Michael C, Agapiou A, Papanastasiou P, Fatta-Kassinos D (2020) Physicochemical and structural characterization of biochar derived from the pyrolysis of biosolids, cattle manure and spent coffee grounds. J Energy Inst 93(5):2063–2073

Subedi R, Taupe N, Pelissetti S, Petruzzelli L, Bertora C, Leahy JJ (2016) Green house gas emission and soil properties following amendment with manure derivedbiochars: influence of pyrolysis temperature and feedstock type. J Envrion Manag 166:73–83

Sumalinog DAG, Capareda SC, de Luna MDG (2018) Evaluation of the effectiveness and mechanisms of acetaminophen and methylene blue dye adsorption on activated biochar derived from municipal solid wastes. J Environ Manag 210:255–262

Sun D, Lan Y, Xu EG, Meng J, Chen W (2016) Biochar as a novel niche for culturing microbial communities in composting. Waste Manag 54:93–100

Tan Z, Lin CSK, Ji X, Rainey TJ (2017) Returning biochar to fields: a review. Appl Soil Ecol 116:1–11

Tang J, Zhu W, Kookana R, Katayama A (2013) Characteristics of biochar and its application in remediation of contaminated soil. J Biosci Bioeng 116(6):653–659

Tian L, Jinjin C, Ji R, Ma Y, Yu X (2022) Microplastics in agricultural soils: sources, effects, and their fate. Curr Opion Envrion Sci Health 25:100311

Tomczyk A, Sokolowska Z, Boguta P (2020) Biochar physiochemical properties: pyrolysis temperature and feedstock kind effects. Rev Environ Sci Biotechnol 19:191–215

Uchiyama M, Hiradate S, Antal MJ (2015) Dissolved phosphorous speciation of flash carbonization, slow pyrolysis and fast pyrolysis biochar. ACS Sustain Chem Eng 3(7):1642–1649

Vagsholm I, Arzoomand NS, Boqvist S (2020) Food security, safety and sustainability- getting the trades off right. Front Sustain Food Syst 4:16

Vamvuka D, Sfakiotakis S (2011) Effects of heating rate and water leaching of perennial energy crop on pyrolysis characteristics and kinetics. Renew Energy 36:2433–2439

Viglasova E, Galambos M, Dankovaa L, Krivosudsky L, Lengauer CL, Hood-Nowonty R, Soja G, Rompel A, Matik M, Briancin J (2018) Production, characterization and adsorption studies on bamboo-based biochar / montmorillonite composite for nitrate removal. Waste Manag 79:385–394

Vuppaladadiyam AK, Vuppaladadiyam SSV, Sahoo A, Murugavelh S, Anthony E, Bhaskar T, Zheng Y, Zhao M, Daun H, Zhao Y, Antunes E, Sarmah AK, Leu SY (2023) Bio-oil and biochar from pyrolytic conversion of biomass: a current and future perspective on the trade-off between economic, environmental and technical indicators. Sci Total Environ 857:159155
Wang Q, Chen L, He LY, Shen XF (2016) Increased biomass and reduced heavy metal accumulation of edible tissues of vegetable crops in presence of plant growth promoting *Neorhizobiumhautlense* T1–17 and biochar. Agril Ecosyst Environ 228:9–18
Wang B, Gao B, Fang J (2017) Recent advances in engineered biochar productions and applications. Crit Rev Environ Sci Technol 47:2158–2207
Wang B, Gao B, Wan W (2019a) Comparative study of calcium alginate ball-milled biochar and their composites on aqueous methylene blue absorption. Environ Sci Pollut Res 26:11535–11541
Wang C, Tan H, Liu H, Wu B, Xu F, Xu H (2019b) A nanoscale ferroferic oxide coated biochar derived from mushroom waste to rapidly remove Cr(VI) and mechanism study. Bioresour Technol Rep 7:100253
Wijitkosum S, Jiwnok P (2019) Elemental composition of biochar obtained from agricultural waste for soil amendment and carbon sequestration. Appl Sci 9(19):3980
Woods WI, Teixeria WG, Lehmann J, Steiner C, Winkler Pirns A, Rebelato L (2009) Amazonian dark earths: Wim Sombroek's vision
Wu X, Wang D, Riaz M, Zhang L, Jiang C (2019) Investigating the effect of biochar on the potential of increasing cotton yield, potassium efficiency and soil environment. Ecotoxicol Environ Saf 182:109451
Wu Z, Chen X, Yuan B, Fu M-L (2020) A facile foaming polymerization strategy to prepare 3D MnO_2 modified biochar based porous hydrogels for efficient removal of Cd(II) and Pb(II). Chemosphere 239:124745
Xiang Y, Yang X, Xu Z, Hu W, Zhou Y, Wan Z, Yang Y, Wei Y, Yang J, Tsang DCW (2020) Fabrication of sustainable manganese ferrite modified biochar from vinasse for enhanced adsorption of fluroquinone antibiotics: effects and mechanisms. Sci Total Environ 709:136079
Xiong B, Zhang Y, Hou Y, Arp HPH, Reid BJ, Cai C (2017) Enhanced biodegradation of PAHs in historically contaminated soil by *M. gilvum* inoculate biochar. Chemosphere 182:316–324
Xu T, Lou L, Luo L, Cao R, Duan D, Chen Y (2012) Effect of bamboo biochar on pentachlorophenol leachability and bioavailability in agricultural soil. Sci Total Environ 412:727–731
Xu K, Lin F, Dou X, Zheng M, Tan W, Wang C (2018) Recovery of ammonium and phosphate from urine as value added fertilizer using wood waste biochar using magnesium oxides. J Clean Prod 187:205–214
Xu Z, Xiang Y, Zhou H, Yang J, He Y, Zhu Z, Zhou Y (2021) Manganese ferrite modified biochar from vinasse for enhanced adsorption of levofloxacin: effects and mechanisms. Environ Pollut 272:115968
Yadav SPS, Bhandari S, Bhatta D, Poudel A, Bhattarai S, Yadav P, Ghimire N, Paudel P, Shresta J, Oli B (2023) Biochar application: a sustainable approach to improve soil health. J Agril Food Res 11:100498
Yao Y, Gao B, Chen J, Zhang M, Inyang M, Li Y, Alva A, Yang L (2013) Engineered carbon (biochar) prepared by direct pyrolysis of Mg-accumulated tomato tissue: characterization and phosphate removal potential. Bioresour Technol 138:8–13
Yao Z, You S, Ge T, Wang CH (2018) Biomass gasification for syngas and biochar co-production: energy application and economic evaluation. Appl Energy 209:43–55
Yu Z, Zhou L, Huang Y, Song Z, Qui Z (2015) Effects of manganese oxide modified biochar composites on adsorption of arsenic in red soil. J Environ Manag 163:155–162
Yun P, Wang J, Pan Y, Shen B, Wu C (2019) Review of biochar for the management of contaminated soil: preparation, application and prospect. Sci Total Environ 659:473–490
Zama EF, Zhu YG, Reid BJ, Sun GX (2017) The role of biochar properties in influencing the sorption and desorption of Pb(II), Cd(II) and As(III) in aqueous solution. J Clean Prod 148:127–136
Zhai S, Li M, Wang D, Zhang L, Yang Y, Fu S (2019) In situ loading metal oxide biochar particles on biochars: reusable materials for efficient removal of methylene blue from wastewater. J Clean Prod 220:460–474

Zhang J, Lu M, Wan J, Sun Y, Lan H, Deng X (2018) Effects of pH, dissolved humic acid and Cu^{2+} on adsorption of norfloxacin on montmorillonite-biochar composite derived from wheat straw. Biochem Eng J 130:104–112

Zhang X, Sun P, Wei K, Huang X, Zhang X (2020) Enhanced H_2O_2 activation and sulfamethoxazole degradation by Fe – impregnated biochar. Chem Eng J 385:123921

Zhao SX, Na T, Wang XD (2017) Effect of temperature on structure and physiochemical properties of biochar with apple tree branches as feedstock material. Energies 10:1293

Zhao C, Wang B, Theng BKG, Wu P, Liu F, Wang S, Lee X, Chen M, Li L, Zhang X (2021) Formation and mechanisms of nanometal oxide-biochar composite for pollution removal: a review. Sci Total Environ 726:14305

Zhao FJ, Tang Z, Song JJ, Huang XY, Wang P (2022) Toxic metal and metalloids: uptake, transport, detoxification, phytoremediation, and crop improvement for safer food. Mol Plant 15(1):27–44

Zheng W, Guo M, Chow T, Bennett DN, Rajagopalan N (2010) Sorption properties of green waste biochar for two triazine pesticides. J Hazard Mater 181:121–126

Zheng H, Wang Z, Deng X, Zhao J, Luo Y, Noval J (2013) Characteristics and nutrient values of biochars produced from giant reed at different temperature. Bioresour Technol 130:463–471

Zheng H, Wang X, Luo X, Wang Z, Xing B (2018) Biochar – induced negative carbon mineralization priming effects in a coastal wetland soil: role of soil aggregates and microbial modulation. Sci Total Environ 610:951–960

Zhou L, Huang Y, Qiu W, Sun Z, Liu Z, Song Z (2017) Adsorption properties of nano-MnO_2-biochar composites for copper in aqueous solution. Molecules 22:173

Zhou Q, Liao B, Lin L, Qiu W, Song Z (2018) Adsorption of Cu(II) and Cd(II) from aqueous solution by ferromanganese binary oxide biochar composites. Sci Total Environ 515:115–122

Zhu X, Chen B, Zhu L, Xing B (2017) Effects and mechanisms of biochar – microbe interactions in soil improvement and pollution remediation: a review. Environ Pollut 227:98–115

Chapter 4
Enhancing Strawberry Production in Different Systems: The Role of Nutrient Solutions and Biofertilizers

Durdigul Botirova, Shoxnoza Zakirova, Nigora Kurbanova, and Dilfuza Egamberdieva

Abstract Hydroponics emerges as a sustainable solution, offering precise nutrient management that enhances both growth and fruit quality. This chapter explores the impact of various nutrient solutions and biofertilizers on the growth and fruit yield of strawberries cultivated in soil and soilless systems. It highlights the potential of organic amendments to improve soil health and boost crop yields. Nutrient management and biofertilizers play a vital role in enhancing strawberry growth, yield, and fruit quality while promoting sustainable farming practices. Balanced fertilization with essential nutrients such as nitrogen (N), phosphorus (P), potassium (K), calcium (Ca), and magnesium (Mg) supports vegetative growth, root development, flowering, and fruit quality, while micronutrients like boron (B) and zinc (Zn) aid enzyme activities and physiological processes. Biofertilizers, including nitrogen-fixing bacteria (*Rhizobium*, *Azotobacter*), phosphate- and potassium-solubilizing microorganisms (*Bacillus*, *Pseudomonas*), mycorrhizal fungi (*Glomus*), and plant growth-promoting rhizobacteria (PGPR), improve nutrient availability, stimulate growth through phytohormone production, enhance stress tolerance, and protect against pathogens. Findings indicate that optimal combinations of nutrient solutions and biofertilizers significantly enhance growth parameters and fruit quality. This work provides valuable insights for growers seeking environmentally friendly and economically viable practices in fruit cultivation.

Keywords Hydroponics · Nutrient solutions · Biofertilizers · Growth parameters

---

D. Botirova · S. Zakirova · N. Kurbanova · D. Egamberdieva (✉)
National Research University TIIAME, Tashkent, Uzbekistan

N. K. Arora et al. (eds.), *Environmental Sustainability and Biotechnology: Opportunities and Challenges*, Microorganisms for Sustainability 46,
https://doi.org/10.1007/978-981-96-9025-1_4

4.1 Introduction

Strawberries are among the most widely consumed fruits globally, renowned for their numerous health benefits, including reducing cancer risk, improving heart health, enhancing insulin sensitivity, and supporting a healthy immune system (Nemzer et al. 2020; Ngouana et al. 2023). This perennial herb, characterized by its fibrous root system, ranks as one of the most prevalent fruit crops worldwide (Lewers et al. 2017; Nakro et al. 2023). They are exceptional for their high antioxidant content compared to other foods and are also a rich source of essential minerals such as potassium, magnesium, and calcium (Tulipani et al. 2011). However, the nutritional profile of strawberries can vary considerably depending on factors such as growing conditions and plant variety (Perin et al. 2019). They are popular for their taste, sweetness, and nutrient density but are highly perishable due to their soft texture and susceptibility to damage (Bose et al. 2019). To extend shelf life and manage postharvest decay, various fungicides are employed, though these often leave residues that can be harmful to both humans and the environment. Globally appreciated for their vibrant color, pleasant aroma, and nutritional value, strawberries help promote health by detoxifying free radicals, participating in antioxidant defense, and preventing DNA damage due to their rich content of bioactive compounds like flavonoids, anthocyanins, and phenolic acids (Giampieri et al. 2015). However, as a nonclimacteric fruit harvested at maturity, strawberries rapidly lose quality post-harvest due to their high susceptibility to microbial contamination and vigorous metabolic activity (Xu et al. 2022; Yıldırım-Yalçın et al. 2022). Their soft and juicy nature also makes them prone to damage during picking, storage, and transport, which increases their vulnerability to fungal pathogens (Zhao et al. 2023).

Given its advantages in efficiency and economics, hydroponic cultivation of alpine strawberries is gaining traction among growers. This soilless system facilitates easier harvesting and results in the earlier production of larger, more uniform fruits that meet market preferences (Caruso et al. 2011). Hydroponics allows for precise control over nutrient delivery, which impacts water and salt relations and, in turn, plant growth and fruit quality. It has been observed that while salinity can hinder vegetative growth, moderate levels can enhance fruit quality and antioxidant levels (Naresh et al. 2024).

The use of organic manures and biofertilizers presents a viable and environmentally friendly alternative to inorganic chemicals and fertilizers. Such practices not only boost crop yields but also enhance soil quality and fertility. Organic manures (e.g., vermicompost and poultry manure) provide essential nutrients, improve soil physical properties (e.g., water retention, aeration, and aggregation), prevent soil degradation, and support beneficial soil microorganisms (Negi et al. 2021). Biofertilizers, derived from naturally occurring microorganisms in the plant rhizosphere or cultivated soil, include beneficial bacteria such as *Acinetobacter, Alcaligenes, Arthrobacter, Bacillus, Burkholderia, Pseudomonas, and Serratia.* Potassium-mobilizing microbes enhance the availability of essential nutrients for strawberries, promoting better vegetative growth and higher fruit yields (Abdullaeva et al. 2023; Egamberdieva et al. 2017a, b, 2019, 2023). Mycorrhizal fungi establish symbiotic relationships with strawberry roots, improving water and nutrient uptake, leading to robust root systems and overall plant health (Robinson Boyer et al. 2016). Biofertilizers enhance the accumulation of nutrients in fruits, such as vitamin C, anthocyanins, and flavonoids, improving both nutritional quality and antioxidant activity. They also improve the plant's ability to cope with abiotic stresses such as drought, salinity, and temperature fluctuations. The mechanisms behind growth stimulation by biofertilizers in strawberries involve a combination of biochemical, physiological, and ecological interactions. These mechanisms enhance nutrient availability, hormone regulation, stress tolerance, and soil health (Negi et al. 2011).

4.2 Strawberry Cultivation Technologies

4.2.1 Hydroponics

Hydroponics is a method of growing plants without soil, utilizing nutrient-rich solutions, with or without substrates like gravel, sand, pumice, perlite, vermiculite, rockwool, peat moss, coco coir, or sawdust (Lee and Lee 2015; Niu and Masabni 2022). This technique lessens reliance on agricultural land and pesticides and can be used in areas with poor soil quality. Hydroponics enhances crop yield and productivity by conserving water, energy, and space, making it a viable option for modern sustainable agriculture (Rajendran et al. 2024). It promotes sustainability through efficient nutrient use, reduced water consumption, and the ability to grow crops year-round in controlled environments (Chowdhury and Asiabanpour 2024). A study by Sahoo et al. (2024) demonstrated that strawberries grown hydroponically showed superior growth, adaptability, and yield compared to those grown in soil. The research concluded that hydroponic systems, with the appropriate nutrient concentrations, could replace traditional soil-based cultivation methods.

4.2.2 Greenhouse

Greenhouse farming is recognized as a precise and sustainable method of smart agriculture. Although greenhouses can support crops year-round by creating controlled environments, they require careful management of parameters like temperature, humidity, and light, especially in harsh climates (Hosny et al. 2024). Modern advancements in greenhouse technology allow for optimal plant production by adjusting internal conditions such as temperature, humidity, light intensity, and CO_2 levels, thereby improving crop yield and quality while addressing global food security issues (Badji et al. 2022).

Greenhouses enable year-round strawberry cultivation regardless of external weather conditions, but they demand considerable energy for heating and cooling, which can be costly. To mitigate energy consumption, energy-efficient technologies and computer simulations are used to optimize greenhouse conditions (Ogunlowo et al. 2023). Research by Khammayom et al. (2022) found that greenhouses can create ideal conditions for strawberries with minimal additional heating needed during the day due to adequate solar radiation. Watering and nutrient delivery can be optimized through drip irrigation and hydroponic systems, reducing resource waste. CO_2 enrichment within the greenhouse enhances photosynthesis and accelerates plant growth. Regular monitoring for pests, diseases, and proper airflow ensures healthy growth. While initial costs and energy use can be challenges, these can be mitigated with renewable energy and modular greenhouses (Dong et al. 2020).

4.2.3 Aquaponics

Aquaponics is an emerging sustainable agricultural method that integrates aquaculture and hydroponics in a symbiotic system, adhering to circular economy principles. This approach reduces the need for external resources and improves their efficiency (Fernández-Cabanás et al. 2022). In aquaponics, fish waste serves as an organic nutrient source for plants, while the plants help purify water, creating a mutually beneficial relationship (Chowdhury and Asiabanpour 2024). However, the fish waste may lack sufficient micronutrients for optimal plant growth, so additional nutrient supplements, such as iron chelate, and buffers like $CaCO_3$ and/or K_2CO_3 for calcium and potassium, are often necessary (Ru et al. 2017; Yang and Kim 2020). Aquaponics often results in higher productivity and resource efficiency compared to traditional land-based systems (Van Ginkel et al. 2017). The system minimizes pollution, such as wastewater, while enhancing production efficiency and stability, which can increase revenue (Tyson et al. 2011).

4.2.4 Open Field

Strawberry cultivation in open fields is a widely practiced method that harnesses natural conditions to grow this popular fruit. Open-field agriculture does not control environmental factors such as sunlight, air composition, and temperature, which influence plant growth (Singh and Tiwari 2010). Open-field cultivation allows for large-scale production with a relatively lower initial investment compared to controlled environments like greenhouses. However, achieving optimal growth and high yields requires careful management of environmental factors such as soil quality, water availability, temperature, and pest control. Strawberry cultivation in open fields requires substantial labor, which is becoming increasingly difficult to find. It is considered low-input or extensive agriculture compared to greenhouse production, as it involves minimal energy consumption and results in a lower environmental impact, although this comes with lower productivity (Nemecek et al. 2011). By implementing best practices, such as proper field preparation, mulching, drip irrigation, and integrated pest management (IPM), open-field strawberry cultivation can produce high-quality fruits while maintaining sustainability and cost-efficiency.

4.2.5 Nutrient(s) Requirement

The essential nutrients required for plant growth are fairly consistent across different terrestrial plant species, particularly for vegetative growth. Nutrient solutions must be formulated for the specific crop and provide all essential nutrients while maintaining optimal pH and electrical conductivity (EC) levels (Kumar and Saini 2020). In hydroponic systems, nutrient solutions are aqueous mixtures primarily composed of inorganic ions from soluble salts that provide essential elements for plant development. Occasionally, organic compounds, such as iron chelates, may also be included. Essential elements are crucial for plant physiology, and their absence can disrupt the entire plant life cycle (Esitken et al. 2010). In hydroponics, nutrient solutions, which are liquid fertilizers prepared in specific compositions, are crucial for plant growth. Plants require nutrients in their ionic form, along with proper oxygen levels and temperature. Productivity in hydroponics depends on both environmental factors and the nutrient solution. Nutrient supply should match crop requirements, with the application frequency adjusted based on crop type, age, media used, and environmental conditions (Kumar and Saini 2020).

Some hydroponic growers may use more nutrients than necessary to prevent deficiencies. However, the key to a successful hydroponic system lies in using the correct nutrient solution and carefully managing nutrient levels to regulate plant uptake. Despite being a sustainable method, hydroponics still relies on chemical fertilizers derived from finite and nonrenewable resources. There is growing interest in organic hydroponics due to the expanding organic food market, with some research suggesting that vegetables can be grown with organic nutrient solutions. Optimizing the use of organic waste in hydroponics requires a solubilization process to break down organic matter and release nutrients (Ezziddine et al. 2021). A study by Sundar et al. (2021) has explored using wastewater as a nutrient source for hydroponics, aiming to both reuse water and manage wastewater effectively. For example, second-cycle rinse water from washing machines has been found effective for indoor hydroponic cultivation without additional fertilizers, offering benefits such as water conservation, reduced sewage disposal, ecosystem protection, and potentially cost-effective food production.

It impacts physiological processes in plant roots, such as nutrient and water absorption, and regulating the solution temperature can enhance the plant physiological functions. Higher temperatures can increase nutrient solubility and uptake, improving the efficiency of photosynthesis (Xu and Huang 2006; Yamori et al.

2006). Cold nutrient solutions generally lead to decreased nutrient uptake and water absorption but can enhance root growth. Regulating nutrient solution temperatures is vital, particularly in controlled environments during winter, as optimizing the solution temperature has shown benefits for various crops (Sethi and Sharma 2007). Studying optimal nutrient temperatures for different crops in adverse conditions is crucial (Meselmani 2023). Strawberry plants have high water requirements due to their shallow roots, large leaf area, and high fruit water content. They are sensitive to water deficits and osmotic stress (Keutgen and Pawelzik 2009; Khayyat et al. 2009). Increasing the nutrient solution strength can improve mineral availability but may also expose plants to mild water stress (Caruso et al. 2011).

4.2.5.1 Nitrogen

Nitrogen (N) is vital for plant growth, runner production, and fruit bud formation. Nitrogen deficiency causes small, pale green leaves that may turn yellow, and older leaves may become reddish. Deficient plants also exhibit reduced fruit size and reddish calyxes around the fruit (Ullio 2010). Nitrogen affects strawberry fruit firmness, quality, and shelf life, with 225 kg ha^{-1} of N being effective in maintaining fruit quality for up to 21 days in storage compared to higher nitrogen levels (Mukkun et al. 2001). Neuweiler et al. (2003) found that the availability of nitrogen in the rooting zone largely influences vegetative development, with an optimal nitrogen supply being crucial during the flowering and harvest periods. Overfertilization with nitrogen at early stages can harm initial plant development, so nitrogen fertilization must be carefully managed.

4.2.5.2 Phosphorus

Phosphorus (P) is essential for plant energy transfer, photosynthesis, and nutrient movement and is present in all plant cells (including DNA and RNA). Phosphorus can be a limiting factor in strawberry production, so maintaining a pH around 6.5 helps optimize its uptake. Continuous monitoring of plant nutrient needs through tissue analysis, soil tests, and leaf observations is recommended. Phosphorus deficiency in strawberries results in small, yellowish-green leaves that turn red with age, as well as reduced fruit size (Domoto 2011).

4.2.5.3 Potassium

Potassium (K) is crucial for cell elongation, water regulation, and carbohydrate synthesis. An adequate potassium supply enhances sugar synthesis, resulting in sweeter strawberries. K is a key nutrient influencing fruit quality and marketability, with its availability often limited by various environmental and soil factors (Jákli et al. 2016).

4.2.5.4 Microelements

Micronutrients are essential for plant growth, acting as catalysts in various organic reactions. Deficiencies can limit fruit crop productivity. The application of zinc sulfate and boric acid can improve fruit yield, with molybdenum enhancing fruit quality. However, high concentrations of zinc sulfate can prolong fruit shelf life, while excessive boric acid can be toxic, hindering growth and quality (Ahmad et al. 2010).

4.2.6 Biofertilizers

4.2.6.1 Organic Fertilizers

Fresh chicken manure has been shown to increase nitrate-nitrogen, available phosphorus, potassium, and organic matter content in soil, lower soil pH, and enhance strawberry yield (Zhang et al. 2024). Combining biofertilizers with organic manures like farmyard manure and vermicompost can further improve strawberry production sustainably, especially in organic farming systems (Negi et al. 2021). Research by Altieri et al. (2014) demonstrated that strawberry varieties exhibit different tolerances to high electrical conductivity from composted organic matter. In spring production, all varieties showed increased sweetness and acidity, with standard nitrate levels and no negative effects on fruit firmness.

A study by Singhalage et al. (2019) demonstrated the effectiveness of biofilmed biofertilizers composed of *Aspergillus* sp. and *Enterobacter* sp. improved fruit yield. Fruit quality analysis indicated that strawberries treated with biofertilizers showed a significant increase in total soluble solids and ascorbic acid content. Given the shallow root systems of strawberry plants, effective and balanced nutrient management is essential throughout the growing period. Total sugars and aroma are crucial for evaluating strawberry fruit quality. Yeast as a biofertilizer has been shown to enhance the production of volatiles and soluble sugars in strawberries. Previous research indicated that yeast biofertilizers could increase lettuce dry weight by 107% and enhance sugar content in sugar beets (Duan et al. 2021).

Combining biofertilizers with organic fertilizers and appropriate growth regulators improves plant growth, fruit quality, and strawberry yield. Studies have demonstrated that specific microorganisms and nitrogen fertilizers positively affect plant growth, and NPK fertilization after microbial inoculation can further enhance plant growth (Singh et al. 2023). Recent research by Panigrahi (2020) has focused on using biofertilizers to influence various aspects of fruit crop growth, including flowering, fruit production, quality parameters such as soluble solids, ascorbic acid, and sugar content.

4.2.6.2 Microbial Inoculants

Biofertilizers provide a renewable source of plant nutrients and support sustainable agriculture. These biofertilizers consist of various microbes, such as algae, bacteria, and fungi, which promote plant growth through rhizosphere colonization and increased nutrient availability. They can be applied to seeds, plant surfaces, or soil (Riaz et al. 2020; Mącik et al. 2014). Biofertilizers are effective in supplying essential nutrients, such as nitrogen, phosphorus, and zinc, and promoting overall plant growth (Szilagyi-Zecchin et al. 2016). They help maintain soil fertility, improve soil structure, enhance plant–water relations, protect against drought, reduce soil-borne diseases, and decrease pest incidence (Goyal and Manoharachary 2014; Rashid et al. 2016).

Microbial inoculants are a green alternative to chemical fertilizers and pesticides, composed of active microorganisms that stimulate microbial activity and enhance nutrient mobility in the soil (Ignatova et al. 2022; Shurigin et al. 2024). They help reduce the negative effects of chemical inputs, thereby improving both the quantity and quality of agricultural produce. Plant growth-promoting rhizobacteria (PGPR) have shown several beneficial effects on strawberry cultivation, particularly regarding plant health, yield, and defense mechanisms. They are effective tools for boosting crop yield, particularly under stressful conditions. These inoculants can include bacteria, fungi, and algae (Kar 2024; Bhattacharya et al. 2024). Bioinoculants are highly efficient and can serve as biopesticides without adverse effects on plant health (Kaminsky et al. 2019). For instance, arbuscular mycorrhizal fungi (AMF) can boost phosphorus, sulfur, and nitrogen fixation along with protecting the plants from phytopathogens, hence improving crop yields (Berruti et al. 2015; Prakash et al. 2022). AMF and PGPR are microbial inoculants that improve plant development, productivity, and the quality of fruits and vegetables. They also positively affect the handling and postharvest storage of produce (Violante and Pérez 2022). Research by Dukare et al. (2021) indicated that these microbial inoculants support plant growth through effective root colonization and the induction of growth-promoting mechanisms. They help plants cope with stress by altering root system architecture, inducing systemic resistance, and regulating stress-responsive genes (Rad 2022). Studies indicate that PGPR, such as *Azospirillum brasilense* and *Pseudomonas brassicacearum,* can enhance the nutritional state of strawberries, promoting growth and potentially improving resistance to pests like spider mites. These rhizobacteria have also been linked to the induction of systemic resistance (ISR), which strengthens the plant's ability to fend off herbivores and pathogens by activating defense-related compounds (Singh and Kumar 2024). In addition to enhancing growth, PGPR applications can increase the concentration of polyphenolic compounds in strawberry plants, which play a role in stress resistance. However, results can vary depending on the bacterial strains used, plant varieties, and environmental factors (Hosseini et al. 2021). In a study on organically grown strawberries, different combinations of PGPR and other microbial biostimulants were tested, though no significant effects on fruit yield or disease suppression were observed in some trials, suggesting the need for further investigation into effective application methods. Furthermore, PGPR's ability to improve nutrient uptake in strawberries, such as nitrogen fixation, can lead to increased fruit production and better overall plant vigor. Studies report that treatments with PGPR can increase the concentrations of beneficial compounds like polyphenols, which boost plant resilience against pathogens (Pii et al. 2018). Pırlak and Köse (2009) observed an increase in "Selva" strawberry yield with the application of PGPR through both foliar and root treatments (Seneviratne et al. 2011). Studies demonstrate that applying a combination of *Azotobacter,* AMF, vermicompost, and phosphate solubilising bacteria (PSB) can maximize strawberry yield, with improvements in plant height, leaf number, and fruit size. In another study, biofertilizers combined with organic manures (like poultry manure and farmyard manure) significantly improved the vegetative growth and fruit production of the strawberry variety "Sweet Charlie." Soni et al. (2018) observed that *Azotobacter* and *Azospirillum* enhanced plant growth parameters such as plant height, number of leaves, and crown diameter, contributing to greater fruit yields. Güneş et al. (2009) reported similar findings, noting that phosphate-solubilizing microorganisms such as *Bacillus* FS-3 and *Aspergillus* FS9 positively impacted strawberry yield and nutrient content. Rana (2003) found that a biofertilizer containing *Azotobacter chroococcum* and *Azospirillum brasilense* improved strawberry growth. Plants treated with this biofertilizer had higher root weight, stem weight, and leaf area compared to those treated with chemical fertilizers and also yielded more fruit. Similarly, Rueda et al. (2016) found the combination of *Azospirillum* spp. and *Azotobacter* spp. improved strawberry growth parameters, such as plant height, chlorophyll content, and root weight. Table 4.1 summarizes some of the key findings from recent studies that highlight the positive impacts of biofertilizers on strawberry growth and development.

Table 4.1 Effect of biofertilizers on strawberry growth and development

Fertilizer	Positive effect	Reference
Chicken manure	Yield enhancement	Zhang et al. (2024)
Manure, vermicompost, *Azotobacter* and *Pseudomonas*	Fruit weight	Negi et al. (2021)
Manure	Soil nutrients	Ombita et al. (2024)
Compost	Sweetness and acidity	Altieri et al. (2014)
Yeast	Stimulation of sugar production	Duan et al. (2021)
A. chroococcum and *A. brasilense*	Yield enhancement	Rana (2003)
Azospirillum spp. and *Azotobacter* spp.	Vitamin C	Rueda et al. (2016)
Biofertilizers	Plant growth promotion	Singh et al. (2023)
Azotobactor and *Aspirillum*	Yield and quality	Panigrahi (2020)
Aspergillus sp. and *Enterobactor* sp.	Fruit quality	Singhalage et al. (2019)
Bacillus and *Aspergillus*	P uptake and yield enhancement	Güneş et al. (2009)

4.2.6.3 Strawberry Disease and Management

Strawberry plants are susceptible to various diseases caused by fungi, bacteria, viruses, and environmental conditions. These diseases can affect different parts of the plant, including leaves, roots, fruits, and crowns, significantly reducing yield and fruit quality. Timely and accurate disease classification during strawberry planting can help growers manage diseases more effectively and minimize losses (Guofeng et al. 2022). Intensive monocropping systems, where the same crop is repeatedly grown in the same plot, often lead to an increase in soil-borne diseases (De Cal et al. 2005). For strawberries, high disease incidence has led to the widespread use of soil disinfection practices, such as methyl bromide fumigation, which has been banned due to environmental concerns. To mitigate soil-borne diseases, soilless and hydroponic systems have gained popularity, as they offer better control of plant health and nutrient supply through fertigation (Martínez et al. 2017). Hydroponic systems enable controlled water and nutrient delivery, resulting in high growth rates and reduced chemical inputs (Mattson and Lieth 2019; Askari-Khorasgani and Pessarakli 2020). *Anthracnose,* caused by *Colletotrichum nymphaeae*, and gray mold, caused by *Botrytis cinerea*, are major fungal diseases impacting strawberry production. Motallebi and Negahban (2024) suggested that using formulated neem seed extracts could serve as an alternative to chemical fungicides for controlling these diseases. Zhang et al. (2024) identified a strain of *Bacillus cereus* from soil that could be used as a biocontrol agent against *Neopestalotiopsis clavispora,* a disease affecting strawberries in northern China. By integrating biological control into modern strawberry farming practices, growers can reduce chemical pesticide use, improve soil health, and ensure higher yields and fruit quality. As research and technology advance, biological control will continue to play a critical role in sustainable strawberry production systems.

4.3 Conclusion

This chapter underscores the significant influence of nutrient solutions and biofertilizers on the growth and fruit yield of strawberries. The incorporation of organic manures and biofertilizers not only enriches the nutrient profile but also promotes beneficial microbial activity in the rhizosphere, enhancing the overall resilience and sustainability of the cultivation system. The results indicate that specific combinations of nutrient solutions and biofertilizers lead to improved growth parameters, increased fruit yield, and superior quality attributes. As the global demand for strawberries continues to rise, the adoption of hydroponic cultivation methods presents a promising avenue for enhancing productivity while addressing challenges such as perishability and postharvest losses. Hydroponics facilitates precise nutrient management, allowing for tailored approaches that optimize plant health and fruit quality. This approach not only meets market preferences for larger, uniform fruits but also supports environmental sustainability by reducing reliance on synthetic fertilizers and chemicals. Overall, the research highlights the potential of hydroponic strawberry production as a viable strategy for growers seeking to enhance both economic returns and environmental stewardship. Integrating nutrient management with biofertilizers improves fruit size, sweetness, color, and shelf life while enhancing soil fertility and structure. Practices like using organic amendments, crop rotation, and mulching further optimize nutrient use efficiency, ensuring sustainable strawberry production with minimal environmental impact. Future studies should continue to explore the long-term impacts of these practices on soil health and crop sustainability, paving the way for more innovative and resilient agricultural systems.

References

Abdullaeva Y, Mardonova G, Eshboev F, Cardinale M, Egamberdieva D (2023) Chickpea-associated endophytes and their potential in plant health and fitness. AIMS Microbiol 10(3):489–506. https://doi.org/10.3934/microbiol.2024024

Ahmad I, Khan MA, Qasim M, Ahmad R, Randhawa AM (2010) Growth, yield and quality of Rosa hybrid L. as influenced by various micronutrients. Pak J Agric Sci 47:5–12

Altieri R, Esposito A, Baruzzi G, Nair T (2014) Corroboration for the successful application of humified olive mill waste compost in soilless cultivation of strawberry. Int Biodeterior Biodegradation 88:118–124. https://doi.org/10.1016/j.ibiod.2013.12.006

Askari-Khorasgani O, Pessarakli M (2020) Tomato (*Solanum lycopersicum*) culture in vermi-aquaponic systems: I. Cultural practices. J Plant Nutr 43:1712–1725. https://doi.org/10.1080/01904167.2020.1739306

Badji A, Benseddik A, Bensaha H, Boukhelifa A, Hasrane I (2022) Design, technology, and management of greenhouse: a review. J Clean Prod 373:133753. https://doi.org/10.1016/j.jclepro.2022.133753

Berruti A, Lumini E, Balestrini R, Bianciotto V (2015) Arbuscular Mycorrhizal fungi as natural biofertilizers: let's benefit from past successes. Front Microbiol 6:1559. https://doi.org/10.3389/fmicb.2015.01559

Bhattacharya A, Mishra P, Mishra I, Arora P, Arora NK (2024) Microbe-based biostimulants: latest developments and future perspectives. In: Arora NK, Bouizgarne B (eds) Microbial Biotechnology for Sustainable Agriculture volume 2. Microorganisms for Sustainability, vol 34. Springer, Singapore. https://doi.org/10.1007/978-981-97-2355-3_2

Bose SK, Howlader P, Jia X, Wang W, Yin H (2019) Alginate oligosaccharide postharvest treatment preserve fruit quality and increase storage life via Abscisic acid signaling in strawberry. Food Chem 283:665–674. https://doi.org/10.1016/j.foodchem.2019.01.060

Caruso G, Villari G, Melchionna G, Conti S (2011) Effects of cultural cycles and nutrient solutions on plant growth, yield and fruit quality of alpine strawberry (Fragaria vesca L.) grown in hydroponics. Sci Hortic (Amsterdam) 129:479–485. https://doi.org/10.1016/j.scienta.2011.04.020

Chowdhury H, Asiabanpour B (2024) Influencing factors for the plant growth patterns in hydroponic and aquaponic environments: a subgroup analysis for sustainable agricultural production. Green Technol Sustain 2:100084. https://doi.org/10.1016/j.grets.2024.100084

De Cal A, Martinez-Treceño A, Salto T, López-Aranda JM, Melgarejo P (2005) Effect of chemical fumigation on soil fungal communities in Spanish strawberry nurseries. Appl Soil Ecol 28:47–56. https://doi.org/10.1016/j.apsoil.2004.06.005

Domoto P (2011) Recognizing and correcting nutrient deficiencies in strawberries. Iowa State University-Extension

Dong Y, Xu F, Liu L, Du X, Ren B, Guo A, Geng Y, Ruan C, Ye H, Huang W, Zhu Y (2020) Automatic system for crop pest and disease dynamic monitoring and early forecasting. IEEE J Sel Top Appl Earth Obs Remote Sens 13:4410–4418. https://doi.org/10.1109/jstars.2020.3013340
Duan W, Peng L, Zhang H, Han L, Li Y (2021) Microbial biofertilizers increase fruit aroma content of Fragaria × ananassa by improving photosynthetic efficiency. Alex Eng J 60:5323–5330. https://doi.org/10.1016/j.aej.2021.04.014
Dukare A, Paul S, Kumar R, Sharma V (2021) Microbial-based inoculants in sustainable agriculture: current perspectives and future prospects. In: Biofertilizers. Elsevier, pp 167–181
Egamberdieva D, Wirth S, Alqarawi AA, Abd-Allah EF, Hashem A (2017a) Phytohormones and beneficial microbes: essential components for plants to balance stress and fitness. Front Microbiol 8:2104. https://doi.org/10.3389/fmicb.2017.02104
Egamberdieva D, Wirth S, Behrendt U, Parvaiz A, Berg G (2017b) Antimicrobial activity of medicinal plants correlates with the proportion of antagonistic endophytes. Front Microbiol 8:199. https://doi.org/10.3389/fmicb.2017.00199
Egamberdieva D, Wirth S, Bellingrath-Kimura SD, Mishra J, Arora NK (2019) Salt-tolerant plant growth promoting rhizobacteria for enhancing crop productivity of saline soils. Front Microbiol 10:2791. https://doi.org/10.3389/fmicb.2019.02791
Egamberdieva D, Eshboev F, Shukurov O, Alaylar B, Arora NK (2023) Bacterial bioprotectants: biocontrol traits and induced resistance to phytopathogens. Microbiol Res 14(2):689–703. https://doi.org/10.3390/microbiolres14020049
Esitken A, Yildiz HE, Ercisli S, Figen Donmez M, Turan M, Gunes A (2010) Effects of plant growth promoting bacteria (PGPB) on yield, growth and nutrient contents of organically grown strawberry. Sci Hortic 124:62–66. https://doi.org/10.1016/j.scienta.2009.12.012
Ezziddine M, Liltved H, Seljåsen R (2021) Hydroponic lettuce cultivation using organic nutrient solution from aerobic digested aquacultural sludge. Agronomy 11:1484. https://doi.org/10.3390/agronomy11081484
Fernández-Cabanás VM, Delgado A, Lobillo-Eguíbar JR, Pérez-Urrestarazu L (2022) Early production of strawberry in aquaponic systems using commercial hydroponic bands. Aquac Eng 97:102242. https://doi.org/10.1016/j.aquaeng.2022.102242
Giampieri F, Forbes-Hernandez TY, Gasparrini M, Alvarez-Suarez JM, Afrin S, Bompadre S, Quiles JL, Mezzetti B, Battino M (2015) Strawberry as a health promoter: an evidence based review. Food Funct 6:1386–1398. https://doi.org/10.1039/c5fo00147a
Goyal A, Manoharachary C (2014) Future challenges in crop protection against fungal pathogens. Springer, New York
Güneş A, Ataoğlu N, Turan M, Eşitken A, Ketterings QM (2009) Effects of phosphate-solubilizing microorganisms on strawberry yield and nutrient concentrations. J Plant Nutr Soil Sci 172:385–392. https://doi.org/10.1002/jpln.200800121
Guo-feng Y, Yong Y, Zi-kang HE, Xin-yu Z, Yong HE (2022) A rapid, low-cost deep learning system to classify strawberry disease based on cloud service. J Integr Agric 21:460–473. https://doi.org/10.1016/s2095-3119(21)63604-3
Hosny KM, El-Hady WM, Samy FM (2024) Technologies, protocols, and applications of internet of things in greenhouse farming: a survey of recent advances. Inf Process Agric 12:91. https://doi.org/10.1016/j.inpa.2024.04.002
Hosseini A, Hosseini M, Schausberger P (2021) Plant growth-promoting rhizobacteria enhance defense of strawberry plants against spider mites. Front Plant Sci 12:783578. https://doi.org/10.3389/fpls.2021.783578
Ignatova L, Usmanova A, Brazhnikova Y, Omirbekova A, Egamberdieva D, Mukasheva T, Kistaubayeva A, Savitskaya I, Karpenyuk T, Goncharova A (2022) Plant probiotic Endophytic Pseudomonas flavescens D5 strain for protection of barley plants from salt stress. Sustain For 14:15881. https://doi.org/10.3390/su142315881
Jákli B, Tränkner M, Senbayram M, Dittert K (2016) Adequate supply of potassium improves plant water-use efficiency but not leaf water-use efficiency of spring wheat. J Plant Nutr Soil Sci 179:733–745. https://doi.org/10.1002/jpln.201600340

Kaminsky LM, Trexler RV, Malik RJ, Hockett KL, Bell TH (2019) The inherent conflicts in developing soil microbial inoculants. Trends Biotechnol 37:140–151. https://doi.org/10.1016/j.tibtech.2018.11.011

Kar S (2024) Role of microbial inoculants on vegetable and fruit quality. In: Bio-inoculants in Horticultural Crops. Elsevier, pp 49–77

Keutgen AJ, Pawelzik E (2009) Impacts of NaCl stress on plant growth and mineral nutrient assimilation in two cultivars of strawberry. Environ Exp Bot 65:170–176. https://doi.org/10.1016/j.envexpbot.2008.08.002

Khammayom N, Maruyama N, Chaichana C, Hirota M (2022) Impact of environmental factors on energy balance of greenhouse for strawberry cultivation. Case Stud Therm Eng 33:101945. https://doi.org/10.1016/j.csite.2022.101945

Khayyat M, Tafazoli E, Rajaee S, Vazifeshenas M, Mahmoodabadi MR, Sajjadinia A (2009) Effects of NaCl and supplementary potassium on gas exchange, ionic content, and growth of salt-stressed strawberry plants. J Plant Nutr 32:907–918. https://doi.org/10.1080/01904160902870689

Kumar P, Saini S (2020) Nutrients for hydroponic systems in fruit crops. In: Urban Horticulture – necessity of the future. IntechOpen

Lee S, Lee J (2015) Beneficial bacteria and fungi in hydroponic systems: types and characteristics of hydroponic food production methods. Sci Hortic 195:206–215. https://doi.org/10.1016/j.scienta.2015.09.011

Lewers KS, Fleisher DH, Daughtry CST (2017) Low tunnels as a strawberry breeding tool and season-extending production system. Int J Fruit Sci 17:233–258. https://doi.org/10.1080/15538362.2017.1305941

Mącik M, Gryta A, Frąc M, Malusá E, Vassilev N (2014) Biofertilizers in agriculture: an overview on concepts, strategies and effects on soil microorganisms. In: Advances in Agronomy. Elsevier, pp 6599–6607

Martínez F, Oliveira JA, Calvete EO, Palencia P (2017) Influence of growth medium on yield, quality indexes and SPAD values in strawberry plants. Sci Hortic (Amsterdam) 217:17–27. https://doi.org/10.1016/j.scienta.2017.01.024

Mattson N, Lieth JH (2019) Liquid culture hydroponic system operation. In: Soilless Culture. Elsevier, pp 567–585

Meselmani M (2023) Nutrient solution for hydroponics. In: Recent Research and Advances in Soilless Culture. IntechOpen

Motallebi P, Negahban M (2024) Neem (*Azadirachta indica*) seed extract formulation for managing anthracnose and gray mold diseases in strawberry. S Afr J Bot 169:66–71. https://doi.org/10.1016/j.sajb.2024.04.027

Mukkun L, Singh Z, Phillips D (2001) Nitrogen nutrition affects fruit firmness, quality and shelf life of strawberry. Acta Hortic 69–71:69. https://doi.org/10.17660/actahortic.2001.553.5

Nakro A, Bamouh A, Bouslama H, Bautista S, Ghaouti A (2023) The effect of potassium-nitrogen balance on the yield and quality of strawberries grown under soilless conditions. Horticulturae 9. https://doi.org/10.3390/horticulturae9030304

Naresh R, Jadav SK, Singh M, Patel A, Singh B, Beese S, Pandey SK (2024) Role of hydroponics in improving water-use efficiency and food security. Int J Environ Clim Change 14:608–633. https://doi.org/10.9734/ijecc/2024/v14i23976

Negi YK, Prabha D, Garg SK, Kumar J (2011) Genetic diversity among cold-tolerant fluorescent *Pseudomonas* isolates from Indian Himalayas and their characterization for biocontrol and plant growth-promoting activities. J Plant Growth Regul 30:128–143. https://doi.org/10.1007/s00344-010-9175-7

Negi YK, Sajwan P, Uniyal S, Mishra AC (2021) Enhancement in yield and nutritive qualities of strawberry fruits by the application of organic manures and biofertilizers. Sci Hortic 283:110038. https://doi.org/10.1016/j.scienta.2021.110038

Nemecek T, Huguenin-Elie O, Dubois D, Gaillard G, Schaller B, Chervet A (2011) Life cycle assessment of Swiss farming systems: II. Extensive and intensive production. Agric Syst 104:233–245. https://doi.org/10.1016/j.agsy.2010.07.007

Nemzer BV, Kalita D, Yashin AY, Yashin YI (2020) Bioactive compounds, antioxidant activities, and health beneficial effects of selected commercial berry fruits: a review. J Food Res 9:78. https://doi.org/10.5539/jfr.v9n5p78

Neuweiler R, Bertschinger L, Stamp P, Feil B (2003) The impact of ground cover management on soil nitrogen levels, parameters of vegetative crop development, yield and fruit quality of strawberries. Eur J Hort Sci 86(4):183–191

Ngouana LST, Tonfack LB, Temegne CN, Agendia AP, Youmbi E (2023) Current status of strawberry (*Fragaria* spp.) cultivation and marketing in Cameroon. J Agric Food Res 14:100761. https://doi.org/10.1016/j.jafr.2023.100761

Niu G, Masabni J (2022) Hydroponics. In: Kozai T, Niu G, Masabni J (eds) Plant factory Basics, Applications and Advances. Elsevier, Amsterdam, pp 153–166

Ombita SN, Mwendwa SM, Mureithi SM (2024) Influence of organic fertilization on growth and yield of strawberry (Fragaria × ananassa) in Kabete and Mbooni areas. Kenya Heliyon 10:e25324. https://doi.org/10.1016/j.heliyon.2024.e25324

Ogunlowo QO, Akpenpuun TD, Na H, Adesanya MA, Rabiu A, Dutta P, Kim HT, Lee HW (2023) Simulation of greenhouse energy and strawberry (Seolhyang sp.) yield using TRNSYS DVBES: A base case, Solar Energy 266:112196. https://doi.org/10.1016/j.solener.2023.112196

Panigrahi PB (2020) Effect of bio-fertilizers on growth, yield and quality of strawberry (Fragaria × ananassa Duch.) cv. Nabila under net tunnel. Pharma Innov J 9:442–446

Perin EC, Messias R d S, Galli V, Borowski JM, Souza ER, Avila LO d, Bamberg AL, Rombaldi CV (2019) Mineral content and antioxidant compounds in strawberry fruit submitted to drough stress. Food Sci Technol 39:245–254. https://doi.org/10.1590/fst.09717

Pii Y, Graf H, Valentinuzzi F, Cesco S, Mimmo T (2018) The effects of plant growth-promoting rhizobacteria (PGPR) on the growth and quality of strawberries. Acta Hortic:231–238. https://doi.org/10.17660/actahortic.2018.1217.29

Pırlak L, Köse M (2009) Effects of plant growth promoting rhizobacteria on yield and some fruit properties of strawberry. J Plant Nutr 32:1173–1184. https://doi.org/10.1080/01904160902943197

Prakash J, Egamberdieva D, Arora NK (2022) A novel *Bacillus safensis*-based formulation along with mycorrhiza inoculation for controlling *Alternaria alternata* and simultaneously improving growth, nutrient uptake, and steviol glycosides in *Stevia rebaudiana* under field conditions. Plan Theory 11(14):1857. https://doi.org/10.3390/plants11141857

Rad AK (2022) Effects of microbial inoculants on growth, yield, and fruit quality under stress conditions. In: Sustainable Horticulture. Elsevier, pp 1–38

Rajendran S, Domalachenpa T, Arora H, Li P, Sharma A, Rajauria G (2024) Hydroponics: exploring innovative sustainable technologies and applications across crop production, with emphasis on potato mini-tuber cultivation. Heliyon 10:e26823. https://doi.org/10.1016/j.heliyon.2024.e26823

Rana RK (2003) Effect of biofertilizers and nitrogen on growth, yield and fruit quality of strawberry. Prog Horti 35:25–30

Rashid MI, Mujawar LH, Shahzad T, Almeelbi T, Ismail IMI, Oves M (2016) Bacteria and fungi can contribute to nutrients bioavailability and aggregate formation in degraded soils. Microbiol Res 183:26–41. https://doi.org/10.1016/j.micres.2015.11.007

Riaz U, Mehdi SM, Iqbal S, Khalid HI, Qadir AA, Anum W, Ahmad M, Murtaza G (2020) Biofertilizers: eco-friendly approach for plant and soil environment. In: Bioremediation and biotechnology. Springer International Publishing, Cham, pp 189–213

Robinson Boyer L, Feng W, Gulbis N, Hajdu K, Harrison RJ, Jeffries P, Xu X (2016) The use of arbuscular mycorrhizal fungi to improve strawberry production in coir substrate. Front Plant Sci 7:1237. https://doi.org/10.3389/fpls.2016.01237

Ru D, Liu J, Hu Z, Zou Y, Jiang L, Cheng X, Lv Z (2017) Improvement of aquaponic performance through micro- and macro-nutrient addition. Environ Sci Pollut Res Int 24:16328–16335. https://doi.org/10.1007/s11356-017-9273-1

Rueda D, Valencia G, Soria N, Rueda B, Manjunatha B, Kundapur R, Selvanayagam M (2016) Effect of *Azospirillum* spp. and *Azotobacter* spp. on the growth and yield of strawberry (*Fragaria vesca*) in hydroponic system under different nitrogen levels. J Appl Pharm Sci:048–054. https://doi.org/10.7324/japs.2016.600108

Sahoo S, Sahoo D, Sahoo KK (2024) Optimization of an efficient hydroponic cultivation method for high yield of strawberry plant. S Afr J Bot 167:429–440. https://doi.org/10.1016/j.sajb.2024.02.033

Seneviratne G, Jayasekara APDA, De Silva MSDL, Abeysekera UP (2011) Developed microbial biofilms can restore deteriorated conventional agricultural soils. Soil Biol Biochem 43:1059–1062. https://doi.org/10.1016/j.soilbio.2011.01.026

Sethi VP, Sharma SK (2007) Greenhouse heating and cooling using aquifer water. Energy (Oxf) 32:1414–1421. https://doi.org/10.1016/j.energy.2006.10.022

Shurigin V, Li L, Alaylar B, Egamberdieva D, Abdugheni R, Mohamad OA, Liu YH, Li WJ (2024) Endophytic bacteria associated with fennel (*Foeniculum vulgare* Mill.) and their plant beneficial properties. AIMS Microbiol 10(2):449–467. https://doi.org/10.3934/microbiol.2024022

Singh V, Kumar B (2024) A review of agricultural microbial inoculants and their carriers in bioformulation. Rhizosphere 29:100843. https://doi.org/10.1016/j.rhisph.2023.100843

Singh RD, Tiwari GN (2010) Energy conservation in the greenhouse system: a steady state analysis. Energy (Oxf) 35:2367–2373. https://doi.org/10.1016/j.energy.2010.02.003

Singh T, Rajan R, Vamsi T, Kumar A, Ramprasad RR, Gopichand GB, Chundurwar K (2023) A review on the impact of organic and biofertilizer amendments on growth, yield, and quality of strawberry. Int J Plant Soil Sci 35:147–158. https://doi.org/10.9734/ijpss/2023/v35i143030

Singhalage ID, Seneviratne G, Madawala HMSP, Wijepala PC (2019) Profitability of strawberry (*Fragaria ananassa*) production with biofilmed biofertilizer application. Sci Hortic 243:411–413. https://doi.org/10.1016/j.scienta.2018.08.033

Soni SH, Kanawjia A, Chaurasiya R, Chauhan PS, Kumar R (2018) Effect of organic manure and biofertilizers on growth, yield and quality of strawberry (Fragaria X ananassa Duch). J Pharm Phytochem 2:128–132

Sundar P, Jyothi K, Sundar C (2021) Indoor hydroponics: a potential solution to reuse domestic rinse water. Biosci Biotechnol Res Asia 18:373–383. https://doi.org/10.13005/bbra/2924

Szilagyi-Zecchin VJ, Mógor ÁF, Figueiredo GGO (2016) Strategies for characterization of agriculturally important bacteria. In: Microbial Inoculants in Sustainable Agricultural Productivity. Springer India, New Delhi, pp 1–21

Tulipani S, Marzban G, Herndl A, Laimer M, Mezzetti B, Battino M (2011) Influence of environmental and genetic factors on health-related compounds in strawberry. Food Chem 124:906–913. https://doi.org/10.1016/j.foodchem.2010.07.018

Tyson RV, Treadwell DD, Simonne EH (2011) Opportunities and challenges to sustainability in aquaponic systems. HortTechnology 21:6–13. https://doi.org/10.21273/horttech.21.1.6

Ullio L (2010) Strawberry fertiliser guide. State of New South Wales through Department of Industry and Investment (Industry & Investment NSW). State of New South Wales through Department of Industry and Investment 941:1–9

Van Ginkel SW, Igou T, Chen Y (2017) Energy, water and nutrient impacts of California-grown vegetables compared to controlled environmental agriculture systems in Atlanta, GA. Resour Conserv Recycl 122:319–325. https://doi.org/10.1016/j.resconrec.2017.03.003

Violante H, Pérez M (2022) Preharvest and postharvest application of microbial inoculants influencing postharvest storage technology in horticultural crops. In: Sustainable Horticulture. Elsevier, pp 399–436

Xu Q, Huang B (2006) Seasonal changes in root metabolic activity and nitrogen uptake for two cultivars of creeping bentgrass. Hort Sci 41:822–826. https://doi.org/10.21273/hortsci.41.3.822

Xu C, Zhang X, Liang J, Fu Y, Wang J, Jiang M, Pan L (2022) Cell wall and reactive oxygen metabolism responses of strawberry fruit during storage to low voltage electrostatic field treatment. Postharvest Biol Technol 192:112017. https://doi.org/10.1016/j.postharvbio.2022.112017

Yamori W, Noguchi K, Hanba YT, Terashima I (2006) Effects of internal conductance on the temperature dependence of the photosynthetic rate in spinach leaves from contrasting growth temperatures. Plant Cell Physiol 47:1069–1080. https://doi.org/10.1093/pcp/pcj077

Yang T, Kim H-J (2020) Characterizing nutrient composition and concentration in tomato-, basil, and lettuce-based aquaponic and hydroponic systems. Water (Basel) 12. https://doi.org/10.3390/w12051259

Yıldırım-Yalçın M, Şeker M, Sadıkoğlu H (2022) Effect of grape derivatives and cross-linked maize starch coatings on the storage life of strawberry fruit. Prog Org Coat 167:106850. https://doi.org/10.1016/j.porgcoat.2022.106850

Zhang S, Wu J, Chen J, Jun S, Yuan Y, Dai X, Wang F, Ma Y (2024) The biological control effect of Bacillus cereus on strawberry leaf spot disease caused by *Neopestalotiopsis clavispora*. Sci Hortic 327:112841. https://doi.org/10.1016/j.scienta.2024.112841

Zhao L, Zhou Y, Liang L, Godana EA, Zhang X, Yang X, Wu M, Song Y, Zhang H (2023) Changes in quality and microbiome composition of strawberry fruits following postharvest application of *Debaryomyces hansenii*, a yeast biocontrol agent. Postharvest Biol Technol 202:112379. https://doi.org/10.1016/j.postharvbio.2023.112379

Chapter 5
Salicylic Acid in Sustainable Agriculture by Mitigating Salt Stress: Impact on Membrane Lipids

Ravina Pai and P. K. Sharma

Abstract Salinity is a significant threat to agriculture, posing a risk to global food security. Overcoming salinity with efficient and cost-effective solutions is imperative for sustainable agriculture, and the use of phytohormones is one such possibility. Salicylic acid (SA), a naturally occurring phenolic plant hormone, plays a crucial role as a signaling molecule to regulate growth and development. In the past two decades, the defensive role of salicylic acid against abiotic stress has gained much attention, and we have previously reported the role of salicylic acid in alleviating salt stress by maintaining water status, ion homeostasis, increase in the rate of photosynthesis, osmolyte accumulation, and reduction in oxidative damage by an increase in antioxidants. In this chapter, we focus on the membrane lipids with regard to the application of SA and its possible function in mitigating salt stress to increase crop productivity.

Keywords Lipid peroxidation · Membrane lipids · Oxidation · Salicylic acid · Salinity

5.1 Introduction

Soil salinity poses a critical threat to global agriculture, with over 6.74 million hectares of salt-affected land in India alone, causing annual losses of 16.84 million tons of crop production (Mandal et al. 2018). Projections indicate a dire escalation, with salinized areas expected to reach 16.2 million hectares by 2050 due to a 10% annual increase in soil degradation (Kumar and Sharma 2020). At the heart of this crisis lies salt stress, which imposes dual challenges on plants: *osmotic stress* and *ionic toxicity*. When exposed to high NaCl concentrations, plants undergo osmotic shock,

R. Pai · P. K. Sharma (✉)
School of Biological Sciences and Biotechnology, Goa University, Taleigao Plateau, Goa, India
e-mail: pksharma@unigoa.ac.in

N. K. Arora et al. (eds.), *Environmental Sustainability and Biotechnology: Opportunities and Challenges*, Microorganisms for Sustainability 46,
https://doi.org/10.1007/978-981-96-9025-1_5

mimicking drought conditions by disrupting cellular water balance (Shavrukov 2013). This leads to reduced turgor pressure, plasma membrane shrinkage, and cell wall damage (Zhao et al. 2021), severely impairing nutrient uptake and metabolic activity. Over time, excessive Na^+ and Cl^- accumulation disrupts ion homeostasis, displacing essential nutrients like K^+, Ca^{2+}, and Mg^{2+} (Hussain et al. 2019), while triggering oxidative stress through reactive oxygen species (ROS) overproduction (Mahmud et al. 2017). These cascading effects compromise membrane integrity, organelle function, and ultimately plant survival.

To combat salinity, plants deploy multifaceted adaptations, including osmolyte synthesis (proline and glycine betaine) (Yancey 2005; Ghosh et al. 2021), selective ion exclusion via Salt Overly Sensitive pathways (Gupta and Huang 2014), vacuolar Na^+ sequestration (Blumwald et al. 2000), and antioxidant systems such as superoxide dismutase (SOD) and ascorbate peroxidase (APX) (Sharma et al. 2012). However, conventional breeding approaches face limitations due to the genetic complexity of salt tolerance and species-specific responses (Yildirim et al. 2008; Jini and Joseph 2017). This underscores the need for innovative, cost-effective interventions. Among these, phytohormones like salicylic acid (SA) have emerged as promising tools. SA modulates stomatal closure, enhances osmolyte production, scavenges ROS, and activates stress-responsive genes (Pai and Sharma 2022, 2024), offering a holistic approach to salinity mitigation.

A critical yet underexplored aspect of SA-mediated resilience lies in its interaction with membrane lipids. Lipids are not merely structural components of cell membranes but also govern signaling, energy storage (via triacylglycerols) (Yang and Benning 2018), and stress adaptation (Kirchhoff et al. 2007). Under salinity, membrane lipid composition shifts dynamically, influencing fluidity, permeability, and organelle function. Phospholipids and sphingolipids further serve as precursors for stress-signaling molecules (Lynch and Dunn 2004), while SA's ability to stabilize lipid membranes under ionic and oxidative stress could play a pivotal role in preserving cellular homeostasis.

This chapter synthesizes current knowledge on SA's role in modulating lipid metabolism under salt stress, with a focus on membrane integrity, signaling, and energy dynamics. By elucidating how SA enhances lipid-mediated stress adaptation, we aim to bridge the gap between biochemical mechanisms and sustainable agricultural practices. The goal is to provide actionable insights for utilizing SA to improve crop resilience, optimize nutrient use efficiency, and mitigate yield losses in saline environments—key steps toward achieving global food security in a changing climate.

5.2 Salicylic Acid: Biosynthesis, Structure, and Functional Roles

Salicylic acid (SA) is a seven-carbon, endogenous growth regulator belonging to a group of phenols. It participates in regulating physiological processes and is considered a signaling molecule known to take part in defense responses under abiotic stress like salinity stress (Jini and Joseph 2017). The name salicylic acid originated from the Latin word Salix which was given to the active ingredient in willow bark (*Salix* sp.) by Raffaele Piria in 1838 (Khan et al. 2015). Salicylic acid, also called beta-hydroxy acid (BHA), is lipophilic in nature and is a colorless, odorless, and crystalline organic acid.

Salicylic acid in plants is synthesized from chorismate by two pathways, namely the phenylamine ammonia-lyase (PAL) pathway in the cytoplasm and the isochorismate (ICS) pathway in the chloroplast (Mustafa et al. 2009). PAL pathway is most common and is induced under biotic and abiotic stresses as a key regulator of the phenylpropanoid pathway (Chen et al. 2009). The status of these biosynthesis pathways varies in different plant species and even different plant parts. In rice, the PAL pathway is significant, while in *Arabidopsis,* ICS pathway is the most important, whereas in soybeans, both pathways are equally crucial in the biosynthesis of SA (Lefevere et al. 2020). Salicylic acid is transported through the phloem by the process of translocation, moving from the site of synthesis to the rest of the plants.

5.3 Role in Biotic and Abiotic Stress Tolerance

The role of SA in response to biotic stresses and pathogenesis is well established (Verberne et al. 2000; Cui et al. 2019). SA triggers a series of events leading to the production of pathogenesis-related proteins aiding in resistance against pathogens. Moreover, SA also plays an important role in systemic acquired resistance (SAR), characterized by an enhanced ability to localize a subsequent attack by a broad range of viral, bacterial, or fungal pathogens in regions both proximal to the initial infection site and throughout the plant body serving as a protective mechanism against future attacks. Studies have shown that SA is the signal compound that induces SAR providing resistance to a variety of pathogens (Horvath and Chua 1994). In the past two decades, researchers have demonstrated the role of SA in mitigating abiotic stress responses such as drought, high and low temperatures, salinity, ozone, UV radiation, and heavy metals. However, the exact mechanism remains elusive (Jayakannan et al. 2015; Hoque et al. 2020). The effect of SA application depends on numerous factors, such as the species, developmental stage of the plant, mode of application, and concentration of applied and endogenous SA levels (Ahmad and Prasad 2011; Pai and Sharma 2024). SA, when applied at lower concentrations, acts as a signaling molecule with antioxidative capacity but in larger amounts may cause oxidative stress (Pai and Sharma 2022).

SA is involved in the signal transduction system, inducing the synthesis of enzymes that take part in proline metabolic pathways to produce defense compounds like compatible solutes, establishing an antioxidant defense system, and improving plant water relations under stress conditions, thereby protecting the plant (Jini and Joseph 2017). SA is effective in the induction of glycine betaine and its accumulation as a part of systematic acquired resistance that is important for plants experiencing salt stress; it also increases the methionine content and suppresses excess ethylene formation under salinity stress. The induction of glycine betaine

(GB) may have other effects, such as the activation of protein kinase when aroused by hyperosmotic stress (Khan et al. 2015). SA-mediated increase in GB level can improve the overall plant growth. SA induces the activity of proline metabolism under abiotic stresses; increased proline production is argued to improve nitrogen assimilation and modulate calcium-mediated oxidative burst defense response in plants (Khan et al. 2015). SA increases proline metabolism that acts as a cytoplasmic osmoticum and may function as a carbon and nitrogen source for post-stress recovery and growth, as a stabilizer for membrane and protein synthesis machinery, as a scavenger of free radicals, as a sink for energy to regulate the redox potential, and also serves to protect the protein against denaturation (Jaiswal et al. 2014). An increase in proline on the application of SA reduced the malondialdehyde (MDA) content and alleviated membrane damage (Li et al. 2014; Pai and Sharma 2022, 2024).

SA is reported to play a crucial role in controlling the sodium entry into the roots and its long-distance transport into the shoots, enhancing the activity of H^+ATPase in roots, preventing the leakage of stress-induced K^+ from roots through the depolarization-activated potassium outward rectifying channel (KOR) and ROS-activated nonselective cation channels, increasing K^+ concentration in shoots during salt stress (Jayakannan et al. 2015). SA also enhanced the activity of H^+ATPase, decreased NaCl-induced membrane depolarization, and minimized NaCl-induced K^+ leakage from the cell within the first hour of salt stress by counteracting salt-induced membrane depolarization and by decreasing K^+ efflux via guard cell outward rectifying K^+ channels (GORK) (Jayakannan et al. 2013) along with the accumulation of N, Mg, Fe, Mn, and Cu (Gunes et al. 2007).

SA also plays a vital role in seed germination, plant growth, photosynthesis, transpiration, and nutrient acquisition during salt stress. Initially, it was thought that hydrogen peroxide (H_2O_2) functions downstream of SA; later, it was found that an increase in H_2O_2 acts upstream of SA, inducing endogenous SA (Leon et al. 1995). Reports have shown the regulation of ROS balance by SA as a mechanism by which SA modulates germination during salt stress because SA and hydrogen peroxide form a "self-amplifying feedback loop" during stress conditions, hydrogen peroxide inducing the accumulation of SA and SA inducing hydrogen peroxide concentration, thus composing of positive feedback at specific concentrations (Lee et al. 2010; Jayakannan et al. 2015).

SA is known to reduce the effects of salinity by improving the ability of the photosynthetic pigments to function along with the antioxidative enzymes (Mahmud et al. 2017). SA increases antioxidative enzymes like SOD, leading to the restoration of oxidative balance by a decrease in H_2O_2 content (Nazar et al. 2015) and minimizing the cellular damage caused by secondary oxidative stress (Jaiswal et al. 2014). However, a higher concentration of SA can be attributed to the negative interaction of antioxidant enzymes (Pirasteh-Anosheh and Emam 2018) as it leads to oxidative burst and cell death as the plant produces more ROS and at the same time reduces its capacity of scavenging H_2O_2 as a result of accumulated ROS that activates cell death under stress conditions (Mittler 2002). Both endogenous and exogenous SA are evidenced to play roles in antioxidant metabolism and have tight control over cellular ROS. However, coordination of SA-dependent and SA-independent signaling components with ROS signaling provides an appropriate defense response (Khan et al. 2015).

In addition, SA is reported to mediate ROS detoxification and activate plant defense responses, protecting membrane lipids from peroxidation by inhibiting lipoxygenase activity and preventing oxidative chain reactions (Panche et al. 2016). An increase in proline on the application of SA reduced lipid peroxidation and alleviated membrane damage (Li et al. 2014). SA is also reported to lower the values of electrolyte leakage (Yildirim et al. 2008) and improve the membrane stability index (MSI), resulting in the repair of plasmalemma injury caused by ROS with an increase in the antioxidants like peroxidase, catalase, SOD, and glutathione reductase (Hayat et al. 2010). SA is also reported to effectively ameliorate the negative effects of stress by the induction of stress-responsive gene expression and induction of expression of biosynthetic enzymes such as dehydrin-like proteins, chaperons, protein kinase activity, chlorophyll, and rubisco (Khalvandi et al. 2021). SA applications can be related to rubisco concentration and activity, which help to keep the integrity of the light-harvesting apparatus and thus productivity in which lipids and proteins play a major role.

Furthermore, SA is reported to lower the values of electrolyte leakage (Yildirim et al. 2008) and improve the membrane stability index (MSI), resulting in the repair of plasmalemma injury caused by ROS with an increase in antioxidants (Hayat et al. 2010). Application of SA regulates many aspects of plant responses under stressed and normal conditions through signaling cross-talks with phytohormones such as auxin, cytokinins, gibberellins, abscisic acid (ABA), ethylene, nitric oxide (NO), and brassinosteroids that can be synergistic or antagonistic. SA triggers the accumulation of ABA under both normal and salinity-stressed plants that help in osmotic adaptation, improved photosynthetic pigments, and growth attributes. SA promotes ethylene, which promotes cell death via superoxide oxidation stress. SA interacts with NO, which plays a role as a second messenger and controls stomatal movement in higher plants (Khan et al. 2015).

5.4 Effect of Salt Stress on Membrane Lipids

Salt stress triggers lipid peroxidation, a destructive process initiated by reactive oxygen species (ROS) that attack polyunsaturated fatty acids (PUFAs) and phospholipid ester bonds (Gill and Tuteja 2010). This three-step reaction—initiation, propagation, and termination—generates reactive intermediates like conjugated diene hydroperoxides, lipid peroxy radicals, and aldehydes such as malondialdehyde (MDA) (Gaschler and Stockwell 2017). MDA covalently binds to lysine residues in proteins and DNA bases, forming mutagenic adducts that destabilize membranes and impair physiological processes like photosynthesis. The extent of peroxidation reflects ROS-induced cellular damage, serving as a biomarker for stress susceptibility (Sharma et al. 2012). For instance, salt-stressed plants exhibit reduced membrane fluidity and integrity due to altered lipid composition, particularly in photosynthetic membranes where galactolipids like monogalactosyldiacylglycerol (MGDG) and digalactosyldiacylglycerol (DGDG) dominate. MGDG (~50% of thylakoid lipids) stabilizes PSII and supports xanthophyll cycle enzymes, while DGDG (~25–30%) maintains membrane stacking (Dörmann and Benning 2002). However, salinity reduces MGDG and DGDG levels, as seen in studies where glycolipid decline correlated with chlorophyll loss and chloroplast ultrastructure damage (Liu et al. 2023). This disruption aligns with the force from lipids (FLP) principle, where altered lipid composition impacts membrane pressure gradients, protein function, and cellular behavior (Reszczyńska and Hanaka 2020).

Plants employ species-specific lipid remodeling to counteract salt stress. Halophytes like *Mesembryanthemum crystallinum* increase phospholipids (e.g., phosphatidylcholine, PC) to enhance membrane fluidity and osmotic balance (Barkla et al. 2018), while C3 plants (*Salicornia perennans*) shift from granal to agranal thylakoids, altering the PC/PE ratio to reduce membrane permeability (Rozentsvet et al. 2022; Wu et al. 2005). In contrast, C4 plants (*Climacoptera crassa*) show minimal thylakoid restructuring, relying instead on sulfolipid (SQDG) and phosphatidylglycerol (PG) adjustments. For example, *Thellungiella* and *Suaeda salsa* increase unsaturated fatty acids in PG and SQDG under salinity, enhancing PSII photoinhibition resistance (Sui and Han 2014; Sui et al. 2010). Similarly, *Spartina patens* maintains plasma membrane stability by elevating the PC/PE ratio, though reduced protein-to-lipid ratios may compromise permeability (Wu et al. 2005). Halophytes like *Salicornia* also accumulate SQDG to stabilize ATPase complexes and PSII (Ramani et al. 2004), while glycophytes like tobacco suffer irreversible DGDG loss (Wang et al. 2014).

Lipid unsaturation and compositional ratios play critical adaptive roles in plants' ability to cope salt stress (Upchurch 2008). Maintaining the MGDG/DGDG ratio with increased unsaturation suppresses chloroplast damage, as shown in studies where stable ratios preserved thylakoid architecture (Bejaoui et al. 2016). Similarly, PG enrichment in *Arabidopsis* under moderate salinity protects ATP synthesis and light-harvesting complexes (Hagio et al. 2002; Sui et al. 2010). Mesophyll cells, with higher MGDG/DGDG content, exhibit greater salinity sensitivity than bundle sheath cells, which prioritize sulfolipid adjustments (Omoto et al. 2015). Salt stress also reduces phosphatidylcholine (PC) and phosphatidylglycerol (PG) in plasma membranes while increasing phosphatidylserine (PS) and phosphatidylinositol (PI), altering membrane asymmetry and signaling (Salama and Mansour 2015). These dynamics highlight the dual role of lipids as structural components and signaling precursors—phospholipids and sphingolipids, for instance, generate secondary messengers that regulate stress responses (Lynch and Dunn 2004).

Ultimately, lipid remodeling under salinity involves trade-offs: halophytes optimize fluidity and photoprotection via SQDG/PG accumulation and galactolipid stabilization (Guo et al. 2019), while glycophytes succumb to peroxidation-driven membrane collapse (Munns and Tester 2008; Parida and Das 2005). Table 5.1 highlights lipid remodeling as a key adaptive mechanism, varying with the photosynthetic type and salt tolerance capacity. Understanding these mechanisms—such as

Table 5.1 Effect on membrane lipids in response to salinity stress in different plant species

Plant species	NaCl stress	Effect on membrane lipids	References
Cucumis melo L. (C3)	250 mM	↑ MGDG/DGDG ratio Irregular stability in photosynthetic membrane	Liu et al. (2023)
Salicornia perennans Willd. (C3)	33%	↑MGDG ↑DGDG ↑SQDG ↑PG ↑PC PE and PI unchanged	Rozentsvet et al. (2022)
Climacoptera crassa (Bied.) Botsch (C4)	4%	Membrane lipids unchanged	
Mesembryanthemum crystallinum L. (facultative CAM)	200 mM	↑PG ↑PI ↑PC ↑PA ↓TG ↓MGDG ↓DGDG ↓chloroplastic lipids	Barkla et al. (2018)
Sulla carnosa (C3) *Sulla coronaria* (C3)	200 mM	↓DGDG ↓PG ↓SQDG ↑MGDG/DGDG MGDG unchanged ↓ MGDG	Bejaoui et al. (2016)
Zea mays (C4)	3% NaCl	In mesophyll cells: ↓MGDG ↓SQDG ↑PC DGDG, PG, PI unchanged In bundle sheath cells: MGDG unchanged ↓SQDG ↑PC DGDG, PG, PI unchanged	Omoto et al. (2015)
Triticum aestivum (C3)	150 mM	↓PC ↓PG ↓DPG ↑PI ↑PE ↑PS	Salama and Mansour (2015)
Thellungiella halophila (C3)	100, 200 mM	↑MGDG ↑DGDG ↑PG ↑SQDG	Sui and Han (2014)
Suaeda salsa L. (C3)	300 mM	↑PG ↑DGDG/MGDG ↓MGDG ↓DGDG ↓SQDG ↓PC	Sui et al. (2010)
Brassica napus (C3-C4 intermediate)	150, 200 mM	↓GL/PL ↑sterol/PL	Zamani et al. (2010)
Spartina patens (C4)	340 mM	↑PC ↑PI ↑PS	Wu et al. (2005)
	510 mM	↑GL ↓PL ↓PC ↓PE PC/PE unaffected Sterol/phospholipid unaffected ↓root plasma membrane fluidity	
Crithmum maritinum	200 mM	↑SQDG	Hamed et al. (2005)
Catharanthus roseus (C3)	100 mM	↑PC ↑PE ↑SQDG ↓MGDG ↓DGDG	Elkahoui et al. (2004)
Aster tripolium (C3)	517 mM	↑ Sulfolipids ↑SQDG	Ramani et al. (2004)
Sesuvium postulacastrum L. (C4)	864 mM	↑ Sulfolipids ↑SQDG	
Arabidopsis thaliana (C3)	100 mM	Sulfolipids unchanged	

↑ Upregulation, ↓ Downregulation, *DGDG* digalactosyldiacylglycerol, *DPG* diphosphatidylglycerol, *GL* glycolipids, *MGDG* monogalactosyldiacylglycerol, *PA* phosphatidic acid, *PC* phosphatidylcholine, *PE* phosphatidylethanolamine, *PG* phosphatidylglycerol, *PI* phosphatidylinositol, *PL* phospholipid, *PS* phosphatidylserine, *SQDG* sulfoquinovosyl diacylglycerol, *TG* triglycerides

MGDG's role in PSII stability or SQDG's ATPase protection—provides actionable targets for improving the crop resilience (Dörmann and Benning 2002; Demé et al. 2014; Lu et al. 2020). By preserving lipid homeostasis or engineering species-specific adaptations (e.g., enhancing PC/PE ratios or fatty acid unsaturation), agricultural strategies can mitigate salt-induced yield losses, addressing the escalating threat of soil salinization to global food security.

5.5 Possible Influence of SA on Lipid Membrane

As mentioned in the above sections, SA has emerged as a promising candidate to enhance tolerance to environmental stresses, including salt stress by modulating membrane lipids. Earlier, it has been observed that SA reduces the salt-induced ROS generation in rice and sorghum mainly by the upregulation of antioxidants like SOD and APX, indicating the role of SA in modulating the cell redox balance mainly observed as reduced lipid peroxidation and increased membrane stability (Pai and Sharma 2022, 2024). Nazar et al. (2011) also observed a reduction in ROS generation on the application of 0.5 mM SA in *V. radiata* against salt stress, leading to reduced lipid peroxidation and electrolyte leakage. Kaur et al. (2022) reported that SA priming in the genotypes of chickpeas exposed to salt stress reduced ROS concentration along with a reduction in MDA and EL, suggesting that SA might shield cells and subcellular compartments from the harmful effects of ROS by preventing cell membranes from depolarizing due to excessive concentration of Na^+ or Cl^- ions. Zorova et al. (2018) reported that salicylate collapsed the transmembrane electrochemical potential in pea mitochondria to prevent basal and oxygen consumption and the activity of ATPase. Belkadhi et al. (2015) reported an increase in PC and PE in *Linum usitatissimum* seeds pretreated with SA against cadmium stress. SA treatment is also reported to increase the levels of PA and PC, thus contributing to membrane structure stability and signaling (Mishra and Tanna 2017). However, studies on the effects of SA on individual membrane lipids under salinity stress are limited. Hence, with known findings, in Fig. 5.1, we tried to elucidate the possible mechanism by which SA may affect membrane proteins under salinity stress (Fig. 5.1).

Other reports propose SA to be a membrane stabilizer due to its activity to protect lipid membranes against oxidative stress (Belkadhi et al. 2015). SA-induced changes in MGDG help in scavenging ROS with its antioxidant property along with conserving photosynthetic energy and photoprotection, which improves photosynthetic activities as seen in earlier studies (Pai and Sharma 2024). SA is reported to influence glycolipids in biotic stress (Al-Hakimi and Alghalibi 2007), which may also aid in abiotic stresses for osmotic protection and maintain hydration levels with their hydrophilic headgroups interacting with water molecules, preventing

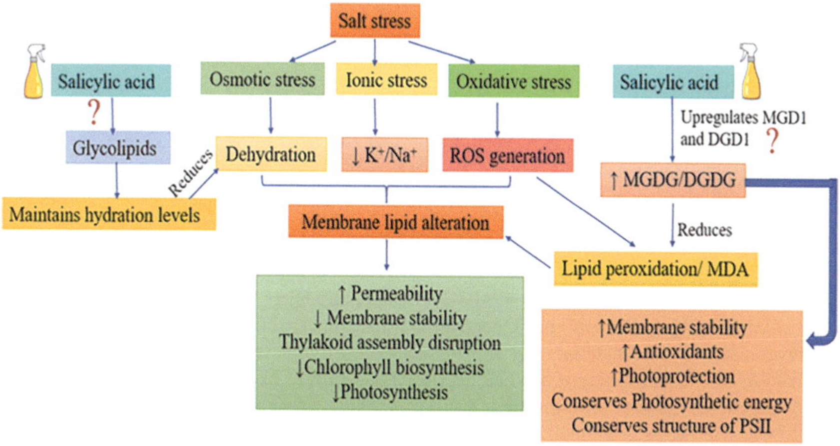

Fig. 5.1 Possible protective role of salicylic acid in mitigating salt stress-induced membrane lipid alterations in plants. Salt stress induces osmotic, ionic, and oxidative damage, leading to membrane lipid alteration, lipid peroxidation, and impaired photosynthesis. SA mitigates these effects possibly by enhancing glycolipid-mediated hydration maintenance, promoting galactolipid (MGDG/DGDG) biosynthesis, and reducing ROS-induced lipid peroxidation. These adjustments stabilize membrane integrity, promote antioxidant defenses, protect photosynthetic energy, and conserve PSII structure under salt stress

dehydration during salt-induced osmotic stress (Perlikowski et al. 2016), thus optimizing photosynthetic efficiency by maintaining the structure and function of PSII (Pai and Sharma 2024). SA may also influence unsaturated fatty acids in the membrane, affecting the fluidity of the membrane and activating the functioning of ion channels/transporters protecting the photosystem and selectively permitting Na^+/K^+/Ca^{2+} aiding in ion homeostasis (Li et al. 2020).

5.6 Conclusion

SA mitigates salt stress through multifaceted mechanisms: reducing oxidative damage via antioxidant upregulation, stabilizing lipid composition (PC, PE, and MGDG), and enhancing membrane fluidity and ion transport. Its role in maintaining glycolipid hydration and PSII photoprotection underscores its potential in sustainable agriculture. However, gaps persist in understanding SA's species-specific effects on individual lipids and molecular signaling. Future studies should explore lipidomic profiles under SA treatment to unravel its full mechanistic potential in salinity resilience.

References

Ahmad P, Prasad MNV (eds) (2011) Abiotic stress responses in plants: metabolism, productivity and sustainability. Springer Science & Business Media

Al-Hakimi AMA, Alghalibi S (2007) Thiamin and salicylic acid as biological alternatives for controlling broad bean rot disease. Saudi J Biol Sci 14:201

Barkla BJ, Garibay-Hernández A, Melzer M, Rupasinghe TW, Roessner U (2018) Single cell-type analysis of cellular lipid remodelling in response to salinity in the epidermal bladder cells of the model halophyte Mesembryanthemum crystallinum. Plant Cell Environ 41(10):2390–2403. https://doi.org/10.1111/pce.13352

Bejaoui F, Salas JJ, Nouairi I, Smaoui A, Abdelly C, Martínez-Force E, Youssef NB (2016) Changes in chloroplast lipid contents and chloroplast ultrastructure in Sulla carnosa and Sulla coronaria leaves under salt stress. J Plant Physiol 198:32–38. https://doi.org/10.1016/j.jplph.2016.03.018

Belkadhi A, De Haro A, Obregon S, Chaïbi W, Djebali W (2015) Exogenous salicylic acid protects phospholipids against cadmium stress in flax (*Linum usitatissimum* L.). Ecotoxicol Environ Saf 120:102–109. https://doi.org/10.1016/j.ecoenv.2015.05.028

Blumwald E, Aharon GS, Apse MP (2000) Sodium transport in plant cells. Biochim Biophys Acta Biomembr 1465(1–2):140–151. https://doi.org/10.1016/S0005-2736(00)00135-8

Chen H, Seguin P, Jabaji SH (2009) Differential expression of genes encoding the phenylpropanoid pathway upon infection of soybean seedlings by *Rhizoctonia solani*. Can J Plant Pathol 31(3):356–367. https://doi.org/10.1080/07060660909507610

Cui N, Lu H, Wang T, Zhang W, Kang L, Cui F (2019) Armet, an aphid effector protein, induces pathogen resistance in plants by promoting the accumulation of salicylic acid. Philos Trans R Soc B 374(1767):20180314. https://doi.org/10.1098/rstb.2018.0314

Demé B, Cataye C, Block MA, Maréchal E, Jouhet J (2014) Contribution of galactoglycerolipids to the 3-dimensional architecture of thylakoids. FASEB J 28(8):3373–3383. https://doi.org/10.1096/fj.13-247395

Dörmann P, Benning C (2002) Galactolipids rule in seed plants. Trends Plant Sci 7(3):112–118. https://doi.org/10.1016/S1360-1385(01)02216-6

Elkahoui S, Smaoui A, Zarrouk M, Ghrir R, Limam F (2004) Salt-induced lipid changes in Catharanthus roseus cultured cell suspensions. Phytochemistry 65(13):1911–1917. https://doi.org/10.1016/j.phytochem.2004.06.021

Gaschler MM, Stockwell BR (2017) Lipid peroxidation in cell death. Biochem Biophys Res Commun 482(3):419–425. https://doi.org/10.1016/j.bbrc.2016.10.086

Ghosh UK, Islam MN, Siddiqui MN, Khan MAR (2021) Understanding the roles of osmolytes for acclimatizing plants to changing environment: a review of potential mechanism. Plant Signal Behav 16(8):1913306. https://doi.org/10.1080/15592324.2021.1913306

Gill SS, Tuteja N (2010) Reactive oxygen species and antioxidant machinery in abiotic stress tolerance in crop plants. Plant Physiol Biochem 48(12):909–930. https://doi.org/10.1016/j.plaphy.2010.08.016

Gunes A, Inal A, Alpaslan M, Eraslan F, Bagci EG, Cicek N (2007) Salicylic acid induced changes on some physiological parameters symptomatic for oxidative stress and mineral nutrition in maize (*Zea mays* L.) grown under salinity. J Plant Physiol 164(6):728–736. https://doi.org/10.1016/j.jplph.2005.12.009

Guo Q, Liu L, Barkla BJ (2019) Membrane lipid remodeling in response to salinity. Int J Mol Sci 20(17):4264. https://doi.org/10.3390/ijms20174264

Gupta B, Huang B (2014) Mechanism of salinity tolerance in plants: physiological, biochemical, and molecular characterization. Int J Genomics 2014:1. https://doi.org/10.1155/2014/701596

Hagio M, Sakurai I, Sato S, Kato T, Tabata S, Wada H (2002) Phosphatidylglycerol is essential for the development of thylakoid membranes in *Arabidopsis thaliana*. Plant Cell Physiol 43(12):1456–1464. https://doi.org/10.1093/pcp/pcf185

Hamed KB, Youssef NB, Ranieri A, Zarrouk M, Abdelly C (2005) Changes in content and fatty acid profiles of total lipids and sulfolipids in the halophyte Crithmum maritimum under salt stress. J Plant Physiol 162(5):599–602. https://doi.org/10.1016/j.jplph.2004.11.010

Hayat S, Hasan SA, Hayat Q, Irfan M, Ahmad A (2010) Effect of salicylic acid on net photosynthetic rate, chlorophyll fluorescence, and antioxidant enzymes in *Vigna radiata* plants exposed to temperature and salinity stresses. Plant Stress 4:62–71

Hoque TS, Sohag AAM, Burritt DJ, Hossain MA (2020) Salicylic acid-mediated salt stress tolerance in plants. In: Plant phenolics in sustainable agriculture: volume 1. Springer, Singapore, pp 1–38. https://doi.org/10.1007/978-981-15-4890-1_1

Horvath DM, Chua NH (1994) The role of salicylic acid in systemic acquired resistance. Curr Opin Biotechnol 5(2):131–136

Hussain S, Bai Z, Huang J, Cao X, Zhu L, Zhu C, Khaskheli MA, Zhang C, Jin Q, Zhang J (2019) 1-Methylcyclopropene modulates physiological, biochemical, and antioxidant responses of rice to different salt stress levels. Front Plant Sci 10:124. https://doi.org/10.3389/fpls.2019.00124

Jaiswal A, Pandurangam V, Sharma SK (2014) Effect of salicylic acid in soybean (*Glycine Max* L. Meril) under salinity stress. Bioscan 9(2):671–676

Jayakannan M, Bose J, Babourina O, Rengel Z, Shabala S (2013) Salicylic acid improves salinity tolerance in *Arabidopsis* by restoring membrane potential and preventing salt-induced K^+ loss via a GORK channel. J Exp Bot 64(8):2255–2268. https://doi.org/10.1093/jxb/ert085

Jayakannan M, Bose J, Babourina O, Rengel Z, Shabala S (2015) Salicylic acid in plant salinity stress signalling and tolerance. Plant Growth Regul 76:25–40. https://doi.org/10.1007/s10725-015-0028-z

Jini D, Joseph B (2017) Physiological mechanism of salicylic acid for alleviation of salt stress in rice. Rice Sci 24(2):97–108. https://doi.org/10.1016/j.rsci.2016.07.007

Kaur H, Hussain SJ, Kaur G, Poor P, Alamri S, Siddiqui H, Khan MIR (2022) Salicylic acid improves nitrogen fixation, growth, yield and antioxidant defence mechanisms in chickpea genotypes under salt stress. J Plant Growth Regul 41:2034–2047. https://doi.org/10.1007/s00344-022-10592-7

Khalvandi M, Siosemardeh A, Roohi E, Keramati S (2021) Salicylic acid alleviated the effect of drought stress on photosynthetic characteristics and leaf protein pattern in winter wheat. Heliyon 7(1):e05908. https://doi.org/10.1016/j.heliyon.2021.e05908

Khan MIR, Fatma M, Per TS, Anjum NA, Khan NA (2015) Salicylic acid-induced abiotic stress tolerance and underlying mechanisms in plants. Front Plant Sci 6:135066. https://doi.org/10.3389/fpls.2015.00462

Kirchhoff H, Haase W, Wegner S, Danielsson R, Ackermann R, Albertsson PA (2007) Low-light-induced formation of semicrystalline photosystem II arrays in higher plant chloroplasts. Biochemistry 46(39):11169–11176. https://doi.org/10.1021/bi700748y

Kumar P, Sharma PK (2020) Soil salinity and food security in India. Front Sustain Food Syst 4:533781. https://doi.org/10.3389/fsufs.2020.533781

Lee S, Kim SG, Park CM (2010) Salicylic acid promotes seed germination under high salinity by modulating antioxidant activity in *Arabidopsis*. New Phytol 188(2):626–637. https://doi.org/10.1111/j.1469-8137.2010.03378.x

Lefevere H, Bauters L, Gheysen G (2020) Salicylic acid biosynthesis in plants. Front Plant Sci 11:338. https://doi.org/10.3389/fpls.2020.00338

Leon J, Lawton MA, Raskin I (1995) Hydrogen peroxide stimulates salicylic acid biosynthesis in tobacco. Plant Physiol 108(4):1673–1678. https://doi.org/10.1104/pp.108.4.1673

Li T, Hu Y, Du X, Tang H, Shen C, Wu J (2014) Salicylic acid alleviates the adverse effects of salt stress in *Torreya grandis* cv. Merrillii seedlings by activating photosynthesis and enhancing antioxidant systems. PLoS One 9(10):e109492. https://doi.org/10.1371/journal.pone.0109492

Li J, Liu LN, Meng Q, Fan H, Sui N (2020) The roles of chloroplast membrane lipids in abiotic stress responses. Plant Signal Behav 15(11):1807152. https://doi.org/10.1080/15592324.2020.1807152

Liu T, Amanullah S, Xu H, Gao P, Du Z, Hu X et al (2023) RNA-Seq identified putative genes conferring photosynthesis and root development of melon under salt stress. Genes 14(9):1728. https://doi.org/10.3390/genes14091728

Lu J, Xu Y, Wang J, Singer SD, Chen G (2020) The role of triacylglycerol in plant stress response. Plan Theory 9(4):472. https://doi.org/10.3390/plants9040472

Lynch DV, Dunn TM (2004) An introduction to plant sphingolipids and a review of recent advances in understanding their metabolism and function. New Phytol 161(3):677–702. https://doi.org/10.1111/j.1469-8137.2004.00992.x

Mahmud S, Sharmin S, Chowdhury BL, Hossain MA (2017) Effect of salinity and alleviating role of methyl Jasmonate in some rice varieties. Asian J Plant Sci:87–93. https://doi.org/10.3923/ajps.2017.87.93

Mandal S, Raju R, Kumar A, Kumar P, Sharma PC (2018) Current status of research, technology response and policy needs of salt-affected soils in India–a review. J Indian Soc Coast Agric Res 36(2)

Mishra A, Tanna B (2017) Halophytes: potential resources for salt stress tolerance genes and promoters. Front Plant Sci 8:252363. https://doi.org/10.3389/fpls.2017.00829

Mittler R (2002) Oxidative stress, antioxidants and stress tolerance. Trends Plant Sci 7(9):405–410. https://doi.org/10.1016/S1360-1385(02)02312-9

Munns R, Tester M (2008) Mechanisms of salinity tolerance. Annu Rev Plant Biol 59(1):651–681. https://doi.org/10.1146/annurev.arplant.59.032607.092911

Mustafa NR, Kim HK, Choi YH, Erkelens C, Lefeber AW, Spijksma G, Verpoorte R (2009) Biosynthesis of salicylic acid in fungus elicited *Catharanthus roseus* cells. Phytochemistry 70(4):532–539. https://doi.org/10.1016/j.phytochem.2009.01.009

Nazar R, Iqbal N, Syeed S, Khan NA (2011) Salicylic acid alleviates decreases in photosynthesis under salt stress by enhancing nitrogen and sulfur assimilation and antioxidant metabolism differentially in two mungbean cultivars. J Plant Physiol 168(8):807–815. https://doi.org/10.1016/j.jplph.2010.11.001

Nazar R, Umar S, Khan NA, Sareer O (2015) Salicylic acid supplementation improves photosynthesis and growth in mustard through changes in proline accumulation and ethylene formation under drought stress. S Afr J Bot 98:84–94. https://doi.org/10.1016/j.sajb.2015.02.005

Omoto E, Iwasaki Y, Miyake H, Taniguchi M (2015) Salinity induces membrane structure and lipid changes in maize mesophyll and bundle sheath chloroplasts. Physiol Plant 157(1):13–23. https://doi.org/10.1111/ppl.12404

Pai R, Sharma PK (2022) Exogenous application of salicylic acid mitigates salt stress in rice seedlings by regulating plant water status and preventing oxidative damage. Environ Exp Biol 20(4):193–204. https://doi.org/10.22364/eeb.20.18

Pai R, Sharma PK (2024) Exogenous supplementation of salicylic acid ameliorates salt-induced membrane leakage, ion homeostasis and oxidative damage in Sorghum seedlings. Biologia 79(1):23–43. https://doi.org/10.1007/s11756-023-01554-9

Panche AN, Diwan AD, Chandra SR (2016) Flavonoids: an overview. J Nutr Sci 5:e47. https://doi.org/10.1017/jns.2016.41

Parida AK, Das AB (2005) Salt tolerance and salinity effects on plants: a review. Ecotoxicol Environ Saf 60(3):324–349. https://doi.org/10.1016/j.ecoenv.2004.06.010

Perlikowski D, Krajewski P, Kosmala A (2016) Remodeling of leaf cellular glycerolipid composition under drought and re-hydration conditions in grasses from the Lolium-Festuca complex. Front Plant Sci 7:179370. https://doi.org/10.3389/fpls.2016.01027

Pirasteh-Anosheh H, Emam Y (2018) Modulation of oxidative damage due to salt stress using salicylic acid in *Hordeum vulgare*. Arch Agron Soil Sci 64(9):1268–1277. https://doi.org/10.1080/03650340.2018.1423556

Ramani B, Zorn H, Papenbrock J (2004) Quantification and fatty acid profiles of sulfolipids in two halophytes and a glycophyte grown under different salt concentrations. Zeitschrift für Naturforschung C 59(11–12):835–842. https://doi.org/10.1515/znc-2004-11-1212

Reszczyńska E, Hanaka A (2020) Lipids composition in plant membranes. Cell Biochem Biophys 78(4):401–414. https://doi.org/10.1007/s12013-020-00947-w

Rozentsvet O, Shuyskaya E, Bogdanova E, Nesterov V, Ivanova L (2022) Effect of salinity on leaf functional traits and chloroplast lipids composition in two C3 and C4 Chenopodiaceae halophytes. Plan Theory 11(19):2461. https://doi.org/10.3390/plants11192461

Salama KH, Mansour MMF (2015) Choline priming-induced plasma membrane lipid alterations contributed to improved wheat salt tolerance. Acta Physiol Plant 37:1–7

Sharma P, Jha AB, Dubey RS, Pessarakli M (2012) Reactive oxygen species, oxidative damage, and antioxidative defense mechanism in plants under stressful conditions. J Bot 2012:1. https://doi.org/10.1155/2012/217037

Shavrukov Y (2013) Salt stress or salt shock: which genes are we studying? J Exp Bot 64(1):119–127. https://doi.org/10.1093/jxb/ers316

Sui N, Han G (2014) Salt-induced photoinhibition of PSII is alleviated in halophyte *Thellungiella halophila* by increases of unsaturated fatty acids in membrane lipids. Acta Physiol Plant 36:983–992. https://doi.org/10.1007/s11738-013-1477-5

Sui N, Li M, Li K, Song J, Wang BS (2010) Increase in unsaturated fatty acids in membrane lipids of Suaeda salsa L. enhances protection of photosystem II under high salinity. Photosynthetica 48:623–629. https://doi.org/10.1007/s11099-010-0080-x

Upchurch RG (2008) Fatty acid unsaturation, mobilization, and regulation in the response of plants to stress. Biotechnol Lett 30:967–977. https://doi.org/10.1007/s10529-008-9639-z

Verberne MC, Verpoorte R, Bol J, Mercado-Blanco J, Linthorst HJ (2000) Overproduction of salicylic acid in plants by bacterial transgenes enhances pathogen resistance. Nat Biotechnol 18(17):779–783. https://doi.org/10.1038/77347

Wang S, Uddin MI, Tanaka K, Yin L, Shi Z, Qi Y, Mano J, Matsui K, Shimomura N, Sakaki T, Deng X, Zhang S (2014) Maintenance of chloroplast structure and function by overexpression of the rice Monogalactosyldiacylglycerol synthase gene leads to enhanced salt tolerance in tobacco. Plant Physiol 165(3):1144–1155. https://doi.org/10.1104/pp.114.238899

Wu J, Seliskar DM, Gallagher JL (2005) The response of plasma membrane lipid composition in callus of the halophyte *Spartina patens* (Poaceae) to salinity stress. Am J Bot 92(5):852–858. https://doi.org/10.3732/ajb.92.5.852

Yancey PH (2005) Organic osmolytes as compatible, metabolic and counteracting cytoprotectants in high osmolarity and other stresses. J Exp Biol 208(15):2819–2830. https://doi.org/10.1242/jeb.01730

Yang Y, Benning C (2018) Functions of triacylglycerols during plant development and stress. Curr Opin Biotechnol 49:191–198. https://doi.org/10.1016/j.copbio.2017.09.003

Yildirim E, Turan M, Guvenc I (2008) Effect of foliar salicylic acid applications on growth, chlorophyll, and mineral content of cucumber grown under salt stress. J Plant Nutr:593–612. https://doi.org/10.1080/01904160801895118

Zamani S, Bybordi A, Khorshidi MB, Nezami T (2010) Effects of NaCl salinity levels on lipids and proteins of canola (*Brassica napus* L.) cultivars. Adv Environ Biol:397–405

Zhao S, Zhang Q, Liu M, Zhou H, Ma C, Wang P (2021) Regulation of plant responses to salt stress. Int J Mol Sci 22(9):4609. https://doi.org/10.3390/ijms22094609

Zorova LD, Popkov VA, Plotnikov EY, Silachev DN, Pevzner IB, Jankauskas SS, Babenko VA, Zorov SD, Balakireva AV, Juhaszova M, Sollott SJ, Zorov DB (2018) Mitochondrial membrane potential. Anal Biochem 552:50–59. https://doi.org/10.1016/j.ab.2017.07.009

Chapter 6
Bioprospecting of Plant Growth-Promoting Fungi (PGPF) for Remediating Plant Stress

Smriti Chouhan, Nandini Singh, and Anil Prakash

Abstract Plant growth-promoting fungi (PGPF) are rhizosphere-colonizing fungi that enhance plant health and productivity through mutualistic interactions. This diverse group encompasses various taxonomic genera including *Aspergillus*, *Fusarium*, *Penicillium*, *Piriformospora*, *Phoma*, *Talaromyces*, and *Trichoderma*, which thrive across ecological niches while benefiting their hosts. By co-evolving with plants, PGPF employ direct and indirect mechanisms to stimulate growth and development. Direct benefits include the mobilization of essential micro- and macronutrients, thereby promoting robust plant growth. Indirectly, PGPF synthesize defense-related enzymes, volatile organic compounds, and phytohormones, which suppress phytopathogens and bolster plant resilience against biotic and abiotic stressors. This chapter reviews recent advances in understanding plant–PGPF symbioses, emphasizing their role in enhancing nutrient acquisition, modulating phytohormone signaling, and inducing systemic resistance (ISR) to combat pathogen incursion. Furthermore, it explores the translational potential of PGPF in sustainable agriculture, highlighting their capacity to improve crop productivity and environmental stress tolerance through phytostimulation and integrated stress mitigation strategies. By bridging ecological insights with agricultural applications, this discussion underscores PGPF as pivotal agents in developing resilient, low-input agroecosystems.

Keywords PGPF · Plant growth promotion · Biotic and abiotic stress · Sustainable agriculture

---

S. Chouhan (✉)
Babulala Gaur Govt. P.G. College, BHEL, Bhopal, MP, India
e-mail: hegpgcbhelbho@mp.gov.in

N. Singh · A. Prakash
Department of Microbiology, Barkatullah University, Bhopal, MP, India

N. K. Arora et al. (eds.), *Environmental Sustainability and Biotechnology: Opportunities and Challenges*, Microorganisms for Sustainability 46,
https://doi.org/10.1007/978-981-96-9025-1_6

6.1 Introduction

By 2050, the global population is projected to reach 9.7 billion—a 24% increase from 2023—necessitating a 70% surge in food production to avert widespread insecurity (Tripathi et al. 2019; Leisinger et al. 2002). Conventional agricultural intensification, reliant on synthetic fertilizers and pesticides, has temporarily boosted yields but at profound ecological costs. These practices degrade soil health, contaminate water systems, disrupt microbial ecosystems, and diminish long-term farmland productivity (Wimalawansa and Wimalawansa 2014). Furthermore, pre- and post-harvest crop losses, exacerbated by climate change and pathogen resistance, underscore the urgency for sustainable alternatives that reconcile productivity with planetary health.

Rhizosphere microorganisms, particularly fungi, have emerged as keystones in this transition. Their metabolic versatility enables symbiotic relationships with plants, mediated by root exudates that recruit beneficial microbial communities (Vacheron et al. 2013). Among these, plant growth-promoting fungi (PGPF)—including *Trichoderma*, *Penicillium*, *Phoma*, and *Fusarium* species—enhance nutrient bioavailability, stimulate stress resilience, and suppress pathogens through eco-friendly mechanisms (Hyakumachi 1994; Zhou et al. 2018). Unlike chemical inputs, PGPF align with organic farming principles, offering dual benefits of yield enhancement and ecological preservation (Naziya et al. 2019).

PGPF employ a synergistic matrix of direct and indirect strategies to optimize plant performance. Direct mechanisms include mineral solubilization (e.g., phosphorus and iron), phytohormone synthesis (auxins and gibberellins), and enzymatic activation of soil nutrients. Indirectly, they induce systemic resistance (ISR) via volatile organic compounds (VOCs), antagonize pathogens through antibiosis, and mitigate abiotic stressors like drought and salinity (Hossain et al. 2017). Despite their potential, fewer than 5% of identified PGPF strains are commercially utilized, hindered by challenges in field efficacy, formulation stability, and compatibility with existing agricultural practices (Lyn et al. 2010). Developing robust bio-inoculants—viable across diverse soils and climates—requires innovations in microbial consortia design, encapsulation technologies, and omics-driven strain optimization.

This chapter examines recent advances in PGPF research, focusing on their mechanisms of nutrient mobilization, systemic resistance induction, and rhizosphere colonization. It explores their long-term impacts on soil biodiversity, carbon sequestration, and yield stability, while addressing translational barriers such as host specificity, bioformulation challenges, and scalability in precision agriculture systems. By bridging lab-to-field gaps, the discussion underscores PGPF's potential to redefine agroecosystem management, offering a sustainable roadmap to replace chemical dependencies with microbiome-driven solutions.

6.2 PGPF and Plant Growth Promotion

Plant growth-promoting fungi are recognized for their broad-spectrum beneficial associations with diverse plant taxa, mediated by multifunctional mechanisms that enhance host fitness (devi et al. 2020). Studies demonstrate that PGPF significantly improve key growth metrics, including biomass accumulation, nutrient uptake efficiency, and photosynthetic capacity, through both direct and indirect modes of action (Murali et al. 2021; Adedayo and Babalola 2023). Symbiotic interactions between PGPF and plants modulate critical physiological and developmental processes, such as seed germination, seedling vigor index, root architecture plasticity (e.g., lateral root initiation and root hair density), shoot apical dominance, flowering phenology, and yield-related traits. Mechanistically, specific PGPF strains secrete bioactive metabolites, including phytohormones (e.g., auxins and gibberellins) and volatile organic compounds (VOCs), which act as signaling molecules to upregulate growth-related gene expression and metabolic pathways (Harman et al. 2004; Naznin et al. 2013). Concurrently, PGPF enhance nutrient bioavailability via phosphate solubilization and siderophore-mediated iron chelation, mitigate abiotic stress through osmolyte synthesis (e.g., proline and trehalose), and suppress phytopathogens via antibiosis (e.g., antifungal metabolites) and competitive niche exclusion (Wakelin et al. 2007; Hossain et al. 2014). These functional outcomes arise not from singular pathways but from synergistic interactions between PGPF-derived biochemical agents, host genotype-specific responses, and environmental variables such as soil microbiota composition and edaphic conditions (Prashar and Vandenberg 2017). Consequently, the efficacy of PGPF-mediated growth promotion is contingent upon compatibility between fungal functional traits, host physiological thresholds, and agroecological contexts (Verma et al. 2019).

6.3 Antagonistic Behavior of PGPF

Plant growth-promoting fungi employ antagonistic strategies as a primary defense mechanism to curtail the proliferation and dissemination of phytopathogens. This antagonism manifests through microbial predation, niche competition, and the biosynthesis of antimicrobial metabolites, including antibiotics and lytic enzymes (Fig. 6.1) (Mohiddin et al. 2017). Central to this mechanism is the enzymatic degradation of pathogen cell walls, facilitated by chitinases, proteases, and β-1,3-glucanases, which disrupt pathogen integrity and viability (Benhamou and Chet 1997). Several studies highlight the efficacy of diverse PGPF strains, such as *Trichoderma harzianum*, *Phoma* sp., *Fusarium equiseti*, and *Penicillium simplicissimum*, in suppressing pathogens like *Rhizoctonia solani*, *Pythium irregulare*, and *Fusarium oxysporum* through direct antagonism (Meera et al. 1994; Thambugala et al. 2020).

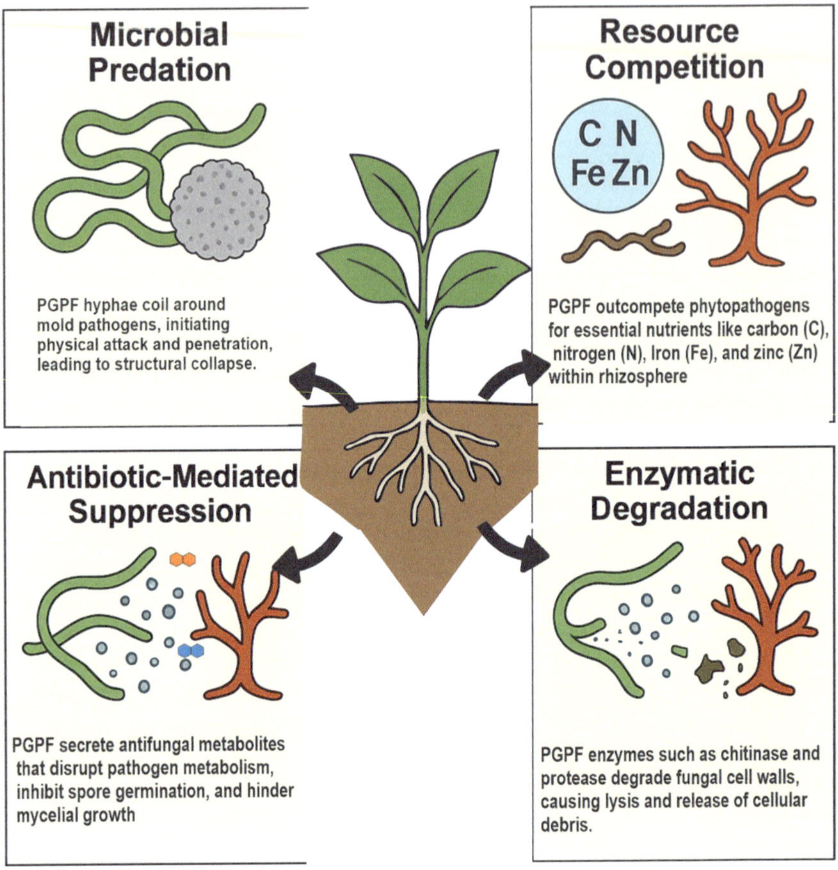

Fig. 6.1 Antagonistic behavior of PGPF

At the molecular level, PGPF-induced predation involves the activation of chitinase genes, which impair pathogen growth by hydrolyzing chitin-rich cell walls (Mohiddin et al. 2017). Additionally, antibiotic production serves as a critical antagonistic tool. For instance, *Gliocladium virens* synthesizes gliovirin, a potent antifungal compound that selectively inhibits *Pythium* spp. (Howell and Stipanovic 1983). Nonpathogenic *Fusarium oxysporum* strains further exemplify resource-mediated competition, suppressing pathogenic *Fusarium* strains by occupying root colonization sites, sequestering nutrients, and inhibiting chlamydospore germination (Kaur et al. 2011). Similarly, *T. harzianum* demonstrates broad-spectrum antagonism against *Pythium* sp., *R. solani*, and *F. oxysporum*, underscoring its biocontrol versatility (Murali et al. 2012; Kaur et al. 2011).

Notably, rhizosphere fungi like *Aspergillus fischeri* exhibit antagonistic activity against *Botrytis cinerea*, further validating the ecological role of PGPF in pathogen suppression (Moreno et al. 2003). While extensive research corroborates the antagonistic potential of PGPF against pathogens (Tóth et al. 2020; Bhatia et al. 2016; Vinale et al. 2008), antagonism is not an exclusive prerequisite for induced systemic resistance (ISR) (discussed later). Non-antagonistic PGPF can also prime plant defenses, suggesting that ISR involves synergistic interactions beyond direct microbial antagonism, such as signaling molecule modulation or host metabolic reprogramming (Hyakumachi and Kubota 2004; Naziya et al. 2020). Thus, while antagonism remains a basis of PGPF-mediated biocontrol, its role exists within a broader framework of plant–microbe cross talk essential for sustainable disease management.

6.4 Resistance Induction by PGPF

Induced resistance (IR) represents a physiological strategy wherein plants amplify their innate defense mechanisms upon exposure to biotic stressors, orchestrated by elicitors or microbial inducers. Broadly categorized into systemic acquired resistance (SAR) and induced systemic resistance (ISR), these pathways are distinguished by their signaling molecules and defense regulation mechanisms. SAR is typically triggered by pathogen infection or salicylic acid (SA)-mediated signaling, leading to the accumulation of pathogenesis-related (PR) proteins, while ISR, often mediated by jasmonic acid (JA) and ethylene (ET) pathways, is primed by beneficial microbes like PGPF (Sánchez-Montesinos et al. 2020; Pieterse et al. 2014).

PGPF enhance plant resilience by systemically activating defense responses in tissues distal to the initial infection site. For instance, root colonization by *T. harzianum*, *Penicillium simplicissimum*, or nonpathogenic *F. oxysporum* strains primes shoots and leaves for heightened resistance against airborne and soilborne pathogens, such as *B. cinerea* and *P. irregulare* (Tucci et al. 2011; Hossain et al. 2007; Gajera and Vakharia 2012). This systemic priming is linked to the elevated activity of defense enzymes like phenylalanine ammonia-lyase (PAL), polyphenol oxidase (PPO), and β-1,3-glucanase (GLU), which fortify cell walls and degrade pathogen components (Loganathan et al. 2010; Murali and Amruthesh 2015).

6.4.1 *Induced Systemic Resistance Activation*

Nonpathogenic PGPF strains, including *Phoma* spp., *Aspergillus ustus*, and *Talaromyces flavus*, suppress diseases like Verticillium wilt and anthracnose by competing for nutrients, colonizing root niches, and secreting elicitors that upregulate defense genes (Koike et al. 2001; Halifu et al. 2019; Yoshioka et al. 2012). VOCs from *Cladosporium* sp. and *Ampelomyces* sp. induce ISR in *Arabidopsis* against *Pseudomonas syringae*, highlighting the role of airborne signaling (Kanchiswamy et al. 2015).

6.4.2 Systematic Acquired Resistance

Systemic acquired resistance is a plant-wide defense response triggered by localized pathogen infection, typically involving cell death at the infection site (Hunt et al. 1996). This response leads to the activation of defense-related genes throughout the plant, providing long-lasting and broad-spectrum resistance against diverse pathogens (Ulaganathan et al. 2004). Unlike ISR, which is usually mediated by nonpathogenic microbes and relies on jasmonic acid and ethylene signaling, SAR is primarily associated with salicylic acid (SA) accumulation and pathogenesis-related (PR) protein expression, making it a distinct and robust form of plant immunity. Studies have shown that similar to beneficial microbes, PGPF also contribute to SAR. For example, *Trichoderma asperelloides* enhances SAR in *Arabidopsis* under low-nitrogen conditions, while *T. harzianum* triggers PR proteins in pepper plants infected with *Phytophthora capsici* (Saxena et al. 2020; Vernooij et al. 1994). Glycoprotein elicitors from *Phytophthora* spp. and *Cladosporium fulvum* further stimulate PR protein synthesis, conferring resistance against *Cercosporanicotianae* and tobacco mosaic virus (Patel et al. 2020; Bartnicki-Garcia 1966).

Fungal-derived elicitors, such as cerebrosides from *F. oxysporum* and the Sm1 protein from *Trichoderma virens*, directly activate defense pathways, reducing anthracnose severity in chili and tomato wilt (Yamagiwa et al. 2011; Djonovic et al. 2007). Bioformulations of *Trichoderma asperellum* and *T. harzianum* demonstrate practical efficacy, suppressing damping-off in sugar beet, bacterial wilt in tomato, and Fusarium wilt in pumpkin through ISR-mediated enzyme activation (Loganathan et al. 2010; Kishimoto et al. 2010; Yoshioka et al. 2012). Similarly, chitin-based formulations combined with *Trichoderma* spp. mitigate head rot and root-knot diseases in cabbage (Kakvan et al. 2013).

While oligosaccharides and glycosphingolipids from PGPF are recognized as potent elicitors, the precise molecular interplay between fungal signals and plant receptors remains unclear (Miao et al. 2012; Ezziyyani et al. 2017). Nevertheless, PGPF-mediated IR exemplifies a sustainable biocontrol strategy, leveraging both antagonism and systemic priming to enhance crop resilience and yield (Naziya et al. 2020; Konappa et al. 2018).

6.5 PGPF and Abiotic Stress Management

Abiotic stressors—including salinity, drought, extreme temperatures, heavy metal toxicity, and flooding—pose significant threats to global crop productivity, with yield losses exceeding 70% in afflicted regions (Jin et al. 2019; He et al. 2018). Plant growth-promoting fungi mitigate these challenges by enhancing plant physiological resilience through multifaceted mechanisms such as osmotic adjustment, antioxidant defense activation, and stress-responsive phytohormone modulation.

6.5.1 Osmotic Balance and Ion Homeostasis

Plant growth-promoting fungi such as *T. harzianum* and *Penicillium* spp. have emerged as critical allies in mitigating abiotic stress in plants. These fungi enhance salt tolerance in crops like cucumber, sesame, and Indian mustard by upregulating key antioxidant enzymes—including superoxide dismutase (SOD), catalase, and ascorbate peroxidase (APX)—which collectively scavenge reactive oxygen species (ROS) and mitigate oxidative damage (Guler et al. 2016). Additionally, they promote the accumulation of osmoprotectants such as proline and betaine, which help maintain cellular osmotic balance under salt stress conditions (Radhakrishnan et al. 2014; Qi and Zhao 2013).

T. asperellum has been reported to enhance drought resilience in maize by elevating antioxidant defense mechanisms and significantly reducing ROS accumulation, thereby preserving cellular integrity during water scarcity (Saxena et al. 2015). In the context of heavy-metal stress, *T. virens* plays a pivotal role in alleviating cadmium toxicity in tobacco plants. It does so by reducing lipid peroxidation and promoting glutathione recycling—two key processes that prevent membrane damage and support redox homeostasis (Montero-Barrientos et al. 2010).

6.5.2 Phytohormone Regulation

Ethylene, a key hormone, while essential for stress responses, can become deleterious at elevated levels under abiotic stress. PGPF alleviate this problem by producing 1-aminocyclopropane-1-carboxylate (ACC) deaminase, an enzyme that hydrolyzes ACC, the immediate precursor of ethylene, into α-ketobutyrate and ammonia (Viterbo et al. 2010). This activity effectively lowers ethylene accumulation in plant tissues, thus mitigating its inhibitory effects on root growth and overall plant development.

For instance, *Trichoderma longibrachiatum* has been reported to enhance wheat growth under saline conditions by downregulating ethylene biosynthesis through ACC deaminase activity (Zhang et al. 2019). Concurrently, it upregulates the expression of Na^+/H^+ antiporter genes (e.g., NHX1), which facilitate the sequestration of excess Na^+ into vacuoles, thereby minimizing cytosolic sodium toxicity and maintaining ionic balance (Zhang et al. 2016; Ortíz-Castro et al. 2008).

Beyond salinity, similar phytohormonal modulation by other *Trichoderma* species has been observed under drought and heavy-metal stress, where ethylene suppression helps sustain root elongation and nutrient uptake (Jaroszuk-Ściseł et al. 2019). This hormone-regulatory mechanism, coupled with enhanced antioxidant defenses, positions PGPF as crucial bio-inoculants for promoting plant vigor and resilience under diverse stress conditions (Rauf et al. 2021). Such integrative physiological responses highlight the potential of PGPF-based formulations in climate-resilient agriculture and stress-adaptive crop production systems. While

PGPF-mediated ROS scavenging and ACC deaminase activity are well-documented, the molecular interplay between fungal elicitors (e.g., cerebrosides and Sm1 protein) and plant stress receptors remains unclear. Additionally, optimizing bioformulations for diverse agroclimatic conditions requires further exploration (Nishimura et al. 2003; Vieira et al. 2017).

6.5.3 Heavy-Metal Detoxification

PGPF play a pivotal role in mitigating heavy-metal toxicity in plants through dual strategies of extracellular sequestration and intracellular detoxification. Fungi such as *Aspergillus* spp., *Penicillium* spp., and *Microsphaeropsis* spp. immobilize metals like zinc (Zn), copper (Cu), and cadmium (Cd) via biosorption to fungal cell walls or chelation by metallothioneins and phytochelatins (Ahmad et al. 2015; Chibuike and Obiora 2014). In few instances, they play dual role as metal stress remediation and plant hormone producer. In a study, *Penicillium resedanum* not only reduces metal bioavailability but also stimulates gibberellin (GA) biosynthesis in chili plants, enhancing root elongation and biomass production under metal-stressed conditions (Mastouri et al. 2010).

Phytochelatins (PCs) play a critical role in plant heavy-metal detoxification by acting as chelators, binding to toxic metals such as Cd, Zn, Cu, and lead (Pb) and sequestering them into vacuolar compartments, thereby preventing cellular damage. Central to this process is the enzyme phytochelatin synthase (PCS), which catalyzes the synthesis of phytochelatins from glutathione (GSH), a sulfur-rich tripeptide. The activity of PCS is directly induced by the presence of heavy metals, triggering a surge in phytochelatin production to neutralize metal toxicity. PGPF and arbuscular mycorrhizal fungi (AMF) enhance this defense mechanism by modulating plant tolerance to heavy metals. PGPF indirectly promotes phytochelatin synthesis by improving sulfur uptake, a nutrient essential for GSH biosynthesis, and may also directly regulate PCS gene expression or enzymatic activity. This symbiotic relationship not only boosts glutathione availability but also amplifies the plant's capacity to compartmentalize metals, effectively reducing oxidative stress. By strengthening phytochelatin-mediated detoxification, PGPF significantly improve plant resilience in metal-contaminated environments, enabling survival and growth under otherwise toxic conditions.

6.6 Ecological and Agricultural Relevance of PGPF

PGPF exhibit remarkable adaptability to saline ecosystems, surviving NaCl concentrations as high as 20% through a combination of osmolyte biosynthesis and membrane stabilization strategies. Under salinity stress, PGPF such as *Aspergillus* and *Penicillium* spp. accumulate compatible solutes like glycerol, trehalose, and proline, which maintain cellular osmotic balance by counteracting ion-induced water loss (Gunde-Cimerman et al. 2000). Concurrently, ergosterol and other sterols in fungal membranes enhance lipid bilayer stability, preserving membrane integrity and functionality under ionic stress (Moura et al. 2010; Turk et al. 2004). Beyond mere survival, these fungi actively bolster crop resilience in saline and drought-prone environments. For instance, *Trichoderma hamatum* enhances cocoa drought tolerance by upregulating aquaporin genes and improving root hydraulic conductivity, while *Penicillium chrysogenum* primes sesame plants for salinity resistance by elevating antioxidant enzymes (SOD, CAT) and reducing lipid peroxidation (Qi and Zhao 2013). Hydroponic trials reveal that PGPF can reduce fertilizer demand under salt stress. For instance, in a study, *Penicillium olsonii* strain A3, isolated from the rhizosphere of the halophyte *Aeluropus littoralis*, as a novel PGPF under 1 M NaCl salt stress, significantly improved plant growth, chlorophyll content, and antioxidant enzyme activities (CAT, SOD), while reducing Na^+ accumulation and increasing K^+ uptake in tobacco plant (Tarroum et al. 2022). A3's culture filtrate (A3CFF) also upregulated key stress-responsive genes (e.g., *NtSOS1*, *NtNHX1*, and *NtHKT1*) and growth-related genes involved in nitrogen and hormone biosynthesis. Notably, A3CFF allowed for a 50% reduction in chemical fertilizer use without compromising plant performance, highlighting its potential as a sustainable biofertilizer (Tarroum et al. 2022).

Another study by Sheteiwy et al. (2022) demonstrates that JA priming significantly enhances wheat tolerance to high salinity stress by activating a suite of physiological and molecular defense mechanisms. Application of 60 μM JA improved growth, yield, and photosynthetic efficiency in wheat cultivars ZM9 and YM25 while also preserving chloroplast ultrastructure under saline conditions. JA priming enhanced osmotic adjustment through increased proline accumulation, maintained water balance, and promoted ion homeostasis by reducing Na^+ and enhancing K^+ levels, partly via upregulation of transporter genes like SOS1, NHX2, and HVP1. Additionally, JA modulated hormonal balance by increasing cytokinins and auxin while reducing abscisic acid levels (Sheteiwy et al. 2022).

Hence, all these approaches not only improve crop yields in marginal soils but also reduce the reliance on chemical fertilizers, positioning PGPF as keystones in sustainable agriculture for climate-vulnerable regions.

6.7 Conclusion

PGPF offer a sustainable and cost-effective strategy to enhance crop productivity, soil health, and disease resistance while reducing reliance on agrochemicals. By synthesizing phytohormones, defense enzymes (e.g., chitinases, glucanases), and antibiotics, PGPF prime systemic resistance against biotic stressors and mitigate abiotic stress through ROS scavenging, osmolyte accumulation, and ethylene modulation. Their role in nutrient solubilization and rhizosphere engineering further improves plant vigor and yield. Deciphering PGPF–plant signaling pathways—particularly during pathogen challenges—reveal actionable insights for optimizing induced resistance mechanisms. As eco-friendly alternatives to synthetic inputs, PGPF-based bioformulations demonstrate scalable potential for long-term agricultural resilience, balancing productivity with environmental stewardship.

References

Adedayo AA, Babalola OO (2023) Fungi that promote plant growth in the rhizosphere boost crop growth. J Fungi 9(2):239. https://doi.org/10.3390/jof9020239

Ahmad P, Hashem A, Abd-Allah EF, Alqarawi AA, John R, Egamberdieva D, Gucel S (2015) Role of *Trichoderma harzianum* in mitigating NaCl stress in Indian mustard (*Brassica juncea* L) through antioxidative defense system. Front Plant Sci 6:868. https://doi.org/10.3389/fpls.2015.00868

Bartnicki-Garcia S (1966) Chemistry of hyphal walls of Phytophthora. Microbiology 42(1):57–69. https://doi.org/10.1099/13500872-42-1-57

Benhamou N, Chet I (1997) Cellular and molecular mechanisms involved in the interaction between *Trichoderma harzianum* and *Pythium ultimum*. Appl Environ Microbiol 63(5):2095–2099. https://doi.org/10.1128/aem.63.5.2095-2099.1997

Bhatia P, Harsh NSK, Dubey RC, Kaushik PK (2016) Antagonistic activity of rhizosphere fungi against *Fusarium solani* causing wilt in *Dalbergia sissoo*. J Sci Res Adv 3:352–357

Chibuike GU, Obiora SC (2014) Heavy metal polluted soils: effect on plants and bioremediation methods. Appl Environ Soil Sci 2014(1):752708. https://doi.org/10.1155/2014/752708

Devi R, Kaur T, Kour D, Rana KL, Yadav A, Yadav AN (2020) Beneficial fungal communities from different habitats and their roles in plant growth promotion and soil health. Microb Biosyst 5(1):21–47. https://doi.org/10.21608/mb.2020.32802.1016

Djonovic S, Vargas WA, Kolomiets MV, Horndeski M, Wiest A, Kenerley CM (2007) A proteinaceous elicitor Sm1 from the beneficial fungus *Trichoderma virens* is required for induced systemic resistance in maize. Plant Physiol 145(3):875–889. https://doi.org/10.1104/pp.107.103689

Ezziyyani M, Hamdache A, Egea-Gilabert C, Requena ME, Candela ME, Mater J (2017) Production of pathogenesis-related proteins during the induction of resistance to *Phytophthora capsici* in pepper plants treated with *Burkholderia cepacia* and *Trichoderma harzianum* in combination compatible. J Mater Environ Sci 8:4785–4795

Gajera HP, Vakharia DN (2012) Production of lytic enzymes by *Trichoderma* isolates during in vitro antagonism with *Aspergillus niger*, the causal agent of collar rot of peanut. Braz J Microbiol 43:43–52. https://doi.org/10.1590/S1517-83822012000100005

Guler NS, Pehlivan N, Karaoglu SA, Guzel S, Bozdeveci A (2016) *Trichoderma atroviride* ID20G inoculation ameliorates drought stress-induced damages by improving antioxidant defence in maize seedlings. Acta Physiol Plant 38:1–9. https://doi.org/10.1007/s11738-016-2153-3

Gunde-Cimerman N, Zalar P, de Hoog S, Plemenitaš A (2000) Hypersaline waters in salterns–natural ecological niches for halophilic black yeasts. FEMS Microbiol Ecol 32(3):235–240. https://doi.org/10.1111/j.1574-6941.2000.tb00716.x

Halifu S, Deng X, Song X, Song R (2019) Effects of two *Trichoderma* strains on plant growth, rhizosphere soil nutrients, and fungal community of *Pinus sylvestris* var. mongolica annual seedlings. Forests 10(9):758. https://doi.org/10.3390/f10090758

Harman GE, Howell CR, Viterbo A, Chet I, Lorito M (2004) Trichoderma species—opportunistic, avirulent plant symbionts. Nature reviews microbiology 2(1):43–56. https://doi.org/10.1038/nrmicro797

He M, He CQ, Ding NZ (2018) Abiotic stresses: general defenses of land plants and chances for engineering multistress tolerance. Front Plant Sci 9:1771.https://doi.org/10.3389/fpls.2018.01771

Hossain MM, Sultana F, Kubota M, Koyama H, Hyakumachi M (2007) The plant growth-promoting fungus *Penicillium simplicissimum* GP17-2 induces resistance in Arabidopsis thaliana by activation of multiple defense signals. Plant Cell Physiol 48(12):1724–1736. https://doi.org/10.1093/pcp/pcm144

Hossain MM, Sultana F, Miyazawa M, Hyakumachi M (2014) The plant growth-promoting fungus *Penicillium* spp. GP15-1 enhances growth and confers protection against damping-off and anthracnose in the cucumber. Journal of Oleo Science 63(4):391–400. https://doi.org/10.5650/jos.ess13143

Hossain MM, Sultana F, Islam S (2017) Plant growth-promoting fungi (PGPF): phytostimulation and induced systemic resistance. In: Plant-microbe Interactions in Agro-ecological Perspectives: volume 2: Microbial Interactions and Agro-ecological Impacts. Springer, pp 135–191. https://doi.org/10.1007/978-981-10-6593-4_6

Howell CR, Stipanovic RD (1983) Gliovirin, a new antibiotic from *Gliocladium virens*, and its role in the biological control of *Pythium ultimum*. Can J Microbiol 29(3):321–324. https://doi.org/10.1139/m83-053

Hunt MD, Ryals JA, Reinhardt D (1996) Systemic acquired resistance signal transduction. Crit Rev Plant Sci 15(5–6):583–606. https://doi.org/10.1080/07352689609382371

Hyakumachi M (1994) Plant-growth-promoting fungi from turfgrass rhizosphere with potential for disease suppression. Soil Microorganisms 44:53–68. https://doi.org/10.18946/jssm.44.0_53

Hyakumachi M, Kubota M (2004) Fungi as plant growth promoter and disease suppressor. In: Fungal Biotechnology in Agriculture, Food and Environmental Applications, vol 21. Marcel Dekker, New York, pp 101–110

Jaroszuk-Ściseł J, Tyśkiewicz R, Nowak A, Ozimek E, Majewska M, Hanaka A et al (2019) Phytohormones (auxin, gibberellin) and ACC deaminase in vitro synthesized by the mycoparasitic *Trichoderma* DEMTkZ3A0 strain and changes in the level of auxin and plant resistance markers in wheat seedlings inoculated with this strain conidia. Int J Mol Sci 20(19):4923. https://doi.org/10.3390/ijms20194923

Jin N, Liu SM, Peng H, Huang WK, Kong LA, Wu YH et al (2019) Isolation and characterization of *Aspergillus niger* NBC001 underlying suppression against *Heterodera glycines*. Sci Rep 9(1):591. https://doi.org/10.1038/s41598-018-37827-6

Kakvan N, Heydari A, Zamanizadeh HR, Rezaee S, Naraghi L (2013) Development of new bioformulations using *Trichoderma* and *Talaromyces* fungal antagonists for biological control of sugar beet damping-off disease. Crop Prot 53:80–84. https://doi.org/10.1016/j.cropro.2013.06.009

Kanchiswamy CN, Malnoy M, Maffei ME (2015) Chemical diversity of microbial volatiles and their potential for plant growth and productivity. Front Plant Sci 6:151. https://doi.org/10.3389/fpls.2015.00151

Kaur R, Kaur J, Singh RS (2011) Nonpathogenic Fusarium as a biological control agent. Plant Pathol J 9(3):79–91

Kishimoto K, Kouzai Y, Kaku H, Shibuya N, Minami E, Nishizawa Y (2010) Perception of the chitin oligosaccharides contributes to disease resistance to blast fungus *Magnaporthe oryzae* in rice. Plant J 64(2):343–354. https://doi.org/10.1111/j.1365-313X.2010.04328.x

Koike N, Hyakumachi M, Kageyama K, Tsuyumu S, Doke N (2001) Induction of systemic resistance in cucumber against several diseases by plant growth-promoting fungi: lignification and superoxide generation. Eur J Plant Pathol 107:523–533. https://doi.org/10.1023/A:1011203826805

Konappa N, Krishnamurthy S, Siddaiah CN, Ramachandrappa NS, Chowdappa S (2018) Evaluation of biological efficacy of *Trichoderma asperellum* against tomato bacterial wilt caused by *Ralstonia solanacearum*. Egypt J Biol Pest Control 28:1–11. https://doi.org/10.1186/s41938-018-0069-5

Leisinger KM, Schmitt KM, Pandya-Lorch R (2002) Six billion and counting: population growth and food security in the 21st century. https://hdl.handle.net/10568/156483

Loganathan M, Sible GV, Maruthasalam S, Saravanakumar D, Raguchander T, Sivakumar M, Samiyappan R (2010) *Trichoderma* and chitin mixture based bioformulation for the management of head rot (*Sclerotinia sclerotiorum* (Lib.) deBary)–root-knot (Meloidogyne incognita

Kofoid and White; Chitwood) complex diseases of cabbage. Arch Phytopathol Plant Protect 43(10):1011–1024. https://doi.org/10.1080/03235400802214885

Lyn ME, Burnett D, Garcia AR, Gray R (2010) Interaction of water with three granular biopesticide formulations. J Agric Food Chem 58(3):1804–1814. https://doi.org/10.1021/jf903309s

Mastouri F, Björkman T, Harman GE (2010) Seed treatment with *Trichoderma harzianum* alleviates biotic, abiotic, and physiological stresses in germinating seeds and seedlings. Phytopathology 100(11):1213–1221. https://doi.org/10.1094/PHYTO-03-10-0091

Meera MS, Shivanna MB, Kageyama K, Hyakumachi M (1994) Plant growth promoting fungi from zoysiagrass rhizosphere as potential inducers of systemic resistance in cucumbers. Phytopathology 84(12):1399–1406

Miao F, Yang R, Chen DD, Wang Y, Qin BF, Yang XJ, Zhou L (2012) Isolation, identification and antimicrobial activities of two secondary metabolites of *Talaromyces verruculosus*. Molecules 17(12):14091–14098. https://doi.org/10.3390/molecules171214091

Mohiddin FA, Bashir I, Padder SA, Hamid B (2017) Evaluation of different substrates for mass multiplication of *Trichoderma* species. J Pharmacogn Phytochem 6:563–569

Montero-Barrientos M, Hermosa R, Cardoza RE, Gutiérrez S, Nicolás C, Monte E (2010) Transgenic expression of the *Trichoderma harzianum* hsp70 gene increases Arabidopsis resistance to heat and other abiotic stresses. J Plant Physiol 167(8):659–665. https://doi.org/10.1016/j.jplph.2009.11.012

Moreno AB, Del Pozo AM, Borja M, Segundo BS (2003) Activity of the antifungal protein from *Aspergillus giganteus* against *Botrytis cinerea*. Phytopathology 93(11):1344–1353. https://doi.org/10.1094/PHYTO.2003.93.11.1344

Moura JCMS, Bonine CAV, de Oliveira Fernandes Viana J, Dornelas MC, Mazzafera P (2010) Abiotic and biotic stresses and changes in the lignin content and composition in plants. J Integr Plant Biol 52(4):360–376. https://doi.org/10.1111/j.1744-7909.2010.00892.x

Murali M, Amruthesh KN (2015) Plant growth-promoting fungus *Penicillium oxalicum* enhances plant growth and induces resistance in pearl millet against downy mildew disease. J Phytopathol 163(9):743–754. https://doi.org/10.1111/jph.12371

Murali M, Amruthesh KN, Sudisha J, Niranjana SR, Shetty HS (2012) Screening for plant growth promoting fungi and their ability for growth promotion and induction of resistance in pearl millet against downy mildew disease. J Phytology 4(5):30–36

Murali M, Naziya B, Ansari MA, Alomary MN, AlYahya S, Almatroudi A et al (2021) Bioprospecting of rhizosphere-resident fungi: their role and importance in sustainable agriculture. J Fungi 7(4):314. https://doi.org/10.3390/jof7040314

Naziya B, Murali M, Amruthesh KN (2019) Plant growth-promoting fungi (PGPF) instigate plant growth and induce disease resistance in *Capsicum annuum* L. upon infection with *Colletotrichum capsici* (Syd.) Butler & Bisby. Biomol Ther 10(1):41. https://doi.org/10.3390/biom10010041

Naziya B, Murali M, Amruthesh KN (2020) Agronomic survey and screening of genotypes for Anthracnose infection in *Capsicum annuum* L. Asian J Plant Sci 19(4): 354-360 https://doi.org/10.3923/ajps.2020.354.360

Naznin HA, Kimura M, Miyazawa M, Hyakumachi M (2013) Analysis of volatile organic compounds emitted by plant growth-promoting fungus *Phoma* sp. GS8-3 for growth promotion effects on tobacco. Microbes and Environments 28(1):42–49. https://doi.org/10.1264/jsme2.ME12085

Nishimura MT, Stein M, Hou BH, Vogel JP, Edwards H, Somerville SC (2003) Loss of a callose synthase results in salicylic acid-dependent disease resistance. Science 301(5635):969–972. https://doi.org/10.1126/science.1086716

Ortíz-Castro RANDY, Martínez-Trujillo MIGUEL, López-Bucio JOSE (2008) N-acyl-L-homoserine lactones: a class of bacterial quorum-sensing signals alter post-embryonic root development in *Arabidopsis thaliana*. Plant Cell Environ 31(10):1497–1509. https://doi.org/10.1111/j.1365-3040.2008.01863.x

Patel ZM, Mahapatra R, Jampala SSM (2020) Role of fungal elicitors in plant defense mechanism. In: Molecular aspects of plant beneficial microbes in agriculture. Academic Press, pp 143–158. https://doi.org/10.1016/B978-0-12-818469-1.00012-2

Pieterse CM, Zamioudis C, Berendsen RL, Weller DM, Van Wees SC, Bakker PA (2014) Induced systemic resistance by beneficial microbes. Annu Rev Phytopathol 52(1):347–375. https://doi.org/10.1146/annurev-phyto-082712-102340

Prashar P, Vandenberg A (2017) Genotype-specific responses to the effects of commercial Trichoderma formulations in lentil (*Lens culinaris* ssp. *culinaris*) in the presence and absence of the oomycete pathogen *Aphanomyces euteiches*. Biocontrol Sci Tech 27(10):1123–1144. https://doi.org/10.1080/09583157.2017.1376035

Qi W, Zhao L (2013) Study of the siderophore-producing *Trichoderma asperellum* Q1 on cucumber growth promotion under salt stress. J Basic Microbiol 53(4):355–364. https://doi.org/10.1002/jobm.201200031

Radhakrishnan R, Kang SM, Baek IY, Lee IJ (2014) Characterization of plant growth-promoting traits of *Penicillium* species against the effects of high soil salinity and root disease. J Plant Interact 9(1):754–762. https://doi.org/10.1080/17429145.2014.930524

Rauf M, Awais M, Ud-Din A, Ali K, Gul H, Rahman MM et al (2021) Molecular mechanisms of the 1-aminocyclopropane-1-carboxylic acid (ACC) deaminase producing *Trichoderma asperellum* MAP 1 in enhancing wheat tolerance to waterlogging stress. Front Plant Sci 11:614971. https://doi.org/10.3389/fpls.2020.614971

Sánchez-Montesinos B, Diánez F, Moreno-Gavíra A, Gea FJ, Santos M (2020) Role of *Trichoderma aggressivum* f. *europaeum* as plant-growth promoter in horticulture. Agronomy 10(7):1004. https://doi.org/10.3390/agronomy10071004

Saxena A, Raghuwanshi R, Singh HB (2015) *Trichoderma* species mediated differential tolerance against biotic stress of phytopathogens in *Cicer arietinum* L. J Basic Microbiol 55(2):195–206. https://doi.org/10.1002/jobm.201400317

Saxena A, Mishra S, Ray S, Raghuwanshi R, Singh HB (2020) Differential reprogramming of defense network in *Capsicum annum* L. plants against *Colletotrichum truncatum* infection by phyllospheric and rhizospheric *Trichoderma* strains. J Plant Growth Regul 39:751–763. https://doi.org/10.1007/s00344-019-10017-y

Sheteiwy MS, Ulhassan Z, Qi W, Lu H, AbdElgawad H, Minkina T et al (2022) Association of jasmonic acid priming with multiple defense mechanisms in wheat plants under high salt stress. Front Plant Sci 13:886862. https://doi.org/10.3389/fpls.2022.886862

Tarroum M, Romdhane WB, Al-Qurainy F, Ali AAM, Al-Doss A, Fki L, Hassairi A (2022) A novel PGPF *Penicillium olsonii* isolated from the rhizosphere of *Aeluropus littoralis* promotes plant growth, enhances salt stress tolerance, and reduces chemical fertilizers inputs in hydroponic system. Front Microbiol 13:996054. https://doi.org/10.3389/fmicb.2022.996054

Thambugala KM, Daranagama DA, Phillips AJ, Kannangara SD, Promputtha I (2020) Fungi vs. fungi in biocontrol: an overview of fungal antagonists applied against fungal plant pathogens. Front Cell Infect Microbiol 10:604923. https://doi.org/10.3389/fcimb.2020.604923

Tóth L, Váradi G, Boros É, Borics A, Ficze H, Nagy I et al (2020) Biofungicidal potential of *Neosartorya* (*Aspergillus*) *fischeri* antifungal protein NFAP and novel synthetic γ-core peptides. Front Microbiol 11:820. https://doi.org/10.3389/fmicb.2020.00820

Tripathi ĀD, Mishra R, Maurya KK, Singh RB, Wilson DW (2019) Estimates for world population and global food availability for global health. In: The Role of Functional Food Security in Global Health (pp. 3–24). Academic Press.

Tucci M, Ruocco M, De Masi L, De Palma M, Lorito M (2011) The beneficial effect of *Trichoderma* spp. on tomato is modulated by the plant genotype. Mol Plant Pathol 12(4):341–354. https://doi.org/10.1111/j.1364-3703.2010.00674.x

Turk M, Mejanelle L, Šentjurc M, Grimalt JO, Gunde-Cimerman N, Plemenitaš A (2004) Salt-induced changes in lipid composition and membrane fluidity of halophilic yeast-like melanized fungi. Extremophiles 8:53–61. https://doi.org/10.1007/s00792-003-0360-5

Ulaganathan K, Basha S, Daida P (2004) SAR proteins and SAR protein homologues and their use in developing fungal resistance in plants. Ann Rev Plant Pathol 2(2):475–497
Vacheron J, Desbrosses G, Bouffaud ML, Touraine B, Moënne-Loccoz Y, Muller D et al (2013) Plant growth-promoting rhizobacteria and root system functioning. Front Plant Sci 4:356. https://doi.org/10.3389/fpls.2013.00356
Verma PP, Shelake RM, Das S, Sharma P, Kim JY (2019) Plant growth-promoting rhizobacteria (PGPR) and fungi (PGPF): potential biological control agents of diseases and pests. In: Microbial Interventions in Agriculture and Environment: volume 1: Research Trends, Priorities and Prospects, pp 281–311. https://doi.org/10.1007/978-981-13-8391-5_11
Vernooij B, Friedrich L, Morse A, Reist R, Kolditz-Jawhar R, Ward E et al (1994) Salicylic acid is not the translocated signal responsible for inducing systemic acquired resistance but is required in signal transduction. Plant Cell 6(7):959–965. https://doi.org/10.1105/tpc.6.7.959
Vieira PM, Santos MP, Andrade CM, Souza-Neto OA, Ulhoa CJ, Aragão FJL (2017) Overexpression of an aquaglyceroporin gene from Trichoderma harzianum improves water-use efficiency and drought tolerance in *Nicotiana tabacum*. Plant Physiol Biochem 121:38–47. https://doi.org/10.1016/j.plaphy.2017.10.012
Vinale F, Sivasithamparam K, Ghisalberti EL, Marra R, Barbetti MJ, Li H et al (2008) A novel role for *Trichoderma* secondary metabolites in the interactions with plants. Physiol Mol Plant Pathol 72(1–3):80–86. https://doi.org/10.1016/j.pmpp.2008.05.005
Viterbo A, Landau U, Kim S, Chernin L, Chet I (2010) Characterization of ACC deaminase from the biocontrol and plant growth-promoting agent *Trichoderma asperellum* T203. FEMS Microbiol Lett 305(1):42–48. https://doi.org/10.1111/j.1574-6968.2010.01910.x
Wakelin SA, Gupta VV, Harvey PR, Ryder MH (2007) The effect of Penicillium fungi on plant growth and phosphorus mobilization in neutral to alkaline soils from southern Australia. Canadian journal of microbiology 53(1):106–115. https://doi.org/10.1139/w06-109
Wimalawansa SA, Wimalawansa SJ (2014) Agrochemical-related environmental pollution: effects on human health. Glob J Biol Agric Health Sci 3(3):72–83
Yamagiwa Y, Inagaki Y, Ichinose Y, Toyoda K, Hyakumachi M, Shiraishi T (2011) *Talaromyces wortmannii* FS2 emits β-caryphyllene, which promotes plant growth and induces resistance. J Gen Plant Pathol 77:336–341. https://doi.org/10.1007/s10327-011-0340-z
Yoshioka Y, Ichikawa H, Naznin HA, Kogure A, Hyakumachi M (2012) Systemic resistance induced in *Arabidopsis thaliana* by *Trichoderma asperellum* SKT-1, a microbial pesticide of seedborne diseases of rice. Pest Manag Sci 68(1):60–66. https://doi.org/10.1002/ps.2220
Zhang S, Gan Y, Xu B (2016) Application of plant-growth-promoting fungi *Trichoderma longibrachiatum* T6 enhances tolerance of wheat to salt stress through improvement of antioxidative defense system and gene expression. Front Plant Sci 7:1405. https://doi.org/10.3389/fpls.2016.01405
Zhang S, Gan Y, Xu B (2019) Mechanisms of the IAA and ACC-deaminase producing strain of *Trichoderma longibrachiatum* T6 in enhancing wheat seedling tolerance to NaCl stress. BMC Plant Biol 19:1–18. https://doi.org/10.1186/s12870-018-1618-5
Zhou LS, Tang K, Guo SX (2018) The plant growth-promoting fungus (PGPF) Alternaria sp. A13 markedly enhances *Salvia miltiorrhiza* root growth and active ingredient accumulation under greenhouse and field conditions. Int J Mol Sci 19(1):270. https://doi.org/10.3390/ijms19010270

Part III
Industrial Biotechnology and Environmental Remediation

Advanced Applications in Industry, Waste Management, and Environmental Restoration

Chapter 7
Microbially Derived Chitinases: Multifunctional Enzymes for Biocontrol, Bioremediation, and Sustainable Industrial Applications

Gaytri Mahajan, Varsha Thadiyan, Tanvi Bisht, and Reena Gupta

Abstract Chitinase, an enzyme responsible for hydrolyzing chitin—a β-1,4-linked N-acetylglucosamine polymer abundant in fungal cell walls and arthropod exoskeletons—plays a pivotal role in ecosystem nutrient cycling and biocontrol. By degrading chitin, it disrupts pathogenic fungi and pests, offering an eco-friendly alternative to chemical pesticides. Its application in agriculture suppresses fungal growth, insect infestations, and nematode damage, enhancing crop yield and quality. Microbially derived chitinase, produced by bacteria, fungi, and actinomycetes, is particularly advantageous due to its high yield, scalability, and adaptability. Beyond biocontrol, it facilitates bioremediation by converting chitinous waste into bioavailable nutrients and supports industrial processes to generate valuable products like chitosan and bioactive oligosaccharides. Ecologically, chitinase influences microbial dynamics, symbiotic plant–microbe interactions, and nutrient recycling, maintaining ecosystem balance. This chapter comprehensively reviews the production, optimization, and multifaceted applications of microbially derived chitinase, emphasizing its significance in agriculture, environmental sustainability, and biotechnology.

Keywords Chitin · Chitinase · Chitinous waste · Trichoderma · Biocontrol

7.1 Introduction

Chitinase is an enzyme that catalyzes the hydrolysis of chitin, a linear polymer of β-1,4-linked N-acetylglucosamine units. Chitin is one of the most abundant biopolymers in nature, primarily found in the exoskeletons of arthropods and in the cell walls of fungi (Mahajan et al. 2024). Chitinase plays a crucial role in the

G. Mahajan · V. Thadiyan · T. Bisht · R. Gupta (✉)
Department of Biotechnology, Himachal Pradesh University, Summerhill, Shimla, India

N. K. Arora et al. (eds.), *Environmental Sustainability and Biotechnology: Opportunities and Challenges*, Microorganisms for Sustainability 46,
https://doi.org/10.1007/978-981-96-9025-1_7

degradation and recycling of chitin, thus contributing to the carbon and nitrogen cycles in ecosystems. In the context of biocontrol, chitinase holds significant importance due to its ability to degrade chitin, a major component of the cell walls of many phytopathogenic fungi and the exoskeletons of various plant pests. By targeting chitin, chitinase disrupts the structural integrity of fungal cell walls and insect exoskeletons, leading to their lysis and eventual death (Oyeleye and Normi 2018). This makes chitinase an effective tool for controlling fungal diseases in plants and managing insect pests in agricultural systems. The biocontrol activity of chitinase has been extensively studied and demonstrated in various crop protection strategies (Veliz et al. 2017). When applied exogenously or produced endogenously by plants or microbial agents, chitinase contributes to the suppression of fungal pathogens (Russo et al. 2023). By degrading the chitin-based cell walls of these pathogens, chitinase inhibits their growth, sporulation, and penetration into plant tissues, thereby improving crop yield and quality (Vaghela et al. 2022). Moreover, chitinase-mediated biocontrol extends beyond fungal pathogens to include insect pests and nematodes. Insecticidal chitinase disrupts the chitin-rich exoskeletons of insect pests, causing physical damage and eventual mortality (Unuofin et al. 2024). Additionally, chitinase has shown nematocidal activity against plant-parasitic nematodes, offering potential solutions for nematode management in agriculture (Chen et al. 2015). The significance of chitinase in biocontrol lies not only in its direct antimicrobial and insecticidal properties but also in its eco-friendly nature. As a naturally occurring enzyme, chitinase offers a sustainable alternative to chemical pesticides, reducing environmental pollution and minimizing risks to nontarget organisms, including beneficial insects, pollinators, and soil microbes. Furthermore, the use of microbially derived chitinase allows for the development of biocontrol agents that are compatible with integrated pest management (IPM) strategies, promoting the conservation of biodiversity and ecosystem health.

Microbially derived chitinase holds significant importance in various fields, particularly in agriculture, environmental management, and biotechnology (Hamid et al. 2013). Unlike chitinase derived from other sources, such as plants or animals, microbially derived chitinase offers several advantages, including higher production yields, easier scalability, and broader applicability across diverse environmental conditions (Sharp 2013). Microorganisms, including bacteria, fungi, actinomycetes, and yeast, are prolific producers of chitinase, which can be harnessed for a wide range of applications (Nayak et al. 2021). The importance of microbially derived chitinase stems from its multifaceted roles in biocontrol, bioremediation, and industrial processes (Falade et al. 2021; Manzar et al. 2022). Chitinase plays a crucial role in bioremediation by facilitating the degradation of chitin-containing waste materials, such as crustacean shells, fungal biomass, and insect exuviate (Chakravarty and Edwards 2022). Through enzymatic hydrolysis, chitinase converts insoluble chitin into soluble chito-oligosaccharides, which can be utilized as carbon and nitrogen sources by chitinolytic microorganisms, thereby promoting the biodegradation of organic waste and the recycling of nutrients in ecosystems (Karthik et al. 2017). Industrial applications of chitinase include the production of chitin-derived products, such as chitosan, glucosamine, and bioactive oligosaccharides (Halder and

Mondal 2018). Chitinase-mediated hydrolysis of chitin yields chitosan, a versatile biopolymer with diverse biomedical, pharmaceutical, and agricultural applications (Thomas et al. 2022; Taokaew et al. 2023).

The degradation of chitin by chitinase is essential for nutrient recycling and carbon turnover in ecosystems (Rajput et al. 2022). In addition to its critical role in nutrient cycling, the degradation of chitin through the activity of chitinase carries significant ecological importance. By dismantling chitin-rich structures, including fungal cell walls and insect exoskeletons, chitinase helps shape microbial community dynamics, mediates interactions between fungi and plants, and influences predator–prey relationships (Canter et al. 2018). Such processes are crucial in maintaining ecological balance and enhancing nutrient availability in various ecosystems (Ahmed and Opal 2020; Gooday 1990; Dukat et al. 2024). Furthermore, chitinase plays a crucial role in symbiotic interactions between plants and beneficial microbes, such as mycorrhizal fungi and nitrogen-fixing bacteria (Khokhani et al. 2021). By facilitating chitin degradation, chitinase promotes the establishment and persistence of beneficial microbial communities in the rhizosphere and phyllosphere of plants (Das et al. 2023). Overall, the role of chitinase in chitin degradation is multifaceted and essential for the functioning of ecosystems (Olatunji 2020). This chapter provides a comprehensive overview of microbially derived chitinase, its production, optimization, and application. The chapter also delves into their selected applications in various fields.

7.2 Chitinase: An Overview

Chitinase is an enzyme that catalyzes the hydrolysis of chitin, a linear polysaccharide composed of β-1,4-linked N-acetylglucosamine (GlcNAc) units. Chitin is one of the most abundant biopolymers in nature, second only to cellulose and is found predominantly in the cell walls of fungi, the exoskeletons of arthropods (such as insects, crustaceans, and spiders), and the radulas of mollusks (Crini 2022). Structurally, chitin resembles cellulose but differs in the presence of an acetyl group attached to the C-2 position of the glucosamine units, imparting rigidity and insolubility to the polymer. Chitin molecules form long, unbranched chains through the β-1,4 glycosidic bonds between GlcNAc monomers (Hoell et al. 2010). These chains can further assemble into microfibrils, providing structural support and mechanical strength to the cell walls of fungi and the exoskeletons of arthropods. Depending on the source and degree of acetylation, chitin can vary in its properties, such as crystallinity, solubility, and susceptibility to enzymatic degradation. Chitinase enzymes play a crucial role in the degradation and recycling of chitin in nature. By cleaving the β-1,4 glycosidic bonds between GlcNAc units, chitinase hydrolyzes chitin into soluble oligomers, primarily chito-oligosaccharides and eventually into monomeric GlcNAc units (Itoh et al. 2019).

The enzymatic activity of chitinase is influenced by various factors, including pH, temperature, substrate concentration, and the presence of cofactors or metal ions (Jeong et al. 2023; Renaud et al. 2023). Optimal conditions for chitinase activity typically range from acidic to neutral pH and moderate temperatures (Das et al. 2012; Liang et al. 2023). Additionally, chitinase activity can be modulated by post-translational modifications, such as glycosylation, phosphorylation, and proteolytic cleavage, which affect enzyme stability, substrate specificity, and catalytic efficiency (Unuofin et al. 2024). Understanding the structure and function of chitinase is essential for harnessing its potential in various applications, including agriculture, biotechnology, and environmental remediation (Gomaa et al. 2021).

7.3 Types of Chitinase

Chitinases are classified into two major families based on their amino acid sequences and catalytic mechanisms: glycoside hydrolase family 18 (GH18) and family 19 (GH19). GH18 chitinases are widespread in bacteria, fungi, plants, and animals, while GH19 chitinases are predominantly found in bacteria and some viruses (Orlando et al. 2021). Both families of chitinases share a conserved catalytic domain with a triosephosphate isomerase (TIM) barrel fold, which houses the catalytic residues involved in chitin hydrolysis.

7.3.1 Glycoside Hydrolase Family 18 (GH18) Chitinases

GH18 chitinases are the most widely distributed and extensively studied chitinases found in nature. They are present in a wide range of organisms, including bacteria, fungi, plants, and animals. Structurally, GH18 chitinases typically consist of a catalytic domain with a triosephosphate isomerase (TIM) barrel fold, which houses the active site responsible for chitin hydrolysis (Funkhouser and Aronson 2007). Some GH18 chitinases also contain additional domains, such as chitin-binding domains (CBDs), carbohydrate-binding modules (CBMs), or fibronectin type III-like domains, which enhance substrate binding and enzyme activity (Chen et al. 2020). GH18 chitinases employ a retaining mechanism for catalysis, in which the enzyme forms a covalent intermediate with the substrate during hydrolysis (Alsina et al. 2019). This mechanism involves two conserved catalytic residues, a nucleophile and an acid/base, which facilitate the cleavage of the glycosidic bond between N-acetylglucosamine (GlcNAc) units in chitin (Hussain and Wilson 2013).

7.3.2 Glycoside Hydrolase Family 19 (GH19) Chitinases

GH19 chitinases are primarily found in bacteria and some viruses, although they are less common than GH18 chitinases. They are absent in eukaryotes, including fungi, plants, and animals. Structurally, GH19 chitinases exhibit a $(\beta/\alpha)_8$ TIM barrel fold, similar to GH18 chitinases, but lack the additional domains typically present in GH18 chitinases (Jamek et al. 2017). GH19 chitinases utilize an inverting mechanism for catalysis, in which a water molecule acts as the nucleophile during glycosidic bond cleavage (Orlando et al. 2021). Unlike GH18 chitinases, GH19 chitinases do not form a covalent intermediate with the substrate during hydrolysis.

In addition to these two main families, chitinases can be further classified into subfamilies and isoforms based on their sequence homology, substrate specificity, and biological functions (Huet et al. 2008). For example, some chitinases exhibit strict specificity for chitin as a substrate, while others display broader substrate specificity, including other glycosidic substrates (Yano et al. 2021). Chitinases can be subdivided into endo and exo forms based on their cleavage sites.

Endochitinases (EC 3.2.1.14) randomly split chitin at the internal sites of the chitin microfibril, forming a soluble, low-molecular-mass multimer (Suresh 2019). The multimer products include di-acetylchitobiose, chitotriose, and chitotetraose, with the dimer being the predominant product.

Exochitinases have also been divided into two subcategories: *Chitobiosidases (EC 3.2.1.29)* act on the nonreducing end of the chitin microfibril, releasing the dimer, di-acetylchitobiose, one by one from the chitin chain. Therefore, there is no release of monosaccharides or oligosaccharides in this reaction. *β-1,4-N-acetylglucosaminidases* (EC 3.2.1.30) split the multimer products, such as di-acetylchitobiose, chitotriose, and chitotetraose, into monomers of N-acetylglucosamine (GlcNAc).

7.4 Microbial Sources of Chitinase

The diversity of microorganisms producing chitinase is remarkable, encompassing a wide range of taxonomic groups and ecological niches. Chitinase-producing microorganisms can be found in virtually every habitat on Earth, from terrestrial to aquatic environments and from polar regions to tropical forests (Banerjee and Mandal 2019). This diversity reflects the importance of chitin degradation in nutrient cycling, carbon turnover, and ecological interactions across diverse ecosystems (Chaulagain et al. 2024). Chitinase-producing microorganisms can be divided into different taxonomic groups, including bacteria, fungi, actinomycetes, and other microorganisms (Atheena et al. 2024). Each group exhibits unique chitinolytic capabilities, enzymatic properties, and ecological roles in chitin degradation and nutrient cycling.

7.4.1 Bacteria

Bacteria are among the most prolific producers of chitinase, with representatives found in various phyla, including Proteobacteria, Actinobacteria, Firmicutes, Bacteroidetes, and Cyanobacteria (Banerjee and Mandal 2019). Many soil bacteria, such as members of the genera *Streptomyces, Bacillus, Pseudomonas,* and *Serratia*, are known for their chitinolytic activity, which play a crucial role in the decomposition of chitin-rich organic matter and the recycling of nutrients in soil ecosystems (Nayak et al. 2020). Marine bacteria, including *Vibrio, Alteromonas,* and, *Flavobacterium* species, are also important contributors to chitin degradation in marine environments, where chitin is abundant in the form of crustacean shells, zooplankton, and marine snow (da Silva et al. 2023). The subsequent section provides a discussion on a few representative bacteria known for their chitinolytic activity.

7.4.1.1 *Streptomyces* spp.

Streptomyces species are filamentous, Gram-positive bacteria abundant in soil ecosystems. When it comes to chitinase activity, *Streptomyces* is one of the most extensively studied genera. They produce a wide range of chitinase enzymes, including endochitinases and exochitinases, which play a crucial role in the decomposition of chitin-rich organic matter and the recycling of nutrients in soil. *Streptomyces griseus*, *S.lividans,* and *S. coelicolor* are examples of chitinase-producing *Streptomyces* species (Subramani et al. 2022).

7.4.1.2 *Nocardia* spp.

Nocardia species are aerobic actinomycetes found in soil, water, and decaying organic matter. Few species produce chitinase enzymes that enable them to degrade chitin-rich substrates and compete with other microorganisms for nutrients. *Nocardia asteroids* and *N. brasiliensis* are examples of chitinase-producing *Nocardia* species (Subash et al. 2024).

7.4.1.3 *Micromonospora* spp.

Micromonospora species are soil-dwelling actinomycetes known for their chitinolytic capabilities. They produce chitinase enzymes involved in the degradation of chitin-containing substrates and the recycling of nutrients in soil ecosystems (Teregulova et al. 2023). Recently, a novel pH-stable chitinase (MaChi1) from *M. aurantiaca* was expressed in *E. coli* and characterized as a GH18 enzyme (57 kDa) with optimal activity at pH 5.0/55 °C, stability across pH 3.0–10.0, and showed enhanced activity under Mg^{2+}/DTT by hydrolyzing diverse substrates (chitin, chitosan, and cellulose) via endo-cleavage, efficiently converting colloidal chitin into GlcNAc (227.2 mg/g) and $(GlcNAc)_2$ (505.9 mg/g) (Guo et al. 2024).

7.4.1.4 *Bacillus* spp.

Bacillus species are ubiquitous bacteria found in soil, water, and the rhizosphere of plants. They are renowned for their chitinolytic activity, which enables them to thrive in chitin-rich environments (Chaulagain et al. 2024). Among the major chitinase-producing *Bacillus* species are *B. thuringiensis, B. subtilis*, and *B. cereus* (Paulsen et al. 2016). Bacilli thrive in diverse habitats, including soil and marine environments, where chitin is found in the form of lysed biomass, insect exoskeletons, and crustacean shells, serving as a significant nutrient source for them (Keyhani and Roseman 1999; Strobel et al. 2010). *Bacillus* species efficiently break down chitin into simpler compounds, utilizing them as carbon and nitrogen sources (Poria et al. 2021). This adaptive strategy enhances their survival and competitiveness against non-chitinase-producing microorganisms in their surroundings. Moreover, many *Bacillus* species serve as biocontrol agents against plant pathogens, including fungi and insect pests. Additionally, chitinases derived from *Bacillus* have significant industrial and agricultural applications. They contribute to converting chitin waste into valuable products such as chito-oligosaccharides (COS) and glucosamine (GS), as well as the development of biofungicides and biopesticides (Singh et al. 2013a, b).

7.4.1.5 *Pseudomonas* spp.

Pseudomonas species are highly adaptable bacteria commonly found in soil, water, and on plant surfaces. They produce chitinase, along with other bioactive metabolites, to combat phytopathogenic fungi (Chen et al. 2015). Among them, *Pseudomonas aeruginosa*, *P. fluorescens*, and *P. putida* are notable chitinase producers. *P. aeruginosa* stands out as a representative bacterium and an opportunistic human pathogen, demonstrating significant chitinase activity that facilitates the breakdown of chitin-containing materials in its surroundings (Folders et al. 2001; Vivekanandhan et al. 2022). Meanwhile, *P. fluorescens*, due to its nonpathogenic nature and excellent plant colonization abilities, also produces chitinase. Hence, it plays a crucial role in the biocontrol of fungal pathogens. In soil habitats, other *Pseudomonas* species, such as *P. stutzeri* and *P. putida,* contribute to chitin turnover by participating in chitin degradation processes (Kahlon 2016). These chitinase enzymes from different *Pseudomonas* species hold significant promise for applications in agriculture and biotechnology, particularly in the development of biopesticides and biofungicides (Unuofin et al. 2024).

7.4.1.6 *Serratia* spp.

Studies have confirmed the role of *Serratia marescens* chitinolytic machinery as one of the best-known enzymes secreting the chitinolytic system (Vaaje-Kolstad et al. 2013). Three different GH18 chitinases—ChiA, ChiB, and ChiC—are produced by

S. marescens (Lin et al. 2015). ChiA is an endochitinase that cleaves chitin internally, producing shorter chitin oligomers, and is particularly effective at breaking down crystalline chitin, making it essential for the initial stages of chitin degradation (Horn et al. 2006). ChiB functions as an exochitinase, cleaving chitin from the nonreducing ends and releasing disaccharides such as chitobiose, thus targeting the β-1,4-glycosidic bonds at the ends of chitin chains and gradually reducing the size of chitin polymers (Brurberg et al. 1996). ChiC, also an endochitinase, has a slightly different substrate specificity and mode of action compared to ChiA, helping in the breakdown of chitin into smaller oligomers that can be further processed by other enzymes (Liu et al. 2023). The combined action of ChiA, ChiB, and ChiC allows *Serratia marcescens* to efficiently degrade chitin (Suzuki et al. 2016). Additionally, by breaking down the exoskeletons of insects, *S. marcescens* effectively controls insect pests (El-Sayed et al. 2024).

7.4.2 Fungi

Fungi are prominent producers of chitinase enzymes, as chitin is a major component of their cell walls and structural support systems. Fungi employ chitinase enzymes for a variety of essential functions, such as remodeling their cell walls, acquiring nutrients, and defending against pathogens that contain chitin (Langner and Göhre 2016). These processes are vital for fungal growth, survival, and ecological interactions (Thakur et al. 2023). Fungal filamentous sporulation is aided by the chitinases, which are linked to the fungal cell wall (Dwyer et al. 2023). Few representative genera of chitinase-producing fungi include the following.

7.4.2.1 *Trichoderma* spp.

Trichoderma species are filamentous fungi ubiquitously distributed in soil, plant rhizospheres, and decomposing organic matter. Production of chitinases by them is critical for mycoparasitism, biocontrol of plant diseases, and degradation of chitin-rich substrates such as fungal cell walls and insect exoskeletons (Anwar et al. 2023). Notably, species like *T. reesei*, *T. harzianum*, and *T. viride* are industrially significant due to their ability to synthesize commercially valuable enzymes, including cellulases and chitinases, which are widely applied in agriculture, textiles, biofuel production, and biotechnology (Schuster and Schmoll 2010). Among *Trichoderma*, *T. harzianum* is particularly recognized as a potent biocontrol agent, capable of colonizing plant roots and secreting chitinases that degrade the cell walls of phytopathogenic fungi including *Fusarium oxysporum* and *Rhizoctonia solani* (Harman et al. 2004; Lorito et al. 2010). Enzymatic activity not only disrupts pathogen invasion but also enhances plant immune responses, featuring its agricultural importance (Hermosa et al. 2012; Pandey et al. 2023). Similarly, *T. viride* employs mycoparasitic strategies, producing chitinases alongside glucanases and proteases to lyse fungal pathogens, thereby suppressing diseases like powdery mildew and root rot (Vinale et al. 2008; Abdelraouf et al. 2024). The agricultural significance of *Trichoderma*-derived chitinases is further highlighted by their role in sustainable farming practices. Studies demonstrate that these enzymes protect crops from fungal infections while reducing the reliance on chemical pesticides, aligning with eco-friendly agricultural goals (Hernández-Melchor et al. 2023; Mukherjee et al. 2013).

7.4.2.2 *Aspergillus* spp.

Aspergillus species are ubiquitous fungi commonly found in soil, decaying organic matter, and indoor environments, where they play vital roles in organic decomposition and nutrient cycling (Abdel-Azeem et al. 2019). Chitinase produced by them take part in the degradation of fungal cell walls, facilitate sporulation, and help in acquiring nutrients from chitin-rich substrates such as insect exoskeletons and decaying plant material (Ornela and Guimarães 2024). Prominent chitinase-producing species are *A. niger*, *A. fumigatus*, and *A. flavus*, each with distinct ecological and industrial roles (Paulussen et al. 2017). These enzymes are critical for degrading chitin-containing waste in bioremediation processes and are employed in food, textile, and biofuel industries (Abdel Wahab et al. 2023). Beyond enzymes, *A. niger* is a major producer of citric acid, accounting for over 90% of global production, and is harnessed for detoxifying heavy metals and organic pollutants in contaminated environments (Behera 2020).

In contrast, *A. fumigatus* is an opportunistic pathogen responsible for invasive aspergillosis in immunocompromised individuals. Despite its clinical significance, this species produces chitinases that enable it to hydrolyze environmental chitin, supporting its survival and nutrient acquisition in diverse niches (Tong et al. 2024). Similarly, *A. flavus* exhibits a dual ecological role: it contaminates crops with carcinogenic aflatoxins, posing severe health risks, while its chitinases contribute to the breakdown of agricultural waste and fungal competitors (Song et al. 2023).

7.4.2.3 *Penicillium* spp.

Penicillium species are ubiquitous molds found in soil, air, organic matter, and even extreme environments such as Arctic sediments (Yuliantoro et al. 2023). Their chitinase also plays a critical role in degrading chitin-containing substrates like plant debris, insect exoskeletons, and fungal cell walls. This enzymatic activity not only aids nutrient recycling in ecosystems but also positions *Penicillium* as a potential biocontrol agent against chitin-rich plant pathogens, such as *Fusarium* spp. and *Phytophthora* spp. (Khan et al. 2020). Notable chitinase-producing species include *P. chrysogenum*, *P. expansum*, and *P. roquefortii*, which have been studied for their biotechnological applications in agriculture and waste management (Pacios-Michelena et al. 2023). Furthermore, their enzymatic versatility extends to marine ecosystems, where they assist in decomposing crustacean and algal biomass, emphasizing their ecological significance (Balabanova et al. 2018). Despite their benefits, certain species, such as *P. expansum*, are postharvest pathogens of fruits, necessitating careful management to balance their industrial utility with agricultural risks (Ballester et al. 2017).

7.4.2.4 *Metarhizium anisopliae*

Metarhizium anisopliae is a filamentous fungus belonging to the order Hypocreales. It is recognized for its entomopathogenic properties. This fungus is widely studied and utilized in biological control due to its ability to effectively target a broad range of insect pests (Schrank and Vainstein 2010). One of the key factors contributing to its pathogenicity is its production of chitinases (Leger 2008). By breaking down chitin, *M. anisopliae* can penetrate and degrade the protective barriers of its insect hosts, facilitating infection and ultimately leading to the insect's death (Tiago et al. 2014). The fungus secretes both endo- and exochitinases (e.g., MaChi1 and MaChi2), which synergistically degrade the chitin-rich exoskeletons of insect hosts, facilitating infection by breaching structural barriers (Qazi 2008). Beyond pathogenicity, *M. anisopliae's* chitinases have broad industrial relevance. For instance, they efficiently convert shrimp and crab shell waste into chitosan, a biodegradable polymer with applications in agriculture, water treatment, and biomedical devices (Sbaraini et al. 2021; Golzar et al. 2023). Chitinase production in *M. anisopliae* is highly dependent on aeration strategies, as oxygen availability regulates gene expression and enzymatic activity.

7.4.2.5 *Verticillium* spp.

Verticillium spp., a genus of soil-dwelling filamentous fungi, are ecologically versatile organisms that produce a suite of chitinases, enzymes critical for nutrient cycling, microbial interactions, and pathogenicity (Acharya et al. 2020). These fungi secrete endochitinases (e.g., VdChi1 in *Verticillium dahliae*) and exochitinases (e.g., VaChi2 in *V. albo-atrum*), which hydrolyze chitin in fungal cell walls, insect exoskeletons, and arthropod cuticles (Li et al. 2024). Ecologically, *Verticillium* chitinases facilitate niche competition in soil ecosystems by degrading chitin-rich residues from decaying fungi or insects, liberating carbon and nitrogen for microbial utilization while suppressing rival fungi (Zhang et al. 2022). For example, *V. dahliae* employs chitinases to weaken the cell walls of competing fungi like *Fusarium oxysporum*, enhancing its dominance in rhizosphere communities (de Jonge et al. 2023). As plant pathogens, these enzymes also play dual roles: while Verticillium species primarily infect hosts via xylem colonization, their chitinases may indirectly modulate plant immunity by releasing chitin oligomers that trigger host defense responses, complicating their role in disease progression (Klosterman et al. 2011). Beyond ecology, *Verticillium* chitinases hold biotechnological promise. They are harnessed for biocontrol against fungal pathogens and for converting agricultural waste, such as shrimp shells and spent mushroom substrates, into value-added products like chito-oligosaccharides, which enhance plant growth and soil health (Chen et al. 2021; Zhang et al. 2022).

7.4.3 Viruses

Some viruses, particularly those infecting chitin-containing eukaryotic hosts (such as arthropods or fungi), may encode chitinases as auxiliary metabolic genes, but this is not common in bacteriophages or most mycoviruses (Abulikemu et al. 2021). Some viruses, particularly large DNA viruses or those infecting chitin-containing hosts, encode GH18 family chitinases—occasionally as multimodular proteins. These chitinases may help degrade chitinous barriers like fungal cell walls or insect gut linings, potentially aiding in host invasion or viral dissemination, though their exact roles remain underexplored (Karwelat et al. 2020). For instance, Baculoviruses, such as Autographa californica nucleopolyhedrovirus (AcMNPV), produce a chitinase (v-CHI1) that degrades chitin in host tissues during late infection, promoting host liquefaction and facilitating virion release and environmental dissemination. While it may also affect chitinous structures like the peritrophic matrix, its primary role is in host tissue breakdown rather than initial dissemination (Shrestha et al. 2018; Clem and Passarelli 2013). Beyond pathogenesis, viral chitinases play underappreciated ecological roles in marine ecosystems. Marine bacteriophages encoding chitinases (e.g., Vibrio-infecting phages) contribute to carbon cycling by degrading chitin from viral lysate biomass, releasing oligosaccharides that fuel microbial food webs (Abulikemu et al. 2021). Recent metagenomic studies have identified novel viral chitinases in oceanic "viromes," suggesting their broader role in global chitin turnover (da Silva et al. 2023). Structurally, viral chitinases exhibit unique adaptations. For instance, the *Pseudomonas* phage ΦPA3 encodes a GH18 chitinase with a carbohydrate-binding module (CBM5) that enhances substrate affinity, a feature rare in nonviral homologues (Middelboe et al. 2025). Biotechnologically, these enzymes are gaining traction for applications such as chitin waste valorization—engineered viral chitinases expressed in *E. coli* have achieved higher conversion efficiency of crab shell chitin into N-acetylglucosamine, a precursor for pharmaceuticals (Han et al. 2024). Despite their potential, viral chitinases remain underexplored, with fewer than 20 characterized to date, highlighting a critical gap in understanding viral enzymatic diversity and ecological impact.

7.5 Production Techniques

7.5.1 Submerged Fermentation (SmF)

Submerged fermentation (SmF) is the predominant method for chitinase production, particularly for substrates with high moisture content, due to its superior process control, enhanced mass transfer, and broad substrate adaptability. SmF allows microorganisms to grow in a liquid medium, ensuring uniform nutrient distribution, precise regulation of environmental parameters (e.g., pH and dissolved oxygen), and efficient utilization of diverse chitin substrates, including colloidal chitin, shrimp shells, and fungal biomass (Subramaniyam and Vimala 2012). This method is favored for industrial-scale production due to its scalability and reproducibility, with bacteria and fungi demonstrating high chitinase yields under optimized SmF conditions (Mahajan et al. 2024). *Beauveria bassiana*, isolated from marine sediment, was optimized for chitinase production using SSF on wheat bran, achieving a peak yield of 246.6 U/gIDS at pH 9.20 (Suresh and Chandrasekaran 1999). In the last few years, various bioreactor configurations to maximize enzyme productivity have been reported. In another study, *Trichoderma virens* UKM1 to optimize chitinase production in a 2 L stirred tank reactor with shrimp waste, found optimal agitation and aeration rates at 200 rpm and 0.33 vvm. This yielded maximum chitinase activities of 4.1 U/L and 5.97 U/L per hour, respectively (Abd-Aziz et al. 2008). Another study developed a cost-effective method to produce chitosanase from shrimp shell waste by pretreating it with 0.4 M acetic acid, enhancing enzyme production fourfold by an actinomycetes *Lentzea* sp. OUR-I1. The process was successfully scaled to an air-lift bioreactor, achieving a maximum enzyme activity of 0.974 U/mL, and the waste product also showed potential as an absorbent for industrial dyes, alongside the production of a yellow color (Suyotha et al. 2021). To compare yields in different subtypes of SSF, a study utilized *Paenibacillus taichungensis,* and it was reported that the net-draft tube air-lift reactor (ALNR) was the most effective. Optimal enzyme activity (21.1 mU/mL), 2.6 times higher than that in a stirred-tank reactor, was achieved at a 2 vvm aeration rate (Chen et al. 2010). This was supported by analyses of shear and oxygen transfer rates, alongside investigations into the strain's morphology under these conditions (Chen et al. 2010).

Critical parameters in SmF are pH, temperature, and aeration. The pH is maintained at 5.5–7.0 for bacterial chitinases and 4.5–6.0 for fungal variants, as deviations >1.0 unit can reduce the activity by 50% (Abo Elsoud and El Kady 2019). For thermophilic *Thermomyces lanuginosus* SSBP, optimum growth temperature and pH were reported to be 323.15 K (50 °C) and 6.5, respectively. It has been noticed that chitinases from SmF exhibit higher thermostability and acid tolerance (pH 4.0–6.0), making them suitable for harsh industrial environments (Saqib et al. 2010). For example, *Aspergillus niveus*, when cultured in submerged fermentation using crab shells, produced extracellular antifungal chitinase, achieving maximum production at 192 h with 1% chitin. The enzyme, purified 1.97-fold, exhibited optimal activity at 65 °C and pH 5.0 and showed stability and enhanced activity in the presence of Mn2+ (Alves et al. 2018).

7.5.2 Solid-State Fermentation

Solid-state fermentation (SSF) is a robust and eco-friendly method for chitinase production, utilizing solid substrates with minimal free water to cultivate microorganisms, primarily filamentous fungi, under conditions that mimic their natural habitats such as decomposing organic matter or soil ecosystems (Baldoni et al. 2020). Unlike SmF, SSF employs lignocellulosic or chitin-rich substrates like wheat bran, rice husk, shrimp shells, or sugarcane bagasse, which serve dual roles as nutrient sources and physical matrices for microbial colonization (Dhevagi 2021). This method is particularly advantageous for fungi such as *A. niger, T. harzianum,* and *P. chrysogenum*, which thrive in low-water-activity environments and secrete chitinases to degrade complex substrates into assimilable sugars (López-García et al. 2024).

In SSF, *Streptomyces* sp. SCUT-3 was used to convert shrimp shell waste (SSW) into high-value products via solid-state fermentation and lactic acid treatment, 92.6% protein (as amino acids/oligopeptides), 92% calcium, and high-purity chitin (Lu et al. 2023). This approach offers ten-fold higher profit, 80% cost reduction, and minimal environmental impact compared to traditional methods. SSF systems require precise control of temperature (25–35 °C for mesophiles), humidity (60–80%), and aeration to maintain microbial metabolic activity (Krishna 2005). The solid matrix creates a gradient of moisture and nutrients, promoting fungal growth in biofilms that enhance enzyme stability and prolong catalytic activity (Rudakiya et al. 2019). For example, in a study, *Aspergillus terreus* SB3 (ON738571) produced chitinase (80.8 U/mL/min), optimized via RSM-CCD under SSF to double-yield (198.5 U/mL/min) with controlling key factors: temperature (35 °C), pH (4.5), substrate (20 g), moisture (55%), and ammonium sulfate (0.25 g) (Adnan et al. 2024). SSF offers lower water consumption, reduced energy costs, and the ability to valorize agro-industrial waste. Agro-industrial waste not only provides carbon and nitrogen but also induces chitinase gene expression in *T. viride* through surface contact signaling (Aita et al. 2019). Similarly, industrial waste supplemented with colloidal chitin may boost chitinase production, demonstrating SSF's potential for circular economy integration (Abu Yazid et al. 2017).

Post-fermentation, chitinases are extracted via aqueous leaching (buffer soaking) or solvent extraction (e.g., 50% ethanol), with recovery efficiencies ranging from 60 to 85% depending on substrate porosity (Adnan et al. 2024). However, SSF faces challenges in scale-up, including heat buildup in large trays and heterogeneous substrate colonization. Innovations like rotary drum bioreactors and modular tray systems with automated humidity control have improved scalability, achieving 90% enzyme recovery in pilot-scale trials (Spier et al. 2011).

7.6 Optimization Parameters

Optimization of culture conditions for chitinase production is a critical aspect of maximizing enzyme yield and activity. This process involves systematically adjusting various environmental factors to create the most favorable conditions for the growth of chitinase-producing microorganisms and the synthesis of chitinase enzymes. The key parameters involved in optimizing culture conditions for chitinase production are as follows:

7.6.1 Medium Composition

The culture medium composition is a critical determinant in chitinase production, exerting profound effects on the metabolic activity and growth of microorganisms. The medium serves as the primary source of nutrients, providing essential elements necessary for cell growth, enzyme synthesis, and chitin degradation (Reid and Ogrydziak 1981). It has been observed that for carbon sources, chitin itself served as the best source. For instance, *T. harzianum* produces chitinase most effectively when grown on fungal cell walls (e.g., *Botrytis cinerea*), chitin, or specific glucans like pustulan/nigeran (de la Cruz 1993). Other carbon sources for chitinase production include wheat bran, rice bran, corn cobs, sugarcane bagasse, coconut husk, lignocellulosic biomass, and chitin-rich shrimp shell waste. These substrates have been used exclusively for chitinase production by different microorganisms, such as *Streptomyces* sp., *B. subtilis, A. niger, P. chrysogenum, T. harzianum* (Le and Yang 2019). Wheat bran, corn cobs, and sugarcane bagasse have been utilized as carbon substrates for fungal chitinase production (López-García et al. 2024). The choice of nitrogen source can influence the rate and yield of chitinase production, with organic nitrogen sources often providing more favorable conditions for enzyme synthesis. For instance, in a study on *Bacillus* sp. R2, chitinase production optimized by applying one variable at a time technique (OVAT), identifying 0.5% yeast extract as the optimal nitrogen source and 24 h incubation at 30 °C, pH 7.5, 2.5–3% NaCl, 180 rpm, and 2.5% inoculum (8.9×10^8 CFU/mL) for peak enzyme activity (Cheba et al. 2018). In contrast to this, statistical design-optimized chitinase production was reported when high chitin and ammonium sulfate levels, with no peptone or urea, were used at optimal pH of 5.6 and 28 °C temperature (Kapat et al. 1996). Phosphorus and sulfur incorporation into the culture medium ensures the availability of these vital nutrients, supporting optimal growth and chitinase production (Gomaa 2012).

7.6.2 pH

pH or the measure of acidity or alkalinity of a solution plays a crucial role in microbial growth, enzyme activity, and stability, including in the context of chitinase production (Kielak et al. 2013). For chitinase production, the optimal pH varies depending on the specific microorganism being used and the composition of the culture medium (Singh et al. 2013a, b). Additionally, the components of the culture medium, such as carbon and nitrogen sources, can influence the optimal pH for chitinase production (Mishra et al. 2012). pH may also vary amongst producing organisms. For instance, *Streptomyces* species have been observed to produce maximum chitinase activity within an acidic to neutral pH range of 5.5–6.0 (Sukalkar et al. 2018). Another study highlights pH 6 as critical for both optimal chitinase production (30 °C, 14 days) and peak enzymatic activity (60 °C) in *Bacillus licheniformis* B307, enabling an eightfold increase in yield (14.2 U/mL) under optimized conditions. *B. subtilis* demonstrates optimal chitinase production at a slightly alkaline pH of 7.5–8.0 (Akeed et al. 2020). On the other hand, pH 8 was optimal for chitinase (CHT1) and chitosanase (CHS1) production by *Pseudomonas* sp. TKU015 using shrimp shell waste, while enzyme activity peaks at pH 6 (CHT1) and pH 4 (CHS1), with distinct pH stability ranges (CHT1: pH 5–7; CHS1: pH 3–9 (Wang et al. 2008).

For fungi, diverse isoelectric points (pH 3.2–13) of chitinolytic enzymes reveal pH-dependent functional versatility, enabling activity across acidic to highly alkaline conditions (Ramirez-Coutiño et al. 2010). Chitinase production in *M. anisopliae* is found to be strongly influenced by ambient pH, with maximal expression occurring at an acidic pH of around 5. Northern analysis also confirmed that chitinase gene expression is pH-dependent, showing increased mRNA accumulation only under conditions favorable to its activity. Further, this pH-specific regulation ensures that chitinase is synthesized effectively during the early stages of infection when the insect cuticle environment is still slightly acidic.

Chitinase production in *M. anisopliae* is strongly influenced by ambient pH, with maximal expression occurring at an acidic pH of around 5, which matches its enzymatic optimum (St. Leger Raymond et al. 1998). Northern analysis confirmed that chitinase gene expression is pH-dependent, showing increased mRNA accumulation only under conditions favorable to its activity. This pH-specific regulation ensures that chitinase is synthesized effectively during early stages of infection when the insect cuticle environment is still slightly acidic. These findings feature the importance of pH optimization in enhancing chitinase production for various biotechnological applications.

7.6.3 Temperature

For chitinase production, the optimal temperature generally ranges between 25 °C and 37 °C, although this can vary significantly depending on the specific microorganism involved (Nampoothiri et al. 2004). For a thermophilic chitinase producer, this can go up to 80 °C. For instance, a study highlights 50 °C as the optimal temperature for maximum chitinase activity from *Paenibacillus* sp. D1, with a high thermal stability at 80 °C (Singh and Chhatpar 2011). On the other hand, thermophilic chitinase from *Thermococcus kodakarensis* exhibited optimal activity at a much higher temperature of 80 °C, showcasing its adaptation to extreme heat. Low-temperature adaptability of a 67 kDa chitinase from marine fungus *Plectosphaerella* sp. MF-1 is also reported This chitinase retains activity at 10 °C (with a minimal activity loss compared to 37 °C), ideal for cold-environment applications like marine waste degradation (Velmurugan et al. 2011). In a study, cold-adapted chitinase from Antarctic *Pseudomonas* sp. GWSMS-1 optimized at 20 °C for production (95.41 U/L activity), with enzymatic activity peaking at 35 °C but retaining >50% efficacy even at 0 °C, critical for low-temperature bioprocessing.

7.6.4 Oxygenation and Aeration

Production by chitinase is highly influenced by aeration and oxygen supply (Haran et al. 1996). A study highlights optimal aeration (0.33 vvm) and agitation (200 rpm) as critical for maximizing chitinase production (5.97 U/L/h) by *T. virens* UKM1 in a bioreactor, balancing oxygen transfer (kLa = 8.34 h^{-1}) while avoiding shear stress at a higher agitation (Abd-Aziz 2008). Enhanced aeration boosted oxygen uptake (1.74 mg O_2/g biomass/h), directly linking the dissolved oxygen levels to fungal metabolic efficiency and enzyme yield (Abd-Aziz 2008). Oxygen deprivation may downregulate the chiA gene, critical for chitinolytic activity. For instance, a study showed the role of aeration-linked oxidative environments in regulating *Aspergillus nidulans* chitinase genes, where redox balance (glutathione) and reactive oxygen species (ROS) during autolysis—common under varying aeration—do not alter chiA or chiB expression, though diamide-induced oxidative stress suppresses both (Pusztahelyi et al. 2006). Aeration impacts ROS levels, directly repressing chiC via superoxide accumulation, linking oxygen availability to gene regulation (Pusztahelyi et al. 2006). Apart from the concentration of dissolved oxygen, agitation is also linked with improving oxygen transfer rates (OTR), ensuring a homogeneous distribution of oxygen. Earlier, it has been reported that 3.5-fold higher chitinase under high aeration (400 rpm) can be achieved compared to static conditions in *S. marcescens* (Roberts and Cabib 1982). It has been noticed that over-aeration can also trigger toxicity by creating reactive oxygen species (ROS), damaging cellular machinery (Akimoto et al. 2000).

7.6.5 Cofactors and Additives

In microbial cultures, the synthesis of chitinase can be significantly enhanced by the presence of specific cofactors and additives. For example, metal ions such as calcium (Ca) and Na have been shown to improve the stability and activity of chitinase (Harini et al. 2021). For instance, in a study, marine chitinase Af Chi from *Alcaligenes faecalis* was evaluated in the presence of various cofactors. At a concentration of 1 mM, Na^+ ions significantly enhanced the enzyme's activity, indicating its halophilic nature (Raval and Raval 2024). Other ions such as K^+, Ca^{2+}, Cu^{2+}, Mg^{2+}, Mn^{2+}, and the chelating agent EDTA also increased the activity, though to a lesser extent. Conversely, Fe^{3+}, Zn^{2+}, Co^{2+}, and Pb^{2+} ions inhibited the enzyme's activity. In contrast to this, a study showed that *Aphanocladium album* chitinase, referred to as Chitinase 1, does not require any cofactors for its activity (Kunz et al. 1992). Another study reports that the purified chitinase *Streptomyces luridiscabiei* U05 shows metal ion sensitivity, and its activity does not strictly depend on the addition of exogenous cofactors, but the presence of certain divalent metal ions such as Ca^{2+} has been found to increase the enzyme's activity, likely by helping to stabilize the active conformation or by optimizing substrate binding (Swiontek et al.

2019). Conversely, treatment with chelating agents (such as EDTA) leads to a decrease in activity.

7.7 Purification Techniques

The purification of chitinase enzymes has evolved significantly, driven by the need for high-purity enzymes in research and industrial applications. Modern workflows integrate traditional methods with cutting-edge innovations to maximize yield, purity, and sustainability.

The first critical step is cell disruption, which employs mechanical, chemical, or enzymatic methods to release intracellular chitinases. Traditional techniques like enzymatic digestion (lysozyme), homogenization, and sonication remain foundational. However, advanced methods such as high-pressure homogenization (HPH) and microfluidization now offer superior reproducibility and efficiency, achieving up to 95% cell lysis with minimal enzyme denaturation (dos Santos Aguilar et al. 2023). Subsequent fractionation separates cellular debris from crude enzyme extracts. Differential centrifugation and ultracentrifugation are widely used, but continuous flow centrifuges have improved throughput, enabled large-scale processing while maintaining enzyme stability (Oshima et al. 2022). Precipitation methods, such as ammonium sulfate salting-out and organic solvent (e.g., acetone) precipitation, remain cost-effective for concentrating chitinases. However, recent advancements prioritize green chemistry; for example, ionic liquids and polyethylene glycol (PEG)-based systems reduce environmental toxicity while achieving comparable purity (Tue et al. 2024).

Chromatography is pivotal for high-resolution purification. Ion-exchange chromatography (IEX) benefits from resins like DEAE-Sepharose Fast Flow, which offer higher binding capacities (up to 120 mg/mL) and selectivity for chitinases with distinct isoelectric points (Lee et al. 2019). Affinity chromatography has advanced through recombinant ligands, such as chitin-binding domains (CBDs) fused to cellulose matrices, and immobilized metal ion affinity chromatography (IMAC) using Ni^{2+} or Co^{2+} for His-tagged enzymes (Smith et al. 2020). Innovations in hydrophobic interaction chromatography (HIC), such as phenyl- or butyl-based resins, enhance the separation of chitinase isoforms by exploiting surface hydrophobicity (Berini et al. 2017).

Electrophoretic techniques like SDS-PAGE remain the gold standard for purity assessment, but capillary electrophoresis (CE) now achieves sub-nanogram sensitivity and resolves chitinase isoforms in minutes (Kim et al. 2022). Post-purification, enzymatic activity assays utilize fluorogenic substrates (e.g., 4-methylumbelliferyl-D-N,N′,N′‘-triacetylchitotriose) for real-time kinetic analysis, while high-throughput assays enable rapid screening of enzyme variants (Patel Price and Naumann 2011; Menghiu et al. 2021). Protein quantification increasingly relies on label-free mass spectrometry and nano-drop spectroscopy, which require only 1–2 μL of the sample (Tamhane et al. 2021).

Automation and machine learning (ML) are revolutionizing purification workflows. Robotic liquid handlers minimize human errors in column packing and buffer exchange, while ML algorithms optimize resin selection and elution gradients based on chitinase physicochemical properties (Ding et al. 2023). These innovations, combined with eco-friendly solvents and modular chromatography systems, are making chitinase purification faster, cheaper, and scalable for industrial applications.

7.8 Molecular Characterization and Structural Analysis

The molecular characterization and structural analysis of chitinases are critical for unraveling their biological roles, evolutionary relationships, and biotechnological potential (Unuofin et al. 2024). Chitinase genes are typically characterized through gene cloning, PCR amplification, and sequence analysis to identify conserved domains (e.g., catalytic GH18/19 motifs) and regulatory elements (Felse and Panda 1999; Aghbashlo et al. 2023). Recent advances in next-generation sequencing (NGS) have revolutionized this field by enabling rapid, high-throughput genome and transcriptome sequencing. For example, metagenomic NGS of soil microbiomes uncovered novel chitinase genes from unculturable bacteria, expanding the known diversity of these enzymes (Cretoiu et al. 2015). CRISPR-Cas9 gene editing further refines functional studies by enabling precise knockouts or modifications in chitinase genes (Han et al. 2024). Structural elucidation of chitinases has been transformed by cryo-electron microscopy (cryo-EM) and X-ray crystallography (Brain 2024). Cryo-EM, with its near-atomic resolution (1.8–3.0 Å), captures dynamic conformational changes during substrate binding and hydrolysis (Chen et al. 2023). X-ray crystallography remains indispensable for resolving active-site architectures, such as the conserved DXDXE motif in GH18 chitinases that coordinates catalytic protons (Meekrathok et al. 2017). Integrated bioinformatics pipelines, such as AlphaFold2 and SWISS-MODEL, predict chitinase structures with 90% accuracy to experimental data, accelerating functional annotation (Ezzine et al. 2024). Tools like Pfam and InterProScan identify conserved domains (e.g., chitin-binding modules), while PhyML reconstructs phylogenetic trees to trace evolutionary relationships (Yang et al. 2021).

7.9 Applications of Microbially Derived Chitinase

7.9.1 *Plant Disease Management*

Chitinases play a pivotal role in plant disease management by targeting chitin, a structural component of fungal cell walls, nematode eggshells, and insect exoskeletons (Gortari and Hours 2008). Plants produce chitinases as part of their innate defense mechanisms (Vaghela et al. 2022). Upon detecting fungal chitin fragments (chito-oligosaccharides) via pattern recognition receptors (PRRs), plants activate induced systemic resistance (ISR) (Pusztahelyi 2018). This primes jasmonic acid (JA) and ethylene (ET) signaling pathways, triggering chitinase synthesis and secretion into the apoplast to degrade invading pathogens (Shu et al. 2024). For instance, fungal chitin recognition in *Arabidopsis* activates MAP kinase cascades, leading to transcriptional upregulation of chitinase genes (e.g., PR3), which hydrolyze β-1,4-glycosidic bonds in chitin, destabilizing fungal cell walls (Reyes Zamora 2024).

Chitinases synergize with plant defense hormones like SA and JA. SA-mediated systemic acquired resistance (SAR) enhances chitinase expression during fungal attacks, while JA/ET pathways amplify chitinase activity against chewing insects (Jain et al. 2016). Chitinases are also well known for their nematicidal activity (Fig. 7.1). A study reported the inhibition of chitin biosynthesis in *Meloidogyne incognita* through RNA interference (RNAi) targeting chitin synthase genes exhibited increased resistance to the nematode, resulting in reduced gall formation and egg mass production (Mani et al. 2020). Chitinase efficacy varies with the target species. For instance, *S. marcescens* chitinases degrade Heteroderaglycines (soybean cyst nematode) eggshells but show limited activity against *Pratylenchus penetrans* (root-lesion nematode), highlighting the need for enzyme specificity (Dutta et al. 2021).

Microbial chitinases also serve as biocontrol agents (Veliz et al. 2017). Rhizosphere microbes like *Bacillus* spp. and *Trichoderm*a spp. colonize root surfaces, secreting chitinases that degrade fungal hyphae and nematode eggs (Fig. 7.1) (Gortari and Hours 2008). By using shrimp and crab shell waste, *B. subtilis* NPU 001 produced a 31 kDa chitinase that inhibits *F. oxysporum* and uniquely generates chitotriose as its main product (Chang et al. 2010). Metagenomic study reveals that rhizosphere microbes exhibit enhanced ability to suppress fungal diseases through chitinase production. For instance, a study identified a novel bacterial chitinase, Chi18H8, from suppressive soil using functional metagenomics. Chi18H8 demonstrated antifungal activity against several plant pathogens, including *Colletotrichum gloeosporioides, Fusarium graminearum, F. oxysporum*, and *Alternaria alternata* (Hjort et al. 2014).

Commercial products such as Kodiak® (*B. subtilis*), BioNem (*Paecilomyces lilacinus*), and MycoStop (*Streptomyces griseoviridis*) leverage chitinolytic activity for pest control, reducing the reliance on synthetic pesticides (Rajendran et al. 2024).

7.9.2 Bioremediation of Chitin-Containing Waste

As chitin is abundantly found in insect exoskeletons and crustacean shells, it presents a significant challenge in waste management (Kumari et al. 2023). In the seafood industry, millions of tons of crustacean shell waste are generated annually, posing disposal challenges due to chitin's resistance to natural degradation (Taokaew and Kriangkrai 2023). Conventional methods, such as high-temperature treatments

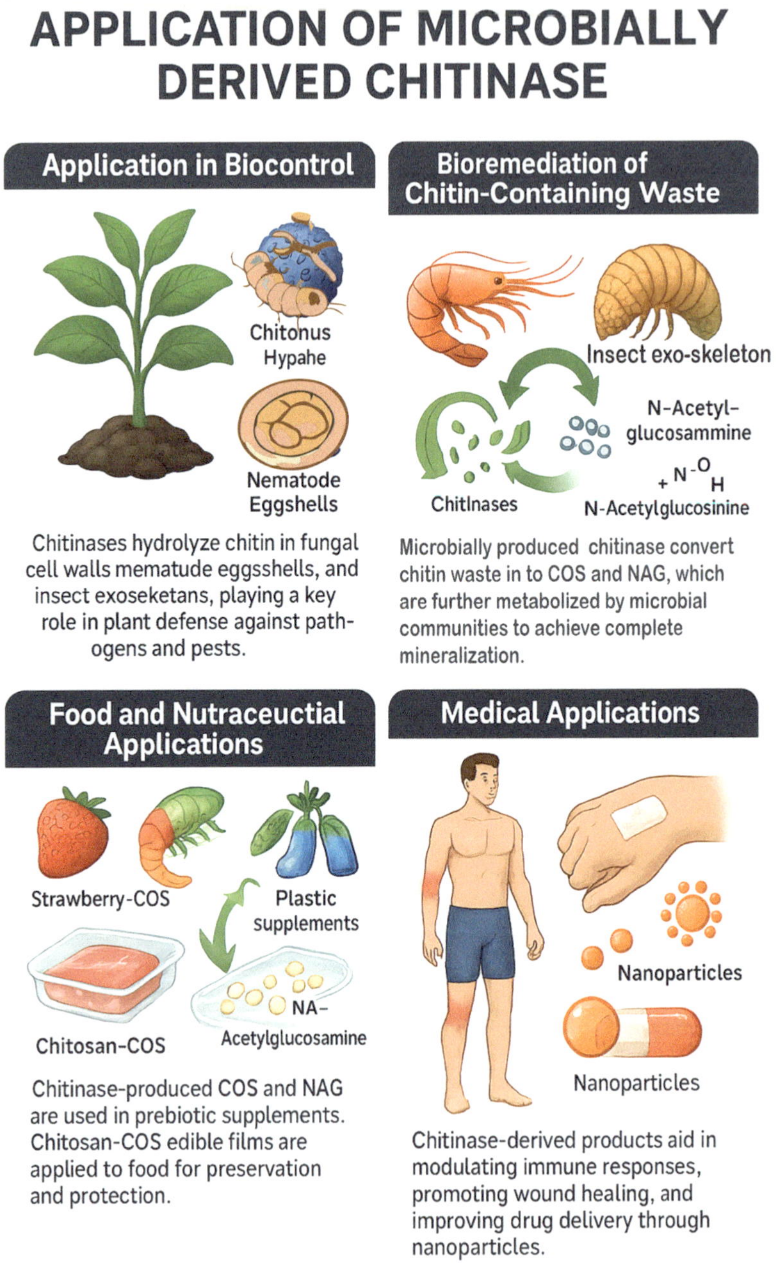

Fig. 7.1 Application of microbially derived chitinase

and chemical processes, are environmentally detrimental and energy-intensive. In contrast, chitinolytic enzymes like chitinases offer a sustainable and efficient solution (Fig. 7.1). These enzymes catalyze the hydrolysis of β-1,4 glycosidic bonds in chitin, producing smaller units like chito-oligosaccharides (COS) and N-acetylglucosamine (NAG), which are further metabolized by microbial communities to achieve complete mineralization (Nour et al. 2024a, b). The bioconversion of chitin waste into COS is particularly advantageous because these oligomers have additional uses. For instance, the COS produced by chitinase A from *S. marcescens* and further expressed in *E. coli* BL21(DE3) demonstrated its utility by forming stable emulsions with various hydrocarbons (Nour et al. 2024a, b). Its emulsification index remained stable across a wide range of salinity, pH, and temperature conditions and facilitated the recovery of motor oil, burned motor oil, and aniline from polluted sand (Nour et al. 2024a, b). Additionally, chitin itself can be employed as a biocarrier in oil spill cleanup. For instance, a hydrocarbon-degrading *Rhodococcus* strain immobilized on shrimp shell chitin showed improved survival and activity, achieving faster crude oil degradation in seawater compared to free cultures (Gentili et al. 2006).

The application of green solvents such as deep eutectic solvents (DESs) and natural deep eutectic solvents (NADESs) has emerged as a complementary technology to enhance chitin bioremediation (Durante et al. 2024). DESs, synthesized from components like quaternary ammonium salts and hydrogen-bond donors, and NADESs, derived from naturally occurring substances like sugars and organic acids, exhibit excellent solubility and biodegradability (Sulthan et al. 2023). These solvents disrupt the crystalline structure of chitin, making it more accessible for enzymatic degradation. Studies revealed that DES treatment can increase enzymatic hydrolysis rates multiple-fold compared to untreated chitin, highlighting their effectiveness in pretreatment processes (Li et al. 2022).

Chitinases and their derived products, such as chitosan and chitin oligomers, demonstrate significant potential in heavy-metal bioremediation by utilizing the trait of biosorption. This adsorption method is advantageous due to its biocompatibility, biodegradability, and cost-effectiveness paired with high efficiency. It requires minimal chemical input, supports regeneration of the biosorbent for reuse, and enables the recovery of adsorbed metals, promoting sustainable resource utilization. A study revealed that chitosan matrix after the adsorption of different concentrations of Cu (II) and Cr (VI) indicates that the biopolymeric matrix can be useful in metal remediation (Anbinder et al. 2019). In another study, chitosan nanoparticle (formed via ionic cross-linking) and calcium-alginate microbeads both adsorb Cu^{2+} via Langmuir isotherms at low concentrations, with alginate exhibiting ~2x higher capacity than chitosan due to denser binding sites (Yu et al. 2013). While chitosan nanoparticles slightly outperform bulk chitosan, alginate's superior adsorption persists even at higher concentrations. Hybrid alginate–chitosan particles show intermediate capacity, balancing their individual properties proportionally (Yu et al. 2013).

These examples highlight the versatility of chitinolytic systems in targeting diverse metal pollutants through the enzymatic degradation of chitin and subsequent biosorption, offering scalable, eco-friendly solutions for environmental detoxification.

7.9.3 Food and Nutraceutical Applications

The enzymatic hydrolysis of chitin into COS and NAG unlocks bioactive compounds with dual food technology and therapeutic value (Fig. 7.1). COS with a degree of polymerization (DP) of 4–6 exhibit superior prebiotic activity, selectively stimulating *Bifidobacterium* and *Lactobacillus* species in the colon (Ji et al. 2022). This modulation enhances short-chain fatty acid (SCFA) production, improving gut barrier function and mitigating inflammatory bowel disease (IBD) (Yang et al. 2023). NAG, a monomeric derivative, is utilized in functional foods for osteoarthritis management due to its role in cartilage repair (Nagaoka 2014).

Chitinase-processed COS are integral to next-generation edible films (Singh et al. 2021). Chitosan–COS nanocomposite coatings on fruits and vegetables are found to improve shelf life and storage at low temperature. For instance, application of foliar chitosan at 250, 500, and 1000 ppm enhances cold stress tolerance in three strawberry varieties (Florida, Festival, and Fortuna), boosting growth, yield, and activating cold-responsive genes and triggered synthesis of novel cold-tolerance proteins (Mahfouze et al. 2024). Another study showed that treating mature green tomatoes with 0.5 g L^{-1} chitosan oligosaccharide during room-temperature storage enhances post-harvest quality by boosting carotenoids (lycopene and β-carotene), vitamin C, and flavor volatiles (e.g., β-ionone). COS upregulated carotenoid biosynthesis genes (SlPSY1 and SlLCY-B2) and accelerated ethylene production via increased expression of ethylene-related genes (SlACS4 and SlETR3), suggesting COS accelerates ripening through ethylene signaling (Shao et al. 2022).

7.9.4 Medical Applications

Chito-oligosaccharides, derived from chitinase-mediated chitin hydrolysis, exhibit potent immunomodulatory effects by interacting with Toll-like receptor 4 (TLR4) and modulating macrophage polarization (Hao et al. 2023). For instance, low-molecular-weight COS (DP 3–5) suppress proinflammatory cytokines (e.g., TNF-α, IL-6) in murine models of colitis, while enhancing anti-inflammatory IL-10 production (Chen et al. 2021). Clinical trials have demonstrated that COS supplementation reduces Crohn's disease relapse rates significantly. For instance, in mouse models of acute and chronic colitis induced by dextran sulfate sodium (DSS) and acetic acid, oral administration of COS significantly reduced mortality and intestinal inflammation, with optimal efficacy observed at doses between 10 and 20 mg/kg/day (Yousef et al. 2012). Additionally, COS treatment suppressed inflammation markers, including NF-κB activation and the proinflammatory cytokines TNF-α and IL-6 in colonic tissues.

However, N-acetylglucosamine (NAG) is applied as a primary chitinase degradation product. For instance, a study utilized short-fiber poly-N-acetyl glucosamine nanofibers (sNAG) and vacuum-assisted closure (VAC) therapy on wound healing in diabetic mice (Erba et al. 2011). The results demonstrated that applying sNAG membranes 24 h before VAC therapy significantly improved wound healing kinetics, increased angiogenesis, and elevated the expression of wound-related growth factors. The findings suggest that integrating sNAG with VAC therapy may offer a synergistic approach to promote wound healing, particularly in diabetic individuals. FDA-approved NAG-based dressings, such as ChitoFlex®, are widely used in burn care, showing improved epithelialization compared to conventional gauze (Sundaram et al. 2021).

COS have emerged as essential players in antitumor and adjuvant therapy. For instance, fungal chitinase (*Aspergillus fumigatus* Chia) induces apoptosis in melanoma (A375) cells by upregulating caspase-3 expression (Pavan et al. 2022). Another study by (Park et al. 2020) showed the role of chitinase in degenerating cancer cell progression. They generated Chi3L1 knockout mice using the CRISPR/Cas9 system to assess the impact of Chi3L1 deficiency on lung tumor formation. The findings revealed that Chi3L1 knockout mice exhibited a significant reduction in lung tumor nodules by increasing the levels of tumor suppressor proteins, including p53, p21, BAX, and cleaved caspase 3, and decreasing the levels of cyclin E1, CDK2, and phosphorylated STAT.

Also, in targeted drug delivery systems chitosan–COS nanocomposites are reported to enable pH-sensitive drug release, improving chemotherapeutic precision. For example, doxorubicin-loaded COS nanoparticles demonstrated tumor regression in murine breast cancer models, with 60% lower cardiotoxicity than free doxorubicin (Goyali et al. 2023). Similarly, COS-functionalized liposomes enhance blood–brain barrier permeability, achieving 5× higher paclitaxel delivery to glioblastoma cells (Haghighi et al. 2020). These are also useful in the preparation of biocompatible medical implants. For instance, in a study, chitinase-processed COS coatings on titanium implants reduce bacterial biofilm formation through electrostatic disruption of microbial membranes (Sasaki et al. 2022).

7.10 Conclusion

Chitinase is a pivotal enzyme that facilitates the hydrolysis of chitin, contributing significantly to nutrient recycling, ecosystem balance, and sustainable biocontrol strategies. Its diverse applications extend to plant disease management, insect pest control, nematode suppression, and bioremediation, underscoring its versatility across agricultural, environmental, and industrial domains. Notably, microbially derived chitinase offers distinct advantages, such as high production yields, scalability, and adaptability to diverse environmental conditions. By disrupting the structural integrity of fungal pathogens and pest exoskeletons, chitinase emerges as a potent biocontrol tool, enhancing crop protection while reducing the reliance on chemical pesticides. Furthermore, its eco-friendly nature aligns with integrated pest management strategies, promoting biodiversity and minimizing risks to nontarget organisms. Beyond biocontrol, chitinase plays a critical role in biodegradation, transforming chitin-based waste into valuable byproducts like chito-oligosaccharides, chitosan, and glucosamine, which have vast industrial and biomedical applications.

References

Abd-Aziz S, Fernandez CC, Salleh MM, Illias RM, Hassan MA (2008) Effect of agitation and aeration rates on chitinase production using *Trichoderma virens* UKM1 in 2-l stirred tank reactor. Appl Biochem Biotechnol 150(2):193–204. https://doi.org/10.1007/s12010-008-8140-4

Abdel Wahab WA, El-Dein AN, Hussein M, Mostafa FA, Saleh SA (2023) Kinetic, thermodynamic and bio-applicable studies on *Aspergillus niger* Mk981235 chitinase. Catal Lett 153(4):1089–1095. https://doi.org/10.1007/s10562-022-04163-9

Abdel-Azeem AM, Abdel-Azeem MA, Abdul-Hadi SY, Darwish AG (2019) *Aspergillus*: biodiversity, ecological significances, and industrial applications. In: Fungal Biology, pp 121–179. https://doi.org/10.1007/978-3-030-10480-1_4

Abdelraouf AMN, Al-Hazmi NE, Naguib DM (2024) *Streptomyces griseus* versus *Trichoderma viride* chitinase as an anti-inflammatory and antifungal agent against human pathogenic fungi. Indian J Microbiol 1-7. https://doi.org/10.1007/s12088-024-01045-6

Abo Elsoud MM, El Kady EM (2019) Current trends in fungal biosynthesis of chitin and chitosan. Bull Natl Res Cent 43(1):59. https://doi.org/10.1186/s42269-019-0105-y

Abu Yazid N, Barrena R, Komilis D, Sánchez A (2017) Solid-state fermentation as a novel paradigm for organic waste valorization: a review. Sustain For 9(2):224. https://doi.org/10.3390/su9020224

Abulikemu S, Yesilyurt A, Gencer D, Usta M, Nalcacioglu R (2021) Comparison of the potential activities of viral and bacterial chitinases. Egypt J Biol Pest Control 31(1):91. https://doi.org/10.1186/s41938-021-00423-3

Acharya B, Ingram TW, Oh Y, Adhikari TB, Dean RA, Louws FJ (2020) Opportunities and challenges in studies of host-pathogen interactions and management of *Verticillium dahliae* in tomatoes. Plan Theory 9(11):1622. https://doi.org/10.3390/plants9111622

Adnan M, Zafar M, Anwar Z (2024) Screening of chitinolytic microfungi and optimization of parameters for hyperproduction of chitinase through solid-state fermentation technique. Appl Biochem Biotechnol 196(4):1840–1862. https://doi.org/10.1007/s12010-023-04663-y

Aghbashlo M, Amiri H, Moosavi Basri SM, Rastegari H, Lam SS, Pan J, Gupta VK, Tabatabaei M (2023) Tuning chitosan's chemical structure for enhanced biological functions. Trends Biotechnol 41(6):785–797. https://doi.org/10.1016/j.tibtech.2022.11.009

Aguilar JGDS (2023) Applications of high-pressure homogenization on microbial enzymes. In: Elsevier eBooks, pp 373–403. https://doi.org/10.1016/b978-0-323-98386-0.00006-3

Ahmed S, Opal SM (2020) Chitin and chitinases: the unfolding biological connections. Nat Rev Microbiol. https://doi.org/10.1038/s41579-020-0364-7

Aita BC, Spannemberg SS, Schmaltz S, Zabot GL, Tres MV, Kuhn RC, Mazutti MA (2019) Production of cell-wall degrading enzymes by solid-state fermentation using agroindustrial residues as substrates. J Environ Chem Eng 7(3):103193. https://doi.org/10.1016/j.jece.2019.103193

Akeed Y, Atrash F, Naffaa W (2020) Partial purification and characterization of chitinase produced by *Bacillus licheniformis* B307. Heliyon 6(5). https://doi.org/10.1016/j.heliyon.2020.e03858

Akimoto C, Aoyagi H, Dicosmo F, Tanaka H (2000) Synergistic effect of active oxygen species and alginate on chitinase production by Wasabia japonica cells and its application. J Biosci Bioeng 89(2):131–137. https://doi.org/10.1016/s1389-1723(00)88726-5

Alsina C, Faijes M, Planas A (2019) Glycosynthase-type GH18 mutant chitinases at the assisting catalytic residue for polymerization of chitooligosaccharides. Carbohydr Res 478:1–9. https://doi.org/10.1016/j.carres.2019.04.001

Alves TB, de Oliveira Ornela PH, de Oliveira AHC, Jorge JA, Guimarães LHS (2018) Production and characterization of a thermostable antifungal chitinase secreted by the filamentous fungus *Aspergillus niveus* under submerged fermentation. 3. Biotech 8(8):369. https://doi.org/10.1007/s13205-018-1397-6

Anbinder PS, Macchi C, Amalvy J, Somoza A (2019) A study of the structural changes in a chitosan matrix produced by the adsorption of copper and chromium ions. Carbohydr Polym 222:114987. https://doi.org/10.1016/j.carbpol.2019.114987

Anwar W, Amin H, Khan HAA, Akhter A, Bashir U, Anjum T, Kalsoom R, Javed MA, Zohaib KA (2023) Chitinase of *Trichoderma longibrachiatum* for control of *Aphis gossypii* in cotton plants. Sci Rep 13(1):13181. https://doi.org/10.1038/s41598-023-39965-y

Atheena PV, Rajesh KM, Raval K, Selvaraj S, Raval R (2024) Identification and characterization of chitinase producing marine microorganism: unleashing the potential of chitooligosaccharides for bioethanol synthesis. Int J Biol Macromol 265:130846. https://doi.org/10.1016/j.ijbiomac.2024.130846

Balabanova L, Slepchenko L, Son O, Tekutyeva L (2018) Biotechnology potential of marine fungi degrading plant and algae polymeric substrates. Front Microbiol 9:1527. https://doi.org/10.3389/fmicb.2018.01527

Baldoni DB, Antoniolli ZI, Mazutti MA, Jacques RJS, Dotto AC, de Oliveira Silveira A, Ferraz RC, Soares VB, de Souza ARC (2020) Chitinase production by *Trichoderma koningiopsis* UFSMQ40 using solid state fermentation. Brazil J Microbiol 51(4):1897–1908. https://doi.org/10.1007/s42770-020-00334-w

Ballester AR et al (2017) Pathogenicity and host specificity of *Penicillium expansum* in apple orchards. Plant Dis 101(4):606–614. https://doi.org/10.1094/PDIS-07-16-1029-RE

Banerjee S, Mandal NC (2019) Diversity of Chitinase-producing bacteria and their possible role in plant pest control. In: Springer eBooks, pp 457–491. https://doi.org/10.1007/978-981-13-8487-5_18

Behera BC (2020) Citric acid from *Aspergillus niger*: a comprehensive overview. Crit Rev Microbiol 46(6):727–749. https://doi.org/10.1080/1040841X.2020.1828815

Berini F, Presti I, Beltrametti F, Pedroli M, Vårum KM, Pollegioni L, Sjöling S, Marinelli F (2017) Production and characterization of a novel antifungal chitinase identified by functional screening of a suppressive-soil metagenome. Microb Cell Factories 16(1):16. https://doi.org/10.1186/s12934-017-0634-8

Brain L (2024) Characterisation of two chitin synthases from *Fusarium graminearum* using yeast as a model. La Trobe. Thesis. https://doi.org/10.26181/26500033.v1

Brurberg MB, Nes IF, Eijsink VGH (1996) Comparative studies of chitinases A and B from *Serratia marcescens*. Microbiology 142(7):1581–1589. https://doi.org/10.1099/13500872-142-7-1581

Canter EJ, Cuellar-Gempeler C, Pastore AI, Miller TE, Mason OU (2018) Predator identity more than predator richness structures aquatic microbial assemblages in *Sarracenia purpurea* leaves. Ecology 99(3):652–660. https://doi.org/10.1002/ecy.2128

Chakravarty J, Edwards TA (2022) Innovation from waste with biomass-derived chitin and chitosan as green and sustainable polymer: a review. Energy Nexus 8:100149. https://doi.org/10.1016/j.nexus.2022.100149

Chang W-T, Chen M-L, Wang S-L (2010) An antifungal chitinase produced by *Bacillus subtilis* using chitin waste as a carbon source. World J Microbiol Biotechnol 26(5):945–950. https://doi.org/10.1007/s11274-009-0244-7

Chaulagain D, Shamabadi NS, Leslie SA, Karig DK (2024) From natural microbe screening to sustained chitinase activity in exogenous hosts. ACS Synth Biol 13(4):1165–1176. https://doi.org/10.1021/acssynbio.3c00637

Cheba BA, Zaghloul TI, El-Mahdy AR, El-Massry MH (2018) Effect of nitrogen sources and fermentation conditions on bacillus sp. R2 chitinase production. Procedia Manufacturing 22:280–287. https://doi.org/10.1016/j.promfg.2018.03.043

Chen H-B, Kao P-M, Huang H-C, Shieh C-J, Chen C-I, Liu Y-C (2010) Effects of using various bioreactors on chitinolytic enzymes production by *Paenibacillus taichungensis*. Biochem Eng J 49(3):337–342. https://doi.org/10.1016/j.bej.2010.01.008

Chen L, Jiang H, Cheng Q, Chen J, Wu G, Kumar A, Sun M, Liu Z (2015) Enhanced nematicidal potential of the chitinase pachi from *Pseudomonas aeruginosa* in association with Cry21Aa. Sci Rep 5(1). https://doi.org/10.1038/srep14395

Chen W, Jiang X, Yang Q (2020) Glycoside hydrolase family 18 chitinases: the known and the unknown. Biotechnol Adv 43:107553. https://doi.org/10.1016/j.biotechadv.2020.107553

Chen L et al (2021) Chito-oligosaccharides from *Verticillium* chitinases improve soybean nodulation and nitrogen fixation. Front Plant Sci 12:734521. https://doi.org/10.3389/fpls.2021.734521

Chen DD, Wang ZB, Wang LX, Zhao P, Yun CH, Bai L (2023) Structure, catalysis, chitin transport, and selective inhibition of chitin synthase. Nat Commun 14(1):4776. https://doi.org/10.1038/s41467-023-40479-4

Clem RJ, Passarelli AL (2013) Baculoviruses: sophisticated pathogens of insects. PLoS Pathog 9(11):e1003729. https://doi.org/10.1371/journal.ppat.1003729

Cretoiu MS, Berini F, Kielak AM, Marinelli F, Van Elsas JD (2015) A novel salt-tolerant chitobiosidase discovered by genetic screening of a metagenomic library derived from chitin-amended agricultural soil. Appl Microbiol Biotechnol 99(19):8199–8215. https://doi.org/10.1007/s00253-015-6639-5

Crini G (2022) 2.Chitin and chitosan. Discoveries and applications for sustainability. https://doi.org/10.1007/978-0-323-96119-6

da Silva DMG, Pedrosa FR, Ângela Taipa M, Costa R, Keller-Costa T (2023) Widespread occurrence of chitinase-encoding genes suggests the Endozoicomonadaceae family as a key player in chitin processing in the marine benthos. ISME Commun 3(1):109. https://doi.org/10.1038/s43705-023-00316-7

Das S, Chilukoti N, Pvsrn S, Jogi M, Pallinti PN, Kaur M, Dutta S, Podile AR (2012) Microbial chitinases for chitin waste management. Microorganisms in Environmental Management. https://doi.org/10.1007/978-94-007-2229-3_6

Das T, Bhattacharyya A, Bhar A (2023) Linking phyllosphere and rhizosphere microbiome to the plant–insect interplay: the new dimension of tripartite interaction. Physiologia 3(1):129–144. https://doi.org/10.3390/physiologia3010011

de Jonge R et al (2023) Competitive interactions between *Verticillium dahliae* and *Fusarium oxysporum* in the tomato rhizosphere. Environ Microbiol 25(3):645–658. https://doi.org/10.1111/1462-2920.16312

de la Cruz J, Rey M, Lora JM, Hidalgo-Gallego A, Domínguez F, Pintor-Toro JA, Llobell A, Benítez T (1993) Carbon source control on *β*-glucanases, chitobiase and chitinase from *Trichoderma harzianum*. Arch Microbiol 159(4):316–322. https://doi.org/10.1007/BF00290913

Dhevagi P, Ramya A, Priyatharshini S, Thanuja KG, Ambreetha S, Nivetha A (2021) Industrially important fungal enzymes: productions and applications. In Fungal biology (pp. 263–309). https://doi.org/10.1007/978-3-030-68260-6_11

Ding Y, Chen S, Liu H, Liu T, Yang Q (2023) Discovery of multitarget inhibitors against insect chitinolytic enzymes via machine learning-based virtual screening. J Agric Food Chem 71(23):8769–8777. https://doi.org/10.1021/acs.jafc.3c00633

Dukat AM, Kuznetsova AM, Klyagin SD, Trushin VO, Klyukina AA, El'cheninov AG, Danilova IV (2024) Chitin degradation by microbial communities of the Kandalaksha Bay, White Sea. Microbiology 93(1):61–78. https://doi.org/10.1134/s0026261723603111

Durante-Salmerón DA, Fraile-Gutiérrez I, Gil-Gonzalo R, Acosta N, Aranaz I, Alcántara AR (2024) Strategies to prepare chitin and chitosan-based bioactive structures aided by deep eutectic solvents: a review. Catalysts 14(6):371. https://doi.org/10.3390/catal14060371

Dutta TK, Phani V, Rao U (2021) Chitinase-mediated disruption of nematode eggshells: A sustainable approach for root-knot nematode control. Biol Control 160:104678. https://doi.org/10.1016/j.biocontrol.2021.104678

Dwyer K, Bentley IS, Fitzpatrick DA, Saleh AA, Tighe E, McGleenan E, Gaffney D, Walsh G (2023) Recombinant production, characterization and industrial application testing of a novel acidic exo/endo-chitinase from *Rasamsonia emersonii*. Extremophiles 27(2):10. https://doi.org/10.1007/s00792-023-01293-4

El-Sayed GM, Emam MT, Hammad MA, Mahmoud SH (2024) Gene cloning, heterologous expression, and in silico analysis of chitinase B from *Serratia marcescens* for biocontrol of *Spodoptera frugiperda* larvae infesting maize crops. Molecules 29(7):1466. https://doi.org/10.3390/molecules29071466

Erba P, Adini A, Demcheva M, Valeri CR, Orgill DP (2011) Poly-N-acetyl glucosamine fibers are synergistic with vacuum-assisted closure in augmenting the healing response of diabetic mice. J Trauma 71(2 Suppl 1):S187–S193. https://doi.org/10.1097/TA.0b013e318225583c

Ezzine A, Ben Hadj Mohamed S, Bezzine S, Aoudi Y, Hajlaoui MR, Baciou L, Smaali I (2024) Improved expression of a thermostable GH18 bacterial chitinase in two different Escherichia coli strains and its potential use in plant protection and biocontrol of phytopathogenic fungi. Mol Biotechnol 66(9):2635–2647. https://doi.org/10.1007/s12033-023-01041-1

Falade AO, Adewole KE, Ekundayo TC (2021) Aptitude of endophytic microbes for production of novel biocontrol agents and industrial enzymes towards agro-industrial sustainability. BJBAS 10(1). https://doi.org/10.1186/s43088-021-00146-3

Felse PA, Panda T (1999) Regulation and cloning of microbial chitinase genes. Appl Microbiol Biotechnol 51(2):141–151. https://doi.org/10.1007/s002530051374

Folders J, Algra J, Roelofs MS et al (2001) Characterization of *Pseudomonas aeruginosa* chitinase, a gradually secreted protein. J Bacteriol 183(10):3040–3048. https://doi.org/10.1128/JB.183.10.3040-3048.2001

Funkhouser JD, Aronson NN Jr (2007) Chitinase family GH18: evolutionary insights from the genomic history of a diverse protein family. BMC Evol Biol 7:96. https://doi.org/10.1186/1471-2148-7-96

Gentili AR, Cubitto MA, Ferrero M, Rodriguéz MS (2006) Bioremediation of crude oil polluted seawater by a hydrocarbon-degrading bacterial strain immobilized on chitin and chitosan flakes. Int Biodeterior Biodegradation 57(4):222–228. https://doi.org/10.1016/j.ibiod.2006.02.009

Golzan SR, Talaei-Hassanloui R, Homayoonzadeh M, Safavi SA (2023) Role of cuticle-degrading enzymes of *Beauveria bassiana* and *Metarhizium anisopliae* in virulence on *Plodia interpunctella* (Lepidoptera, Pyralidae) larvae. J Asia Pac Entomol 26(2):102038. https://doi.org/10.1016/j.aspen.2023.102038

Gomaa EZ (2012) Chitinase production by *Bacillus thuringiensis* and *Bacillus licheniformis*: their potential in antifungal biocontrol. J Microbiol (Seoul, Korea) 50(1):103–111. https://doi.org/10.1007/s12275-012-1343-y

Gomaa EZ (2021) Microbial chitinases: properties, enhancement and potential applications. Protoplasma 258(4):695–710. https://doi.org/10.1007/s00709-021-01612-6

Gooday GW (1990) The ecology of chitin degradation. Adv Microb Ecol 11:387–430. https://doi.org/10.1007/978-1-4684-7612-5_8

Gortari MC, Hours RA (2008) Fungal chitinases and their biological role in the antagonism onto nematode eggs. A review. Mycol Progress 7(4):221–238. https://doi.org/10.1007/s11557-008-0571-3

Goyali JC, Patel R, Kim S (2023) pH-responsive chitosan-COS nanocomposites for targeted delivery of doxorubicin in breast cancer therapy. J Control Release 354:1–14. https://doi.org/10.1016/j.jconrel.2023.03.012

Guo H-Z, Wang D, Yang H-T, Wu Y-L, Li Y-C, Xia G-H, Zhang X-Y (2024) Heterologous expression and characterization of a pH-stable chitinase from *Micromonospora aurantiaca* with a potential application in chitin degradation. Mar Drugs 22(6):287. https://doi.org/10.3390/md22060287

Haghighi H, Licciardello F, Pulvirenti A (2020) COS-functionalized liposomes for enhanced blood-brain barrier permeability in glioblastoma treatment. Acta Biomater 113:393–405. https://doi.org/10.1016/j.actbio.2020.06.037

Halder SK, Mondal KC (2018) Microbial valorization of chitinous bioresources for chitin extraction and production of chito-oligomers and N-acetylglucosamine: trends, perspectives and prospects. In: Microbial Biotechnology (pp. 69–107). https://doi.org/10.1007/978-981-10-7140-9_4

Hamid R, Khan MA, Ahmad M, Ahmad MM, Abdin MZ, Musarrat J, Javed S (2013) Chitinases: an update. J Pharm Bioallied Sci 5(1):21–29. https://doi.org/10.4103/0975-7406.106559

Han J, Ullah M, Andoh V, Khan MN, Feng Y, Guo Z, Chen H (2024) Engineering bacterial chitinases for industrial application: from protein engineering to bacterial strains mutation! A comprehensive review of physical, molecular, and computational approaches. J Agric Food Chem 72(42):23082–23096. https://doi.org/10.1021/acs.jafc.4c06856

Hao W, Li K, Liu S, Yu H, Li P, Xing R (2023) Pleiotropic modulation of chitooligosaccharides on inflammatory signaling in LPS-induced macrophages. Polymers 15(7):1613. https://doi.org/10.3390/polym15071613

Haran S, Schickler H, Chet I (1996) Molecular mechanisms of lytic enzymes involved in the biocontrol activity of *Trichoderma harzianum*. Microbiology 142(9):2321–2331. https://doi.org/10.1099/13500872-142-9-2321

Harini N, Han YX, Sukweenadhi J (2021) The role of calcium ions to improve activity of chitinase isolated from Vibrio sp. Jordan J Biol Sci 14(05):925–931. https://doi.org/10.54319/jjbs/140507

Harman GE et al (2004) *Trichoderma* species—opportunistic, avirulent plant symbionts. Nat Rev Microbiol 2(1):43–56. https://doi.org/10.1038/nrmicro797

Hermosa R et al (2012) Plant-beneficial effects of *Trichoderma* and of its genes. Microbiology 158(1):17–25. https://doi.org/10.1099/mic.0.052274-0

Hernández-Melchor DJ, Guerrero-Chávez AC, Ferrera-Rodríguez MR, Ferrera-Cerrato R, Larsen J, Alarcón A (2023) Cellulase and chitinase activities and antagonism against *Fusarium oxysporum* f. sp. *cubense* race 1 of six *Trichoderma* strains isolated from Mexican maize cropping. Biotechnol Lett 45(3):387–400. https://doi.org/10.1007/s10529-023-03316-5

Hjort K, Presti I, Elväng A, Marinelli F, Sjöling S (2014) Bacterial chitinase with phytopathogen control capacity from suppressive soil revealed by functional metagenomics. Appl Microbiol Biotechnol 98(6):2819–2828. https://doi.org/10.1007/s00253-013-5287-x

Hoell IA, Vaaje-Kolstad G, Eijsink VG (2010) Structure and function of enzymes acting on chitin and chitosan. Biotechnol Genet Eng Rev 27(1):331–366. https://doi.org/10.1080/02648725.2010.10648156

Horn SJ, Sørbotten A, Synstad B, Sikorski P, Sørlie M, Vårum KM, Eijsink VG (2006) Endo/exo mechanism and processivity of family 18 chitinases produced by *Serratia marcescens*. FEBS J 273(3):491–503. https://doi.org/10.1111/j.1742-4658.2005.05079.x

Huet J, Rucktooa P, Clantin B, Azarkan M, Looze Y, Villeret V, Wintjens R (2008) X-ray structure of papaya chitinase reveals the substrate binding mode of glycosyl hydrolase family 19 chitinases. Biochemistry 47(32):8283–8291. https://doi.org/10.1021/bi800655u

Hussain M, Wilson JB (2013) New paralogues and revised time line in the expansion of the vertebrate GH18 family. J Mol Evol 76(4):240–260. https://doi.org/10.1007/s00239-013-9553-4

Itoh T, Kimoto H (2019) Bacterial chitinase system as a model of chitin biodegradation. Adv Exp Med Biol 1142:131–151. https://doi.org/10.1007/978-981-13-7318-3_7

Jain S, Vaishnav A, Kumari S, Varma A, Tuteja N, Choudhary DK (2016) Chitinolytic bacillus-mediated induction of Jasmonic acid and defense-related proteins in soybean (*Glycine max* L. Merrill) plant against Rhizoctonia solani and *Fusarium oxysporum*. J Plant Growth Regul 36(1):200–214. https://doi.org/10.1007/s00344-016-9630-1

Jamek SB, Nyffenegger C, Muschiol J, Holck J, Meyer AS, Mikkelsen JD (2017) Characterization of two novel bacterial type A exo-chitobiose hydrolases having C-terminal 5/12-type carbohydrate-binding modules. Appl Microbiol Biotechnol 101(11):4533–4546. https://doi.org/10.1007/s00253-017-8198-4

Jeong GJ, Khan F, Tabassum N, Kim YM (2023) Chitinases as key virulence factors in microbial pathogens: understanding their role and potential as therapeutic targets. Int J Biol Macromol 249:126021. https://doi.org/10.1016/j.ijbiomac.2023.126021

Ji X, Zhu L, Chang K, Zhang R, Chen Y, Yin H, Jin J, Zhao L (2022) Chitooligosaccahrides: digestion characterization and effect of the degree of polymerization on gut microorganisms to manage the metabolome functional diversity in vitro. Carbohydr Polym 275:118716. https://doi.org/10.1016/j.carbpol.2021.118716

Kahlon RS (2016) Pseudomonas: genome and comparative genomics. In Springer eBooks (pp. 127–191). https://doi.org/10.1007/978-3-319-31198-2_4

Kapat A, Rakshit SK, Panda T (1996) Optimization of carbon and nitrogen sources in the medium and environmental factors for enhanced production of chitinase by *Trichoderma harzianum*. Bioprocess Eng 15(1):13–20. https://doi.org/10.1007/bf00435522

Karthik N, Binod P, Pandey A (2017) Chitinases. In: Current Developments in Biotechnology and Bioengineering. Elsevier, pp 335–368

Karwelat D, Schmeck B, Ringel M, Benedikter BJ, Hübner K, Beinborn I et al (2020) Influenza virus-mediated suppression of bronchial Chitinase-3-like 1 secretion promotes secondary pneumococcal infection. FASEB J 34(12):16432–16448. https://doi.org/10.1096/fj.202001688R

Keyhani NO, Roseman S (1999) Physiological aspects of chitin catabolism in marine bacteria. Biochim Biophys Acta 1473(1):108–122. https://doi.org/10.1016/S0304-4165(99)00153-2

Khan RAA et al (2020) Biocontrol potential of *Penicillium* spp. against plant pathogens. J Fungi 6(4):259. https://doi.org/10.3390/jof6040259

Khokhani D, Carrera Carriel C, Vayla S, Irving TB, Stonoha-Arther C, Keller NP, Ané JM (2021) Deciphering the chitin code in plant symbiosis, defense, and microbial networks. Ann Rev Microbiol 75:583–607. https://doi.org/10.1146/annurev-micro-051921-114809

Kielak AM, Cretoiu MS, Semenov AV, Sørensen SJ, van Elsas JD (2013) Bacterial chitinolytic communities respond to chitin and pH alteration in soil. Appl Environ Microbiol 79(1):263–272. https://doi.org/10.1128/AEM.02546-12

Kim H et al (2022) Capillary electrophoresis for rapid enzyme analysis. Anal Chem 94(8):3565–3573. https://doi.org/10.1021/acs.analchem.1c05234

Klosterman SJ, Subbarao KV, Kang S, Veronese P, Gold SE, Thomma BP, Chen Z, Henrissat B, Lee YH, Park J, Garcia-Pedrajas MD, Barbara DJ, Anchieta A, de Jonge R, Santhanam P, Maruthachalam K, Atallah Z, Amyotte SG, Paz Z, Inderbitzin P et al (2011) Comparative genomics yields insights into niche adaptation of plant vascular wilt pathogens. PLoS Pathog 7(7):e1002137. https://doi.org/10.1371/journal.ppat.1002137

Krishna C (2005) Solid-state fermentation systems—an overview. Crit Rev Biotechnol 25(1–2):1–30. https://doi.org/10.1080/07388550590925383

Kumari R, Kumar M, Vivekanand V, Pareek N (2023) Chitin biorefinery: a narrative and prophecy of crustacean shell waste sustainable transformation into bioactives and renewable energy. Renew Sust Energ Rev 184:113595. https://doi.org/10.1016/j.rser.2023.113595

Kunz C, Sellam O, Bertheau Y (1992) Purification and characterization of a chitinase from the hyperparasitic fungus *Aphanocladium album*. Physiol Mol Plant Pathol 40(2):117–131. https://doi.org/10.1016/0885-5765(92)90039-X

Langner T, Göhre V (2016) Fungal chitinases: function, regulation, and potential roles in plant/pathogen interactions. Curr Genet 62(2):243–254. https://doi.org/10.1007/s00294-015-0530-x

Le B, Yang SH (2019) Microbial chitinases: properties, current state and biotechnological applications. World J Microbiol Biotechnol *35*(9):144. https://doi.org/10.1007/s11274-019-2721-y

Lee SH et al (2019) Ion-exchange chromatography optimization for chitinase purification. J Chromatogr B 1125:121713. https://doi.org/10.1016/j.jchromb.2019.121713

Li Z, Liu C, Hong S, Lian H, Mei C, Lee J, Wu Q, Hubbe MA, Li M (2022) Recent advances in extraction and processing of chitin using deep eutectic solvents. Chem Eng J 446:136953. https://doi.org/10.1016/j.cej.2022.136953

Li Y et al (2024) Chitinase diversity and function in *Verticillium* species: insights into soil ecology and pathogenicity. Fungal Biol Rev 38:100345. https://doi.org/10.1016/j.fbr.2024.100345

Liang B, Song W, Xing R, Liu S, Yu H, Li P (2023) The source, activity influencing factors and biological activities for future development of chitin deacetylase. Carbohydr Polym 321:121335. https://doi.org/10.1016/j.carbpol.2023.121335

Lin FP, Wu CY, Chen HN, Lin HJ (2015) Effects of C-terminal domain truncation on enzyme properties of *Serratia marcescens* chitinase C. Appl Biochem Biotechnol 175(8):3617–3627. https://doi.org/10.1007/s12010-015-1530-5

Liu J, Xu Q, Wu Y, Sun D, Zhu J, Liu C, Liu W (2023) Carbohydrate-binding modules of ChiB and ChiC promote the chitinolytic system of *Serratia marcescens* BWL1001. Enzym Microb Technol 162:110118. https://doi.org/10.1016/j.enzmictec.2023.110118

López-García CL, Guerra-Sánchez G, Santoyo-Tepole F, Olicón-Hernández DR (2024) Chitinase induction in *Trichoderma harzianum*: a solid-state fermentation approach using shrimp waste and wheat bran/commercial chitin for chitooligosaccharides synthesis. Prep Biochem Biotechnol 54(8):1040–1050. https://doi.org/10.1080/10826068.2024.2313631

Lorito M et al (2010) Genes from mycoparasitic fungi as a source for improving plant resistance to fungal pathogens. Proc Natl Acad Sci 107(25):11581–11586. https://doi.org/10.1073/pnas.1009912107

Lu D-L, Zhang M-S, Deng J-J, Lu W-J, Yang Z-D, Li Z-W, Chen Y-C, Luo X-C (2023) Highly efficient shrimp shell recovery by solid-state fermentation with *Streptomyces* sp. SCUT-3. Chem Eng J 458:141256. https://doi.org/10.1016/j.cej.2022.141256

Mahajan G, Thadiyan V, Bisht T, Gupta R (2024) Optimizing microbial chitinase production through submerged fermentation: A comparative study of *Bacillus* and *Streptomyces* strains. Appl Biochem Biotechnol 196(5):2456–2470. https://doi.org/10.1007/s12010-024-04923-5

Mahfouze SA, Mahfouze HA, Helmi RY, Fathallah FB, Aboud KA, Ottai ME (2024) The role of chitosan in improving the cold stress tolerance in strawberry varieties. Jordan J Biol Sci *17*(04):645–655. https://doi.org/10.54319/jjbs/170408

Mani V, Reddy CS, Lee S-K, Park S, Ko H-R, Kim D-G, Hahn B-S (2020) Chitin biosynthesis inhibition of *Meloidogyne incognita* by RNAi-mediated gene silencing increases resistance to transgenic tobacco plants. Int J Mol Sci 21(18):6626. https://doi.org/10.3390/ijms21186626

Manzar N, Kashyap AS, Goutam RS, Rajawat MVS, Sharma PK, Sharma SK, Singh HV (2022) *Trichoderma*: advent of versatile biocontrol agent, its secrets and insights into mechanism of biocontrol potential. Sustain For 14(19):12786. https://doi.org/10.3390/su141912786

Meekrathok P, Kukic P, Nielsen JE, Suginta W (2017) Investigation of ionization pattern of the adjacent acidic residues in the DXDXE motif of GH-18 Chitinases using theoretical pK_a calculations. J Chem Inf Model 57(3):572–583. https://doi.org/10.1021/acs.jcim.6b00536

Menghiu G, Ostafe V, Prodanović R, Fischer R, Ostafe R (2021) A high-throughput screening system based on fluorescence-activated cell sorting for the directed evolution of Chitinase A. Int J Mol Sci 22(6):3041. https://doi.org/10.3390/ijms22063041

Middelboe M, Traving SJ, Castillo D, Kalatzis PG, Glud RN (2025) Prophage-encoded chitinase gene supports growth of its bacterial host isolated from deep-sea sediments. ISME J 19(1):wraf004. https://doi.org/10.1093/ismejo/wraf004

Mishra P, Kshirsagar PR, Nilegaonkar SS, Singh SK (2012) Statistical optimization of medium components for production of extracellular chitinase by *Basidiobolus ranarum*: a novel biocontrol agent against plant pathogenic fungi. J Basic Microbiol 52(5):539–548. https://doi.org/10.1002/jobm.201100446

Mukherjee PK et al (2013) *Trichoderma* research in the genome era. Annu Rev Phytopathol 51:105–129. https://doi.org/10.1146/annurev-phyto-082712-102353

Nagaoka I (2014) Recent aspects of the chondroprotective and anti-inflammatory actions of glucosamine, a functional food. Juntendo Med J 60(6):580–587. https://doi.org/10.14789/jmj.60.580

Nampoothiri KM, Baiju TV, Sandhya C, Sabu A, Szakacs G, Pandey A (2004) Process optimization for antifungal chitinase production by *Trichoderma harzianum*. Process Biochem 39(11):1583–1590. https://doi.org/10.1016/S0032-9592(03)00282-6

Nayak SK, Dash B, Nayak S, Mohanty S, Mishra B (2020) Chitinase producing soil bacteria: prospects and applications. In CRC Press eBooks (pp. 289–298). https://doi.org/10.1201/9780429485794-30

Nayak SK, Nayak S, Mohanty S, Sundaray JK, Mishra BB (2021) Microbial chitinases and their applications: an overview. Microbial Chitinases and Their Applications: An Overview, pp 313–340. https://doi.org/10.1002/9781119525899.ch15

Nour AM, El-Sayed H, Abdelgawad MA (2024a) Chitinase-derived chitooligosaccharides: from immunomodulation to sustainable therapeutics. Biotechnol Adv 72:108305. https://doi.org/10.1016/j.biotechadv.2024.108305

Nour SA, Emam MTH, El-Sayed GM, Sakr EAE (2024b) Utilizing chitooligosaccharides from shrimp waste biodegradation via recombinant chitinase A: a promising approach for emulsifying hydrocarbon and bioremediation. Microb Cell Factories 23(1):126. https://doi.org/10.1186/s12934-024-02388-z

Olatunji O (2020) Aquatic biopolymers. In Springer series on polymer and composite materials. https://doi.org/10.1007/978-3-030-34709-3

Orlando M, Buchholz PCF, Lotti M, Pleiss J (2021) The GH19 engineering database: sequence diversity, substrate scope, and evolution in glycoside hydrolase family 19. PLoS One 16(10):e0256817. https://doi.org/10.1371/journal.pone.0256817

Ornela PH, Guimarães LHS (2024) Purification, characterization and antifungal activity of the *Aspergillus niveus* chitinase produced using shrimp shells. Appl Biosci 3(2):220–232. https://doi.org/10.3390/applbiosci3020015

Oshima K, Nakamura K, Guo H, Smith RL (2022) Mini-review on application of analytical centrifugation, ultracentrifugation and centrifugal devices to phase equilibria and separation processes. Fluid Phase Equilib 558:113457. https://doi.org/10.1016/j.fluid.2022.113457

Oyeleye A, Normi YM (2018) Chitinase: diversity, limitations, and trends in engineering for suitable applications. Biosci Rep 38(4):BSR2018032300. https://doi.org/10.1042/BSR20180323

Pacios-Michelena S, González CNA, Herrera RR, Alvarez-Perez OB, González MLC, Valdés RA, Valdés JAA, Salas MG, Iliná A (2023) Biomass from phytopathogens and culture conditions improve *Penicillium chrysogenum* antimicrobial activity and antifungal compounds production. Environ Qual Manag 33(1):349–358. https://doi.org/10.1002/tqem.22065

Pandey AK, Samota MK, Tanti A, Babu A (2023) *Trichoderma reesei* induces defense-related biochemical markers associated with resistance to Fusarium dieback in tea crop. Biol Control 180:105200. https://doi.org/10.1016/j.biocontrol.2023.105200

Park KR, Yun HM, Yoo K, Ham YW, Han SB, Hong JT (2020) Chitinase 3 like 1 suppresses the stability and activity of p53 to promote lung tumorigenesis. Cell Commun Signal 18:1–13.

Paulsen S, Andersen B, Gram L, Machado H (2016) Biological potential of chitinolytic marine bacteria. Mar Drugs 14(12):230. https://doi.org/10.3390/md14120230

Paulussen C, Hallsworth JE, Álvarez-Pérez S, Nierman WC, Hamill PG, Blain D, Rediers H, Lievens B (2017) Ecology of aspergillosis: insights into the pathogenic potency of *Aspergillus fumigatus* and some other *Aspergillus* species. Microb Biotechnol 10(2):296–322. https://doi.org/10.1111/1751-7915.12367

Pavan SR, Venkatesan J, Kim S-K, Prabhu A (2022) Anticancer effects of Chitooligosaccharides. In: Kim S-K (ed) Chitooligosaccharides: prevention and control of diseases. Springer International Publishing, pp 121–137. https://doi.org/10.1007/978-3-030-92806-3_8

Poria V et al (2021) Current perspectives on chitinolytic enzymes and their agro-industrial applications. Biology 10(12):1319. https://doi.org/10.3390/biology10121319

Price NP, Naumann TA (2011) A high-throughput matrix-assisted laser desorption/ionization-time-of-flight mass spectrometry-based assay of chitinase activity. Anal Biochem 411(1):94–99. https://doi.org/10.1016/j.ab.2010.12.027

Pusztahelyi T (2018) Chitin and chitin-related compounds in plant-fungal interactions. Mycology 9(3):189–201. https://doi.org/10.1080/21501203.2018.1473299

Pusztahelyi T, Molnár Z, Emri T, Klement É, Miskei M, Kerékgyártó J, Balla J, Pócsi I (2006) Comparative studies of differential expression of chitinolytic enzymes encoded bychiA, chiB, chiC andnagA genes in *Aspergillus nidulans*. Folia Microbiol 51(6):547–554. https://doi.org/10.1007/BF02931619

Qazi SS (2008) Regulatory role of ambient pH in the expression of pathogenicity determinant gene products of *Beauveria bassiana* and *Metarhizium anisopliae*. https://harvest.usask.ca/bitstream/10388/etd-03112008-114358/1/quazi.pdf

Rajendran K et al (2024) Chitinase from Streptomyces mutabilis as an effective eco-friendly biocontrol agent. Appl Biochem Biotechnol 196(1):18–31. https://doi.org/10.1007/s12010-023-04567-8

Rajput M, Kumar M, Pareek N (2022) Myco-chitinases as versatile biocatalysts for translation of coastal residual resources to eco-competent chito-bioactives. Fungal Biol Rev 41:52–69. https://doi.org/10.1016/j.fbr.2022.04.001

Ramirez-Coutiño L, Espinosa-Marquez J, Peter MG, Shirai K (2010) The effect of pH on the production of chitinolytic enzymes of *Verticillium fungicola* in submerged cultures. Bioresour Technol 101(23):9236–9240. https://doi.org/10.1016/j.biortech.2010.06.160

Raval KMRK, Raval R (2024) Marine chitinase AfChi: green defense management against *Colletotrichum gloeosporioides* and anthracnose. AMB Express 14(1):128. https://doi.org/10.1186/s13568-024-01786-1

Reid JD, Ogrydziak DM (1981) Chitinase-overproducing mutant of *Serratia marcescens*. Appl Environ Microbiol 41(3):664–669. https://doi.org/10.1128/aem.41.3.664-669.1981

Renaud S, Dussutour A, Daboussi F, Pompon D (2023) Characterization of chitinases from the GH18 gene family in the myxomycete *Physarum polycephalum*. Biochimica et biophysica acta. General subjects 1867(6):130343. https://doi.org/10.1016/j.bbagen.2023.130343

Reyes Zamora O, Troncoso-Rojas R, Báez-Flores ME, Tiznado-Hernández ME, Rascón-Chu A (2024) Signaling of plant defense mediated by receptor-like kinases, receptor-like cytoplasmic protein kinases and MAPKs triggered by fungal chitin in horticultural crops. Horticulturae 10(4):361. https://doi.org/10.3390/horticulturae10040361

Roberts WK, Cabib E (1982) *Serratia marcescens* chitinase: one-step purification and use for chitin determination. Anal Biochem 127(2):402–412. https://doi.org/10.1016/0003-2697(82)90194-4

Rudakiya DM (2019) Strategies to improve solid-state fermentation technology. In Elsevier eBooks (pp. 155–180). https://doi.org/10.1016/b978-0-444-64223-3.00010-2

Russo C, Valle MS, Casabona A, Malaguarnera L (2023) Chitinase signature in the plasticity of neurodegenerative diseases. Int J Mol Sci 24(7):6301. https://doi.org/10.3390/ijms24076301

Saqib AAN, Hassan M, Khan NF, Baig S (2010) Thermostability of crude endoglucanase from *Aspergillus fumigatus* grown under solid state fermentation (SSF) and submerged fermentation (SmF). Process Biochem 45(5):641–646. https://doi.org/10.1016/j.procbio.2009.12.011

Sasaki K, Yamamoto T, Fujita M (2022) Chitinase-processed COS coatings for antibacterial titanium implants. ACS Appl Mater Interfaces 14(15):17176–17188. https://doi.org/10.1021/acsami.2c03245

Sbaraini N, Junges Â, de Oliveira ES, Webster A, Vainstein MH, Staats CC, Schrank A (2021) The deletion of chiMaD1, a horizontally acquired chitinase of *Metarhizium anisopliae*, led to higher virulence towards the cattle tick (*Rhipicephalus microplus*). FEMS Microbiol Lett 368(12):fnab066. https://doi.org/10.1093/femsle/fnab066

Schrank A, Vainstein MH (2010) *Metarhizium anisopliae* enzymes and toxins. Toxicon 56(7):1267–1274. https://doi.org/10.1016/j.toxicon.2010.03.008

Schuster A, Schmoll M (2010) Biology and biotechnology of *Trichoderma*. Appl Microbiol Biotechnol 87(3):787–799. https://doi.org/10.1007/s00253-010-2632-1

Shao Z, Chen H, Hu S, Liu H, Meng F, Li S, Zhang B, Zhu C, Wang G, Liu L, Wang Q (2022) Chitosan oligosaccharide treatment improves quality attributes of tomato fruit stored under room temperature. Postharvest Biol Technol 189:111914. https://doi.org/10.1016/j.postharvbio.2022.111914

Sharp RG (2013) A review of the applications of chitin and its derivatives in agriculture to modify plant-microbial interactions and improve crop yields. Agronomy 3(4):757–793. https://doi.org/10.3390/agronomy3040757

Shrestha A, Bao K, Chen YR, Chen W, Wang P, Fei Z, Blissard GW (2018) Global analysis of Baculovirus autographa californica multiple nucleopolyhedrovirus gene expression in the Midgut of the lepidopteran host *Trichoplusia ni*. J Virol 92(23):e01277–e01218. https://doi.org/10.1128/JVI.01277-18

Shu P, Li Y, Sheng J, Shen L (2024) Recent advances in dissecting the function of ethylene in interaction between host and pathogen. J Agric Food Chem 72(9):4552–4563. https://doi.org/10.1021/acs.jafc.3c07978

Singh AK, Chhatpar HS (2011) Purification and characterization of Chitinase from *Paenibacillus* sp. D1. Appl Biochem Biotechnol 164(1):77–88. https://doi.org/10.1007/s12010-010-9116-8

Singh RK, Kumar DP, Solanki MK, Singh P, Srivastva AK, Kumar S, Kashyap PL, Saxena AK, Singhal PK, Arora DK (2013a) Optimization of media components for chitinase production by chickpea rhizosphere associated *Lysinibacillus fusiformis* B-CM18. J Basic Microbiol 53(5):451–460. https://doi.org/10.1002/jobm.201100590

Singh SP, Thumar JT, Gohel SD, Kikani B, Shukla R, Sharma A, Dangar K (2013b) Actinomycetes from marine habitats and their enzymatic potential. In Elsevier eBooks (pp. 191–214). https://doi.org/10.1533/9781908818355.2.191

Singh A, Mittal A, Benjakul S (2021) Chitosan, chitooligosaccharides and their polyphenol conjugates: preparation, bioactivities, functionalities and applications in food systems. Food Rev Intl 39(4):2297–2319. https://doi.org/10.1080/87559129.2021.1950176

Smith J et al (2020) Advances in affinity chromatography: recombinant ligands and IMAC. Biotechnol Prog 36(5):e3030. https://doi.org/10.1002/btpr.3030

Song Y, Liu X, Feng S, Zhao K, Qi Z, Wu W, Xiao J, Xu H, Ran M, Qin B (2023) Discovery, identification, and insecticidal activity of an *Aspergillus flavus* strain isolated from a saline-alkali soil sample. Microorganisms 11(11):2788. https://doi.org/10.3390/microorganisms11112788

Spier MR, Vandenberghe LPDS, Medeiros ABP, Soccol CR (2011) Application of different types of bioreactors in bioprocesses. Bioreactors: Design, Properties and Applications, pp 53–87

St Leger RJ (2008) Studies on adaptations of *Metarhizium anisopliae* to life in the soil. J Invertebr Pathol 98(3):271–276. https://doi.org/10.1016/j.jip.2008.01.007

St. Leger Raymond J, Joshi L, Roberts D (1998) Ambient pH is a major determinant in the expression of cuticle-degrading enzymes and hydrophobin by metarhizium anisopliae. Appl Environ Microbiol 64(2):709–713. https://doi.org/10.1128/AEM.64.2.709-713.1998

Strobel SA, Driks A, Gutnick DL (2010) Chitin utilization by soil bacteria: mechanisms and ecological implications. Microb Ecol 60(2):435–447. https://doi.org/10.1007/s00248-010-9722-6

Subash HA, Santhosh K, Kannan K, Pitchiah S (2024) Extraction of Keratin degrading enzyme from marine Actinobacteria of *Nocardia* sp and their antibacterial potential against oral pathogens. Oral Oncol Reports 9:100184. https://doi.org/10.1016/j.oor.2024.100184

Subramani AK, Raval R, Sundareshan S, Sivasengh R, Raval K (2022) A marine chitinase from *Bacillus aryabhattai* with antifungal activity and broad specificity toward crystalline chitin degradation. Prep Biochem Biotechnol 52(10):1160–1172. https://doi.org/10.1080/10826068.2022.2033994

Subramaniyam (2012) Solid state and submerged fermentation for the production of bioactive substances: a comparative study. Int J Sci Nat 3:480–486

Sukalkar S, Kadam T, Bhosale H (2018) Optimization of chitinase production from streptomyces macrosporeus M1. Research Journal of Life Sciences Bioinformatics Pharmaceutical and Chemical Sciences 4(1). https://doi.org/10.26479/2018.0401.09

Sulthan R, Reghunadhan A, Sambhudevan S (2023) A new era of chitin synthesis and dissolution using deep eutectic solvents- comparison with ionic liquids. J Mol Liq 380:121794. https://doi.org/10.1016/j.molliq.2023.121794

Sundaram MN, Pradeep A, Varma PK, Jayakumar R (2021) Different forms of chitosan and its derivatives as hemostatic agent and tissue sealants. In: Jayakumar R, Prabaharan M (eds) Chitosan for biomaterials IV: biomedical applications. Springer International Publishing, pp 1–28. https://doi.org/10.1007/12_2021_98

Suresh PV (2019) Enzymatic technologies of chitin and chitosan. In: Enzymatic Technologies for Marine Polysaccharides (pp. 449–466). CRC Press.

Suresh PV, Chandrasekaran M (1999) Impact of process parameters on chitinase production by an alkalophilic marine *Beauveria bassiana* in solid state fermentation. Process Biochem 34(3):257–267. https://doi.org/10.1016/S0032-9592(98)00092-2

Suyotha W, Cheirsilp B, Yano S, Matsuba S, Konno H (2021) Production of chitosanase by Lentzea sp. OUR-I1 using acid-pretreated shrimp Shell in an air-lift bioreactor and the feasibility of utilizing the residual biomass. Waste Biomass Valor 12(5):2445–2458. https://doi.org/10.1007/s12649-020-01191-2

Suzuki K, Shimizu M, Sasaki N, Ogawa C, Minami H, Sugimoto H, Watanabe T (2016) Regulation of the chitin degradation and utilization system by the ChiX small RNA in *Serratia marcescens* 2170. Biosci Biotechnol Biochem 80(2):376–385. https://doi.org/10.1080/09168451.2015.1083399

Swiontek Brzezinska M, Jankiewicz U, Kalwasińska A, Świątczak J, Żero K (2019) Characterization of chitinase from *Streptomyces luridiscabiei* U05 and its antagonist potential against fungal plant pathogens. J Phytopathol 167(7–8):404–412. https://doi.org/10.1111/jph.12809

Tamhane VA, Sant SS, Jadhav AR, War AR, Sharma HC, Jaleel A, Kashikar AS (2021) Label-free quantitative proteomics of *Sorghum bicolor* reveals the proteins strengthening plant defense against insect pest *Chilo partellus*. Proteome Sci 19(1):6. https://doi.org/10.1186/s12953-021-00173-z

Taokaew S, Kriangkrai W (2023) Chitinase-assisted bioconversion of chitinous waste for development of value-added chito-oligosaccharides products. Biology 12(1):87. https://doi.org/10.3390/biology12010087

Teregulova GA, Sineva ON, Markelova NN, Sadikova VS, Uvarov GV, Kovalenko MA, Manucharova NA (2023) Evaluation of the Chitinolytic and antibiotic activity of the *Streptomyces avidinii* INA 01467 and *Micromonospora aurantiaca* INA 01468. Почвоведение 5:594–602. https://doi.org/10.31857/s0032180x22601207

Thakur D, Bairwa A, Dipta B, Jhilta P, Chauhan A (2023) An overview of fungal chitinases and their potential applications. Protoplasma 260(4):1031–1046. https://doi.org/10.1007/s00709-023-01839-5

Thomas R, Fukamizo T, Suginta W (2022) Bioeconomic production of high-quality chitobiose from chitin food wastes using an in-house chitinase from *Vibrio campbellii*. Bioresour Bioprocess 9(1):86. https://doi.org/10.1186/s40643-022-00574-8

Tiago PV, De Oliveira NT, De Luna Alves Lima EA (2014) Biological insect control using *Metarhizium anisopliae*: morphological, molecular, and ecological aspects. Ciência Rural 44(4):645–651. https://doi.org/10.1590/s0103-84782014000400012

Tong L, Li Y, Lou X, Wang B, Jin C, Fang W (2024) Powerful cell wall biomass degradation enzymatic system from saprotrophic *Aspergillus fumigatus*. Cell surface (Amsterdam, Netherlands) 11:100126. https://doi.org/10.1016/j.tcsw.2024.100126

Tue NH, Phuc NH, Hoa PTB, Tien NQD, Loc NH (2024) Partitioning recombinant chitinase from *Nicotiana benthamiana* by an aqueous two-phase system based on polyethylene glycol and phosphate salts. Int J Biol Macromol 269(Pt 2):131924. https://doi.org/10.1016/j.ijbiomac.2024.131924

Unuofin JO, Odeniyi OA, Majengbasan OS, Igwaran A, Moloantoa KM, Khetsha ZP, Iwarere SA, Daramola MO (2024) Chitinases: expanding the boundaries of knowledge beyond routinized chitin degradation. Environ Sci Pollut Res Int 31(26):38045–38060. https://doi.org/10.1007/s11356-024-33728-6

Vaaje-Kolstad G, Horn SJ, Sørlie M, Eijsink VGH (2013) The chitinolytic machinery of Serratia marcescens—a model system for enzymatic degradation of recalcitrant polysaccharides. FEBS J 280:3028–3049. https://doi.org/10.1111/febs.12181

Vaghela B, Vashi R, Rajput K, Joshi R (2022) Plant chitinases and their role in plant defense: A comprehensive review. Enzym Microb Technol 159:110055. https://doi.org/10.1016/j.enzmictec.2022.110055

Veliz EA, Martínez-Hidalgo P, Hirsch AM (2017) Chitinase-producing bacteria and their role in biocontrol. AIMS Microbiol 3(3):689–705. https://doi.org/10.3934/microbiol.2017.3.689

Velmurugan N, Kalpana D, Han JH, Cha HJ, Lee YS (2011) A novel low temperature chitinase from the marine fungus *Plectosphaerella* sp. strain MF-1. Bot Mar 54(1):75–81. https://doi.org/10.1515/bot.2010.068

Vinale F et al (2008) *Trichoderma*–plant–pathogen interactions. Soil Biol Biochem 40(1):1–10. https://doi.org/10.1016/j.soilbio.2007.07.002

Vivekanandhan P, Swathy K, Chordia MA (2022) *Pseudomonas* species-derived chitinase mass multiplication, production cost analysis, and marketing: as a biocontrol agent for crop protection. In: Amaresan N, Dharumadurai D, Cundell DR (eds) Industrial microbiology based entrepreneurship. Microorganisms for Sustainability, vol 42. Springer, Singapore. https://doi.org/10.1007/978-981-19-6664-4_1

Wang S-L, Chen S-J, Wang C-L (2008) Purification and characterization of chitinases and chitosanases from a new species strain *Pseudomonas* sp. TKU015 using shrimp shells as a substrate. Carbohydr Res 343(7):1171–1179. https://doi.org/10.1016/j.carres.2008.03.018

Yang Y, Sossah FL, Li Z, Hyde KD, Li D, Xiao S, Fu Y, Yuan X, Li Y (2021) Genome-wide identification and analysis of Chitinase GH18 gene family in *Mycogone perniciosa*. Front Microbiol 11:596719. https://doi.org/10.3389/fmicb.2020.596719

Yang X, Zeng D, Li C, Yu W, Xie G, Zhang Y, Lu W (2023) Therapeutic potential and mechanism of functional oligosaccharides in inflammatory bowel disease: a review. Food Sci Human Wellness 12(6):2135–2150. https://doi.org/10.1016/j.fshw.2023.03.027

Yano S, Kanno H, Tsuhako H, Ogasawara S, Suyotha W, Konno H, Makabe K, Uechi K, Taira T (2021) Cloning, expression, and characterization of a GH 19-type chitinase with antifungal activity from *Lysobacter* sp. MK9-1. J Biosci Bioeng 131(4):348–355. https://doi.org/10.1016/j.jbiosc.2020.11.005

Yousef M, Pichyangkura R, Soodvilai S, Chatsudthipong V, Muanprasat C (2012) Chitosan oligosaccharide as potential therapy of inflammatory bowel disease: therapeutic efficacy and possible mechanisms of action. Pharmacol Res 66(1):66–79. https://doi.org/10.1016/j.phrs.2012.03.013

Yu K, Ho J, McCandlish E, Buckley B, Patel R, Li Z, Shapley NC (2013) Copper ion adsorption by chitosan nanoparticles and alginate microparticles for water purification applications. Colloids Surf A Physicochem Eng Asp 425:31–41. https://doi.org/10.1016/j.colsurfa.2012.12.043

Yuliantoro ID, Prihatiningrum AE, Sutarman N (2023) Exploration and inhibition test of *Penicillium* sp. in vitro by *Trichoderma*. IOP Conference Series Earth and Environmental Science 1242(1):012012. https://doi.org/10.1088/1755-1315/1242/1/012012

Zhang DD, Dai XF, Klosterman SJ, Subbarao KV, Chen JY (2022) The secretome of *Verticillium dahliae* in collusion with plant defence responses modulates Verticillium wilt symptoms. Biol Rev Camb Philos Soc 97(5):1810–1822. https://doi.org/10.1111/brv.12863

Chapter 8
Unveiling Nature's Bacterial Solution to Radioactive Waste

Birson Ingti, Anupama Moirangthem, Susmita Paul, Anuj Kumar Borah, and Damsai Reang

Abstract Radioactive waste pollution in the environment has increased due to increased human activities. Humans and animals were regularly exposed to radioactive waste through a variety of channels, resulting in long-term health impacts such as cancer and other diseases. The rapid radionuclide discharge into the environment has led to the development of public and academic interest in radionuclide processing. Its removal or transformation from the environment is extremely difficult; nevertheless, bacteria provide an attractive alternative option for radioactive waste cleanup known as "bioremediation." The bacteria used several mechanisms to change refractory organic compounds from radioactive waste into less hazardous, less radioactive, or more useful forms. Genetically engineered microorganisms that can degrade or transform radionuclides in the environment have also increasingly been utilized over the past decades. Thus, native bacteria with the ability to tolerate radionuclides naturally can be studied and used for the bioremediation of these hazardous wastes. Furthermore, genetically modified bacteria, in conjunction with an omics-based strategy, could be employed as a long-term alternative option for removing or transforming radionuclides from the environment.

Keywords Radionuclides · Bioremediation · Omics · Radioactive waste

---

B. Ingti (✉) · A. Moirangthem · S. Paul
Department of Microbiology, Royal School of Biosciences, The Assam Royal Global University, Guwahati, Assam, India

A. K. Borah
Department of Biotechnology, Royal School of Biosciences, The Assam Royal Global University, Guwahati, Assam, India

D. Reang
Department of Environmental Science, Royal School of Environmental and Earth Sciences, The Assam Royal Global University, Guwahati, Assam, India

N. K. Arora et al. (eds.), *Environmental Sustainability and Biotechnology: Opportunities and Challenges*, Microorganisms for Sustainability 46,
https://doi.org/10.1007/978-981-96-9025-1_8

8.1 Introduction

Radioactive wastes containing radionuclides are constantly generated from various radioactive materials, and their concentration in the environment has increased with the global trend toward mining and nuclear power. Numerous episodes of radioactive wastes seeping into soil and groundwater have been witnessed, and their concentration in the environment has increased. Some of the sources of radioactive waste include nuclear power programs, plants, the nuclear industry, nuclear weapons development, and/or accidental disposal. Many radioactive elements such as U-235, U-238, Np-237, Cs-137, Pu-239, Am-241, Tc-99, Sr-90, etc. are present in the chemicals (liquid/effluent) emitted by nuclear facilities or other industries, and their release in the environment imposes a huge adverse effect on human health and the environment. Therefore, the rapid radionuclide discharge into the environment has promoted to the development of public and academic interest in radionuclide processing. However, managing and removing radioactive wastes from the environment has been challenging due to their propensity for bioaccumulation and radiation. Prior approaches for disposing of radioactive waste included deep-digging wells and ocean dumping. However, because radioactive elements such as 237 Np, 239Pu, 241Am, 99Tc, and 90Sr are soluble and susceptible to polluting groundwater, neither of these approaches is suitable for waste disposal. Therefore, the application of bacteria offers an ideal alternative option to control the biogeochemical cycles of radioactive wastes known as "bioremediation" (Vandana et al. 2021). This process transformed resistant organic molecules from radioactive waste into less harmful or more beneficial forms by the microorganisms. Radionuclides can become less hazardous by becoming more soluble, less radioactive, or nonradioactive, bioavailable, and mobile through bioremediation mediated by bacteria. During the last 20 years, several studies have been conducted to investigate the potent function of bacteria in the remediation of radionuclide-contaminated sites, including both land and water, which may be an environmentally beneficial, economically viable, and socially acceptable strategy. Biosorption, biomineralization, bio-reduction, bioaccumulation, and enzymatic reduction (direct and/or indirect) can all influence radionuclide solubility in biogeochemical systems. For bioremediation to be successful, biological, physical, and chemical characteristics of a bacterial cell are essential. The cleanup of radioactive waste using bacteria is economical, practical, and environmentally friendly. The bio-engineering of these species is another significant method for radioactive waste bioremediation (Anand et al. 2023). Recently, cutting-edge technology for cleaning up radionuclide-contaminated environments utilizing genetically modified organisms (GMOs) has become accessible. Radionuclides could also be successfully eliminated from the environment via omics-based approaches or molecular pathways leading to radionuclide microbial transformation. Metagenomics and proteomics are both omics-based methods that can aid in closing knowledge gaps about the variables that influence microbial growth and metabolism (Patel et al. 2022). Thus, this chapter shall briefly describe such species and discuss certain strategies to overcome existing limitations.

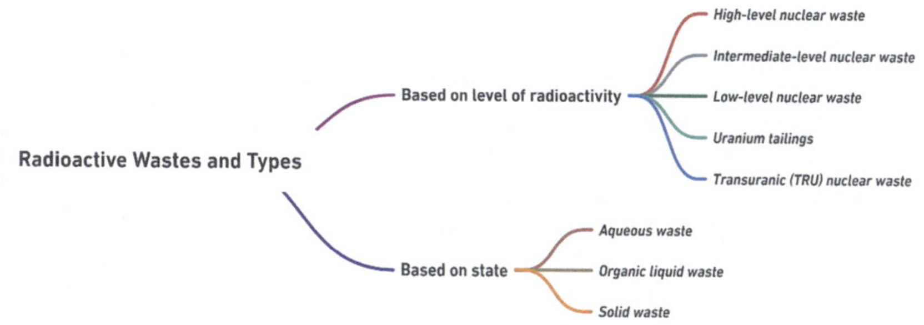

Fig. 8.1 Classification of radioactive waste based on levels of activity and state of radioactive waste

8.2 Radioactive Waste/Nuclear Waste

8.2.1 Types of Nuclear Waste

Nuclear wastes are classified as low, intermediate, or high level depending on their degree of activity or radioactivity, which indicates their potential threat (Fig. 8.1). Ninety percent of all nuclear waste is composed of low-level waste (LLW) or moderately contaminated items, such as tools and work gear, while LLW only contributes 1% of all radiation. Although high-level waste accounts for just 3% of total nuclear waste volume, it accounts for 95% of total radioactivity and is mostly composed of spent nuclear reaction waste or used nuclear fuel (AC03223434 2001; Rossiello and Failla 1994). Radioactive waste can be categorized according to its inherent radioactivity, and the state of such waste is demonstrated in the subsequent section of this state of art.

8.2.1.1 Low-Level Nuclear Waste

The nuclear fuel cycle created the majority of the low-level waste. This trash consists of items that have become radioactive due to exposure to neutron radiation and/or hazardous substances. Most of this nuclear waste is composed of machinery and tools, medical tubes, clothing, wiper rags, swabs, injection needles, luminous dials, mops, filters, and remnants from the reactor's water treatment process. These substances were referred to as "mixed low-level waste" (MLLW). In liquid form, these wastes typically have less than a microcurie of radioactivity per gallon. Even though low-level waste makes up just 1% of total radioactivity, most of it is not particularly radioactive (Waste and Technology 1980).

8.2.1.2 Intermediate-Level Nuclear Waste

Only 1% of all radio activities are accounted by low-level waste, but most of it is not extremely radioactive. This waste may contain up to one curie of radioactivity per gallon, and its radioactivity in liquid form is 100–1000 times higher than that of low-level wastes. The International Atomic Energy Agency (IAEA) classifies radioactive waste into alpha-bearing waste, and short- and long-lived waste under the intermediate-level waste (ILW) and LLW categories. Alpha-bearing waste refers to radioactive materials that contain one or more actinides or other radionuclides that emit alpha radiation (USEPA 2009).

8.2.1.3 High-Level Nuclear Waste

This sort of waste has a liquid content of several hundred to several thousand curies per gallon and comprises highly radioactive transuranic elements and fission products created in the core of a nuclear reactor. It needs cooling since this kind of trash produces a lot of heat. After nuclear fuel is burnt and removed from reactors, this waste is produced, and it accounts for more than 95% of all radiation. To prevent dangerous exposure while handling and moving this type of garbage, it should be covered. High-level waste can instantly kill nearby people and animals because it contains radioactive nuclear fuel rods (Grover 1974).

8.2.1.4 Uranium Tailings

Uranium tailings are radioactive by-products of the process of extracting uranium from its ore. Radon gas and other radioactive materials present in the tailings have the potential to escape into the air. Wind can disseminate radioactive dust from unsealed tailings on open terrain, which could add radionuclides to the ecosystem and food chain. The primary radiation risks of uranium tailings are radon gas release, wind dispersal of radionuclide dust, and gamma radiation due to radium decay (Abdelouas 2006).

8.2.1.5 Transuranic Nuclear Waste

Nuclear weapons production facilities, nuclear testing sites, and power plants all produce transuranic nuclear waste (TRU). There is a lot of plutonium in the TRU waste. Although the emitted radiation have the capability of causing lung issues in people, the modest radiation released makes this type of waste not particularly dangerous (Kok et al. 2016).

8.2.2 *Classification Based on the State of Radioactive Waste*

The radioactive waste is divided into categories based on its condition: solid waste, organic liquid, and aqueous waste (Fig. 8.1) (AC03223434 2001).

8.2.2.1 Aqueous Waste

It is created by radioisotope laboratories, nuclear research facilities, radiopharmaceuticals, and other sources, and it produces 50% or more of the radioactive waste. The aqueous wastes may be very reactive and contain compounds with effluent concentrations, such as acids. The testing of radiochemical neutron activation was caused by a unique kind of reactor that generated aqueous waste. These wastes can be exceedingly active (high concentrations of acid or peroxide), and they can be problematic for people (AC03223434 2001; USEPA 2009).

8.2.2.2 Organic Liquid Waste

Nuclear research facilities, radioisotopes laboratories, radiolabeling and radiopharmaceuticals, scientific research institutions or centers, medical centers, industrial and pilot plants produce organic liquid waste. It contains biological, radiological, and nuclear materials that are dangerous to people (AC03223434 2001; Natarajan et al. 2020). When classifying the liquid wastes, the levels of long-lived radionuclides and tritium are also considered.

8.2.2.3 Solid Waste

Solid waste is produced by businesses, testing facilities, and mines (radioisotopes) by using radioactive material as a fuel for their reactors. Solid wastes that are combustible, non-compactable, and compactable are divided into two groups: low- and intermediate-level wastes people (AC03223434 2001; Natarajan et al. 2020). Solid radioactive wastes are categorized as either compressible or noncompressible and combustible or noncombustible based on their physical characteristics.

8.3 Sources of Radioactive Wastes

Industries and metal mining (industrial uses include radiographic imaging of mechanical parts and irradiation of products for disinfection) are also sources of radioactive waste. In nations with nuclear power reactors, nuclear weapons, or nuclear fuel processing plants, most of the waste comes from the nuclear fuel cycle and reprocessing of nuclear weapons. The other sources, which are also included in medical and industrial wastes, include naturally occurring radioactive materials (NORM) that can be concentrated due to the processing or consumption of coal, oil, and gas, as well as certain minerals.

8.3.1 Nuclear Fuel Cycle

Fission product radioisotopes, irradiation of nonfuel materials, and activation product radioisotopes are utilized in fuel generation. Waste from the uranium extraction process that emits alpha radiation typically comes from the front end of the nuclear fuel cycle. Radium and its decay products are frequently present in it. The concentration of the mineral uranium dioxide (UO_2) is about a thousand times more radioactive than the granite used in construction. Uranium hexafluoride gas (UF_6) is produced by refining yellowcake ($U3O_8$). Enrichment is applied to the gas to raise the U-235 level from 0.7% to roughly 4.4% (LEU). The hard ceramic oxide (UO_2) is then created from it for assembly as reactor fuel elements (Leenaers 2014).

Depleted uranium (DU), mostly the U-238 isotope with a U-235 level of less than 0.3%, is the primary by-product of enrichment. Either UF_6 or $U3O_8$ is used for storage. Some of it is utilized for anti-tank munitions and, occasionally, even a sailboat keel, where its exceptionally high density makes it beneficial. It also weakens or down-blends highly enriched uranium from weapons stockpiles that are already being used as reactor fuel, and it is combined with plutonium to produce mixed oxide fuel (MOX) (Kessler 2011).

The spent fuel rods at the back end of the nuclear fuel cycle contain fission products that release beta and gamma radiation and actinides that release alpha particles, including uranium-234 (half-life: 245,000 years), neptunium-237 (2.144 million years), plutonium-238 (87.7 years), and americium-241 (432 years). Californium (half-life: 898 years) is one example of a neutron emitter. These isotopes are produced in nuclear reactors.

It is critical to distinguish between reprocessing old fuel and processing uranium to generate fuel. The extremely radioactive by-products of fission are found in used fuel. Numerous of these are neutron absorbers or neutron poisons as they are known in this context. These eventually reach a point where they absorb enough neutrons to cause the chain reaction to halt even when all the control rods are removed. Even if there is still a significant amount of uranium-235 and plutonium present at that stage, the fuel in the reactor needs to be replaced with new fuel. This old fuel is often "stored" in the United States, but in other nations like Russia, the United Kingdom, France, Japan, and India, the fuel is reprocessed to eliminate the fission products and can then be used again (Žagar et al. 2011). High-level waste concentrated forms are the chemicals utilized in the process and the fission products removed from the fuel. Although most countries recycle fuel through a single plutonium cycle, Russia has a closed cycle, and India is considering multiple plutonium recycling plans (Kansra 1999).

8.3.1.1 Fuel Structure and Enduring Radioactivity

The activity of U-233 varies across three fuel types. In MOX fuel, the concentration of U-233 increases over the first 650,000 years due to its production through the decay of Np-237, which is initially formed in the reactor when U-235 absorbs neutrons (Nuttin et al. 2005). Activity totals for the three fuel kinds. Radiation from short-lived nuclides is present in region 1, Sr-90 and Cs-137 are present in region 2, and the decay of Np-237 and U-233 may be seen on the far right. The use of different fuels in nuclear reactors, the spent nuclear fuel (SNF), has a variable composition and has different activity curves. U-238, together with other uranium isotopes, other actinides, fission products, and activation products, is the most prevalent substance (Anttila 2005).

A thorough waste management plan for SNF should take long-lived radioactive waste from the back end of the fuel cycle into account. An integrated waste management strategy for SNF should take long-lived radioactive waste at the conclusion of the fuel cycle into consideration. Actinides present in the SNF significantly influence long-term radioactive decay due to their very long half-lives. Actinide composition in the SNF depends on the type of fuel utilized in nuclear reactors (Kansra 1999). The usage of thorium-containing nuclear fuels is one illustration of this effect. Th-232 is a fertile substance that can produce a fissile isotope of U-233 through two beta minus decays and a neutron capture process. U-233 will be present in a cycle with thorium's SNF. Its radioactive decay will have a significant impact on the SNF's long-term activity curve in about a million years. The fuels that were burned were mixed oxide fuel (MOX; no thorium), thorium with weapons-grade plutonium (WGPu), and thorium with reactor-grade plutonium (RGPu). The original amount of U-233 and its degradation over a million years may be seen for RGPu and WGPu. The three fuel types' combined activity curve is impacted (Bunn et al. 2005). Actinides can be removed from spent fuel during nuclear reprocessing so they can be used or destroyed (Sood and Patil 1996).

8.3.1.2 Proliferation Apprehensions

Concerns regarding proliferation have arisen since uranium and plutonium are components of nuclear weapons. Typically, the plutonium found in spent nuclear fuel is reactor grade. As uranium and plutonium are used in nuclear weapons, proliferation concerns have existed. The plutonium contained within spent nuclear fuel is usually reactor grade. It also has large levels of undesirable impurities like plutonium-240, plutonium-241, and plutonium-238 besides plutonium-239, which is highly suitable for nuclear bomb construction. Although there are less expensive ways to generate fissile material (such as enriching uranium or using certain reactors to produce plutonium), it is very difficult to separate these isotopes (Herring et al. 2001).

Highly radioactive fission products, the majority of which have a brief half-life, are abundant in high-level waste. This is an issue because the fission products will gradually decay if waste is stored, perhaps in deep geological storage, reducing the radioactivity of trash and making the plutonium more available.

Recycling plutonium and utilizing it as fuel, for instance, in fast reactors, is a method of addressing this problem. The plutonium and uranium that are separated in pyrometallurgical fast reactors are polluted by actinides and cannot be utilized to make nuclear weapons (Lyman 1994).

8.3.2 Heritage Waste

Numerous places contain or are contaminated with radiation because of historical operations, frequently connected to the radium industry, uranium mining, and military programs. According to the Department of Energy, there are "huge quantities of contaminated soil and water," "millions of gallons of radioactive waste," "thousands of tons of spent nuclear fuel and material," and "radioactive waste" all in the United States alone (Carnes et al. 1998).

8.3.3 Medicine

Medical facilities (used for things like radiation and radiodiagnosis) that are gamma and beta emitters are frequently found in radioactive medical waste. It falls into two primary categories. A variety of short-lived gamma emitters are utilized in diagnostic nuclear medicine. Many of these can be discarded by allowing them to decompose for a little period before being disposed of like regular waste (Schmor 2011).

8.3.4 Coal, Oil, and Gas Industries

There may be alpha, beta, neutron, or gamma emitters in industrial source waste. While neutron emitting devices are utilized in a variety of applications, including oil well logging, gamma emitters are used in radiography (Frankle and Dale 2013).

The amount of radioactive uranium, barium, thorium, and potassium in pure coal is negligibly lower than the typical level of these elements in the Earth's crust.The ash content of "dirty" coals can also indicate a slightly higher-than-average presence of adjacent shale or mudstone layers. Because they do not burn well, the more active ash minerals concentrate in the fly ash. Fly ash has a similar level of radioactivity as that of black shale but lower radioactivity compared to phosphate rocks. However, it is much worse because a proportion of it reaches the air to be breathed into the body (Gabbard 1993).

Industrial waste from oil and gas often contains radium and its decay products. The sulfate scale from an oil well can be very radioactive, but the water, oil, and gas from a well sometimes include radon. Solid radioisotopes produced by radon decay cover the interior of pipes. This is where propane is refined within the oil processing plant. It is one of the most contaminated sections of the plant because radon and propane have very similar boiling points (Cowie et al. 2012).Workers operating in direct contact with the crude oil and brine in some oil wells may be exposed to doses that have detrimental health consequences from radioactive materials. The comparatively large concentration of these components in brine makes its disposal a technological challenge as well. However, since the 1980s, the brine has been exempt from the dangerous waste restrictions in the United States and can be disposed of regardless of its content in radioactive or toxic compounds (Sm 2020).

8.3.5 Rare-Earth Mining

Because rare-earth ore naturally contains radioactive elements like radium and thorium, mining processes also produce waste and mineral deposits that are marginally radioactive (Cooper 2005).

8.4 Effect of Radioactive Waste on the Atmosphere as Well as Living Organisms

Radioactive wastes, such as transuranic waste, contain radionuclides that, when breathed in or absorbed by humans, can cause biological responses. Despite the low levels of natural radioactivity, living organisms have always been exposed to it through their environment, which includes the soil and groundwater. Radioactive waste releases radioactive particles that can pose direct and indirect risks to the environment and living things. Radiation waste can have an immediate negative impact on people by harming bodily cells (somatic effects) or reproductive cells (genetic consequences). Somatic effects will be lessened if waste radiation exposure is spread out over a long period of time (Rao 2001).

Apart from these, an organism's genetic makeup can be impacted by radioactive waste, and this harmful effect can be passed down through the generations. Radioactive waste exposure can result in gastrointestinal injury, intestinal lining destruction, DNA damage, thyroid cancer, cancer in general in young children, and in severe cases, death (AC03223434 2001; Kautsky et al. 2013; Taylor and Taylor 1997). Radioactive waste disposal in terrestrial and aquatic environments will poison the environment and cause serious sickness in both humans and animals. The soil contaminated by radioactive waste and radioactive elements get into the trees or crops that are growing, which may cause plant mortality. Thus, radioactive contamination of plants will have a negative impact on the food chain (Kautsky et al. 2013).

These radioactive wastes including radionuclides or radioisotopes contaminate the water if dumped in a deeper well or the deep ocean.

8.5 Bioremediation of Radionuclides Using the Merit of Bacteria

Bacterial communities have a major impact on cycling, retaining and releasing nutrients and soil carbon and maintaining ecosystems (McKenney et al. 2018; Newton et al. 2011; Van der Gucht et al. 2007). It can have both direct and indirect impacts on the environment when composition changes occur. Numerous studies have established that radionuclide contamination significantly affects the structure and functions of bacterial communities (Fan et al. 2018; Hemme et al. 2015; Narendrula-Kotha and Nkongolo 2017; Xiao et al. 2019; Yan et al. 2016). As bacterial cells tend to be susceptible to radionuclides, constant exposure to them can impact bacterial diversity and composition (Ager et al. 2010). Bacterial cells may be damaged by the radioactive material by changing the cell components, such as membranes, and enzyme viability by protein denaturation. The most vulnerable stage of a bacterial cell is when cell division is taking place actively due to the breakage or DNA alteration caused by the radiation. Nevertheless, there has been the recovery of some bacteria from highly radio-contaminated environments that can resist radioactive materials (Ryabova et al. 2020). These bacteria and archaea are capable of surviving to doses that are a thousand times the dosage that can kill humans. The way bacteria resist radiation, varies among species and requires the coordination of multiple biochemical and genetic processes. The resistant groups against radionuclides include *Deinococcus* spp., *Thermococcus* spp., *Chloroflexi*, Actinobacteria, and Proteobacteria. In some cases, it also includes *Agrobacterium* sp., *Enterobacter* sp., and *Klebsiella* sp. (Yablokov et al. 2009). Over recent years, there has been a significant growth in interest in the biodiversity and activity of these bacteria that inhabit radionuclide-contaminated sites. The composition of these bacterial communities has become known to be specific to the site and related to concentration levels and/or different geological and physiological environmental conditions. The abundance of *Acidithiobacillus*, *Pseudomonas*, *Acinetobacter*, and *Nitrosomonas* was observed in different uranium mine waste sites in Germany, whereas the dominance of *Sphingomonas*, *Acidovorax*, *Acinetobacter*, and *Ralstonia* may be attributed to radioactive waste contaminated with U (Akob et al. 2007; Jaswal et al. 2019a, b; Yan et al. 2016). Radionuclide-contaminated environments are often found with abundant and active Proteobacteria (alpha, beta, delta, and gamma proteobacteria), *Acidobacteria, Actinobacteria, Bacteroidetes,* and *Firmicutes* (Hemme et al. 2010; Jaswal et al. 2019a, b; Kumar et al. 2013; Suriya et al. 2017) (Fig. 8.2).

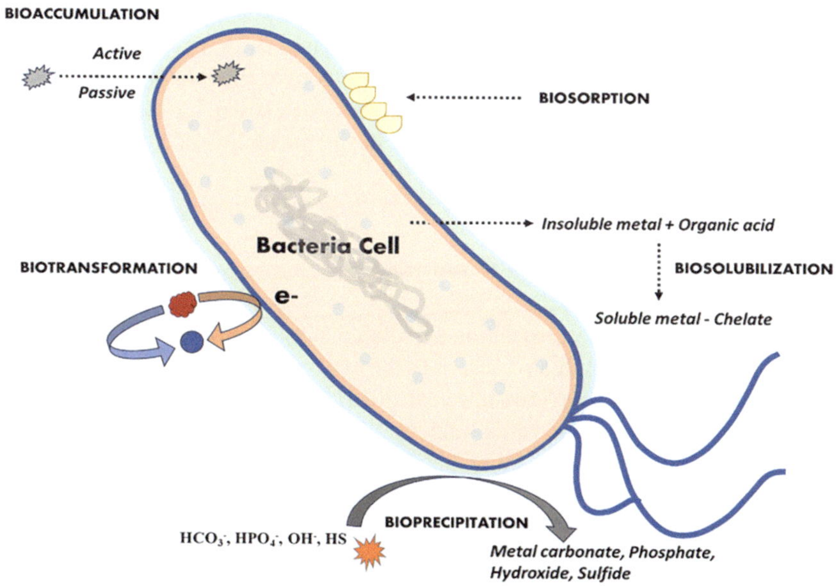

Fig. 8.2 Mechanism of bacterial bioremediation of radionuclides and heavy metals

8.5.1 Mechanism Involved in the Bioremediation of Radionuclide by Bacteria

Different microorganisms can degrade diverse contaminants depending on their concentration and metabolic requirements. Microorganisms are said to play a significant role and show the greatest potential in bioremediation across the globe. Bacteria may interact with radionuclides in numerous ways. It influences the toxicity and transport of metals and radionuclides (McCullough et al. 1999) under the fundamental principles of biotransformation, biosorption, bioaccumulation, bioprecipitation, and biosolubilization. Each provides the possibility of bioremediation of radioactive and metallic impurities in the environment (Koul and Adlakha 2021).

8.5.1.1 Remediation Through Bioaccumulation

The retention and concentration of a substance within an organism is known as bioaccumulation. In this process, solutes are transported from the outside into the cell cytoplasm via the cell membrane, where the metal is sequestered (McCullough et al. 1999). Numerous metal ions can be accumulated inside microbial cells via the bioaccumulation process. Some metals are easily absorbed within the bacterial cell because they are necessary for the cell's operation (Newsome et al. 2014). The adsorption of radionuclides, which happens as a result of the electrostatic contact between the positively charged metals and the negatively charged bacterial cell surface, is the foundation of bioaccumulation. Bioaccumulation may be metabolism-dependent (passive) or metabolism-independent (active) (Shukla et al. 2017). The intracellular accumulation of radionuclides may be caused by the radionuclide transporter and the effect of nuclides on the permeability of cell membranes (Merroun and Selenska-Pobell 2008).

8.5.1.2 Remediation Through Biosorption

Biosorption describes the association of soluble substances with the cell surface (McCullough et al. 1999). It also describes the metabolism-independent sorption of radionuclides into biomass (Newsome et al. 2014). The cell surfaces of the microbes have different cell wall compositions among microorganisms, leading to significant differences in the type and amount of metal ions binding to them (Goyal et al. 2003).

Gram-positive bacteria are comprised of a peptidoglycan layer, whereas Gram-negative bacteria consist of lipopolysaccharides. The radionuclides concentrate at the cell surface and pass through the cell membrane using adsorption or ion exchange. The presence of a phosphate moiety on the cell membrane creates an electronegative charge that attracts positive cations of metals and radionuclides (Goyal et al. 2003). Biosorption can be direct or indirect, where direct biosorption is achieved by interacting between the functional group of the cell wall and cationic nuclide, and indirect absorption is achieved by the interaction with the extracellular polymeric substances, surface layer, or capsule (Merroun and Selenska-Pobell 2008). Biosorption can occur passively by surface complexation onto the cell wall and surface layers in both living cells and dead biomass (Fomina and Gadd 2014).

8.5.1.3 Remediation Through Precipitation

Bacteria have the capability of inducing the precipitation of radionuclides by altering their chemical conditions in their surroundings. Organic and inorganic compounds produced by the bacteria react with radionuclides, leading to precipitation as insoluble compounds, which leads to immobilization. The alkanization reaction on the cell surface enables microbes to precipitate the metals and radionuclides easily as carbonates and hydroxides (Van Roy et al. 1997). The formation of mineral precipitates around radionuclides, through bacterial-assisted processes of biomineralization, can reduce their mobility and bioavailability. The utilization of biomineralization to eliminate uranium contamination, for instance, by *Citrobacter* sp., demonstrates its ability. It can altogether eliminate U from contaminated surroundings through the accumulation of NaUO2PO4 within and upon the cell wall via a system that involves a gradient of entrance of UO4 and gradient of exit of PO4 (Keasling et al. 2000).

8.5.1.4 Remediation Through Biotransformation

Biotransformation is a microbial-mediated structural modification of a chemical compound by organisms or enzyme systems. Comparatively more polar molecules are formed during the biotransformation (Ager et al. 2010). Certain bacterial processes can alter the concentration of radionuclides. The transformation includes oxidation-reduction, which changes the solubility of radionuclides. The presence of electron donors and acceptors highly influences the activities of certain microorganisms. Both aerobic and anaerobic bacteria have a significant role in biotransformations, which may be enzymatically driven or without enzymes. The oxidized form of the radionucleotides and metals such as U, Technetium (Tc), and chromium (Cr) are used by microbes as terminal electron acceptors.

8.5.1.5 Biosolubilization of Radionuclides

It involves the breakdown and solubilization of contaminants by microorganisms by producing enzymes and metabolites that can transform insoluble or less soluble contaminants into more soluble forms. The solubilization process, which involves autotrophic metabolism, causes radionuclides to leak from solid matrices. The autotrophic bacteria take up the Fe and S compounds while solubilizing the nuclides and metals to obtain energy. Several variables affect biosolubilization, such as the oxidation state of the nuclide, pH in the vicinity of bacteria, moisture, and inorganic content of the bacteria (Shukla et al. 2017).

8.5.1.6 Bacterial Reduction of Radionuclides

It is usually found that the oxidized state of radionuclides is more soluble and mobile in an aqueous system enhancing their harmful effects. Reduced forms of radionuclides, on the other hand, are insoluble in an aqueous environment and thus can be removed in precipitate form. These methods of eliminating uranium contamination by enzymatically reducing U (VI) to insoluble forms of U (IV) are demonstrated by microorganisms like *Shewanella putrefaciens* and *Desulfovibrio vulgaris* (Lovley and Phillips 1994; Wildung et al. 2000). It was also noted that the long-lived radioisotope 99Tc was precipitated and remedied after Tc (VII) was reduced by microorganisms to low-valency oxide of Tc (IV). *Desulfovibrio desulfuricans, Geobacter metallireducens,* and *Desulfovibrio fructosovorans* are some metal-reducing bacteria having ^{99}Tc remediation efficacies (Alliot et al. 2009; De Luca et al. 2001; Lloyd et al. 1999). The *Geobacter* sp. and the iron-reducing bacterium *Rhodoferax ferrireducens* have the metabolic capacity to enzymatically reduce other radionuclides of the low to medium category, such as Th, Np, Pu, and Am (Kim et al. 2012; Lloyd and Macaskie 2000). Further, microorganisms that may reduce metal or sulfate can also indirectly reduce soluble pollutants in sedimentary and subterranean environments (Prakash et al. 2013).

8.5.1.7 Biostimulation of Radionuclides

By promoting the activity of dissimilatory sulfate and iron-reducing organisms, bioavailability and mobility of uranium in groundwater can be reduced. *Desulfovibrio* sp., *Geobacter* sp., and *Shewanella* sp. are some important microorganisms capable of radionuclide reduction, but their in situ efficacy is limited because of the presence of nitrate. On the other hand, non-reducible heavy metals retard nitrate removal (Rao et al. 2001). The use of a suitable carbon source (e.g., ethanol) could enhance nitrate reduction and heavy metal remediation ex situ before in situ removal of radionuclides such as U (VI) (Taylor et al. 1997). Evidence revealed that raising pH with the addition of carbon (ethanol) to stimulate the growth of denitrifying and radionuclide-reducing microorganisms has been an effective approach leading to the removal of radionuclides including U (VI) by around 80% (Kautsky et al. 2013).

8.5.1.8 Bioaugmentation of Radionuclides

Bioaugmentation is an emerging and promising technology to remediate radionuclides, which involves the introduction of specific microorganisms (natural or genetically modified) to enhance the radionuclide removal capabilities of the existing microbial community. For instance, microorganisms of the genus *Deinococcus* that are highly resistant to radiation can be introduced into a radio waste-contaminated environment toward U(VI), Cr(VI), Tc(VII), and other radionuclide remediation (Jeong and Choi 2020). In such processes, the right microorganism is chosen based on the radionuclide(s) present and the environmental factors at the polluted site. The survival of the injected microorganisms in the contaminated environment is one of the main challenges of bioaugmentation. Native or indigenous bacteria might defeat newly imported species in a race for scarce nutrients. Consequently, while ex situ bioaugmentation has been in common practice for bioremediation of municipal wastewater treatment, in situ bioaugmentation is still in its developing stage and currently showing less effects (McCullough et al. 1999).

8.5.1.9 Removing Radionuclides Through Biofilm Formation

Biofilm formation has the potency to immobilize and sequester radio wastes in contaminated environments. Extant shreds of evidence reveal that *Exiguobacterium profundum*, *Pseudomonas putida*, and *Bacillus marisflavi* biofilms can carry out biosorption of uranium (Manobala et al. 2019a, b). Biofilms of *Desulfovibrio desulfuricans* G20, a sulfate-reducing bacteria, are reported to reduce dissolved U(VI) and form complexes with carbonates toward remediating uranium contamination (Marsili et al. 2007). Similarly, *Geobacter sulfurreducens* biofilms display prompt resistance to uranium toxicity and quite ably reduce U(VI) to U(IV). While most biofilms opt to perform permeation and biomineralization in the cell envelope, *Geobacter sulfurreducens* biofilms reduce the uranium in the top stratum, preserving cellular functions and catalytic capacity to remediate more and more uranium from contaminated environments (Cologgi et al. 2014). As an alternate strategy, ethanol-stimulated biofilm remediation can successfully lower high U(VI) concentrations by 87% (Marsili et al. 2007). Because biofilms, the surface-attached microbial consortia, wherein cells are self-immobilized in extracellular polymeric matrices, do not need to be separated from the bulk liquid waste, they are thought to be the ideal tool for bioremediation.

8.5.2 Other Emerging Bioremediation Technologies Radionuclide Removal

For radionuclide remediation at contaminated sites, vitrification, chemical immobilization, electrokinetic remediation, and soil excavation are the most often utilized physical and chemical approaches, with capping and washing being the most advised methods. Nevertheless, these traditional approaches to pollution management are more costly and less successful in reducing pollutants across large areas. Thus, bioremediation, an environment-friendly and less costly measure, has emerged as a mainstay in getting rid of radionuclides and their hazardous effects from contaminated environments. Consequently, numerous technologies and strategies have been developed to actuate bioremediations of radioactive wastes. The subsequent section of the current state of the art aims to briefly illuminate the biological measures employed to remove radionuclides.

8.5.2.1 Use of Chelating Agents

Radionuclides can be complexed with chelating agents produced biologically to enable them to be removed from contaminated environments. Chelating agents that can be utilized to form biodegradable complexes (aerobic and anaerobic degradation) with radionuclides are citric acid, hydroxyl-acetic acid, oxalic acid, tartaric acid, EDTA, diethylenetriamine pentaacetic acid (DTPA), nitrilotriacetic acid (NTA), and N-hydroxy ethylenediamine triacetic acid (HEDTA). This led to the precipitated ions falling out as water-insoluble hydroxides or oxides, which prevented them from migrating into the groundwater (Prakash et al. 2013).

8.5.2.2 Biochar and Radionuclide Removal

Biochar, which is a form of charcoal made from organic materials (biomass) using pyrolysis, has been proposed as a green adsorbent for the elimination of caesium (Cs), copper (Cu), nickel (Ni), and lead (Pb) wastes from different sources (Mészároš et al. 2023). Biochar has revealed efficacy in the adsorption and subsequent removal of Uranium (U-232) from water even in picomolar quantities (Philippou et al. 2022). The affinity and adsorption ability of biochar is subjected to improvement with surface modification such as oxidation. Further, biochar (labeled as PMBN3) was found to enhance Sr(II) (strontium) removal from test water samples (Guo et al. 2022).

8.5.2.3 Phytoremediation

Plants can absorb, accumulate, and immobilize radionuclides contributing to bioremediation in a process known as phytoremediation. Green plants have been widely used for accumulation (phytoaccumulation/phytoextraction), filtration (rhizofiltration), stabilization (phytostabilization), volatilization (phytovolatilization), and degradation (phytodegradation and rhizodegradation) of radioactive wastes (Yan et al. 2021). Sunflower (*Helianthus annuus* L) and bean (*Phaseolus vulgaris)* are, for example, reported to remediate uranium contamination from soil and water through rhizofiltration (Lee and Yang 2010). Sunflowers also possess significant efficacies in the removal of ^{137}Cs, ^{90}Sr, ^{125}I, and ^{226}Ra from contaminated water and soil. *Brassica campestris* (Chou et al. 2005), *Zea mays* (Stojanović et al. 2010), *Amaranthus retroflexus* (Fuhrmann et al. 2002), and *Cyperus iria* (Yan 2016) are some other plants noted for their phytoremediation potential. Further, it was discovered that high antioxidant capacity plants have high remediation efficiency and therefore high tolerance to element toxicity. Moreover, plants that are radionuclide-tolerant, have a big biomass, and have a well-developed root system are suitable for phytoremediation.

8.5.2.4 Microbial Fuel Cells and Radionuclide Removal

A microbial fuel cell (MFC) is a device that uses microorganisms' metabolic activities to produce power. Since MFCs are capable of transforming organic matter into energy while eliminating contaminants from wastewater, such as nuclides like uranium and U(VI), they can be employed in wastewater treatment facilities (Roy et al. 2023). Phosphatase enzyme produced by denitrifying bacteria such as *Pseudomonas* species present as part of microbial consortia at the MFC leads to the generation of inorganic phosphate from glycerol-3-phosphates. The inorganic phosphates so generated combine with U (VI) to culminate in insoluble uranyl phosphate with efficiency as high as 90%. Thus, MFC provides not only an efficient setup as a source of power but also prompt removal of U (VI) and nitrate contamination from water samples (Vijay et al. 2020).

8.5.2.5 Bioinspired Material for Radionuclide Removal

Extant pieces of evidence reveal that natural materials like peat and mosses are quite effective in radionuclide sorption. This has been an inspiration for the scientific community to design and create a synthetic or modified form of natural biological or non-biological materials popularly called bioinspired material for the remediation of radionuclides. For instance, multifunctional mineral MXene hydrogels (MMHs) have been developed for the biomineralization of radionuclides (Li et al. 2022). Similarly, 2-aminobenzamidoxime containing bioinspired nano-traps have been synthesized for uranium extraction (Sun et al. 2018).

8.5.2.6 Radionuclide Removal Using Nanotechnology

Radionuclides can be selectively adsorbate and immobilized using nanoscale materials like nanoparticles and nanocomposites. Nanomaterials are broadly of three types—nanoadsorbents, nanomembranes, and nanocatalysts. These materials are efficient at removing radionuclides because of their high surface area and high tensile strength, reactivity, and excellent biocompatibility (Kumar et al. 2020). Shanshan Feng's group has shown that magnetic Prussian blue nanospheres (MPBNs) help remediate cesium from contaminated wastewater (Feng et al. 2022). Numerous types of nanomaterials, such as metal nanoparticles, carbon-based nanomaterials, cellulose nanomaterials, and carbon dots (CDs), have been explored for the efficient detection and remediation of uranium and reviewed by Neeru Rani's research group (Rani et al. 2023). Fungus-Fe3O4 bio-nanocomposite is reported to cause biochemical reductive precipitation of ^{238}U (Li et al. 2015). Through physicochemical adsorption, ZnO-based nanoadsorbents are reported to remediate Th(IV) (Kaynar et al. 2015). Hydroxyl-apatite (HAp) nanoparticles, similarly, could remediate ^{90}Sr, Co(II), and Eu(III) by more than 90% (Sheha et al. 2016). Sodium titanate nanobelt (Na-TNB) nanocrystals are reported to remove ^{137}Cs and ^{90}Sr contamination through ion exchange (Wen et al. 2016). While the effectiveness of the majority of nanomaterials is quite recommendable, their industrial scale-up and ecotoxicity minimization are yet to be met and thus need upgradation.

8.5.2.7 Bioaccumulation in Aquatic Systems

Certain aquatic organisms are reported to efficiently accumulate radionuclides from water. Filamentous algae growing in the residual water of tailings ponds (in proximity to the uranium industry) are reported to bioaccumulate very high concentrations of ^{226}Ra (9850 Bq/kg) (Jha et al. 2010). Microalga *Vacuoliviride crystalliferum* is reported to harbor cesium accumulation properties (Iwamoto and Shiraiwa 2017). For instance, marine brown macroalgae can accumulate U-238 up to concentrations as high as 3.8–11 Bq/kg (Ishikawa et al. 2004). Extant evidence reveals that marine mollusk like *Littorina littorea* reportedly accumulates U-238 (1.36–18.9 Bq/kg),

Th-232 (0.007–2.5 Bq/kg), Pb-210 (1.11–68 Bq/kg), and Po-210 (13.1–399 Bq/kg) (McDonald et al. 1991). Similarly, marine Pisces like coastal elasmobranchs reportedly is known for U-238 and Po-210 accumulation and biomagnification (Pentreath et al. 1980). Gillian M. Stewart's group has studied a large number of different marine animals with radioactive bioaccumulation efficiency including U-238, Th-234/228/232, Ra-226, etc. These marine organisms belong to a wide range of organisms including phytoplankton, microalgae, macroalgae, zooplankton, salps, crustaceans, mollusks, cephalopods, fishes, and others (Stewart et al. 2008). The radionuclides so accumulated through aquatic organisms can be successfully removed from the aquatic environment by using these organisms, which are safe to collect and dispose of. The specific radionuclide, the surrounding environment, and the intended application can all impact the effectiveness of these newly developed bioremediation techniques. Hence, in-depth investigation and testing are required to ascertain the viability and effectiveness of these approaches in different contexts. Nonetheless, microbial bioremediation looks a more promising and favorable outlook among all the available approaches toward the remediation of radionuclides. Microorganisms can take up various routes such as bioaccumulation, biosorption, bioimmobilization, and biofilm toward reducing radio wastes load. Further, microbes have the added advantage of being easily genetically altered to enhance their capacity to sequester radionuclides, a topic that is briefly covered in the section that follows in this state of the art.

8.5.2.8 Genetically Modified Bacterial Bioremediation and Omics-Integrated Bioremediation

Genetically engineering and recombinant DNA technology (RDT) have led to the identification and culturing of bacteria with specific and enhanced ability to remediate selected radionuclides through mechanisms such as biosorption and bioaccumulation. Such genetically altered bacteria possess the ability to target, bind, and transform/degrade radionuclides. Through RDT, transgenic bacterial cells may be generated with new proteins or enzymes that facilitate metal/radionuclide binding and subsequent sequestration. It has been found that bacterial strains with higher levels of metal-binding peptides and metal transporter proteins are better at collecting radionuclides. By adding the metal transporter gene nixA from *Helicobacter pylori* and the metal binding protein gene merTP from *Serratia marcescens*, a recombinant *Escherichia coli* strain was created. These *E. coli* strains showed five times greater U(VI) sorption capacity than a typical strain (Beckwith et al. 2001; Prakash et al. 2013). The radio-tolerant microorganism *Deinococcus radiodurans* known for its Cr(VI), U(VI), and Tc(VII) detoxification has undergone various genetic engineering processes toward achieving enhanced remediation capabilities. *Deinococcus radiodurans* R1 does not naturally form biofilms, and hence large culture of the bacterium is required making downstream processing tedious (Manobala et al. 2019a, b). Through genetic engineering, a recombinant strain of the bacterium *D. radiodurans* (DR1-bf+) with the ability to build biofilms has been developed.

The strain showed significant uranium removal efficiency in the presence of calcium. Further, it was observed that remediation efficacy got better with biofilm age (Manobala et al. 2019a, b). Similarly, a recombinant *D. radiodurans* strain is now available with a trans-gene merA from *E. coli* that makes *D. radiodurans* catabolize toluene and mercury (Brim et al. 2000). The introduction of YieF protein and PhoN (a phosphatase) into *D. radiodurans* R1 through RDT enhances the bacterium's ability to precipitate uranium along with chromium (VI) reduction (Xu et al. 2018). *Deinococcus geothermalis*, another extreme heat and radiation-resistant microorganism, has been engineered to obtain a Hg(II)-resistant strain capable of reducing not only Hg(II) but U(VI) and Cr(VI) at high temperatures (55 °C) (Brim et al. 2003). Genetically engineered microbes (GEMs) are promising as their radionuclide remediation efficiency is considered and rightfully has demonstrated excellent ex situ bioremediation activity. However, the in situ applicability of GEMs is currently at an early state demanding scientific attention and intervention. Further, GEMs may also need to compete for survival with the natural preexisting microbial community before demonstrating the intended action.

Advanced genomics and proteomics methods can also be employed to detect and investigate possible genes and proteins involved in radionuclide metabolism, a method commonly referred to as omics-based approach. Production of desired proteins and enzymes into radionuclide-resistant organisms for bioremediation is now possible through such a method toward improved management of radioactive waste. Based on recent genome-wide studies, *G. sulfurreducens* contains hundreds of c-type cytochrome genes that contribute to radionuclide reduction (Methe et al. 2003). Based on comparative genomics studies, radioresistant *Thermococcus gammatolerans* contains genes enabling it to tolerate radionuclide stress such as thioredoxin reductase (tgo180), glutaredoxin-like protein (tg1302), and peroxiredoxins (tg1253 and tg1220). The radioresistant *Thermococcus gammatolerans* are known to possess genes including thioredoxin reductase (tgo180), glutaredoxin-like protein (tg1302), and peroxiredoxins (tg1253 and tg1220) that allow the organism to withstand radionuclide stress, according to comparative genomics research (Zivanovic et al. 2009). Similarly, it was shown that when organisms such as *Rhodopseudomonas palustris* CGA009 and *Novosphingobium aromaticivorans* F-199 were exposed to radioactive cobalt, the expression of the NiCoT gene increased, suggesting that the NiCoT gene may have a role in managing radionuclides (Raghu et al. 2008). An essential understanding of radionuclide removal is provided by knowing the genes and the processes involved in their translation products. Some of the genes, their source organisms, and the brief function of such genes with regard to the bioremediation of radionuclides are depicted in Table 8.1. The identification of certain bacteria that are advantageous for in situ bioremediation may result from the integration of omics-based methodologies. This offers adequate assistance in creating transgenic organisms with excellent ability to exist in heterogenous radionuclide waste areas and perform bioremediation of the same through multiple processes such as bioreduction, biosorption, and biodegradation.

Table 8.1 Genes/proteins of genetically modified microorganisms implicated in radionuclide remediation

Sl. no	Gene/ protein name	Remark on bioremediation of radionuclides	Reference
1.	NiCoT	NiCoT-expressing *E. coli* cells demonstrated 85% removal of Cobalt (two-cycle treatment). The NiCoT gene was obtained from *Rhodopseudomonas palustris* CGA0009 and *Novosphingobium aromaticivorans* F-199 (NA)	Raghu et al. (2008)
2.	phoK	*Sphingomonas* species were used to isolate the phoK gene, which codes for an alkaline phosphatase. BSAR-1 and then recombinant *E. coli* overexpressing phoK was genetically engineered. Such *E. coli* demonstrated more phoK activity compared to the parent *Sphingomonas* sp. and precipitated more than 90% of uranium in a comparatively short span of time	Nilgiriwala et al. (2008)
3.	phoN	Using *Salmonella enterica* serovar Typhi's phoN gene, *Deinococcus radiodurans* demonstrated improved radionuclide removal. In only over 6 hours, this genetically modified *D. Radiodurans* precipitated more than 90% of the added uranium from a 0.8 mM uranyl nitrate solution	Appukuttan et al. (2006)
4.	*mer*A	The merA gene encoding mercury resistance was transferred from *E. coli* to *Deinococcus geothermalis* through RDT procedures. *Deinococcus geothermalis* so generated now having merA expression led to biotransformation of Hg (II) to the less toxic volatile Hg (I)	Brim et al. (2003)
5.	*ctyc*3	*Desulfovibrio desulfuricans* strain G20 bacterium lacking the gene ctyc3 displayed reduced potential of bioreduction of uranium (VI) to uranium (IV) compared to the wild-type counterpart	Payne et al. (2002)
6.	ChrR6	Recombinant E. coli with ChrR6 gene expression displayed enhanced U(VI) to U(IV) reduction activity along with chromate reductase activity	Barak et al. (2006)

8.6 Challenges and Limitations of Bacterial Bioremediation and Degradation of Radionuclides

Given its potential to endanger both human health and ecosystems, radionuclide contamination is one of the biggest environmental issues. Bioremediation is still the most important method for remediating such hazardous waste due to its higher benefit-to-risk ratio among the several techniques and technical breakthroughs made in this area. However, using microorganisms and microbiological processes to bioremediate radionuclides is the most appealing measure and numerous such processes have been covered in the existing state of art. Despite the advancements, there are still a number of obstacles to be addressed in the field of using microorganisms for radioactive bioremediation. Long-term impacts require optimizing the remediation setup and circumstances for the effective bioremediation of radionuclides, which are also accompanied by competing anions, toxic metals, organic compounds, and chelating agents. This has been a fundamental concern in many bacterial bioremediation advancements. While ex situ bioremediation has been satisfactory, in situ removal of radionuclides, through bioaugmentation and/or biostimulation has been challenging. Firstly, the introduced microbial species (wild type/genetically modified) need to acclimatize to the treatment site environment and compete with preexisting consortia of microorganisms. Secondly, it might be difficult to increase metabolic activity while preserving the necessary development parameters (pH, temperature, nutrition, etc.) in situ. Thirdly, the diverse nature of contaminants itself is a challenge for in situ remediation. Lastly, while intended bioremediation is achieved, it is of utmost necessity to understand that the natural local microflora of the treatment site should not be disturbed (Prakash et al. 2013).

Although quite effective bacterial bioremediations have some inherent challenges and limitations. Low bacterial concentration and slow growth may inhibit bacterial growth and subsequently radionuclide removal. Certain bacteria could be useful for eliminating one class of radionuclides but ineffective for eliminating another type of radioactive waste-demanding alternative. Many a time, large-scale application of bacterial remediation may not replicate the observations made in lab-scale setup. Finally, it can be challenging to sustain the bacterial population over time and meet the expenses, particularly if the pollution persists or if other environmental variables hinder the bacteria's development or the environment's ability to remove radionuclides.

Integration of omics and genetic engineering has drastically enhanced the efficacy of radionuclide removal from contaminated sites. However, such approaches have their limitations. Firstly, the generation of GEMs and subsequent approval from the concerned authority itself is a laborious process. Secondly, compatibility—both with the preexisting organisms at the treatment site and with the growth parameters—is a prerequisite for the release of GEMs.

Thirdly, after remediation is complete, the GEMs should be withdrawn from the treatment site, which is again a complex process. Finally, there shouldn't be any ecotoxicity caused by GEMs. Once these challenges and limitations are neutralized, bacterial bioremediation can be considered the most reliable and sound method for radionuclide removal ex situ as well as in situ.

8.7 Conclusion

The rapid accumulation of radioactive waste in the environment is a big concern these days because of its serious impact on animals and human health and the ecosystem. Its long-term management remains tough since several new nuclear energy firms are sprouting up in various nations and continue to expand to meet energy demand. As a result, there is a need to develop viable, eco-friendly, and cost-effective technology to prevent these dangerous radionuclides from being discharged into the environment or to develop innovative ways to remove radionuclides to keep the environment and humanity safe. Thus, many bacteria have the ability to tolerate radionuclides naturally and can be properly explored and used for the bioremediation of these toxic wastes. Further, genetically modified bacteria with the potential to remediate the same could be used as a sustainable alternative option. Recent advancements in omics-based innovation such as genomics, transcriptomics, and proteomics can be further utilized to assist the development of the bioremediation system.

References

Abdelouas A (2006) Uranium mill tailings: geochemistry, mineralogy, and environmental impact. Elements 2(6):335–341

AC03223434, A (2001) Methods for the minimization of radioactive waste from decontamination and decommissioning of nuclear facilities. Internat. Atomic Energy Agency

Ager D, Evans S, Li H, Lilley AK, Van Der Gast CJ (2010) Anthropogenic disturbance affects the structure of bacterial communities. Environ Microbiol 12(3):670–678

Akob DM, Mills HJ, Kostka JE (2007) Metabolically active microbial communities in uranium-contaminated subsurface sediments. FEMS Microbiol Ecol 59(1):95–107

Alliot I, Alliot C, Vitorge P, Fattahi M (2009) Speciation of technetium (IV) in bicarbonate media. Environ Sci Technol 43(24):9174–9182

Anand R, Mohan L, Bharadvaja N (2023) Phyto-& microbial-remediation of radioactive waste, microbial technologies in industrial wastewater treatment. Springer:215–241

Anttila M (2005) Radioactive characteristics of the spent fuel of the Finnish nuclear power plants. Posiva Oy

Appukuttan D, Rao AS, Apte SK (2006) Engineering of *Deinococcus radiodurans* R1 for bioprecipitation of uranium from dilute nuclear waste. Appl Environ Microbiol 72(12):7873–7878

Barak Y, Ackerley D, Dodge C, Banwari L, Alex C, Francis A, Matin A (2006) Analysis of novel soluble chromate and uranyl reductases and generation of an improved enzyme by directed evolution. Appl Environ Microbiol 72(11):7074–7082

Beckwith CS, McGee DJ, Mobley HL, Riley LK (2001) Cloning, expression, and catalytic activity of helicobacter hepaticus urease. Infect Immun 69(9):5914–5920

Brim H, McFarlan SC, Fredrickson JK, Minton KW, Zhai M, Wackett LP, Daly MJ (2000) Engineering *Deinococcus radiodurans* for metal remediation in radioactive mixed waste environments. Nat Biotechnol 18(1):85–90

Brim H, Venkateswaran A, Kostandarithes HM, Fredrickson JK, Daly MJ (2003) Engineering *Deinococcus geothermalis* for bioremediation of high-temperature radioactive waste environments. Appl Environ Microbiol 69(8):4575–4582

Bunn M, Holdren JP, Fetter S, Van Der Zwaan B (2005) The economics of reprocessing versus direct disposal of spent nuclear fuel. Nucl Technol 150(3):209–230

Carnes SA, Schweitzer M, Peelle EB, Wolfe AK, Munro JF (1998) Measuring the success of public participation on environmental restoration and waste management activities in the US Department of Energy. Technol Soc 20(4):385–406

Chou F-I, Chung H-P, Teng S-P, Sheu S-T (2005) Screening plant species native to Taiwan for remediation of 137Cs-contaminated soil and the effects of K addition and soil amendment on the transfer of 137Cs from soil to plants. J Environ Radioact 80(2):175–181

Cologgi DL, Speers AM, Bullard BA, Kelly SD, Reguera G (2014) Enhanced uranium immobilization and reduction by *Geobacter sulfurreducens* biofilms. Appl Environ Microbiol 80(21):6638–6646

Cooper MB (2005) Naturally occurring radioactive materials (NORM) in Australian industries-review of current inventories and future generation. Report ERS-006 to the Radiation Health and Safety Advisory Council

Cowie M, Mously K, Fageeha O, Nassar R (2012) NORM Management in the oil and gas industry. Ann ICRP 41(3–4):318–331

De Luca G, De Philip P, Dermoun Z, Rousset M, Verméglio A (2001) Reduction of technetium (VII) by *Desulfovibrio fructosovorans* is mediated by the nickel-iron hydrogenase. Appl Environ Microbiol 67(10):4583–4587

Fan Y-Y, Li B-B, Yang Z-C, Cheng Y-Y, Liu D-F, Yu H-Q (2018) Abundance and diversity of iron reducing bacteria communities in the sediments of a heavily polluted freshwater lake. Appl Microbiol Biotechnol 102:10791–10801

Feng S, Ni J, Cao X, Gao J, Yang L, Jia W, Chen F, Feng S, Zhang Y, Ma F (2022) Separation and removal of radionuclide cesium from water by biodegradable magnetic Prussian blue nanospheres. PRO 10(12):2492

Fomina M, Gadd GM (2014) Biosorption: current perspectives on concept, definition and application. Bioresour Technol 160:3–14

Frankle CM, Dale GE (2013) Unconventional neutron sources for oil well logging. Nucl Instrum Methods Phys Res, Sect A 723:24–29

Fuhrmann M, Lasat MM, Ebbs SD, Kochian LV, Cornish J (2002) Uptake of cesium-137 and strontium-90 from contaminated soil by three plant species; application to phytoremediation. J Environ Qual 31(3):904–909

Gabbard A (1993) Coal combustion: nuclear resource or danger. Oak Ridge National Laboratory Review 26(3&4)

Goyal N, Jain S, Banerjee U (2003) Comparative studies on the microbial adsorption of heavy metals. Adv Environ Res 7(2):311–319

Grover JR (1974) Management of high level radioactive wastes. Ann Nucl Sci Eng 1:329–331. https://doi.org/10.1016/0302-2927(74)90013-0

Guo Y, Hong Nhung NT, Dai X, He C, Wang Y, Wei Y, Fujita T (2022) Strontium ion removal from artificial seawater using a combination of adsorption with biochar and precipitation by blowing CO2 nanobubble with neutralization. Front Bioeng Biotechnol 10:819407

Hemme CL, Deng Y, Gentry TJ, Fields MW, Wu L, Barua S, Barry K, Tringe SG, Watson DB, He Z (2010) Metagenomic insights into evolution of a heavy metal-contaminated groundwater microbial community. ISME J 4(5):660–672

Hemme CL, Tu Q, Shi Z, Qin Y, Gao W, Deng Y, Nostrand JDV, Wu L, He Z, Chain PS (2015) Comparative metagenomics reveals impact of contaminants on groundwater microbiomes. Front Microbiol 6:1205

Herring JS, MacDonald PE, Weaver KD, Kullberg C (2001) Low cost, proliferation resistant, uranium–thorium dioxide fuels for light water reactors. Nucl Eng Des 203(1):65–85

Ishikawa Y, Kagaya H, Saga K (2004) Biomagnification of 7Be, 234Th, and 228Ra in marine organisms near the northern Pacific coast of Japan. J Environ Radioact 76(1–2):103–112

Iwamoto K, Shiraiwa Y (2017) Accumulation of cesium by aquatic plants and algae. Impact of cesium on plants the environment:171–185

Jaswal R, Pathak A, Chauhan A (2019a) Metagenomic evaluation of bacterial and fungal assemblages enriched within diffusion chambers and microbial traps containing uraniferous soils. Microorganisms 7(9):324

Jaswal R, Pathak A, Edwards B III, Lewis R III, Seaman JC, Stothard P, Krivushin K, Blom J, Rupp O, Chauhan A (2019b) Metagenomics-guided survey, isolation, and characterization of uranium resistant microbiota from the Savannah River Site, USA. Genes 10(5):325

Jeong S-W, Choi YJ (2020) Extremophilic microorganisms for the treatment of toxic pollutants in the environment. Molecules 25(21):4916

Jha V, Tripathi R, Sethy N, Sahoo S, Shukla A, Puranik V (2010) Bioaccumulation of 226Ra by plants growing in fresh water ecosystem around the uranium industry at Jaduguda, India. J Environ Radioactivity 101(9):717–722

Kansra V 1999. Status of power reactor fuel reprocessing in India

Kautsky U, Lindborg T, Valentin J (2013) Humans and ecosystems over the coming millennia: overview of a biosphere assessment of radioactive waste disposal in Sweden. Ambio 42:383–392

Kaynar ÜH, Ayvacıklı M, Hiçsönmez Ü, Kaynar SÇ (2015) Removal of thorium (IV) ions from aqueous solution by a novel nanoporous ZnO: isotherms, kinetic and thermodynamic studies. J Environ Radioact 150:145–151

Keasling J, Van Dien S, Trelstad P, Renninger N, McMahon K (2000) Application of polyphosphate metabolism to environmental and biotechnological problems. Biochemistry (Mosc) 65(3):324–331

Kessler G 2011. Proliferation-proof uranium/plutonium fuel cycles: safeguards and non-proliferation

Kim S-J, Koh D-C, Park S-J, Cha I-T, Park J-W, Na J-H, Roh Y, Ko K-S, Kim K, Rhee S-K (2012) Molecular analysis of spatial variation of iron-reducing bacteria in riverine alluvial aquifers of the Mankyeong River. J Microbiol 50:207–217

Kok KD, Heckman J, Devarakonda M (2016) Waste disposal: transuranic waste, high-level waste and spent nuclear fuel, and low-level radioactive waste. In: Nuclear engineering handbook. CRC Press, pp 521–556

Koul B, Adlakha K (2021) Bioremediation of radionuclides by plant-microbe system: current progress and challenges. Microbe Mediated Remediation of Environmental Contaminants:303–320

Kumar R, Nongkhlaw M, Acharya C, Joshi SR (2013) Uranium (U)-tolerant bacterial diversity from U ore deposit of Domiasiat in North-East India and its prospective utilisation in bioremediation. Microbes Environ 28(1):33–41

Kumar V, Katyal D, Nayak SS (2020) Removal of heavy metals and radionuclides from water using nanomaterials: current scenario and future prospects. Environ Sci Pollut Res 27:41199–41224

Lee M, Yang M (2010) Rhizofiltration using sunflower (Helianthus annuus L.) and bean (Phaseolus vulgaris L. var. vulgaris) to remediate uranium contaminated groundwater. J Hazard Mater 173(1–3):589–596

Leenaers A (2014) Surface-engineered low-enriched uranium-molybdenum fuel for research reactors. University of Ghent-SCK• CEN

Li L, Xu M, Chubik M, Chubik M, Gromov A, Wei G, Han W (2015) Entrapment of radioactive uranium from wastewater by using fungus-Fe3O4 bio-nanocomposites. RSC Adv 5(52):41611–41616

Li X, Shan G, Ma R, Shek C-H, Zhao H, Ramakrishna S (2022) Bioinspired mineral MXene hydrogels for tensile strain sensing and radionuclide adsorption applications. Front Phys 17(6):63501

Lloyd JR, Macaskie LE (2000) Bioremediation of radionuclide-containing wastewaters. Environ Microbe-Metal Interactions:277–327

Lloyd J, Ridley J, Khizniak T, Lyalikova N, Macaskie L (1999) Reduction of technetium by Desulfovibrio desulfuricans: biocatalyst characterization and use in a flowthrough bioreactor. Appl Environ Microbiol 65(6):2691–2696

Lovley DR, Phillips EJ (1994) Reduction of chromate by Desulfovibrio vulgaris and its c 3 cytochrome. Appl Environ Microbiol 60(2):726–728

Lyman ES (1994) A perspective on the proliferation risks of plutonium mines. US Dep. Energy Plutonium Stabilization and Immobilization Workshop, Dec

Manobala T, Shukla SK, Rao TS, Kumar MD (2019a) Uranium sequestration by biofilm-forming bacteria isolated from marine sediment collected from southern coastal region of India. Int Biodeterior Biodegrad 145:104809

Manobala T, Shukla SK, Subba Rao T, Dharmendira Kumar M (2019b) A new uranium bioremediation approach using radio-tolerant *Deinococcus radiodurans* biofilm. J Biosci 44:1–9

Marsili E, Beyenal H, Palma LD, Merli C, Dohnalkova A, Amonette JE, Lewandowski Z (2007) Uranium immobilization by sulfate-reducing biofilms grown on hematite, dolomite, and calcite. Environ Sci Technol 41(24):8349–8354

McCullough J, Hazen T, Benson S (1999) Bioremediation of metals and radionuclides: what it is and how it works. Lawrence Berkeley National Lab. (LBNL), Berkeley

McDonald P, Cook G, Baxter M (1991) Natural and artificial radioactivity in coastal regions of UK, radionuclides in the study of marine processes. Springer:329–339

McKenney E, Koelle K, Dunn R, Yoder A (2018) The ecosystem services of animal microbiomes. Mol Ecol 27(8):2164–2172

Merroun ML, Selenska-Pobell S (2008) Bacterial interactions with uranium: an environmental perspective. J Contam Hydrol 102(3–4):285–295

Mészároš L, Šuránek M, Melichová Z, Frišták V, Ďuriška L, Kaňuchová M, Soja G, Pipíška M (2023) Green biochar-based adsorbent for radiocesium and Cu, Ni, and Pb removal. J Radioanal Nucl Chem 332:1–15

Methe B, Nelson KE, Eisen J, Paulsen I, Nelson W, Heidelberg J, Wu D, Wu M, Ward N, Beanan M (2003) Genome of *Geobacter sulfurreducens*: metal reduction in subsurface environments. Science 302(5652):1967–1969

Narendrula-Kotha R, Nkongolo KK (2017) Bacterial and fungal community structure and diversity in a mining region under long-term metal exposure revealed by metagenomics sequencing. Ecol Genetic Genomic 2:13–24

Natarajan V, Karunanidhi M, Raja B (2020) A critical review on radioactive waste management through biological techniques. Environ Sci Pollut Res 27:29812–29823

Newsome L, Morris K, Lloyd JR (2014) The biogeochemistry and bioremediation of uranium and other priority radionuclides. Chem Geol 363:164–184

Newton RJ, Jones SE, Eiler A, McMahon KD, Bertilsson S (2011) A guide to the natural history of freshwater lake bacteria. Microbiol Mol Biol Rev 75(1):14–49

Nilgiriwala KS, Alahari A, Rao AS, Apte SK (2008) Cloning and overexpression of alkaline phosphatase PhoK from *Sphingomonas* sp. strain BSAR-1 for bioprecipitation of uranium from alkaline solutions. Appl Environ Microbiol 74(17):5516–5523

Nuttin A, Heuer D, Billebaud A, Brissot R, Le Brun C, Liatard E, Loiseaux J-M, Mathieu L, Meplan O, Merle-Lucotte E (2005) Potential of thorium molten salt reactors detailed calculations and concept evolution with a view to large scale energy production. Prog Nucl Energy 46(1):77–99

Patel R, Mugunthan J, Singh P, Mukherjee S, Koka R (2022) Microbial bioremediation and biodegradation of radioactive waste contaminated sites, microbes and microbial biotechnology for green remediation. Elsevier:733–746

Payne RB, Gentry DM, Rapp-Giles BJ, Casalot L, Wall JD (2002) Uranium reduction by *Desulfovibrio desulfuricans* strain G20 and a cytochrome c 3 mutant. Appl Environ Microbiol 68(6):3129–3132

Pentreath R, Woodhead D, Harvey B, Ibbett R (1980) A preliminary assessment of some naturally-occurring radionuclides in marine organisms (including deep sea fish) and the absorbed dose resulting from them. Proceedings of the third NEA seminar on marine radioecology. Tokyo 1–5 October 1979

Philippou M, Pashalidis I, Theocharis CR (2022) Uranium isotope (U-232) removal from waters by biochar fibers: an adsorption study in the sub-picomolar concentration range. Molecules 27(19):6765

Prakash D, Gabani P, Chandel AK, Ronen Z, Singh OV (2013) Bioremediation: a genuine technology to remediate radionuclides from the environment. Microb Biotechnol 6(4):349–360

Raghu G, Balaji V, Venkateswaran G, Rodrigue A, Maruthi Mohan P (2008) Bioremediation of trace cobalt from simulated spent decontamination solutions of nuclear power reactors using E. coli expressing NiCoT genes. Appl Microbiol Biotechnol 81:571–578

Rani N, Singh P, Kumar S, Kumar P, Bhankar V, Kamra N, Kumar K (2023) Recent advancement in nanomaterials for the detection and removal of uranium: a review. Environ Res 11:116536

Rao K (2001) Radioactive waste: the problem and its management. Curr Sci 81(12):1534–1546

Roy H, Rahman TU, Tasnim N, Arju J, Rafid MM, Islam MR, Pervez MN, Cai Y, Naddeo V, Islam MS (2023) Microbial fuel cell construction features and application for sustainable wastewater treatment. Membranes 13(5):490

Rossiello LA, Failla L (1994) Classification of radioactive wastes in Italy. American Industrial Hygiene Association Journal 55(7):658–659

Ryabova A, Kozlova O, Kadirov A, Ananeva A, Gusev O, Shagimardanova E (2020) DetR DB: a database of ionizing radiation resistance determinants. Genes 11(12):1477

Schmor P (2011) Review of cyclotrons for the production of radioactive isotopes for medical and industrial applications. Rev Accel Sci Technol 4(01):103–116

Sheha R, Moussa S, Attia M, Sadeek S, Someda H (2016) Novel substituted hydroxyapatite nanoparticles as a solid phase for removal of Co (II) and Eu (III) ions from aqueous solutions. J Environ Chem Eng 4(4):4808–4816

Shukla A, Parmar P, Saraf M (2017) Radiation, radionuclides and bacteria: an in-perspective review. J Environ Radioact 180:27–35

Sm S (2020) DeSmog the Syrian job : Uncovering the Oil Industry's Radioactive Secret 1–23

Sood D, Patil S (1996) Chemistry of nuclear fuel reprocessing: current status. J Radioanal Nucl Chem 203:547–573

Stewart GM, Fowler SW, Fisher NS (2008) The bioaccumulation of U-and Th-series radionuclides in marine organisms. Radioact Environm 13:269–305

Stojanović MD, Stevanović DR, Milojković JV, Grubišić MS, Ileš DA (2010) Phytotoxic effect of the uranium on the growing up and development the plant of corn. Water Air Soil Pollut 209:401–410

Sun Q, Aguila B, Perman J, Ivanov AS, Bryantsev VS, Earl LD, Abney CW, Wojtas L, Ma S (2018) Bio-inspired nano-traps for uranium extraction from seawater and recovery from nuclear waste. Nat Commun 9(1):1–9

Suriya J, Chandra Shekar M, Nathani NM, Suganya T, Bharathiraja S, Krishnan M (2017) Assessment of bacterial community composition in response to uranium levels in sediment samples of sacred Cauvery River. Appl Microbiol Biotechnol 101:831–841

Taylor DM, Taylor SK (1997) Environmental uranium and human health. Rev Environ Health 12(3):147–158

USEPA EPA (2009) A user-friendly reference document for hazardous waste characteristics. United States Environ Prot Agency Washingt DC 30

Van der Gucht K, Cottenie K, Muylaert K, Vloemans N, Cousin S, Declerck S, Jeppesen E, Conde-Porcuna J-M, Schwenk K, Zwart G (2007) The power of species sorting: local factors drive bacterial community composition over a wide range of spatial scales. Proc Natl Acad Sci 104(51):20404–20409

Van Roy S, Peys K, Dresselaers T, Diels L (1997) The use of an Alcaligenes eutrophus biofilm in a membrane bioreactor for heavy metal recovery. Res Microbiol 148(6):526–528

Vandana UK, Gulzar A, Laskar IH, Meitei LR, Mazumder P (2021) Role of microbes in bioremediation of radioactive waste. Microbial Rejuvenation of Polluted Environment 1:329–352

Vijay A, Khandelwal A, Chhabra M, Vincent T (2020) Microbial fuel cell for simultaneous removal of uranium (VI) and nitrate. Chem Eng J 388:124157

Waste R, Technology M (1980) USNRC technical training center 6-1. Rev 0311:1–18

Wen T, Zhao Z, Shen C, Li J, Tan X, Zeb A, Wang X, Xu A-W (2016) Multifunctional flexible free-standing titanate nanobelt membranes as efficient sorbents for the removal of radioactive 90Sr2+ and 137Cs+ ions and oils. Sci Rep 6(1):20920

Wildung RE, Gorby YA, Krupka KM, Hess NJ, Li S, Plymale AE, McKinley JP, Fredrickson JK (2000) Effect of electron donor and solution chemistry on products of dissimilatory reduction of technetium by *Shewanella putrefaciens*. Appl Environ Microbiol 66(6):2451–2460

Xiao S, Zhang Q, Chen X, Dong F, Chen H, Liu M, Ali I (2019) Speciation distribution of heavy metals in uranium mining impacted soils and impact on bacterial community revealed by high-throughput sequencing. Front Microbiol 10:1867

Xu R, Wu K, Han H, Ling Z, Chen Z, Liu P, Tian F, Zafar Y, Malik K, Li X (2018) Co-expression of YieF and PhoN in *Deinococcus radiodurans* R1 improves uranium bioprecipitation by reducing chromium interference. Chemosphere 211:1156–1165

Yablokov A, Nesterenko V, Nesterenko A (2009) Consequences of the Chernobyl catastrophe for public health and the environment 23 years later. Ann N Y Acad Sci 1181:318–326

Yan X (2016) Uptake of radionuclide thorium by twelve native plants grown in uranium mill tailings soils from south part of China. Nucl Eng Des 304:80–83

Yan X, Luo X, Zhao M (2016) Metagenomic analysis of microbial community in uranium-contaminated soil. Appl Microbiol Biotechnol 100:299–310

Yan L, Van Le Q, Sonne C, Yang Y, Yang H, Gu H, Ma NL, Lam SS, Peng W (2021) Phytoremediation of radionuclides in soil, sediments and water. J Hazard Mater 407:124771

Žagar T, Buršič A, Špiler J, Kim D, Chiguer M, David G, Gillet P (2011) Recycling as an option of used nuclear fuel management strategy. Nucl Eng Des 241(4):1238–1242

Zivanovic Y, Armengaud J, Lagorce A, Leplat C, Guérin P, Dutertre M, Anthouard V, Forterre P, Wincker P, Confalonieri F (2009) Genome analysis and genome-wide proteomics of *Thermococcus gammatolerans*, the most radioresistant organism known amongst the Archaea. Genome Biol 10(6):1–23

Chapter 9
Bacterial Remediation of Radioactive Waste: Current Understanding on Mechanistic Insights and Future Direction

Uddhav Sarania, Priya Borah, Nabajit Kalita, Mridulina Hazarika, Paramita Chakravarty, W. James Singha, Glory Borah, Janayita Biswa Sarma, and Hemen Deka

Abstract The management and remediation of radwaste is a pressing global concern as it is associated with environmental and human health. The physical and chemical methods for radwaste remediation are expensive and not environment friendly. The latest studies have confirmed the potential of indigenous microorganisms, more particularly bacteria in safe and efficient cleanup of radioactive contaminants. As earliest inhabitants of Earth, bacteria have demonstrated remarkable resilience in adapting to extreme environmental conditions. These microorganisms not only endure such hazardous surroundings but also actively participate in the remediation and cleanup of polluted environments. Bacteria isolated from radioactive sites present intriguing subjects that offer a unique perspective on their capabilities and the underlying mechanisms of their interactions. Understanding these mechanisms between radwaste and bacteria is of paramount importance for optimizing and enhancing the efficiency of radwaste treatment methods, which are often regarded as environmentally friendly and cost-efficient. Considering this, the chapter aims to provide a concise overview of the current understanding of the bacterial remediation mechanisms of the radwaste and its underlying radionuclide-bacterial interactions. Additionally, it emphasizes the limitations of traditional methods and underscores the significance of the diversity of indigenous bacteria for cleaning up radwaste-polluted environments. Furthermore, in advancing the technological field of radwaste management using bacteria, this chapter highlights future prospects and opportunities.

U. Sarania · P. Borah · N. Kalita · M. Hazarika · P. Chakravarty · W. J. Singha · G. Borah · J. B. Sarma · H. Deka (✉)
Department of Botany, Ecology and Environmental Remediation Laboratory, Gauhati University, Guwahati, India
e-mail: hemendeka@gauhati.ac.in

N. K. Arora et al. (eds.), *Environmental Sustainability and Biotechnology: Opportunities and Challenges*, Microorganisms for Sustainability 46,
https://doi.org/10.1007/978-981-96-9025-1_9

Keywords Bacterial remediation · Bioaccumulation · Bioreduction · Radioisotopes · Radwaste

9.1 Introduction

The ever-increasing volume of waste and its proper management have emerged as critical global challenges, driven by their detrimental environmental impacts and risks to human health (Zorpas 2020). Among the diverse categories of waste, radioactive waste (radwaste) represents a uniquely hazardous concern due to its persistent radioactivity, long-term ecological consequences, and direct threats to human well-being. The proliferation of nuclear technologies since the mid-twentieth century—spanning energy production, military applications, medical diagnostics, and industrial processes—has resulted in the accumulation of substantial quantities of radwaste, which contains radionuclides that remain hazardous for millennia (Yin and Zou 2021; Reena et al. 2012). Unlike organic pollutants, which degrade into benign byproducts such as CO_2 and water, radionuclides—unstable elements emitting ionizing radiation—cannot be chemically "destroyed" and persist in ecosystems, necessitating strategies that either immobilize or isolate them (Hazen and Tabak 2005; Natarajan et al. 2020). The remediation of these contaminants is further complicated by their environmental interactions, as factors like pH, redox potential, and microbial activity influence their bioavailability and mobility in soil and water systems. Compounding these challenges, the anticipated expansion of nuclear energy and industrial activities is projected to escalate radwaste production, amplifying contamination risks and underscoring the urgency for innovative, sustainable remediation solutions (Ding et al. 2019; Saling 2018). Conventional physical and chemical methods, such as vitrification, ion exchange, and chemical precipitation, often fall short due to their high costs, energy-intensive processes, and secondary environmental impacts, which limit their scalability and ecological compatibility (Czekała et al. 2023). In this context, biological strategies, particularly bacterial remediation, have gained prominence as eco-friendly alternatives. Certain bacteria exhibit intrinsic metabolic and physiological adaptations that enable survival in radionuclide- and heavy metal-contaminated environments, allowing them to interact with pollutants through mechanisms such as biosorption (surface binding via cell wall polymers), bioreduction (enzymatic conversion of toxic forms to less soluble states, e.g., U(VI) to U(IV)), biomineralization (precipitation into stable mineral phases like phosphates or sulfides), and bioaccumulation (intracellular sequestration) (Sundar et al. 2023; Reena et al. 2012; Patel et al. 2022). For instance, species within the genera *Pseudomonas*, *Bacillus*, and *Shewanella* demonstrate exceptional tolerance to uranium and technetium, while *Desulfovibrio desulfuricans* facilitate the reductive immobilization of radionuclides through sulfate-reducing metabolisms (Lloyd 2003). Advances in omics technologies—genomics, proteomics, and metabolomics—have further deepened insights into the genetic and molecular pathways underpinning these processes, enabling the identification of stress-response genes, metal-binding proteins, and redox-active enzymes critical for bacterial survival and pollutant transformation (Aziz et al. 2021). Despite these advancements, practical challenges remain, including environmental variability (e.g., temperature, nutrient availability), competition within microbial consortia, and the need for robust field-scale applications (Gonçalves et al. 2022; Bhatt et al. 2021). To address these limitations, interdisciplinary approaches integrating bacterial remediation with nanotechnology (e.g., nanoparticle-enhanced biosorption) or physicochemical treatments (e.g., electrokinetics) are being explored to create synergistic, scalable systems (Tonelli et al. 2023). Future research directions emphasize the discovery of extremophilic bacterial strains from radioactive niches, genetic engineering to amplify metal-binding or reduction capabilities, and AI-driven predictive models to optimize microbial consortia design and deployment in heterogeneous environments (Dangi et al. 2019; Thakur and Kumar 2024). By synthesizing mechanistic insights with technological innovation, this chapter aims to advance the development of efficient, sustainable strategies for mitigating radwaste contamination, ultimately contributing to global efforts to balance nuclear energy advancement with ecological and human safety.

9.2 Radioactive Waste

9.2.1 Classification

According to the International Atomic Energy Agency (IAEA), any material that contains radionuclides at concentrations or radioactivity rates higher than the exempt amount set by the relevant authorities is considered radioactive waste (McGinnes 2007). Radioactive waste can be classified broadly into two categories:

(i) *Qualitative approach:* Qualitative approach is based on the origin and physical state of the radioactive waste which includes solid waste, liquid waste, and gaseous waste (Hosan 2017).
(ii) *Quantitative approach:* In quantitative approach, solid, liquid, and gaseous radioactive wastes have been divided further into different classes (Fig. 9.1) based on radiation dose and concentration level, respectively (Hosan 2017).

The IAEA further classified radioactive waste into six categories based on radionuclide half-life and radioactivity content. The six categories are as follows:

(a) *Exempt waste (EW)*

Exempt waste (EW) contains radionuclides in a very less concentration, which is not considered harmful to people or surrounding environment and therefore no radiation protection provisions are required (Khelurkar et al. 2015). The activity concentration of exempt waste ranges from 10^{-1} Bq/g to 10^{4} Bq/g depending on the specific radionuclides (Darda et al. 2021). Examples of

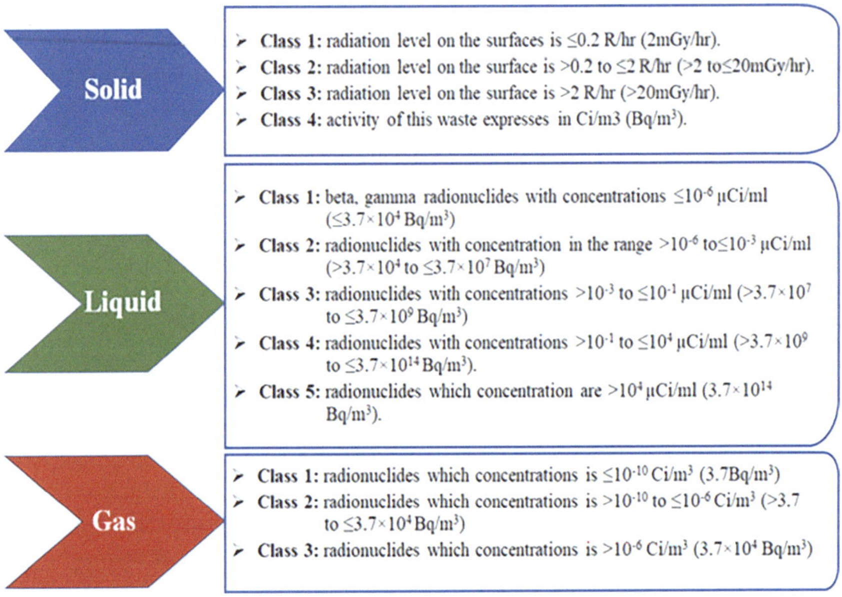

Fig. 9.1 Showing quantitative approaches of classification of radioactive waste

exempt waste include plaster, bricks, and metals collected from nuclear power sites.

(b) *Very short-lived waste (VSLW)*

VSLW contains nuclides with extremely less half-life that undergo radioactive decay rapidly. Therefore, this kind of waste is hazardous for a brief period and is kept in storage until they reach the threshold for exempt waste so that they are no longer hazardous to the environment (Ojovan 2011).

(c) *Very low-level waste (VLLW)*

VLLW is the term used to define waste generated from the nuclear power plants with concentration levels that are either slightly above or close to the thresholds required for waste material to get clearance from the regulatory control (Hosan 2017).

(d) *Low-level waste (LLW)*

LLW is mainly produced by the nuclear fuel cycle, hospitals, and industries. It primarily contains relatively little short-lived radioactive radiation. It can be buried in the shallow landfills and doesn't need to be isolated during handling or transportation (Khelurkar et al. 2015).

(e) *Intermediate-level waste (ILW)*

ILW contains long-lived radionuclides that emit alpha particles and do not decay to a level of activity concentration suitable for near-surface disposal. Therefore, it needs to be disposed of at a greater depth rather than near surface disposal (Ojovan 2011).

(f) *High-level radioactive waste (HLW)*

HLW is described as radioactive waste whose activity concentration is high enough to require shielding during handling operations and produce significant amounts of heat during radioactive decay (Khelurkar et al. 2015). The majority of the radionuclides in HLW are fission products and transuranic elements produced in the reactor core. Because fission products are so radioactive, therefore HLW is considered as the most hazardous nuclear waste (Darda et al. 2021).

9.2.2 *Sources of Radioactive Waste*

Radioactive waste is produced by a variety of processes, such as nuclear fuel cycle and the reprocessing of nuclear weapons, industrial waste, medical waste, and naturally occurring radioactive materials (NORM). Human industrial processes such as phosphate fertilizer production, uranium mining, and oil and gas production can concentrate and expose naturally occurring sources of radioactive materials (Rao et al. 2022). Nuclear fuel cycle is a prime producer of radioactive waste which produces radioactive waste from both front end and back end of the cycle. Generally, the front end produces alpha-emitting radioactive waste and back end produces beta- and gamma-radiating wastes (Khelurkar et al. 2015). Nuclear fuel cycle also produces alpha particles such as Uranium-234 (^{234}U), Neptunium-237(^{237}Np), Plutonium-238 (^{238}Pu), and Americium-241 (^{241}Am). Military weapon development such as nuclear weapons is also a major source of radioactive waste. The reprocessed waste from nuclear weapons is expected to include materials with significantly elevated specific activities, such as Plutonium-238 (^{238}Pu) and alpha-emitting actinides like Plutonium-239 (^{239}Pu), which is a fissile material utilized in bombs (Reed 2017). Additionally, trace amounts of beta or gamma radiation-emitting substances like americium (Am) and tritium (T) are also emitted during nuclear weapon testing processes. Moreover, radioactive materials are used for many purposes in the medical industry, and they are a major cause for radioactive pollution. Beta particles and gamma ray emitters are typically found in medical waste. A few isotopes used in medicine are Yttrium-90 (^{90}Y) for lymphoma treatment, Iodine-131 (^{131}I) for thyroid cancer treatment, Strontium-89 (^{89}Sr) for bone cancer treatment, Iridium-192 (^{192}Ir) for brachytherapy, and Cobalt-60 (^{60}Co) and Caesium-137 (^{137}Cs) for brachytherapy and external radiation (Khelurkar et al. 2015).

Since the infamous disaster of nuclear accidents in Hiroshima and Nagasaki, scientists have especially emphasized the effects of certain notable radioisotopes generated during the production of nuclear power (Nakamura and Kikuchi 2011). Such radioisotopes include Strontium-95 (half life: 27 years), Zinc-65 (half life: 245 days), Iodine-131 (half life: 8 days), and Tritium, H-3 (half life: 12 years) (Eidemüller and Eidemüller 2021). Strontium-90 was first reported to be released into the atmosphere during the atomic bombings of Hiroshima and Nagasaki (Paatero et al. 2010). It was also documented to be released as a byproduct during electricity generation by the nuclear reactors (Semenishchev and Voronina 2020). The human body often mistakenly identifies it as calcium leading to its accumulation in bone structures particularly in young individuals (Mayneord 2011). Likewise, Zinc-65 released from the reactor shells and discharged into the water bodies in supposedly negligible amounts can affect the neurological health and shows potential toxicity when ingested in excess (Rajkhowa et al. 2021). On the other hand, Iodine-131 released in the air as a gas in minute concentration can increase the risk of thyroid cancer (Elkhatib et al. 2024; Leeman-Neill et al. 2013). Moreover, Tritium (H-3) was initially not considered as an environmental contaminant due to its low production levels; however, its combination with oxygen to form heavy water can exhibit properties like toxic radioisotopes (Mosin and Ignatov 2015).

9.2.3 Ecological Implications and Impact on Human Health

Radwaste possesses a significant adverse threat to both living organisms and the overall ecosystem (Adeola et al. 2023). Therefore, it must be properly managed to ensure that exposure of radioactive waste does not surpasses beyond the standard regulations (Poškas et al. 2019). Out of all the wastes either generated by natural or anthropogenic means, radwaste stands out to be the most complex and threatening type of waste concerning both environmental and human health as shown in Fig. 9.2 (Hamblin 2008). Its unpredictable and unstable behavior confers to its complexity and poses a threat to both humankind and the environment (Shiels 2019). Moreover, radioactive elements can pollute wide ranges of regions such as air, water, and soil through radon gas, which is a noble gas generated during radioactive decaying process of thorium or uranium (Blanco-Novoa et al. 2018). Notable quantity of radwaste was found in landfill sites, harbors, and marine sediments (Fang 2002). Improper disposal of such radioactive waste can pollute the soil and that can impact both the plant growth and fertility of soil (Smičiklas and Šljivić-Ivanović 2016). Studies reported that radionuclides can interfere with the biogeochemical cycle in soil impeding the availability of nutrients crucial for plant growth (Ali et al. 2023). Moreover, radwaste exposure to the soil microbiome can lead to alterations in the abundance and density of microbial community (Theodorakopoulos et al. 2017). This observation can be supported by the fact that some microbes are sensitive to radiation resulting in their decline, while others are resistant to radiation leading in their adaptation to such areas (Castillo et al. 2015). A study conducted by Geras' kin et al. (2011) reported the potential impact of radionuclides on soil enzyme activities after the Chernobyl fallout and found out huge alterations in dehydrogenase, catalase, urease, and invertase activities. Another study carried out by McNamara et al. (2003) investigated the impact of radiation generated by gamma isotopes on soil enzyme activities. The findings suggested that the specific responses and alterations on the soil enzyme activities differ considerably based on radiation dose, exposure time, and the total resilience of the soil ecosystem (McNamara et al. 2003). Soil enzymes play a crucial role in assessing the overall health of an ecosystem and significant changes in these parameters can have a profound impact on life within it (Joshi et al. 2018). However, radioisotopes generated from various nuclear power

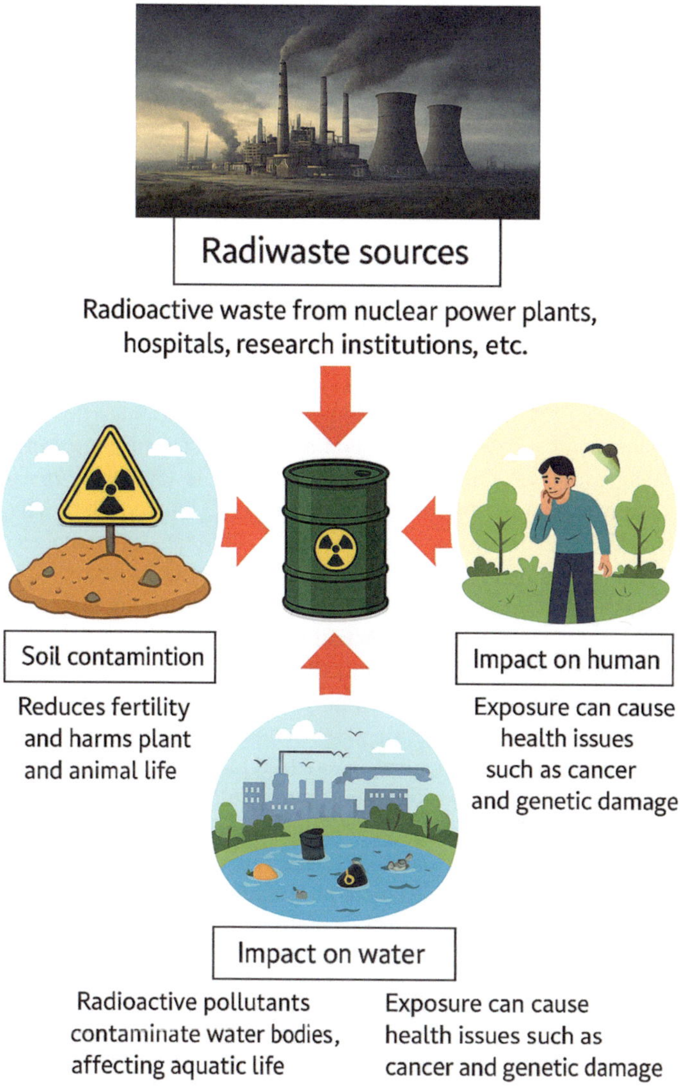

Fig. 9.2 Showing ecological impacts of radwaste

plants discharged into the water bodies accumulated in the food chain through aquatic life forms and eventually reach plants through irrigation with such contaminated water (McBride 2022). Therefore, the necessity for thorough investigation for the potential threat induced by radwaste on the ecosystems has been recognized across the globe, and a series of international regulations has been made in this matter (Environment Canada 1997; IAEA 2006; ICRP 2003, 2008).

Radioactive waste can peculiarly change their properties and forms as they surpass the decaying stage (Adloff 2018). Radwaste decay radioactivity is a natural phenomenon in which radiation energy is released by certain decaying substances (Bagher et al. 2014). The radioactive decay-generated emissions can be mainly divided into three types: alpha particles, beta particles, and gamma rays (Karmaker et al. 2021). The alpha particles cannot penetrate the human skin because it comprises relatively slower moving mass of neutrons and protons Hence, it can be inferred that humans are significantly immune to such radiations (Al-Rubaii 2018), whereas beta particles possess electrons that can move at a very fast speed and can penetrate through the skin or even several inch thicker wood. They are henceforth considered relatively dangerous even when occurred naturally (Dondelinger 2012). Gammy rays are intense energy rays like X-rays that can easily pass through materials of several inches like steel as well as concrete of several feet in thickness. As a direct consequence of this, they are highly dangerous (McCarthy et al. 2003). Although plants and animals possess certain amounts of natural protection against radiation that is outside their bodies, no protection is known against radwaste-generated isotopes that are trapped inside the bodies (Walker 2000; Monteith and Unsworth 2013). The ionizing potential of radiation is a factor that makes it harmful for human health (Burgio et al. 2018). Moreover, prolong exposure of high-intensity radiation can result in extreme sickness and even death (Belpomme et al. 2018). Even though alpha particles are comparatively less harmful, it can still result in severe damage in the body when accidentally eaten or inhaled (Zakariya and Kahn 2014). When beta particles can penetrate the skin, they can cause severe burns (Waghmare 2013). Overexposure of most dangerous type of radiation, namely, gamma radiation, X-rays, and neutrons, may lead to vomiting, nausea, weight loss, weakness, diarrhea, fever, hair loss, and internal bleeding. Another deteriorating impact of ionizable radiation is its ability to induce genetic mutation that can sometimes cause cancer (Samarth et al. 2020). Additionally, human radiation exposure has been linked to several illnesses including acute radiation syndrome and cancer of the lungs, thyroid, and skin as well as cardiac, reproductive, and neurological conditions (Sundar et al. 2023). Besides, some radioisotopes are capable of forming heavy water, which show the tendency to replace normal hydrogen from water molecules and can disrupt important biological processes (Goncharuk 2014).

There is a huge difference in the activity of natural and synthetic radiation on living organisms (Fang 2002). Natural radiation generated from cosmic rays and naturally occurring radioactive elements such as thorium, radium, and uranium can only affect plants and animals from the exterior (Shahbazi-Gahrouei et al. 2013). Meanwhile, man-made radiations cause damage internally (Busby 2013). Radwaste-generated radioisotopes often can enter food chain and become increasingly concentrated as they ascend up the food chain and ultimately reach humans (Xu and Zhang 2010). Once inside the human body, the radioisotopes continue to decay and release radiation for a period of hours to years based on the half-life of the isotopes involved (Spiers 2013; Dondelinger 2012). Several data concerning the speed of decay from the initial encounter showed that the complete decay of many radioactive isotopes can take up to several hundreds to thousand years (Pfützner et al. 2012). Ecological impacts of radwaste have been depicted in Fig. 9.2.

9.3 Radwaste Remediation

9.3.1 Traditional Methods

Since the discovery of radioactive waste, various physical and chemical methods have been explored and put into practice for its management. Some of these methods include incineration, evaporation, dumping, acid digestion, and wet oxidation (Rao et al. 2022). The incineration method involves the burning of solid radwaste at high temperatures in kilns producing S, HCl, CO_2, and H_2O as byproducts. Gas-filtering devices are necessary for this technology to control radioactive emissions (IAEA 2006). The main drawback of incineration method is its ash product containing concentrated radioactive isotope—hazardous for transportation and unsuitable for burial. In order to transform radioactive ash into a condition safe for final disposal, it must be further processed using additional industrial machinery and a variety of technological processes, which require significant energy expenditures (Bobrakov et al. 2014). Again, evaporation is a type of unit operation wherein radioactive waste and other hazardous waste elements, such as salts and HMs, are extracted from effluent wastes (Natarajan et al. 2020). It assists in lowering the amount of radioactive waste from intermediate- and low-level radwaste. The main limitation of evaporation is that it's an energy-intensive technology with significant running costs. Furthermore, the presence of a large number of inactive salts slows down the evaporation process with some organic salts, which increases the risk of an evaporation explosion (IAEA 2006; Natarajan et al. 2020). On the other hand, in the process of acid digestion, organic radwaste is digested by mixtures of strong acids (H_3PO_4 and HNO_3) at high temperature and adequate pressure which breaks down the bonds between the waste's constituent parts and produces inorganic liquids and gases like CO_2 and O_2 (Rao et al. 2022). Consequently, wet oxidation is the method of oxidation of liquid radioactive waste at low temperatures. O_2, H_2O_2, and O_3 are used as oxidizers that degrade or transform the waste through the process of oxidation. Although the process generates clean solvents, the challenge lies in the immobilization of residues and storage of the oxidizing agents (Rao et al. 2022). Again, dumping is the final storage of high-level radioactive waste that cannot be recycled (Clayton et al. 2024). Radwaste (solid or mixed) is generally dumped either in deep ocean repositories or deep underground wells. Before being dumped into the ocean, the waste undergoes a series of treatments that include combustion, evaporation, and compaction to avoid contamination (Thakur et al. 2024). The major disadvantage associated with dumping process is the high possibility of leakage and contamination of ocean and groundwater.

9.3.2 *Biological Methods*

The biological method of radwaste remediation involves the use of living organisms, such as bacteria, fungi, or plants to mitigate or eliminate radioactive contaminants from the polluted environment (Thakare et al. 2021). This approach involves the natural abilities of these organisms to interact with and transform radionuclides by making them less hazardous and promotes their immobilization. This environmentally friendly tactic to remediate radioactive waste which primarily includes bacterial remediation and phytoremediation (Sasikumar et al. 2003; Reena et al. 2012) is briefly addressed below.

9.3.2.1 Phytoremediation

Phytoremediation of radioactive waste is the process of removal or immobilization of radioactive elements from contaminated soil, sludge, or water using plants (Natarajan et al. 2020; Hu et al. 2014). It is an eco-friendly method with low operational cost, and this is more effective to put into practice in accidentally contaminated land or water systems (Rao et al. 2022). Based on the breakdown of radioactive nuclides, phytoremediation can be categorized into the following processes, such as phytoextraction, phytovolatilization, phytostabilization, rhizofiltration, and phytodegradation as shown in Fig. 9.3.

Phytoextraction is the process of extraction of the pollutants from the contaminated soil through the roots, translocating them to the aboveground portion of the plant and then harvesting and destroying safely within the plants (Boonyapookana et al. 2005). Ramaswami et al. (2001) mentioned that terrestrial plants have high uranium extraction capacity in sandy soil as compared to organic soil. Again, phytovolatilization involves the uptake of contaminants from the site and dispersing them into the atmosphere through evaporation (Singh et al. 2023). The radioactive elements can be released into the atmosphere by mixing them with volatile organic solvents produced by plants. Upon volatilization, leaves disperse the toxic metals into the atmosphere for further degradation (Saleh 2012; Rao et al. 2022). However, it should be considered that the continuous release of radioactive substances in the atmosphere may raise the number of radioactive elements in the atmosphere (Shukla et al. 2024). Accordingly, phytostabilization or phytosequestration involve stabilizing and immobilizing contaminants within the soil by utilizing plant roots to induce precipitation, thereby reducing their mobility and bioavailability in the surrounding environment (Latif et al. 2023). The selection of plants for phytostabilization should encompass a combination of both trees and grasses (Thakur et al. 2024). The major advantage of this method is the production of no secondary waste, thereby eliminating the necessity for additional treatment (Bakshe and Jugade 2023). However, this process is a less developed area within the field of phytoremediation research. Again, the process of rhizofiltration can be defined as the filtration of radioactively polluted water through a mass of roots for the removal of toxic elements. This

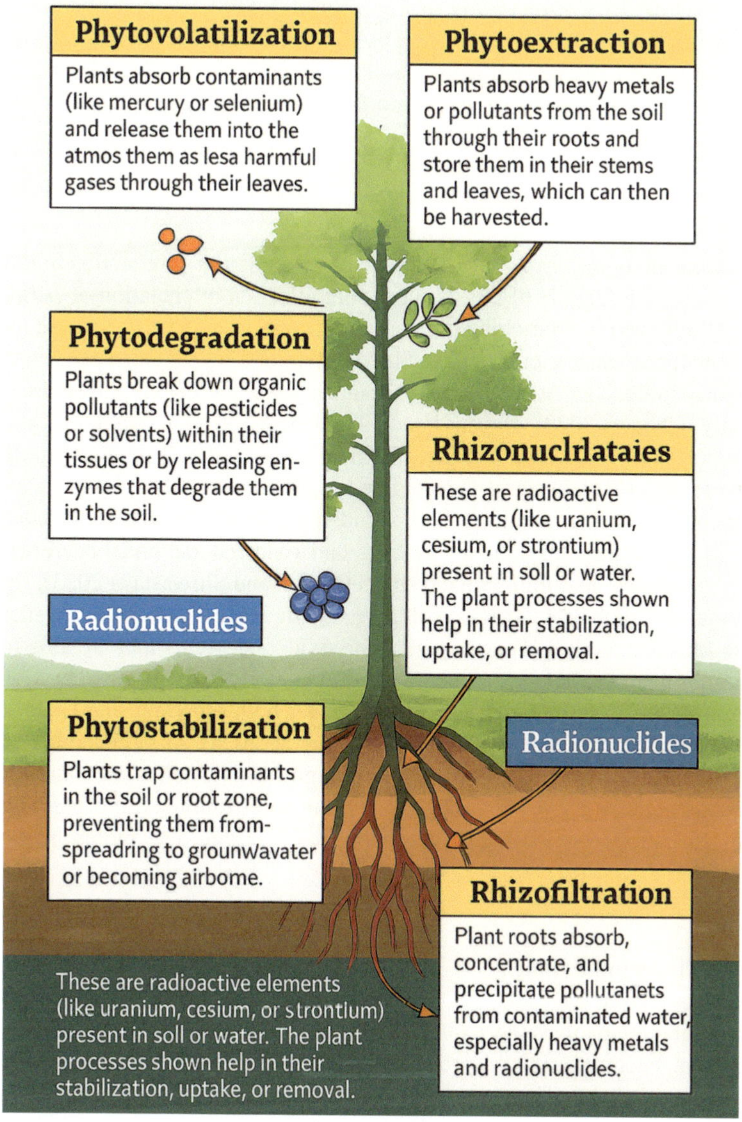

Fig. 9.3 Showing different phytoremediation mechanisms against radwaste pollutants

cost-effective technique is efficient in eliminating only low-level pollutants (Rao et al. 2022). Plants having rapidly growing root systems with good accumulation capacity can be best suited for the rhizofiltration method. For example, the water hyacinth (*Eichhornia crassipes*) significantly contributes to the rhizofiltration process by accumulating radioactive nuclides in its roots (Cerne et al. 2011). Moreover, phytodegradation is the utilization of plants in conjugation with microbes to degrade toxic substances mainly suitable for the degradation of hydrophobic organic chemicals (Rao et al. 2022). According to Eapen et al. (2006), phytodegradation is the absorption and breakdown of pollutants by enzymes inside the plant. Plant enzymes such as dehalogenase, nitroreductase, and nitrilase are involved in the degradation of contaminants during phytodegradation process (Kumar et al. 2019).

9.3.2.2 Bacterial Remediation

The mechanism of bacterial remediation involves oxidation or reduction of radionuclides, changes in the pH of the solution, solubilization or chelation of radioactive elements followed by immobilization, and reduction in bioavailability and toxicity of radionuclides (Sarkar et al. 2024; Shukla et al. 2017). Enzymes can effectively immobilize radioactive nuclides with ease in bacteria, whether employed directly or indirectly (Cheema 2023). Depending on site-specific edaphic conditions, radioactive elements may be treated with direct (oxidation-reduction reactions) and indirect enzymatic actions of bacteria (Francis and Nancharaiah 2015). Bacteria execute direct action by utilizing organic compounds to facilitate the release of electrons, which in turn transform soluble, mobile, and oxidized radionuclide forms into reduced, insoluble, and immobilized states (Verma and Shrivastava 2023). Indirect action may be initiated by certain sulfate-reducing bacteria, capable of generating organic acid metabolites, mineral acids, and oxidizing agents that are able to efficiently minimize the pH of the medium (Shukla et al. 2017). Hence, by reducing long-term risks, bacterial bioremediation provides both economic and environmental benefits over traditional techniques (Sarkar et al. 2024). As depicted in Fig. 9.4, the remediation of radioactive elements through bacteria relies on some specific mechanisms, including bioaccumulation, biomineralization or bio-precipitation, biotransformation or bioreduction, bio-solubilization or bioleaching, and biosorption.

9.3.2.3 Bioaccumulation

Bioaccumulation involves the accumulation and amplification of a substance inside an organism, encompassing the processes of biomagnification and bioconcentration (Campbell et al. 2022). This process involves electrostatic binding facilitated by the attraction between bacterial cell wall, which is negatively charged and metal cations. As shown in Fig. 9.5, the mechanisms emphasize generation of different complexes between radioactive nuclide and several ligands or chelating agents including sulfate, hydroxide, and phosphate, which foster the generation of insoluble and less harmful substances inside the cell (Riaz et al. 2024). The process may be passive or active. Unlike active bioaccumulation, which is a slower process demanding more energy, passive bioaccumulation is relatively faster and entails lower energy consumption (Shukla et al. 2017). Again, the accumulation of radionuclides by bacteria can occur either intracellularly or extracellularly. Accumulation through

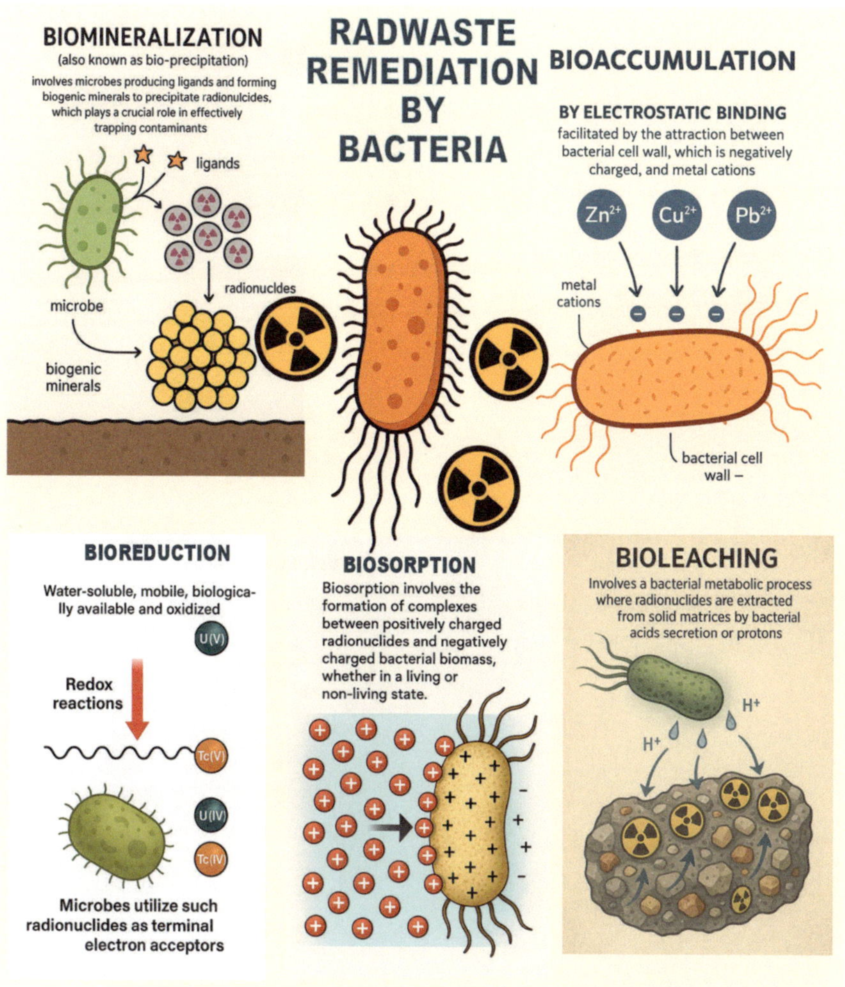

Fig. 9.4 Showing different radwaste remediation mechanisms employing bacteria

intercellular process is regarded as more prominent compared to extracellular, which influences the permeability of cell membrane (Ding et al. 2019). Radionuclides such as uranium (U), cesium (Cs), radium (Ra), plutonium (Pu), strontium (Sr), and thorium (Th) have been reported to exhibit potential bioaccumulation by bacteria. A study by Zhao et al. (2016) elucidated the uranium bioaccumulation process exhibited by *Bacillus* sp. in liquid environment. The XPS analysis further confirmed that the presence of various functional groups on the biomass of *Bacillus* sp. was crucial for uranium bioaccumulation. Additionally, certain cyanobacteria such as *Arthrospira platensis* demonstrate the ability for bioaccumulation of various radionuclides including ^{241}Am, ^{233}U, ^{137}Cs, ^{239}Pu, ^{237}Np, and ^{90}Sr (Zinicovscaia et al. 2020). Besides, findings from Ozdemir et al. (2022) revealed considerable

Fig. 9.5 Different bacterial remediation mechanisms of U(VI) or UO_2^{2+}

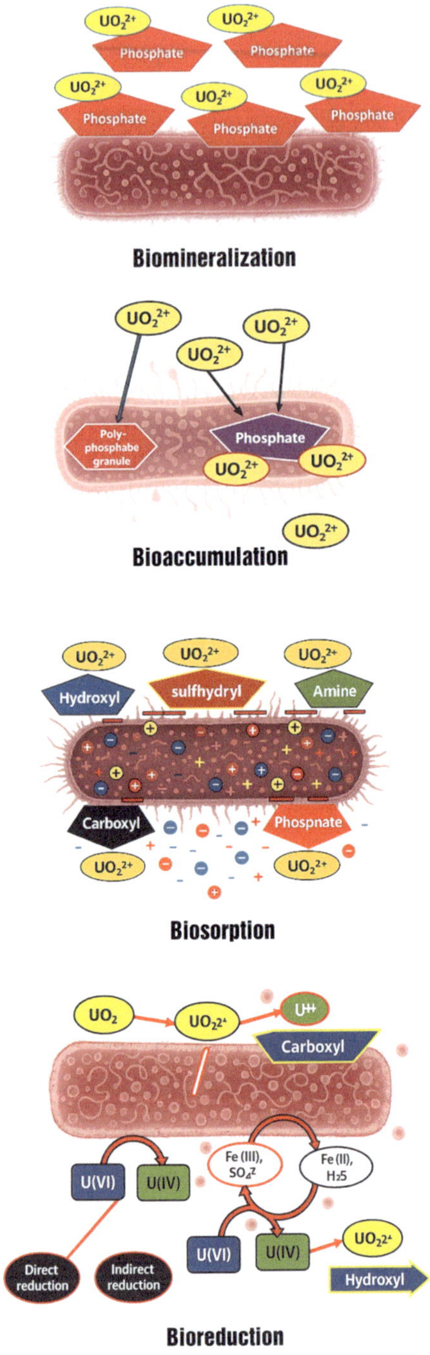

bioaccumulation efficacy of *Bacillus cereus* against U(VI) and Th (IV) after 24 h of application. Additionally, Li et al. (2016) reported that 100 mg of *Streptomyces sporoverrucosus* could accumulate approximately 60% of the total uranium within 12 h of application. Notably, bioaccumulation of uranium in *Pseudomonas* sp. (Dummi Mahadevan and Zhao 2017), bioaccumulation of strontium by *Micrococcus luteus* (Phillips et al. 2002), and bioaccumulation of Cs-137 by *Bacillus licheniformis* and *Bacillus pumilus* (Abdel-Razek et al. 2015) were reported in various recent studies to demonstrate significant bioaccumulation capacity of such bacteria. Bioaccumulation is significantly influenced by the presence of capsules, slime layers, or S-layers in the bacterial surface (Panicker 2022). After the accumulation process, the polyphosphate bodies interact with the accumulated radionuclides and reduce its toxic nature (Shukla et al. 2017). These intracellular chelation mechanisms have been proposed to be liable for the development of tolerance to metals and subsequent radionuclides.

Biomineralization or Bio-Precipitation

The process of biomineralization, also known as bio-precipitation, involves microbes producing ligands and forming biogenic minerals to precipitate radionuclides, which plays a crucial role in effectively trapping contaminants (Jabbar and Wallner 2015). Microbes can enzymatically produce highly localized ligands like sulfides, carbonates, phosphates, and hydroxides around their cells that serve as nucleation sites for precipitation (Jiang et al. 2020). In contrast to conventional direct precipitation, the biogenic mineral method entails the microbial-facilitated formation of a metal biomineral, serving as a scaffold for the sequential capture and deposition of the targeted radionuclides (Jabbar and Wallner 2015). The process of biomineralization depends on pH, temperature, humid substances, ions, and enzyme activity (Jiang et al. 2020). The secretions and metabolic activities of bacteria can induce alterations in the pH of their immediate environment, thereby modifying the pH in the vicinity of the metal during the process (Shukla et al. 2017). Two types of biomineralization, i.e., biological-induced mineralization (BIM) and biological-controlled mineralization (BCM), is naturally exhibited by microorganisms (Thakare et al. 2021). BIM involves microbes altering their environment for extracellular mineral precipitation, while BCM emphasizes microbial control during the process. BIM frequently takes place through redox reactions, interactions with metabolites such as phosphate, or adhering onto cell surfaces (Ding et al. 2019). These processes ultimately lead to the formation of mineral precipitates containing radionuclides. Efficient uranium biomineralization was observed in bacteria such as *Citrobacter* sp. and *Serratia* sp. (Ding et al. 2019). In addition, prior instances of uranium biomineralization were also noted when *Pseudomonas* species were provided with tributyl phosphate (Thomas and Macaskie 1996). Newsome et al. (2014) isolated bacterial cell coated in uranium phosphate from contaminated soil, revealing inherent biomineralization process in nature. The rapid development of metal phosphate creates a defensive barrier, potentially hindering cellular metabolism

(Vandana et al. 2021). Wang et al. (2019) demonstrated the potential biomineralization of U(VI) employing *Kocuria* sp. Additionally, *Serratia* sp. has been identified as relying on the synthesis of crystalline hydroxyapatite, a process strategically employed for the subsequent recovery of Eu and Sr (Gangappa et al. 2016). The biomineralization approach has also been explored as a method for the removal of strontium (Francis and Nancharaiah 2015) and uranium (Xu et al. 2018). Huang et al. (2017) also revealed successful U(VI) bio-precipitation employing *Shewanella putrefaciens*. The induction of biomineralization takes place by the binding of radionuclides to biofilms generated by microorganisms. Keasling et al. (2000) revealed the deposition of polycrystalline $NaUO_2PO_4$ inside and in the vicinity of the outer membrane of *Citrobacter*. Both the adsorption of uranium onto lipopolysaccharides (LPS) and the acid phosphatase activity in the cell wall of *Citrobacter* sp. played a crucial role in influencing the biomineralization process (Prakash et al. 2013). Another study by Zhong et al. (2021) revealed the U(VI) immobilization capacity of *Bacillus* spp. in the uranium-contaminated site in Southeast China. Such bacteria can survive in high U(VI) concentrations and effectively remove it from water by forming a stable uranium-phosphate compound. The removal mechanisms include enzymatic bio-precipitation, cell surface binding, and the use of extracellular polymeric substances suggesting their potential for uranium bioremediation. The bio-precipitation of majority of metal ions is facilitated by the presence of organic metabolites such as oxalic acid, citric acid, hydroxyl-acetic acid, and similar compounds (Koul and Adlakha 2021). It is crucial to understand that this process does not alter the oxidation states of the precipitated metals or radionuclides unlike other processes like biotransformation (Ruiz-Fresneda et al. 2023).

Biotransformation or Bioreduction

Enzymatic conversion of soluble radionuclides into insoluble forms by certain bacteria effectively immobilizes such radionuclides and minimizes their exposure to living organisms by restricting their mobility (Tan et al. 2024). Microbes play a key role in reducing water-soluble, mobile, biologically available, and oxidized radionuclides like hexavalent uranium, U(VI), and hexavalent technetium, Tc (VII) through redox reactions (Francis and Nancharaiah 2015). The microbes utilize such radionuclides as terminal electron acceptors by making them insoluble and less biologically active. The bioreduction or biotransformation process of radionuclides might be done by direct and indirect pathways as shown in Fig. 9.5. Numerous bacterial genera including dissimilatory ferric iron-reducing bacteria such as *Shewanella* and *Geobacter* spp., sulfate-reducing bacteria like *Desulfovibrio* and *Desulfosporosinus* spp., and extremophiles such as the thermophilic *Thermoanaerobacter* and the haloalkaliphilic *Halomonas* have been shown to exhibit bioreduction activities (Tan et al. 2024). In the case of direct method, anaerobic bacteria can enzymatically reduce the higher oxidation state of radionuclides to lower oxidation state (Cheema 2023). For instance, bacterial isolates such as *Desulfovibrio* and *Shewanella* can act as electron acceptors and directly use the electrons provided by some electron donors like acetic acid, lactic acid, ethanol, and hydrogen to reduce U(VI) into U(IV) and Tc (VII) into Tc(IV) (Ding et al. 2019). The indirect reduction of radionuclides is carried out by some sulfate-reducing bacteria to form less toxic and less soluble substrate by decreasing their bioavailability. For example, sulfate-reducing bacteria such as *Geobacter sulfurreducens*, *Anaeromyxobacter dehalogenans,* and *Shewanella putrefaciens* are capable of indirect bioreduction of radionuclides (Ding et al. 2019). Similarly, Renshaw et al. (2005) affirmed the process of hexavalent uranium reduction employing *G. sulfurreducens*. Lovley et al. (1991) were first to show that *Shewanella oneidensis* and *Geobacter metallireducens* might effectively harness energy for anaerobic growth through the reduction of U(VI). Additional potential bacteria like *Desulfovibrio vulgaris*, *Clostridium* sp., and *D. desulfuricans* can reduce uranium to form less bioactive state (Roh et al. 2015). Similarly, Pu (IV) biotransformation mechanism of *Pseudomonas fluorescens* in the environment having abundant citric acid was elucidated by Francis et al. (2006). Again, the bioreduction capacity of *D. desulfuricans* to reduce Np(V) has been reported by Jabbar and Wallner (2015). As documented by Francis (1998), U(VI) was reduced to its trivalent form, U(III), by some sulfate-reducing bacteria as well as *Micrococcus alactolyticus* having cell-free abstract. Prakash et al. (2013) revealed the significant involvement of cytochrome C_3 in the reduction of U(VI) by *D. desulfuricans*. Likewise, the involvement of the homologous cytochrome (PpcA) in the reduction of U(VI) was also verified in *G. sulfurreducens* (Ding et al. 2019). Francis (2007) also recorded the anaerobic reduction-induced transformation of tetravalent plutonium, Pu(IV) into its trivalent state, Pu(III), facilitated by *Serratia marcescens, Clostridium* sp., and *Bacillus* sp. As identified by Chabala and Chirwa (2010), four bacterial strains such as *Klebsiella* sp., *Pseudomonas* sp., *Pantoea* sp., and *Enterobacter* sp., are capable of reducing U(VI) by individual application and in bacterial consortia. They achieved a complete reduction of U(VI) at concentrations up to 200 mg/L and demonstrated an 88% reduction at 400 mg/L concentration. The radioactive elements are immobilized, precipitated, and insoluble in their reduced state. In contaminated waste, bioreduction provides a strong assurance for controlling the soluble content and motility of target radionuclides (Patel et al. 2022).

Biosorption

Biosorption involves the formation of complexes between positively charged radionuclides and negatively charged biomass whether in a living or nonliving state (Prakash et al. 2013). The biosorption process entails capturing and immobilizing a specific radioactive nuclide exhibiting species specificity. The specificity could be determined by environmental parameters, including pH, temperature, and aeration. Additionally, it relies upon the characteristic features of microbes including the cell growth pattern, the production of metabolites and extracellular polymers, and the composition on the cell surface, in addition to the kind of radionuclide whether positively or negatively charged (Ding et al. 2019). Biosorption occurs through direct interaction, involving the binding of radionuclide as cations to functional groups present in bacterial cell wall as anions that are depicted in Fig. 9.5. Alternatively, it can come of indirectly along interactions with extracellular polymeric substances (EPS), S-layer, or capsule structures (Shukla et al. 2017). Radionuclides such as Pu, Np, U, and Th have the ability to attach to surface of bacterial cell through the assistance of ligands including hydroxyl, amine, phosphate, sulfhydryl, and carboxyl groups (Dummi Mahadevan and Zhao 2017). Biosorption typically serves as the initial stage for subsequent microbe interactions, including bioaccumulation, biomineralization, and bioreduction. Previous reports indicated that both Gram-positive and Gram-negative bacteria can effectively adsorb radionuclides from the contaminated water. Gram-negative bacteria that were found to have radionuclide biosorption activity included *Pseudomonas*, *Stenotrophomonas*, *Acidovorax*, and *Shewanella*, whereas *Streptomyces*, *Kocuria*, *Bacillus*, *Microbacterium*, and *Cellulosimicrobium* were included under Gram-positive bacteria showing radionuclide biosorption activity (Tan et al. 2024). Additionally certain bacterial strains such as *Firmicutes* and *Citrobacter freundii* possess notable biosorption capabilities against the radionuclides (Riaz et al. 2024). The cell wall of *Citrobacter freundii* possesses functional groups like carboxyl compounds that emerge as one of the active sites for the microbial absorption, exerting a substantial influence on the biosorption process (Cheema 2023). Moreover, a *Pseudomonas* strain is capable of engaging in biosorption of thorium and uranium ions by employing sequestration of intracellular parts as demonstrated by Natarajan et al. (2020). The sorption of multiple metal ions has been facilitated by utilizing extracellular polymeric substances (EPS) derived from *Rhodococcus rhodochrous* and *Rhodococcus opacus* as biosorbents (Shukla et al. 2017). The findings of Icgen and Yilmaz (2018) unveiled the elimination of Sr^{2+} from water solutions could be efficiently achieved through the utilization of *Micrococcus luteus* isolate. This particular isolate demonstrated complete sorption of Sr^{2+} (774.5 μg ml^{-1}) within approximately 30 h. Bacterial species such as *B. cereus* and *Bacillus polymyxa* also exhibited significant biosorption ability against the toxic radionuclides (Long et al. 2017; Shevchuk and Klimenko 2009). Similarly, the biosorption of U(VI) by *Bacillus amyloliquefaciens* (Liu et al. 2019) and *P. fluorescens* (Zheng et al. 2022) have also been reported in previous studies. It was also disclosed that dead cell biomass exhibits enhanced sorption capability due to the effects of various functional groups instead of living cells in entire process (Xie et al. 2008). However, a significant drawback of this process is the swift saturation of radioactive nuclides and competitive displacement, resulting from other cations competing for site of attachment on the cell surface of bacteria (Jabbar and Wallner 2015). Therefore, this technology is usually employed for the elimination of radionuclides at minor concentrations (Shukla et al. 2017). However, the efficacy of biosorption for radionuclide remediation could enhance through sophisticated approaches, including the application of recombinant DNA technology and the stimulation of microbial growth in the contaminated soil (Prakash et al. 2013).

Bio-Solubilization

Bio-solubilization, also known as biomining or bioleaching, involves a metabolic process where radionuclides are extracted from solid matrices (Qiu et al. 2019). The process encompasses leaching conducted by either autotrophic or heterotrophic microorganisms, the generation of complex agents or metabolites and various chemical transformations (Gavrilescu 2022). Autotrophic leaching involves the aerobic oxidation of mineral ores facilitated by autotrophic bacteria, predominantly thiobacilli like *Thiobacillus ferrooxidans* (Frutos 1998). This method is widely employed on an industrial scale to extract metals and uranium from ores, in addition to recovering metals from waste materials. Heterotrophic bacteria illustrate critical involvement in heterotrophic leaching, a process where the solubilization of radionuclides occurs through the generation of protons, organic acids, and chelating agents (Ye et al. 2024). Consequently, the development of complexes could lead to enhanced solubility of radionuclides as it involves the liberation of the radionuclide followed by its conversion into hydroxides, oxides, or other compounds with reduced solubility (Williamson et al. 2024). Physical factors such as the pH in the bacterial environment, moisture levels, nuclide oxidation state, and inorganic substrate content can significantly influence the bio-solubilization process (Kaksonen et al. 2017). Typically, such varieties of bacteria exhibit mesophilic and acidophilic characteristics (Srichandan et al. 2019). Francis and Nancharaiah (2015) revealed that the existence of citric acid as vital metabolite of bacteria results in an extended enhancement of nuclide solubility. Bacteria such as *Leptospirillum ferrooxidans, Acidithiobacillus thiooxidans, Sulfolobus, Acidithiobacillus ferrooxidans,* and *Sulfolobus* (Shukla et al. 2017) have been enumerated for their ability to enhance metal solubilization. Additionally, studies have indicated that *Sulfolobus, Acetobacter* sp., and *A. ferrooxidans* are some bacteria having the ability to facilitate the solubilization of metals (Reitz et al. 2015; Qu et al. 2019; Mao et al. 2015). Study by Kim et al. (2021) investigated the feasibility of bioleaching using *A. thiooxidans* to remediate soil contaminated with Co (cobalt) and Cs (cesium) near the nuclear power plant. The bacteria exhibited tolerance to high levels of Co, Cs, and γ-irradiation, enabling efficient removal up to 95% of both radionuclides from the contaminated soil. Moreover, the study by Abdulla et al. (2018) revealed the bioleaching of uranium from Egyptian ores using native actinomycetes. The results showed that selected isolates, particularly *Streptomyces bacillaris*, effectively solubilized uranium from sandstone with bioleaching demonstrating potential as an eco-friendly and cost-effective alternative to conventional chemical methods for uranium extraction. Table 9.1 provides a comprehensive summary of various methods employed for the remediation of known radioactive elements.

Table 9.1 Bacterial remediation mechanisms of different radionuclides

Radioactive elements and their isotopes	Half-life (years)	Mechanisms	Bacteria involved	References
Uranium (U) 238^{U}, 235^{U}	4.41×10^{9} years	Bioaccumulation	*Pseudomonas* sp.	Dummi Mahadevan and Zhao (2017)
			Bacillus cereus	Ozdemir et al. (2022)
			Streptomyces sporoverrucosus	Li et al. (2016)
		Biomineralization	*Kocuria* sp.	Wang et al. (2019)
			Shewanella putrefaciens	Huang et al. (2017)
			Citrobacter sp.	Ding et al. (2019)
		Bioreduction	*Micrococcus alactolyticus*	Francis (1998)
			Geobacter sulfurreducens	Renshaw et al. (2005)
			Desulfovibrio	Ding et al. (2019)
		Biosorption	*Bacillus amyloliquefaciens*	Liu et al. (2019)
			Pseudomonas fluorescens	Zheng et al. (2022)
Neptunium (Np) 237^{Np}	2.14×10^{6} years	Bioreduction	*Desulfovibrio desulfuricans and Shewanella putrefaciens*	Jabbar and Wallner (2015)
			S. Oneidensis and *Geobacter metallireducens*	Neu et al. 2005
		Biosorption	*Acetobacterium* sp. *and Halomonas* sp.	Francis (2007)
Thorium (Th) 232Th	14.1×10^{9} years	Bioaccumulation	*Bacillus cereus*	Ozdemir et al. (2022)
Plutonium (Pu) 239^{Pu}	24,100 years	Biotransformation	*Pseudomonas fluorescens*	Francis et al. (2006)
			Bacillus sp., *Serratia marcescens,* and *clostridium*	Francis (2007)
		Bioreduction	*Shewanella putrefaciens* and *Geobacter metallireducens*	Natarajan et al. (2020)

(continued)

Table 9.1 (continued)

Radioactive elements and their isotopes	Half-life (years)	Mechanisms	Bacteria involved	References
Cesium (Cs) 137^{Cs}	30 years	Bioaccumulation	*Bacillus pumilus* and *B. Licheniformis*	Abdel-Razek et al. (2015)
			Pseudomonas aeruginosa	Kang et al. (2017)
		Biosorption	*Synechococcus* sp.	Yu et al. (2020)
Strontium (Sr) 90^{Sr}	29 years	Biosorption	*Micrococcus luteus*	Icgen and Yilmaz (2018)
			Pseudomonas sp.	Natarajan et al. (2020)
		Biomineralization	*Serratia* sp.	Gangappa et al. (2016)

9.3.2.4 Limitation of Biological Approaches

Bacterial-based remediation of radioactive waste has been a topic of research for many years. While it has shown promise, this method faces constraints due to the complexity involved in utilizing microorganisms for the treatment of radioactive liquid waste. Moreover, its efficacy is significantly influenced by various environmental elements like biological, non-biological, and climatic factors. Some of the abiotic factors that affect the bioremediation include physicochemical factors like ionic strength, absence or presence of electron acceptor or donor, solubility, and pH of the radwaste (Vandana et al. 2021; Anand et al. 2023). For instance, the ionic strength of the ligands is dependent on the pH value of the radwaste that can modify its frequency of absorption. Besides, solubility also increases with decrease in pH. Moreover, it can alter the cell surface charge of the bacteria by modifying its isoelectric point. Besides, the efficiency of the bioremediation is also dependent on the type of nuclide, treatment time, temperature, nutrient availability, and initial concentration (Ma et al. 2023; Patel et al. 2022). Furthermore, the conversion rates of radioactive waste tend to be slow and bioremediation can take a long time to complete. While bacteria can sequester radioactive elements through biosorption or bioleaching and bio-precipitation processes or through indirect and direct redox transformation, these processes may not be effective enough to remove all the radioactive waste. For instance, in the case of biosorption technique, the cell surface of the microbes can become saturated and prevent further bindings of the cations (Vandana et al. 2021).

9.4 Future Directions

Nuclear energy is a vital global resource, but its waste management poses challenges. Nuclear energy remains a critical global resource, but its sustainable future hinges on overcoming the challenges of radioactive waste management. Conventional methods, which rely on costly chemical treatments to reduce waste volume, often necessitate long-term storage or environmentally risky practices like deep-sea burial. To address these limitations, innovative biological solutions are emerging as promising alternatives. Microbial remediation offers transformative potential by leveraging the natural ability of microorganisms to regulate the solubility, mobility, and activity of radionuclides. For instance, recent studies have demonstrated the efficacy of *Deinococcus radiodurans*—a radiation-resistant bacterium—in sequestering uranium (U) and technetium (Tc) through enzymatic reduction, even in high-radiation environments (Sarkar et al., 2024). Similarly, engineered strains of *Shewanella oneidensis* have shown enhanced bioreduction capabilities, converting soluble U(VI) to insoluble U(IV) nanoparticles under anaerobic conditions, effectively immobilizing the contaminant (Sheng et al. 2011). Besides, recent advances in synthetic biology including the use of CRISPR-Cas9 for targeted genome editing have further enhanced the ability of such bacteria to utilize and transform the radioactive waste (Sahoo et al. 2022). The exploration of indigenous microbial communities with inherent capabilities to transform radioactive waste into less toxic forms also provides valuable insights into natural attenuation mechanisms (Shukla et al. 2019).

9.5 Conclusion

The transition from conventional radioactive waste management to biologically driven strategies represents a paradigm shift toward sustainability and efficiency. While traditional methods remain entrenched in cost and ecological trade-offs, bacterial remediation—enhanced by genetic engineering and interdisciplinary innovation—provides a viable pathway to mitigate the environmental and health risks posed by radwaste. By harnessing the metabolic versatility of microbes, particularly recombinant strains tailored for extreme conditions, the scientific community can achieve efficient detoxification and immobilization of hazardous substances. Moreover, the potential repurposing of remediated radionuclides for commercial use underscores the economic and environmental value of these approaches. Success, however, demands sustained collaboration across scientific disciplines, robust regulatory frameworks, and public engagement to balance nuclear energy's benefits with its responsibilities. Ultimately, the integration of microbial solutions into waste management practices promises not only to safeguard ecosystems but also to redefine humanity's relationship with nuclear technology in the Anthropocene era.

References

Abdel-Razek AS, Shaaban MT, Mahmoud SA, Kandeel EM (2015) Bioaccumulation of $_{137}$Cs by immobilized bacterial species isolated from radioactive wastewater. Appl Environ Microbiol 3:112–118

Abdulla HM, Taher HS, Ibrahim HA (2018) Bioleaching of uranium from Egyptian rocks using native actinomycete strains. Geomicrobiol J 35(2):91–99

Adeola AO, Iwuozor KO, Akpomie KG, Adegoke KA, Oyedotun KO, Ighalo JO et al (2023) Advances in the management of radioactive wastes and radionuclide contamination in environmental compartments: a review. Environ Geochem Health 45(6):2663–2689

Adloff JP (2018) Fundamentals of radiochemistry. CRC Press

Ali S, Baloch SB, Bernas J, Konvalina P, Onyebuchi EF, Naveed M et al (2023) Phytotoxicity of radionuclides: A review of sources, impacts and remediation strategies. Environ Res 117479

Al-Rubaii TAJY (2018) Calculation of the stopping power and range for alpha particles in some materials and tissues. Doctoral dissertation, University of Baghdad

Anand R, Mohan L, Bharadvaja N (2023) Phyto-& microbial-remediation of radioactive waste. In: Microbial technologies in industrial wastewater treatment. Springer, Singapore, pp 215–241

Aziz A, Yasmeen T, Tariq M, Arif MS, Shahzad SM, Riaz M, Ali S, Rizwan M (2021) Genomics in understanding bioremediation of inorganic pollutants. In: Handbook of bioremediation. Academic Press, pp 397–410

Bagher AM, Vahid M, Mohsen M (2014) Introduction to radioactive materials. Int J Renew Sustain Energy 3(3):59–67

Bakshe P, Jugade R (2023) Phytostabilization and rhizofiltration of toxic heavy metals by heavy metal accumulator plants for sustainable management of contaminated industrial sites: A comprehensive review. J Hazardous Mater Adv 10:100293

Belpomme D, Hardell L, Belyaev I, Burgio E, Carpenter DO (2018) Thermal and non-thermal health effects of low intensity non-ionizing radiation: an international perspective. Environ Pollut 242:643–658

Bhatt P, Bhatt K, Sharma A, Zhang W, Mishra S, Chen S (2021) Biotechnological basis of microbial consortia for the removal of pesticides from the environment. Crit Rev Biotechnol 41(3):317–338

Blanco-Novoa O, Fernández-Caramés TM, Fraga-Lamas P, Castedo L (2018) A cost-effective IoT system for monitoring indoor radon gas concentration. Sensors 18(7):2198

Bobrakov AN, Kudrinskii AA, Kulygin VM, Pereslavtsev AV, Polkanov MA, Shiryaevskii VL, Artemov AV (2014) Russian experience in solid radioactive waste processing: achievements and prospects. Russ J Gen Chem 84:1041–1049

Boonyapookana B, Parkpian P, Techapinyawat S, DeLaune RD, Jugsujinda A (2005) Phytoaccumulation of lead by sunflower (*Helianthus annuus*), tobacco (*Nicotiana tabacum*), and vetiver (*Vetiveria zizanioides*). J Environ Sci Health 40(1):117–137

Burgio E, Piscitelli P, Migliore L (2018) Ionizing radiation and human health: reviewing models of exposure and mechanisms of cellular damage. An epigenetic perspective. Int J Environ Res Public Health 15(9):1971

Busby C (2013) Aspects of DNA damage from internal radionuclides. In: New research directions in DNA repair. IntechOpen

Campbell PG, Hodson PV, Welbourn PM, Wright DA (2022) Contaminant bioaccumulation: kinetics and modelling uptake mechanisms. Ecotoxicology 61

Castillo H, Schoderbek D, Dulal S, Escobar G, Wood J, Nelson R, Smith G (2015) Stress induction in the bacteria *Shewanella oneidensis* and *Deinococcus radiodurans* in response to below-background ionizing radiation. Int J Radiat Biol 91(9):749–756

Cerne M, Smodis B, Strok M (2011) Uptake of radionuclides by a common reed (Phragmites australis (Cav.) Trin. ex Steud.) grown in the vicinity of the former uranium mine at Zirovskivrh. Nucl Eng Des 241(4):1282–1286

Chabalala S, Chirwa EM (2010) Uranium (VI) reduction and removal by high performing purified anaerobic cultures from mine soil. Chemosphere 78(1):52–55

Cheema HA (2023) A review on bioremediation-an emerging technology for treatment of radionuclide waste. J Mod Agric Biotechnol 2(1):2

Clayton R, Kirk J, Banford A, Stamford L (2024) A review of radioactive waste processing and disposal from a life cycle environmental perspective. Clean Techn Environ Policy:1–18

Czekała W, Drozdowski J, Łabiak P (2023) Modern technologies for waste management: a review. Appl Sci 13(15):8847

Dangi AK, Sharma B, Hill RT, Shukla P (2019) Bioremediation through microbes: systems biology and metabolic engineering approach. Crit Rev Biotechnol 39(1):79–98

Darda SA, Gabbar HA, Damideh V, Aboughaly M, Hassen I (2021) A comprehensive review on radioactive waste cycle from generation to disposal. J Radioanal Nucl Chem 329(1):15–31

Ding C, Cheng W, Nie X (2019) Microorganisms and radionuclides. In: Interface science and technology, vol 29. Elsevier, pp 107–139

Dondelinger RM (2012) Radiation. Biomed Instrum Technol 46(3):219–223

Dummi Mahadevan G, Zhao F (2017) A concise review on microbial remediation cells (MRCs) in soil and groundwater radionuclides remediation. J Radioanal Nucl Chem 314:1477–1485

Eapen S, Singh S, Thorat V, Kaushik CP, Raj K, D'Souza SF (2006) Phytoremediation of radiostrontium (90Sr) and radiocesium (137Cs) using giant milky weed (Calotropis gigantea R. Br.) plants. Chemosphere 65(11):2071–2073

Eidemüller D, Eidemüller D (2021) Types of radioactive substances. Nuclear Power Explained:77–86

Elkhatib H, Abdelmaksoud A, Othman M, Abdelaziz M, Kotb H (2024) Modeling and assessment of radioactive iodine dispersion inside Egyptian radioisotope production facility. Appl Radiat Isot 203:111082

Environment Canada (1997) Environmental assessments of priority substances under the Canadian environmental protection act: guidance manual, version 1.0. Report EPS/2/CC/3E, vol 2. Chemical Evaluation Division, Commercial Chemicals Evaluation Branch, p 111

Fang HY (2002) Radioactive nuclear waste. Pract Period Hazard Toxic Radioact Waste Manag 6(2):102–111

Francis AJ (1998) Biotransformation of uranium and other actinides in radioactive wastes. J Alloys Compd 271:78–84

Francis AJ (2007) Microbial mobilization and immobilization of plutonium. J Alloys Compd 444:500–505

Francis AJ, Nancharaiah YV (2015) In situ and ex situ bioremediation of radionuclide-contaminated soils at nuclear and norm sites. In: Environmental remediation and restoration of contaminated nuclear and norm sites. Woodhead Publishing, pp 185–236

Francis AJ, Dodge CJ, Gillow JB (2006) Biotransformation of plutonium complexed with citric acid. Radiochim Acta 94(9–11):731–737

Frutos FG (1998) Bacterial leaching of minerals. In: Mineral processing and the environment. Springer Netherlands, Dordrecht, pp 43–72

Gangappa R, Yong P, Singh S, Mikheenko I, Murray AJ, Macaskie LE (2016) Hydroxyapatite biosynthesis by a *Serratia sp.* and application of nanoscale bio-HA in the recovery of strontium and europium. Geomicrobiol J 33(3–4):267–273

Gavrilescu M (2022) Microbial recovery of critical metals from secondary sources. Bioresour Technol 344:126208

Geras' kin S, Oudalova A, Dikareva N, Spiridonov S, Hinton T, Chernonog E, Garnier-Laplace J (2011) Effects of radioactive contamination on Scots pines in the remote period after the Chernobyl accident. Ecotoxicology 20:1195–1208

Gonçalves CR, da Silva Delabona P (2022) Strategies for bioremediation of pesticides: challenges and perspectives of the Brazilian scenario for global application—a review. Environ Adv 8:100220

Goncharuk VV (2014) Drinking Water: Physics, Chemistry and Biology. Springer Cham, pp 426

Hamblin JD (2008) Gods and devils in the details: marine pollution, radioactive waste, and an environmental regime circa 1972. Dipl Hist 32(4):539–560

Hazen TC, Tabak HH (2005) Developments in bioremediation of soils and sediments polluted with metals and radionuclides: 2. Field research on bioremediation of metals and radionuclides. Rev Environ Sci Biotechnol 4:157–183

Hosan MI (2017) Radioactive waste classification, management and environment. Eng Int 5(2):63–72

Hu N, Ding D, Li G (2014) Natural plant selection for radioactive waste remediation. In: Radionuclide Contamination and Remediation Through Plants (Eds. Gupta DK, Walther C). Springer Cham, pp 33–53

Huang W, Nie X, Dong F, Ding C, Huang R, Qin Y et al (2017) Kinetics and pH-dependent uranium bioprecipitation by *Shewanella putrefaciens* under aerobic conditions. J Radioanal Nucl Chem 312:531–541

IAEA (2006) Fundamental safety principles safety fundamentals, jointly sponsored by Euratom, FAO, IAEA, ILO, OECD/NEA, PAHO, WHO; IAEA safety standards no. SF-1. International Atomic Energy Agency, Vienna

Icgen B, Yilmaz F (2018) Biosorption of strontium from aqueous solutions by Micrococcus luteus Sr02. Geomicrobiol J 35(4):284–293

ICRP (2003) A framework for assessing the impact of ionising radiation on nonhuman species. In: ICRP publication 91, annals of the ICRP 33(3). Pergamon Press, Oxford, p 201

ICRP (2008) The concept and use of reference animals and plants for the purposes of environmental protection. In: Valentin J (ed) International Commission on Radiological Protection Publication 108, Annals of the ICRP 38(4e6). Elsevier, Oxford, p 76

Jabbar T, Wallner G (2015) Biotransformation of radionuclides: trends and challenges. Radionuclides in the Environment: Influence of chemical speciation and plant uptake on radionuclide migration. Radionuclides Environ:169–184

Jiang L, Liu X, Yin H, Liang Y, Liu H, Miao B et al (2020) The utilization of biomineralization technique based on microbial induced phosphate precipitation in remediation of potentially toxic ions contaminated soil: a mini review. Ecotoxicol Environ Saf 191:110009

Joshi S, Mohapatra B, Mishra JPN (2018) Microbial soil enzymes: implications in the maintenance of rhizosphere ecosystem and soil health. In: Advances in soil microbiology: recent trends and future prospects: Volume 1. Soil-Microbe Interaction, pp 179–192

Kaksonen AH, Boxall NJ, Usher KM, Ucar D, Sahinkaya E (2017) Biosolubilisation of metals and metalloids. Sustainable Heavy Metal Remediation: Volume 1: Principles and Processes, pp 233–283

Kang SM, Jang SC, Heo NS, Oh SY, Cho HJ, Rethinasabapathy M et al (2017) Cesium-induced inhibition of bacterial growth of *Pseudomonas aeruginosa* PAO1 and their possible potential applications for bioremediation of wastewater. J Hazard Mater 338:323–333

Karmaker N, Maraz KM, Islam F, Haque MM, Razzak M, Mollah MZI et al (2021) Fundamental characteristics and application of radiation. GSC Adv Res Rev 7(1):064–072

Keasling JD, Van Dien SJ, Trelstad P, Renninger N, McMahon K (2000) Application of polyphosphate metabolism to environmental and biotechnological problems. Biochemistry 65(3):324–331

Khelurkar N, Shah S, Jeswani H (2015) A review of radioactive waste management. In: In 2015 International Conference on Technologies for Sustainable Development (ICTSD). IEEE, pp 1–6

Kim YG, Jeong D, Um W, Kim KW, Ko MS (2021) The bioleaching assessment for nuclear power plant-soil contaminated with Co and Cs using A. Thiooxidans sp. J Environ Chem Eng 9(1):104791

Koul B, Adlakha K (2021) Bioremediation of radionuclides by plant-microbe system: current progress and challenges. Microbe Mediated Remediation of Environmental Contaminants, pp 303–320

Kumar N, Jeena N, Gangola S, Singh H (2019) Phytoremediation facilitating enzymes: an enzymatic approach for enhancing remediation process. In: Smart bioremediation technologies. Academic Press, pp 289–306

Latif A, Abbas A, Iqbal J, Azeem M, Asghar W, Ullah R et al (2023) Remediation of environmental contaminants through phytotechnology. Water Air Soil Pollut 234(3):139

Leeman-Neill RJ, Brenner AV, Little MP, Bogdanova TI, Hatch M, Zurnadzy LY et al (2013) RET/PTC and PAX8/PPARγ chromosomal rearrangements in post-Chernobyl thyroid cancer and their association with iodine-131 radiation dose and other characteristics. Cancer 119(10):1792–1799

Li X, Ding C, Liao J, Du L, Sun Q, Yang J et al (2016) Bioaccumulation characterization of uranium by a novel *Streptomyces sporoverrucosus* dwc-3. J Environ Sci 41:162–171

Liu L, Liu J, Liu X, Dai C, Zhang Z, Song W, Chu Y (2019) Kinetic and equilibrium of U (VI) biosorption onto the resistant bacterium *Bacillus amyloliquefaciens*. J Environ Radioact 203:117–124

Lloyd JR (2003) Microbial reduction of metals and radionuclides. FEMS Microbiol Rev 27(2–3):411–425

Long J, Li H, Jiang D, Luo D, Chen Y, Xia J, Chen D (2017) Biosorption of strontium (II) from aqueous solutions by *Bacillus cereus* isolated from strontium hyperaccumulator *Andropogon gayanus*. Process Saf Environ Prot 111:23–30

Lovley DR, Phillips EJ, Gorby YA, Landa ER (1991) Microbial reduction of uranium. Nature 350(6317):413–416

Ma H, Shen M, Tong Y, Wang X (2023) Radioactive wastewater treatment technologies: A review. Molecules 28(4):1935

Mao XL, Ding ZX, Yuan SB (2015) Dissolution of 238U from low-level contaminated soil by *Acidithiobacillus ferrooxidans*. Appl Mech Mater 700:225–229

Mayneord WV (2011) Radiation and health. Routledge, pp 141

McBride C (2022) The investigation of the chemical degradation of the acetate ion under operating conditions relevant to pressurised water reactors. Doctoral dissertation, University of Manchester

McCarthy ME, Thomas B, DiNardi SR (2003) Ionizing radiation. In: Fairfax VA (ed) The occupational environment: its evaluation and control, 2nd edn. American Industrial Hygiene Association, pp 572–610

McGinnes DF (2007) Waste sources and classification. Radioactivity in the Environment 9:8–40

McNamara NP, Black HIJ, Beresford NA, Parekh NR (2003) Effects of acute gamma irradiation on chemical, physical and biological properties of soils. Appl Soil Ecol 24(2):117–132

Monteith J, Unsworth M (2013) Principles of environmental physics: plants, animals, and the atmosphere. Academic Press

Mosin O, Ignatov I (2015) Isotope purification of drinking water from heavy isotopes deuterium (2H), tritium (3H) and oxygen (18O). *Вода: гигиена и экология* 3(3–4):9–21

Nakamura A, Kikuchi M (2011) What we know, and what we have not yet learned: triple disasters and the Fukushima nuclear fiasco in Japan. Public Adm Rev 71(6):893–899

Natarajan V, Karunanidhi M, Raja B (2020) A critical review on radioactive waste management through biological techniques. Environ Sci Pollut Res 27:29812–29823

Neu MP, Icopini GA, Boukhalfa H (2005) Plutonium speciation affected by environmental bacteria. Radiochim Acta 93(11):705–714

Newsome L, Morris K, Lloyd JR (2014) The biogeochemistry and bioremediation of uranium and other priority radionuclides. Chem Geol 363:164–184

Ojovan MI (2011) Radioactive waste characterization and selection of processing technologies. In: Handbook of advanced radioactive waste conditioning technologies. Woodhead Publishing, pp 1–16

Ozdemir S, Acer O, Kılınç E (2022) Bioaccumulation, tolerance, and removal of U (VI) and Th (IV) by a novel thermophilic *Bacillus cereus* ST14 isolated from hot spring mud samples in Afyonkarahisar, Turkey. Biomass Conversion and Biorefinery 14:1–13

Paatero J, Saxén R, Buyukay M, Outola I (2010) Overview of strontium-89, 90 deposition measurements in Finland 1963–2005. J Environ Radioact 101(4):309–316

Panicker SG (2022) Microbial diversity at the polluted sites. In: Microbial diversity in hotspots. Academic Press, pp 135–154

Patel R, Mugunthan J, Singh P, Mukherjee S, Koka R (2022) Microbial bioremediation and biodegradation of radioactive waste contaminated sites. In: Microbes and microbial biotechnology for green remediation. Elsevier, pp 733–746

Pfützner M, Karny M, Grigorenko LV, Riisager K (2012) Radioactive decays at limits of nuclear stability. Rev Mod Phys 84(2):567

Phillips RW, Wiegel J, Berry CJ, Fliermans C, Peacock AD, White DC, Shimkets LJ (2002) *Kineococcus radiotolerans* sp. nov., a radiation-resistant, gram-positive bacterium. Int J Syst Evol Microbiol 52(3):933–938

Poškas P, Kilda R, Šimonis A, Jouhara H, Poškas R (2019) Disposal of very low-level radioactive waste: Lithuanian case on the approach and long-term safety aspects. Sci Total Environ 667:464–474

Prakash D, Gabani P, Chandel AK, Ronen Z, Singh OV (2013) Bioremediation: a genuine technology to remediate radionuclides from the environment. Microb Biotechnol 6(4):349–360

Qiu L, Feng J, Dai Y, Chang S (2019) Mechanisms of strontium's adsorption by *Saccharomyces cerevisiae*: contribution of surface and intracellular uptakes. Chemosphere 215:15–24

Qu Y, Li H, Wang X, Tian W, Shi B, Yao M, Zhang Y (2019) Bioleaching of major, rare earth, and radioactive elements from red mud by using indigenous chemoheterotrophic bacterium *Acetobacter* sp. Fortschr Mineral 9(2):67

Rajkhowa S, Sarma J, Das AR (2021) Radiological contaminants in water: pollution, health risk, and treatment. In: Contamination of water. Academic Press, pp 217–236

Ramaswami A, Carr P, Burkhardt M (2001) Plant-uptake of Uranium: hydroponic and soil system studies. Int J Phytoremediation 3(2):189–201

Rao TS, Panigrahi S, Velraj P (2022) Transport and disposal of radioactive wastes in nuclear industry. In: Microbial biodegradation and bioremediation. Elsevier, pp 419–440

Reed BC (2017) An examination of the potential fission-bomb weaponizability of nuclides other than 235U and 239Pu. Am J Phys 85(1):38–44

Reena R, Majhi MC, Arya AK, Kumar R, Kumar A (2012) BioRadBase: a database for bioremediation of radioactive waste. Afr J Biotechnol 11(35):8718–8721

Reitz T, Rossberg A, Barkleit A, Steudtner R, Selenska-Pobell S, Merroun ML (2015) Spectroscopic study on uranyl carboxylate complexes formed at the surface layer of *Sulfolobus acidocaldarius*. Dalton Trans 44(6):2684–2692

Renshaw JC, Butchins LJ, Livens FR, May I, Charnock JM, Lloyd JR (2005) Bioreduction of uranium: environmental implications of a pentavalent intermediate. Environ Sci Technol 39(15):5657–5660

Riaz S, Sohail M, Sahar R (2024) Microbial bioremediation of metals and radionuclides: approaches and advancements. Microbes Based Approaches for the Management of Hazardous Contaminants, pp 242–256

Roh C, Kang C, Lloyd JR (2015) Microbial bioremediation processes for radioactive waste. Korean J Chem Eng 32:1720–1726

Ruiz-Fresneda MA, Martinez-Moreno MF, Povedano-Priego C, Morales-Hidalgo M, Jroundi F, Merroun ML (2023) Impact of microbial processes on the safety of deep geological repositories for radioactive waste. Front Microbiol 14:1134078

Sahoo S, Routray SP, Lenka S, Bhuyan R, Mohanty JN (2022) CRISPR/Cas-mediated functional gene editing for improvement in bioremediation: an emerging strategy. In: Kumar V, Thakur IS (eds) Omics insights in environmental bioremediation. Springer, Singapore. https://doi.org/10.1007/978-981-19-4320-1_2

Saleh HM (2012) Water hyacinth for phytoremediation of radioactive waste simulate contaminated with cesium and cobalt radionuclides. Nucl Eng Des 242:425–432

Saling J (2018) Radioactive waste management. Routledge

Samarth R, Kumar M, Matsumoto Y, Manda K (2020) The effects of ionizing and non-ionizing radiation on health. Recent Trends Adv Environ Health 179

Sarkar S, Ghosh B, Ghosh S (2024) Microbial interventions in bioremediation of nuclear waste. In: Radioactive pollutant: sources, issues and remediation. Cham, Springer Nature Switzerland, pp 325–352

Sasikumar CS, Papinazath T (2003, December) Environmental management: bioremediation of polluted environment. In: Proceedings of the third international conference on environment and health. Department of Geography, University of Madras, Chennai and Faculty of Environmental Studies, York University, New York, pp 465–469

Semenishchev VS, Voronina AV (2020) Isotopes of strontium: properties and applications. Strontium Contamination in the Environment, pp 25–42

Shahbazi-Gahrouei D, Gholami M, Setayandeh S (2013) A review on natural background radiation. Adv Biomed Res 2:65

Sheng L, Szymanowski J, Fein JB (2011) The effects of uranium speciation on the rate of U (VI) reduction by Shewanella oneidensis MR-1. Geochim Cosmochim Acta 75(12):3558–3567

Shevchuk IA, Klimenko NI (2009) Biological features of sorption of U (VII) and strontium ions by *Bacillus polymyxa* IMV 8910 cells. J Water Chem Technol 31:324–328

Shiels JA (2019) Synthesis and characterisation of highly active nuclear waste simulants. Doctoral dissertation, University of Leeds

Shukla A, Parmar P, Saraf M (2017) Radiation, radionuclides and bacteria: an in-perspective review. J Environ Radioact 180:27–35

Shukla A, Parmar P, Saraf M et al (2019) Isolation and screening of bacteria from radionuclide containing soil for bioremediation of contaminated sites. Environ Sustain 2:255–264. https://doi.org/10.1007/s42398-019-00068-y

Shukla S, Vaishnavee VHS, Dedha A, Phutela S, Shukla SK (2024) Role of plants in remediation of radioactive pollutant from the environment. Radioactive Pollutant: Sources, Issues and Remediation, pp 81–112

Singh G, Bhadange S, Bhawna F, Shewale P, Dahiya R, Aggarwal A et al (2023) Phytoremediation of radioactive elements, possibilities and challenges: special focus on agricultural aspects. Int J Phytoremediation 25(1):1–8

Smičiklas I, Šljivić-Ivanović M (2016) Radioactive contamination of the soil: assessments of pollutants mobility with implication to remediation strategies. In: Soil contamination–current consequences and further solutions. InTech [Internet], Rijeka, pp 253–276

Spiers FW (2013) Radioisotopes in the human body: physical and biological aspects. Elsevier

Srichandan H, Mohapatra RK, Parhi PK, Mishra S (2019) Bioleaching: A bioremediation process to treat hazardous wastes. Soil Microenvironment for Bioremediation and Polymer Production, pp 115–129

Sundar H, Padmini S, Devi PB (2023) Bioremediation of nuclear waste effluent using different communities of microbes. Metagenomics to Bioremediation, pp 403–444

Tan JP, Clyde CW, Ng CC, Yeap SK, Yong CY (2024) Advancements in microbial-mediated radioactive waste bioremediation: A review. J Environ Radioact 280:107530

Thakare M, Sarma H, Datar S, Roy A, Pawar P, Gupta K et al (2021) Understanding the holistic approach to plant-microbe remediation technologies for removing heavy metals and radionuclides from soil. Curr Res Biotechnol 3:84–98

Thakur A, Kumar A (2024) Emerging paradigms into bioremediation approaches for nuclear contaminant removal: from challenge to solution. Chemosphere 352:141369

Thakur A, Prasad B, Kumar A, Sharma P, Singh V, Gupta S (2024) Biological techniques in the mitigation of radioactive pollutants. In: Radioactive pollutant: sources, issues and remediation. Cham, Springer, pp 175–193

Theodorakopoulos N, Février L, Barakat M, Ortet P, Christen R, Piette L et al (2017) Soil prokaryotic communities in Chernobyl waste disposal trench T22 are modulated by organic matter and radionuclide contamination. FEMS Microbiol Ecol 93(8)

Thomas RA, Macaskie LE (1996) Biodegradation of tributyl phosphate by naturally occurring microbial isolates and coupling to the removal of uranium from aqueous solution. Environmental Science & Technology 30(7):2371–2375

Tonelli FMP, Roy A, Murthy HA (eds) (2023) Green nanoremediation: sustainable management of environmental pollution. Springer Nature

Vandana UK, Gulzar ABM, Laskar IH, Meitei LR, Mazumder PB (2021) Role of microbes in bioremediation of radioactive waste. Microbial Rejuvenation of Polluted Environment 1:329–352

Verma R, Shrivastava A (2023) Radiation tolerant life forms and methods used to remediate radioactive wastes from soil. Nat Environ Pollut Technol 22(3):1363

Waghmare CM (2013) Radiation burn—from mechanism to management. Burns 39(2):212–219

Walker JS (2000) Permissible dose: a history of radiation protection in the twentieth century. Univ of California Press

Wang Y, Nie X, Cheng W, Dong F, Zhang Y, Ding C et al (2019) A synergistic biosorption and biomineralization strategy for *Kocuria* sp. to immobilizing U (VI) from aqueous solution. J Mol Liq 275:215–220

Williamson AJ, Binet M, Sergeant C (2024) Radionuclide biogeochemistry: from bioremediation toward the treatment of aqueous radioactive effluents. Crit Rev Biotechnol 44(4):698–716

Xie S, Yang J, Chen C, Zhang X, Wang Q, Zhang C (2008) Study on biosorption kinetics and thermodynamics of uranium by *Citrobacter freundii*. J Environ Radioact 99(1):126–133

Xu Q, Xi H, Zhang Y (2010) Marine pollution in water, sediment and biota. In: Impact, monitoring and management of environmental pollution. Nova Science Pub Inc, Hauppauge NY, p 152191

Xu R, Wu K, Han H, Ling Z, Chen Z, Liu P et al (2018) Co-expression of YieF and PhoN in *Deinococcus radiodurans* R1 improves uranium bioprecipitation by reducing chromium interference. Chemosphere 211:1156–1165

Ye J, Chen W, Zhu J, Ren Q, Peng K, Wu H et al (2024) Separate or simultaneous treatment of nitrate and uranium in groundwater via bioremediation: review and prospect. Chem Eng J 15:155326

Yin M, Zou K (2021) The implementation of the precautionary principle in nuclear safety regulation: challenges and prospects. Sustain For 13(24):14033

Yu R, Chai H, Yu Z, Wu X, Liu Y, Shen L et al (2020) Behavior and mechanism of cesium biosorption from aqueous solution by living *Synechococcus* PCC7002. Microorganisms 8(4):491

Zakariya NI, Kahn MTE (2014) Benefits and biological effects of ionizing radiation. Sch Acad J Biosci 2(9):583–591

Zhao C, Liu J, Li X, Li F, Tu H, Sun Q et al (2016) Biosorption and bioaccumulation behavior of uranium on Bacillus sp. dwc-2: investigation by box-Behenken design method. J Mol Liq 221:156–165

Zheng X, Hu P, Yao R, Cheng J, Chang Y, Mei H et al (2022) Biosorption behavior and biomineralization mechanism of low concentration uranium (VI) by *Pseudomonas fluorescens*. J Radioanal Nucl Chem 331(11):4675–4684

Zhong J, Hu X, Liu X, Cui X, Lv Y, Tang C et al (2021) Isolation and identification of uranium tolerant phosphate-solubilizing *Bacillus* spp. and their synergistic strategies to U (VI) immobilization. Front Microbiol 12:676391

Zinicovscaia I, Safonov A, Zelenina D, Ershova Y, Boldyrev K (2020) Evaluation of biosorption and bioaccumulation capacity of cyanobacteria *Arthrospira* (*spirulina*) *platensis* for radionuclides. Algal Res 51:102075

Zorpas AA (2020) Strategy development in the framework of waste management. Sci Total Environ 716:137088

Chapter 10
Food Waste Biotransformation: Advancing Enzyme Production Through Microbial Processes

Neela Satheesh, Messenbet G. Kassa, and Shiv Kumar Dwivedi

Abstract Increased industrialization and improper disposal of food waste have emerged as significant global issues. A substantial amount of food waste, including household kitchen waste, often goes untreated or ends up in landfills. Food waste generally contains a considerable fraction of organic materials such as proteins, fats, carbohydrates, and lipids. To promote sustainable waste management, practices like composting and recycling are key, focusing on reducing the environmental impact of food waste. For decades, fermentative microorganisms have been instrumental in enhancing the quality, texture, and value of food products. A promising yet underutilized approach involves leveraging these microorganisms to convert waste streams into valuable substances like enzymes—such as amylases, cellulases, proteases, and xylanases. However, due to the heterogeneous nature, high moisture content, and limited calorific value of food waste, specific strategies are necessary for its effective conversion into valuable enzymes. This book chapter delves into the utilization of various food wastes as substrates for producing industrially important enzymes using different microorganisms. It also discusses optimal fermentation conditions, substrate selection, and the influence of process parameters on enzymatic yields.

Keywords Circular economy · Food waste · Green catalysts · Microbial enzymes · Waste utilization

---

N. Satheesh (✉)
Department of Food Science and Technology, School of Health Sciences, Faculty of Science, Amity University Punjab, Mohali, Punjab, India

M. G. Kassa
College of Agriculture, Food, Climate Science, Injibara University, Injibara, Ethiopia

S. K. Dwivedi
Department of Environmental Science, School of Earth and Environmental Sciences, Babasaheb Bhimrao Ambedkar University, Lucknow, Uttar Pradesh, India

N. K. Arora et al. (eds.), *Environmental Sustainability and Biotechnology: Opportunities and Challenges*, Microorganisms for Sustainability 46,
https://doi.org/10.1007/978-981-96-9025-1_10

10.1 Introduction

Every stage of the food system, from the production of raw materials, selection, processing, transportation, storage to disposal, plays a role in the creation of waste (Sawaya 2017). The United Nations Food and Agriculture Organization (FAO) estimated roughly 1.3 billion tons of food losses and food waste (FW) annually.[1] Category-wise FW can be segregated into dairy products, fresh fruits and vegetables, meat, and baked goods. In developed countries, major FW generated per capita mainly consists of household and food sector (Fig. 10.1). The per capita FW by consumers in Europe and North America is between 95 and 115 kilograms a year.[2]

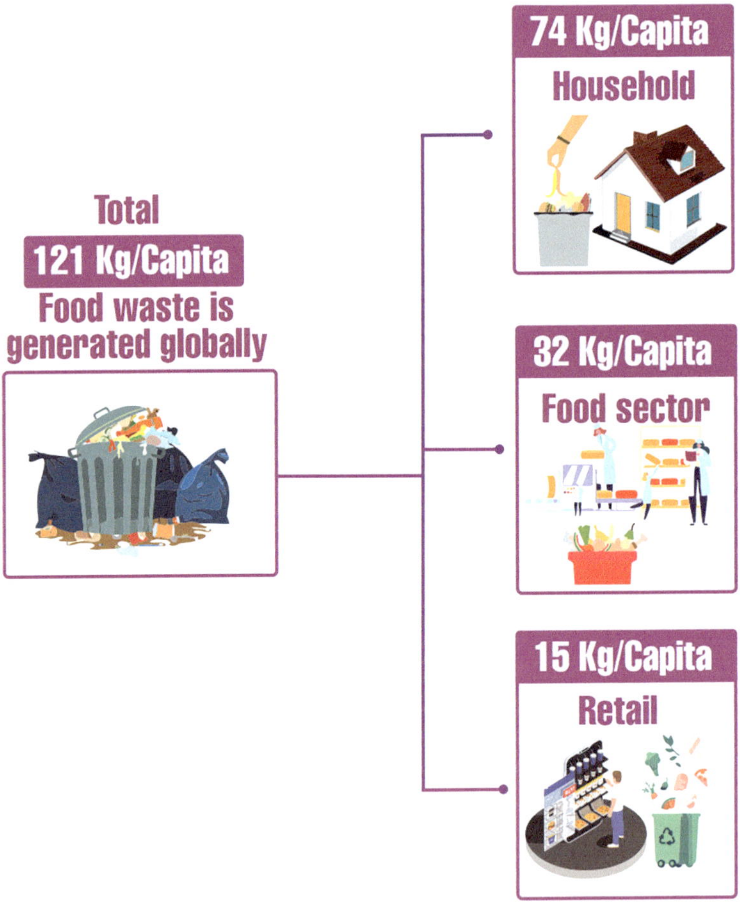

Fig. 10.1 Category-wise distribution of food waste

[1] https://www.fao.org/platform-food-loss-waste/flw-data/en/

[2] https://tabledebates.org/research-library/global-food-losses-and-food-waste-extent-causes-and-prevention

Similarly, for sub-Saharan Africa and South/Southeast Asia, the figure is between 6 and 11 kilograms a year.[3] In less developed countries, significant food losses are reported primarily during postharvest and processing stages. Factors such as rapid population growth, lifestyle changes, and economic surges, particularly in Asian countries, are driving substantial increases in FW (Sharma et al. 2024). It is projected that by 2025, urban areas in developing nations will generate approximately 416 million tons of FW annually (Thi et al. 2015).

FW generated in restaurants, cafeterias, household and commercial kitchens, and food processing plants predominantly consists of organic material (Fig. 10.1). It also constitutes a significant portion of global municipal solid waste. FW includes proteins, lipids, carbohydrates (such as cellulose, hemicelluloses, and starch), lignin, and organic acids. However, its composition can vary widely based on several factors (Lebersorger and Schneider 2011). Compared to vegetable waste, fruit waste typically contains higher quantities of total solids, volatile solids, and a greater carbon to nitrogen ratio (Esparza et al. 2020). The total solids content can be influenced by the type of waste and the moisture content of the source materials (Vera Zambrano et al. 2019). Similarly, the source of waste also impacts levels of volatile solids, carbon, and nitrogen; for instance, fruit waste has significantly higher levels of these components than vegetable waste (Vera Zambrano et al. 2019).

For decades, FW has been managed through various processes, including anaerobic digestion, incineration, composting, landfilling, and dumping. It has also been utilized as livestock feed and for energy generation. Although FW can be converted into valuable products, offering monetary benefits, its sustainable management remains a serious issue (Ghosh et al. 2016). In many Asian countries, a common disposal method involves dumping FW in landfills or dumpsites with other domestic waste. This practice generates odors and noxious gases and can contaminate soil and water resources. Furthermore, since FW often has a moisture content exceeding 80%, incineration is an energy-intensive and environmentally inefficient operation that causes air pollution (Peer et al. 1993).

Composting is an effective method to manage FW, converting it into soil conditioner and fertilizer. This approach is cheaper and more environmentally friendly compared to other treatment methods. However, challenges such as a limited market, the presence of pathogenic microbes, and the potential for substantial leachate discharge from compost remain concerns (O'Connor et al. 2021; Palaniveloo et al. 2020). While FW can be used as animal feed, its high moisture content and variable composition can lead to microbial spoilage. Dehydrating FW before feeding it to animals addresses this issue, but the drying process requires energy proportional to the FW's moisture content and can emit greenhouse gases (Georganas et al. 2020; Salemdeeb et al. 2017). FW is also promising as a substrate in biorefineries for producing chemicals (Dugmore et al. 2018) and value-added products (Ong et al. 2018). The concept of an FW biorefinery has recently gained traction, with ongoing research into FW valorization methods (Esteban and Ladero 2018).

[3] https://africacheck.org/fact-checks/reports/do-americans-and-europeans-waste-15-times-more-food-people-africa

Due to its chemical heterogeneity, alternative FW treatment approaches focus on producing high-value products such as organic acids, biofuels, biodiesel, and enzymes. Microbial enzymes play a crucial role in many industrial operations, including food processing, animal feed production, biofuels, paper and pulp manufacturing, pharmaceuticals, textiles, polymer synthesis, and detergents (Adrio and Demain 2014). Commercially valuable enzymes like cellulases, xylanases, amylases, lipases, proteases, and pectinases are extensively used in various industrial processes, especially in the food industry (Ramesh et al. 2020). FW is an excellent source of proteins, lipids, carbohydrates, organic acids, and essential minerals. It can be used to add value and provide an economical substrate for microbial fermentation (Sabater et al. 2020).

Microbes exhibit the remarkable capability to rapidly utilize substantial quantities of organic waste to produce enzymes through fermentation (Table 10.1). Due to their amenability to genetic engineering and the diverse array of biochemical processes they possess, microbes-based fermentation of organic waste has emerged as a sustainable means for commercial production of enzymes (Rao et al. 1998; Vachher et al. 2021). The contemporary demand for sustainable practices underscores the need for novel approaches in both pretreatment and bioconversion methods to convert FW into fermentable substrates. This shift from conventional methodologies to a sustainable paradigm is essential for generating value-added products, notably enzymes, from FWs. In the following sections, the utilization of different types of FW including cereals, fruits, oil cakes, dairy, and meat for production of valuable enzymes is critically discussed.

10.2 Enzymes Produced Form the Cereals Waste

A significant portion of both human and animal diets consists of cereals and their by-products. Although a substantial quantity of solid waste is produced during grain processing, which can serve as animal feed or manure, these by-products also have nutritive value for human consumption due to their unique biological, physicochemical, and functional characteristics (Luithui et al. 2019). Cereal by-product extracts are utilized in baked goods (bread, biscuits, cookies), extruded items (breakfast cereals, pasta), and other value-added food products (Ahmed et al. 2018). However, the current methods of utilizing various cereal by-products from pre- and post-processing stages are not optimal for industrial sustainability. The primary by-products of milling cereals include steep liquor, grain pieces (germ, bran, husk, and middling), spent grains, and the endospermic tissue aleurone layer. These components are rich sources of fiber, vitamins, minerals, phytoestrogens, and phenolic compounds (Papageorgiou and Skendi 2018). Bran from corn, rice, and wheat comprises the outer kernel layer—pericarp and aleurone—with minimal flour content but high levels of fiber and protein. Additionally, bran contains compounds such as sterols, polyphenols, arabinoxylans, ferulic acid, and vitamins that show potential in mitigating modern-day ailments like cancer, diabetes, and cardiovascular diseases

Table 10.1 Various enzymes produced through fermentation of food waste

S. no	Enzyme	Microorganism used	Waste used	References
1	Amylases	*Candida guilliermondii* CGL-A10	Date wastes	Chandrasekaran and Bahkali (2013)
		Bacillus sp.	Mango, banana, potato, and lemon waste	Hasan et al. (2017)
		Penicillium expansum MT-1	Waste loquat (*Eriobotrya japonica Lindley*) kernels	Erdal and Taskin (2010)
		Bacillus sp.	Groundnut shell and cassava waste	Selvam et al. (2016)
		Aspergillus oryzae	Edible oil cakes	Sivaramakrishnan Sivaramakrishnan and Gangadharan (2009)
		Bacillus subtilis	Banana peel	Krishna (1999)
		B. subtilis (CBTK 106)	Banana waste	Krishna and Chandrasekaran (1996)
2	Levansucrase	*Bacillus megaterium*	Banana peel, orange peel, and lemon waste	Homann et al. (2007)
3	Glucoamylase	*Aspergillus* sp. HA-2	Rice flake manufacturing waste	Anto et al. (2006)
		Aaspergillus awamori and A. oryzae	Bread and cake waste and vegetable waste	Melikoglu et al. (2015)
		Aspergillus niger	Food waste from the student canteens	Wang et al. (2008)
4	Endoglucanase	*Bacillus pumilus* EB3	Oil palm fruit bunch	Ariffin et al. (2008)
5	Polygalacturonase	*A. awamori* MTCC 9166	Fruit skins and peels (apple, banana, orange, jackfruit, mango, and pineapple)	Padma et al. (2012)
6	Protease	*Penicillium digitatum*	Agriculture industrial waste residues (apple pomace, lemon peel)	Aissaoui et al. (2014)

(continued)

Table 10.1 (continued)

S. no	Enzyme	Microorganism used	Waste used	References
7	L-asparaginase	*A. niger*	Cereal processing wastes like meal and bran	Mishra (2006)
8	Keratinolytic protease	*B. subtilis PF1*	Chicken feathers waste, rapeseed cake, and potato peel	Bhange et al. (2016)
9	Ligninase	*Phanerochaete chrysosporium*	Oil palm biomass	Faison and Kirk (1985)
10	Alkaline protease	*B. subtilis*	*Imperata cylindrica* grass and potato peel	Mukherjee et al. (2008)
11	Lipase	*Staphylococcus pasteuri*	Coconut oil mill waste	Kanmani et al. (2015)
		A. niger	Rice bran and Jatropha seed cake	Veerabhadrappa et al. (2014)
		Yarrowia lipolytica	Mango agro-industrial wastes	Fickers et al. (2011)
		B. subtilis strain NL-39	Edible oil waste like oilseed meal and cereal bran	Mazhar et al. (2018)
		Micrococcus roseus	Oil cakes like groundnut, coconut, gingili, mustard, and castor	Lawrence et al. (1967)
		Penicillium simplicissimum	Castor bean waste	Sztajer et al. (1992)
		Bacillus sp. SK II-5	*Moringa oleifera*	Irfansyah et al. (2021)
		Yarrowia lipolytica	Sugarcane molasses and waste cooking oil	Haba et al. (2000)
		Stenotrophomonas maltophilia	Agriculture wastes from markets and mills	Hasan-Beikdashti et al. (2012)
		A. oryzae (NCIM 1212) and *Aspergillus japonicas* (MTCC no. 1975)	Oil cake	Jain and Naik (2018)
		Bacillus licheniformis	Edible oil industry waste	Nthangeni et al. (2001)
		Pseudomonas, Bacillus, Candida, Rhodococcus, and Staphylococcus	Fried oil as selective medium	Haba et al. (2000)

(continued)

Table 10.1 (continued)

S. no	Enzyme	Microorganism used	Waste used	References
12	Lipase (thermostable)	*Bacillus cereus* ASSCRC-P1	Waste frying oil	Awad et al. (2015)
13	Pectinase	*B. subtilis* EFRL 01	Date syrup	Abdul Sattar Qureshi (2012)
		Penicillium citrinum	Orange waste	Sandhya and Kurup (2013)
		Aspergillus foetidus (NRRL 341, ATCC 16878)	Citrus waste, fruit wastes from cashew apple, banana, pineapple, and grape	Cavalitto et al. (1996)
		Penicillium sp.	Using banana peel, wheat bran, sugarcane bagasse, and orange bagasse	Patil and Chaudhari (2010)
		A. niger	Citrus waste peel	Patil and Chaudhari (2010)
		Bacillus sp. MFW7	Cassava waste	Kalaichelvan (2012)
		Pseudomonas sp. *and Bacillus* sp. *A. niger, Aspergillus flavus and P. chrysogenum*	Orange peel, citrus peel, and potato peel	Hildebrand (1971)
		Saccharomyces cerevisiae	Different fruit peels (mango, grape, tamarind kernel, sapota, banana, apple pomace, papaya, pomegranate) and oil cake (groundnut, palm oil, sunflower, coconut, rice bran, and wheat bran)	Poondla et al. (2015)
		B. subtilis strain BK-3	Peel waste (pineapple, citrus, watermelon, Chico, and banana)	Prajapati et al. (2021)
14	Phenylalanine ammonia lyase	*Rhodosporidium toruloides*	Waste cooking oil	Babich et al. (2013)
15	Endopectinase	*A. niger* PC5	Date pomace	Bari et al. (2010)

(continued)

Table 10.1 (continued)

S. no	Enzyme	Microorganism used	Waste used	References
16	Cellulase	*A. japonicus* URM5620	Passion fruit peel waste hydrolysate	Herculano et al. (2011)
		Trichoderma viride VKF3	Oil cake (coconut oil, groundnut, neem, rice bran) and cassava starch	Nathan et al. (2014)
		Rhizopus sp.	Fruit peels (mango, melon, orange, and watermelon)	Tang et al. (2012)
		A. niger and *S. cerevisiae*	Pineapple and orange peels (fruit wastes)	Sohail et al. (2009)
		A. niger and *T. viride*	Agriculture industry waste (pineapple, orange, and banana)	Mandels et al. (1971)
		Various microbe	Palm oil mill effluent	Hii et al. (2012)
		B. subtilis (CBTK 106)	Banana wastes	Krishna and Chandrasekaran (1996)
17	Cellulase and hemicellulase	*Trichoderma reesei*	Horticultural waste	Häkkinen et al. (2014)
18	Xylanase	*Trichoderma harzianum* 1073 D3	Fruit waste of orange, lemon, apple, pear, banana, melon, and hazelnut shell	Seyis and Aksoz (2005)
		B.licheniformis P11(C)	Oat-spelt xylan or cereal bran (wheat, rice, rice, bajra, barley, corn) as carbon source	Bajaj and Manhas (2012)
		A. flavus	Peel from passion fruit	Chen et al. (2019)
19	Endoxylanase	*A. japonicus*	Fruits peels from different fruits (pineapple, banana, mango, orange, tangerine, nectarine, peach)	da Silva et al. (2019)
20	Esterase	*Bacillus altitudinis* AP-MSU	Fish-processing wastes	Esakkiraj et al. (2009)

(continued)

Table 10.1 (continued)

S. no	Enzyme	Microorganism used	Waste used	References
21	β-glucosidase	*A. niger* CECT2915 and CECT2088, *A. ibericus* MUM 03.49 and MUM 04.86, and *R. oryzae* MUM 10.260	Winery industry waste, olive pomace, and brewer's spent grain	Ahmed et al. (2017)
		Penicillium verruculosum	Passion fruit peel	Volkov et al. (2020)
22	Laccase	*Trametes hirsuta*	Kiwi fruit wastes	Abadulla et al. (2000)
		Pycnoporus sanguineus	Oil palm biomass	Vite-Vallejo et al. (2009)
		Pleurotus eryngii	Agro-industrial peach waste	Munoz et al. (1997)
		Bacillus sp. MSK-01	Fruit juice waste	Sondhi et al. (2021)
23	Laccase and manganese peroxidase	*Lentinus edodes*	Brewing industry waste products	Hatvani and Mécs (2001)
24	Tannase	*Penicillium montanense* URM 6286	Agro-industrial waste barbados cherry (*Malpighia emarginata*) and mangaba fruit (*Hancornia speciosa* Gomes)	de Lima et al. (2014)
25	Carboxymethyl cellulase	*Paenibacillus polymyxa*	Mango peel	Kumar et al. (2012)
26	Glucoamylase	*A. oryzae*	Husk and bran (rice, wheat) cottonseed powder, corn steep solid, bagasse powder, oil cake (coconut oil, groundnut) corn steep liquor, and soybean meal	Zambare (2010)
27	Polyhydroxyalkanoate	*Bacillus thermoamylovorans*	Waste cooking oil	Sangkharak et al. (2021)
28	Endoglucanase	*B. pumilus* EB3	Oil palm bunches	Ariffin et al. (2008)
29	Ligninolytic enzymes	*Pleurotus eryngii*	Apricot and pomegranate wastes	Akpinar and Urek (2014)
30	Phytase	*Thermoascus aurantiacus* SL16W	Different 16 types of the different lignocellulosic wastes	Tanruean et al. (2021)

(continued)

Table 10.1 (continued)

S. no	Enzyme	Microorganism used	Waste used	References
31	Invertase	*A. flavus*	Fruit peel waste (orange, pineapple, and pomegranate)	Uma et al. (2010)
32	Carboxymethyl cellulase	*Penicillium funiculosum* 515, *Myrothecium verrucaria* 9095, and *A. niger* TMF-15	Grape waste material and pressed apple fruit pulp	George et al. (2001)
	Multiple enzymes			
33	Cellulase, amylase, and tannase	*B. megaterium* and *saccharomyces cerevisiae*	Fruit waste	Unban et al. (2020)
34	Lipase, protease, amylase, cellulase, invertase, phytase, and tannase	Filamentous fungi (*Trichoderma virens*, *Fusarium oxysporum*, *A. niger*, *Talaromyces assiutensis* and *Penicillium toxicarium*)	Olive oil mill wastes	Zaier et al. (2021)
35	Amylases, β-glucosidases, carboxymethyl cellulase (CMCase), and xylanases	*Lichtheimia ramosa*	Fruit wastes from *Bocaiuva, Guavira,* and Pequi	de Andrade Silva et al. (2014)
36	Cellulases, xylanases, and pectinases	*A. awamori*	Grape pomace	Botella et al. (2005)
37	Augmented cellulase and xylanase	*A. niger* and *T. reesei*	Rice straw, pea pods waste, cauliflower waste, and wheat bran	Dhillon et al. (2011)
38	Cellulases, hemicellulases, pectinase, and amylases	*A. niger* P-19	De-oiled rice bran	Laothanachareon et al. (2022)
39	Protease, α-amylase and cellulase	*S. cerevisiae* and Lactic acid bacteria	Fruit peels (orange, pineapple, banana, and mixed fruit)	Janse and Pretorius (1995)
40	Endoglucanase and xylanase	*A. niger*	Food waste from kitchen	Dobrev and Zhekova (2012)
41	Cellulase and xylanase	*A. niger* DWA8	Oil palm waste as substrate	Tai et al. (2019)
42	Amylase, cellulase, and xylanase	*B. licheniformis* KIBGE-IB3	Date fruit wastes as substrate	Aslam et al. (2020)
43	Cellulase and xylanase	*Aspergillus heteromorphus*	Distillery spent wash and rice straw	Bajar et al. (2020)
44	α-Amylase, protease, and polygalacturonate lyase	*B. subtilis* strain 11,089	Orange and potato peels	Mahmood et al. (1998)

(continued)

Table 10.1 (continued)

S. no	Enzyme	Microorganism used	Waste used	References
45	Glucoamylase and protease	*A. awamori* 2B	Waste bread pieces	Negi et al. (2011)
46	Pectinase and pectin lyase	*B. subtilis* SAV-21	Various agriculture residues and fruit peel wastes	Kaur and Gupta (2017)
47	Ligninase and cellulase	*A. niger* 01UAs181, *A. uvarum* MUM08.01, and *A. ibericus* MUM03.49	Brewery spent grain, olive pomace, vine-shoot trimmings, and Mediterranean agro-industrial wastes	Leite et al. (2019)
48	Lignin peroxidase, manganese peroxidase, and laccase	*Stenotrophomonas* sp., *B. subtilis*, and *Aeribacillus* sp.	Palm oil wastes	Olajuyigbe et al. (2018)
49	Cellulase and xylanase	*Streptomyces thermocoprophilus* TC13W	Oil palm bunch	Sinjaroonsak et al. (2020)

(Patil et al. 2022). These by-products are also widely used in enzyme production (Verni et al. 2019).

10.2.1 α-Amylase

Solid-state fermentation (SSF) has proven to be effective in converting dry cereal waste into useful substances including enzymes (Amoah et al. 2020; Prueckler et al. 2014). Numerous agricultural residues including pulses husk (red gram), bran (rice and wheat), straw (rice and jowar straw), and jowar spathe have been investigated for obtaining α-amylase by using SSF from *Gibberella fujikuroi*. In a study, Mulimani et al. (2000) found that in the presence of *Gibberella fujikuroi, Aspergillus oryzae* fermentation of wheat bran generated the maximum amount of α-amylase.

Similarly, the suitability of rice husk and wheat bran fermentation by *Bacillus subtilis* to generate α-amylase was also investigated by Baysal et al. (2003). Also, a notable variation in enzyme production was observed when utilizing wheat bran, with a 7.3-fold increase compared to rice husk. This disparity was linked to the distinct compositional differences between bran and husk. Soni et al. (2003) utilized *Bacillus* sp. in wheat bran fermentation and reported an increased amount of thermostable α-amylase. It has been observed that by incorporating other FW in cereals waste, enzyme yield can be enhanced. For instance, by adding soybean meal, glycerol, and magnesium oxide into wheat bran, a yield of α-amylase was found to increase when *Bacillus* sp. is used as a fermentative organism (Sodhi et al. 2005).

Similarly, Asgher et al. (2007), Baysal et al. (2003), and Shukla and Kar (2006) used industrial waste from potato starch extraction, along with wheat bran and rice husk, to produce affordable carbon substrates that *B. subtilis* could exploit to produce amylase.

10.2.2 Cellulases

With approximately 60% of carbohydrates in the form of starch and cellulose, FWs emerge as a valuable organic resource for creating high-value end products, including cellulases (Salimi et al. 2019). Cellulolytic microorganisms extensively utilize carbohydrates for energy metabolism, while proteins or lipids are not used as energy sources (Lynd et al. 2002). Cellulosic FW might not have as many microbes as other types of FW, such as animal meat and dairy waste. However, it is believed that certain fungi and bacteria can be used effectively and inexpensively to produce stable biozymes from this waste. For instance, enhanced cellulase formation was observed when *Penicillium* sp. was introduced to rice straw, with the generated enzyme exhibiting stability at 65 °C and a pH between 4 and 5 (Picart et al. 2007). Similarly, Annamalai et al. (2014) found that *Bacillus carboniphilus* CAS displayed higher cellulase activity (4040.45 U/mL) at initial pH of 9.0 and 50 °C when yeast extract and rice bran were utilized at 2.52 g/L and 6.27 g/L, respectively. Gutierrez-Correa and Tengerdy (1997) studied the cooperative role of parent *Trichoderma* sp. strain and *Aspergillus* sp. in the fermentation of agricultural waste. The synergy between these two strains resulted in increased production of β-glucosidase, cellulase, and endoglucanase. Ravindran et al. (2018) identified rice hull and bran as superior carbon sources compared to carbohydrates (glucose, fructose, maltose, and sucrose) for synthesizing the cellulase enzyme. Researchers reported the production of carboxymethyl cellulase at excellent concentrations (102 U/mL) by fermenting 0.25% peptone and 2% rice hull (carbon and nitrogen sources) with *Bacillus amyloliquefaciens* DL-3. *Klebsiella* sp. (Gao et al. 2012). In another study, *Klebsiella* PRW-1 exhibited the potential for cellulolytic enzyme production from various agricultural wastes, with sugarcane trash and grass identified as optimal carbon sources (Waghmare et al. 2014).

10.2.3 Xylanases

Xylanases are extensively employed for the breakdown of complex polysaccharides, present in lignocellulose, into simpler forms in various industries. They are produced by lignocellulosic wastes such as soybean hulls, rice straws, and wheat bran. Lignocellulosic wastes serve as a reservoir for complex carbohydrates and other essential metabolites, required for growth of numerous fermentative bacteria, including *Streptomyces* sp., *Thermomonospora* sp., *Bacillus* sp., *Cellulomonas* sp., and *Clostridium* sp. (Buda et al. 2024). Apart from bacteria, fungi are also known to convert lignocellulosic waste in to useful bioproducts including xylanase. Research has shown that *Aspergillus terreus* and *Aspergillus niger* produce high levels of xylanase when fermenting wheat bran (Gawande and Kamat 1999). *Thermoascus aurantiacus* also produces thermostable xylanase with good activity through SSF of grain waste (Milagres et al. 2004). Rice bran fermentation with *Bacillus* sp. yields efficient xylanase production (Virupakshi et al. 2005), while *Cladosporium* sp. prefers rice straw for optimal production (Alokika and Singh 2019). de Castro and Sato (2013) found that wheat bran is the best substrate for simultaneous α-amylase and protease production via SSF. In another study by Azzouz et al. (2021), a strain of *A. niger* was optimized to produce endoxylanase from wheat straw under SSF, enhancing xylanase yield three-fold using response surface methodology. The purified enzyme showed high thermal, pH stability, optimal activity at 50 °C and pH 6, and effectively hydrolyzed wheat arabinoxylan. This makes it suitable for applications in different food processing applications (Juturu and Wu 2012).

10.2.4 Proteases

Cereals and pulses are central to Indian diets, but their harvesting and processing result in significant wastage, including husk/hulls, germs, bran, and broken kernels. Notably, these waste products are rich in fiber, protein, vitamins, and minerals that nourish microorganism growth hence served as a valuable substrate for enzyme production—such as acid phosphatase, lipase, and especially protease (Yadav et al. 2021). Cost-effective commercial production of proteases can be achieved using bran (wheat and rice), husk (lentil and rice), and other agro-industrial wastes. Studies have highlighted cereal waste as an ideal substrate for protease production. For instance, *Bacillus* sp. was found to produce alkaline protease using lentil husk and wheat bran (Uyar and Baysal 2004). Mung bean husk yielded the highest enzyme output for *Bacillus* sp. (Prakasham et al. 2006), while wheat bran was optimal for *Pseudomonas aeruginosa* in SSF (Meena et al. 2013). Fungal strains like *Rhizopus* sp., *Aspergillus* sp., and *Penicillium* sp. are known for their robust protease production (Sharma et al. 2022). Recently, thermophilic fungi such as *Mycothermus thermophilus* and *Neurospora crassa* have shown promise in protease production using wheat and cereal bran via SSF (Sharma et al. 2022).

10.2.5 Laccases

Laccases, extensively applicable in numerous industries, can be produced from agricultural wastes, namely, wheat bran, corncob, maize steep liquor, and soybean husk (Gupte et al. 2023). Rebhun and Hadar (2005) reported that Basidiomycetes like *Cerrena unicolor* and *Cucurbita maxima* demonstrated high laccase output by fermenting wheat bran. *Trametes versicolor* produced more laccase with maize steep liquor as an inducer and source of nitrogen (Wang et al. 2014). Similarly, Risdianto et al. (2012) reported different white-rot fungi, including *Phanerochaete* sp., *Trametes* sp., *Marasmius* sp., and *T. versicolor*, successfully synthesized laccase on substrates like rice straw, corncob, and rice husk. Brewer's spent grain (BSG) which is a common waste from the brewing industry is also known to be used as a substrate by many strains of *T. versicolor* (Dhillon et al. 2012).

10.3 Enzymes from Fruits and Vegetables Waste

A substantial volume of waste in the agri-food sector is from fragments discarded during various stages such as washing, chopping, product preparation, and packing of fruits and vegetables. This waste comprises leaves, peels, pomace, skins, rinds, pits, pulp, stems, seeds, twigs, and damaged produce, constituting the solid portion. The food processing industry also generates considerable wastewater that requires suitable treatment. Notably, the wastewater from the fruits and vegetables sector exhibits low concentrations of both macro- and micronutrients, as noted in studies by Esparza et al. (2020) and Zhu et al. (2023).

The solid FW generated from fruits and vegetables primarily consists of soluble sugars, hydrolyzable materials, and fiber. These FWs are usually generated from the canning, freezing, drying, dehydration, pulping, juice concentration, and related industries (Chaudhary and Dangi 2022). Pectinases and cellulases can be derived from fruit waste, particularly citrus fruits. Mamma et al. (2008) explored the role of dry citrus fruit skins in part of substrate for production of different enzyme cocktails, including pectin degrading, cellulose degrading, and xylan degrading enzymes, through fermentation of temperate fungi such as *A. niger* BTL, *Fusarium oxysporum* F3, *N. crassa* DSM 1129, *Penicillium decumbens, Rhodotorula* sp., and *Mucor mucorales*. Notably, *Rhodotorula* sp. exhibited the highest enzyme yield when orange peelings were used as a substrate. Few commercial important enzymes derived from FW are discussed here.

10.3.1 Pectinase

Researchers investigated the utilization of citrus fruit waste in SSF to study various hydrolytic and oxidative enzymes. For example, *Pleurotus pulmonarius* was employed for orange waste fermentation, with pectinase being the primary enzyme produced, exhibiting the highest activity after more than one month of fermentation (Contato et al. 2020). In terms of oxidative enzymes, laccase took precedence, showing maximal activity around 20 days of fermentation. Toward the end of the 45 days fermentation duration, lower levels of different complex enzymes (manganese peroxidase, β-glucosidase, and β-xylosidase) were detected. Additionally, amylase, CMCase, and xylanase remained comparable and were reported lower than 1.5 IU/mL later one and half of the fermentation durations (Contato et al. 2020).

Furthermore, Oberoi et al. (2010) explored the capability of pulp obtained after juice extracted from kinnow fruit for cellulase production by *Trichoderma reesei* Rut C-30. Likewise, different combinations of fruit waste with other food cereal waste like wheat bran in four parts to one part reported highest endo-1,4-β-glucanase enzyme activity.

The solid by-product of the apple juice extraction processing considered as the apple pomace, is rich in carbohydrates, pectin, crude fiber, and minerals, providing ample nutrients to various microbes (Zhang et al. 2021). Therefore, this apple pomace is considered a suitable raw material for developing various enzymes in microbial fermentation. Hang and Woodams (1994) investigated the suitability of apple pomace as a substrate to produce β-fructofuranosidase by *Aspergillus fumigatus, Aspergillus foetidus*, and *A. niger*. Similarly, Villas-Bôas et al. (2003) reported the SSF (solid-state fermentation) of apple pomace with *Candida utilis* for the production of lignin and cellulose degrading enzymes. Additionally, Sharma et al. (2016) reported the use of apple pomace in production of polygalacturonase by *Lentinus edodes* in SSF.

In mango pulp processing units, substantial quantities of mango peels and kernels are generated as by-products (Aggarwal et al. 2017). These by-products are recognized as excellent sources of edible digestible fiber with significant concentration of extractable antioxidant polyphenolic compounds. Padma et al. (2012) conducted a study on polygalacturonase production using *Aspergillus awamori* MTCC 9166 in SSF, employing various fruit wastes such as fruit peels (apple, banana, citrus, mango, and pineapple) with higher pectin amounts. Furthermore, Kumar et al. (2012) examined the amylase enzyme produced from mango kernel powder as a substrate in conjunction with *Fusarium solani*. The study found that optimal amylase production occurred at a mango kernel powder at 5%, pH of 4, and a temperature of 30 °C on the ninth day of incubation. Saravanan et al. (2012) fine-tuned a culture medium to improve cellulose degrading enzymes production by *T. reesei*, employing mango peel powder as a substrate. This optimization led to increased cellulase activity under conditions that were experimentally confirmed as optimal.

Solid residues (waste) from pineapple processing (canning), including peel, inedible core, crown, and stem, constitute approximately 80% cellulose. These residues present significant potential to produce cellulase and hemicellulase. Maftukhah (2019) investigated the production of cellulose degrading enzymes from pineapple solid waste using *Trichoderma longibrachiatum*, *A. niger*, and *Saccharomyces cerevisiae*. Similarly, Fortkamp and Knob (2014) investigated the production of xylene degradation enzymes by *Trichoderma viride* using pineapple peel as a substrate. Also, Hariharan and Nambisan (2013) studied the optimization of process parameters for the peroxidase (lignin and manganese) and laccase production from *Ganoderma lucidum* by SSF of pineapple waste.

Banana waste serves as a substrate to produce α-amylase, hemicellulases, and cellulases. Krishna and Chandrasekaran (1996) reported the utilization of banana fruit stalk waste in fermentative production of amylases by *B. subtilis*. Commercial production of cellulose, pectin degrading enzymes, and β-glucosidase are successfully produced from the banana peel pretreated with pectinase. The banana peel was employed as a base for producing extracellular laccase by the fungi *Trametes pubescens*. Examination of enzyme production, specifically cellulolytic and ligninolytic, from banana waste in SSF conditions indicated a greater presence of ligninolytic enzymes compared to cellulolytic enzymes. This finding implies that banana waste may be well suited to produce ligninolytic enzymes (Reddy et al. 2003).

Waste from yellow passion fruit demonstrates the capacity to generate hydrolytic and oxidative enzymes through the activity of fungi, including *Pleurotus ostreatus, P. pulmonarius, Macrocybe titans, Ganoderma lucidum*, and *Grifola frondosa*. This enzymatic production occurs in SSF, both when these fungi are cultivated independently and when combined with wheat bran as a supportive substrate (Khuwijitjaru and Klinchongkon 2020). Research has also explored laccase production from waste generated from yellow passion fruit processing industries and wheat bran with *G. frondosa* cultures, with pectinase identified as the primary enzyme produced by yellow passion fruit waste and *M. titans* recognized as the most proficient producer (Zilly et al. 2012). Similarly, passion fruit waste is utilized to produce endoxylanase arylglycosidases (β-glucosidase, β-xylosidase, and β-galactosidase) through the fermentation of *P. ostreatus, P. pulmonarius, and G. lucidum* (Sharma et al. 2016).

Jackfruit generates significant waste quantities, and SSF using *A. niger* NCIM 548 resulted in the pectinase enzymes production from jackfruit waste as a substrate (Wadhwa et al. 2016). Relatedly, Anuradha et al. (2014) demonstrated polygalacturonase production from jackfruit waste by *A. awamori* MTCC 9166, attributing it to the favorable concentrations of pectin. Substantial polygalacturonase production was also reported for various fruit wastes rich in pectin, like different fruits and vegetable peels (orange, jackfruit, carrot, and beetroot) (Almowallad et al. 2022).

The date fruit processing industry generates considerable amounts of by-products like seeds and press cakes. Studies investigated the preparation of endopectinase from date fruit processing waste by using the *A. niger* PC5. Additional study reported yeast *Candida guilliermondii* CGLA10 producing amylase using date wastes (Chandrasekaran and Bahkali 2013). Tomato pomace proves to be a favorable substrate to produce hydrolytic enzymes, such as xylanase, exo-polygalacturonase, cellulase, and α-amylase by *A. awamori*. The highest activities of xylanase and exo-PG were observed in the initial days of culture, while CMCase and α-amylase production remained relatively constant throughout the fermentation process (Umsza-Guez et al. 2011). In a separate study, carrot pomace, a by-product from carrot juice processing, was utilized for optimal polygalacturonase production by *Bacillus* sp. RCPTM1. Researchers also fine-tuned growth conditions to achieve maximum invertase enzyme production in SSF by *Aspergillus* sp. by employing carrot peel (Sharma et al. 2016).

Ananthi et al. (2014) explored SSF of lipid degrading enzymes (lipase) by *Bacillus cereus* MSU AS on different FW as substrates, with onion skin waste yielding maximum lipase. Azzaz et al. (2021) investigated the pectinase production using sugar beet solid waste as the primary organic carbon source for *A. niger*. Due to the higher pectin quantities in the sugar beet waste, it proved effective for producing pectinase enzymes without the addition of pectinaceous material. Heerd et al. (2014) compared the pectinase production capability of the *Aspergillus sojae* ATCC 20235 and *A. sojae* CBS 100928 variants under optimized conditions and concluded that later variant accomplished the greatest enzyme activity.

In a study, dos Santos et al. (2012) reported the SSF of potato peel to produce cellulolytic enzymes production. Similarly, Elayaraja et al. (2011) reported the thermostable alkaline α-amylase production from *Bacillus firmus* CAS 7 variant using potato peel wastage. Potato peel waste has the immense capacity for the laccase production by the *Trametes hirsuta* under SSF, and the usage of the potato's peels reduced considerable production costs. Similarly, Schalchli et al. (2017) employed potato processing wastes as substrate in the production of lignin degradation enzymes by *Anthracophyllum discolor*.

Krishna (1999) employed banana fruit solid wastes like peels as the substrate for the cellulase degrading enzymes by the *B. subtilis* (CBTK 106) fermentation. Also, the high concentration of lignocellulosic wastes like banana peels can be successfully used in the production of an extracellular halotolerant, thermoalkaline cellulase by *Bacillus halodurans* CAS1. *Aspergillus* sp. MPS-002 and *Phyllosticta* sp. MPS-001 were assessed for their ability to produce lignocellulolytic enzymes by banana fruit solid waste.

Botella et al. (2007) highlighted the vast prospective of grape juice processing industry solid waste to produce xylanase by *A. awamori*. Mamma et al. (2008) utilized orange peels to produce high xylanase activities from *A. niger*.

10.3.2 Tannases

Tannases are mainly employed in the food processing and pharmaceutical industries for hydrolysis of tannins into glucose and gallic acid. Agri- and plant-based FWs containing barks, leaves, seeds, and peels may contain significant amount of tannic acid (Sarangi et al. 2023). Commercial preparation of tannase involves hydrolyzing tannin either chemically or by enzymatic means. Commercial preparation of tannase typically involves hydrolyzing tannins either chemically or enzymatically; however, it often results in low yield and the production of impure gallic acid (Dhiman et al. 2018). Tannase production using fruit waste such as peels (banana, pomegranate, apple) and whole fruits waste like *Psidium guajava (guava), Terminalia chebula (myrobalan), Syzygium cumini (black plum), Citrus limon (lemon), Citrus sinensis (orange), and Mangifera indica (mango) via fermentation has been explored as a sustainable alternative (Dhiman et al.* 2022; Lekshmi et al. 2021). Studies also indicate use of Indian gooseberry leaves, tamarind seed, cashew waste, castor bean residues, and coffee husk for extraction of tannase using indigenous fungal strains of *Aspergillus, Penicillium* sp., and *Paecilomyces* sp. (Sutaoney et al. 2023). Kumar et al. (2016) discovered that among various tannin-rich wastes, black tea waste served as the optimal substrate for *Aspergillus glaucus* to produce tannase via SSF. The study revealed that the most efficient bioconversion of tannic acid, at concentrations as high as 200 g/L, occurred at a temperature of 40 °C. Nandini et al. (2014) utilized FW and agricultural residues for tannase synthesis, identifying bacterial isolates that produced maximum yields of tannase and gallic acid. Similarly, de Lima et al. (2014) investigated the use of FW and agro-industrial waste for tannase production by *Penicillium montanense* URM6286 under SSF, noting the highest tannase concentrations after a 3-day incubation period. Furthermore, the fungal strain *Penicillium atramentosum* KM was successfully used for the extracellular production of tannase through submerged fermentation, employing powdered leaves of amla, jamun, ber, keekar, and jamoa as substrates. Optimal tannase production was observed using 2% (w/v) amla leaves and 3% (w/v) keekar leaves. In addition to fungi, certain probiotic bacteria have also been documented for their tannase-producing capabilities by leveraging various agro-industrial residues as substrates (García Méndez et al. 2021).

10.4 Enzymes Produced from the Dairy Industry Waste

The dairy processing industry generates around six tons of solid waste per million liters of raw milk, with the majority being liquid waste (Patra and Duary 2020). These wastes predominantly consist of organic material in liquid form or in suspension from. The milk industry waste required high levels of biological and chemical oxygen demand and composed of the fats, proteins, carbohydrates, nitrogen sources, phosphorous source, and different suspended solids (Patra and Duary 2020). Microbial growth on dairy industry wastes yields various enzymes such as β-galactosidase, α-amylase, manganese peroxidase, protease, penicillin acylase, penicillin amidase, polygalacturonase, cutinase, inulinase, lipase, nattokinase, or α-galactosidase (Sar et al. 2022). This section will cover a few of these enzymes in detail.

10.4.1 Lactase and β-Galactosidase

Traditionally, microorganisms like *Kluyveromyces* sp., *Aspergillus* sp., *Streptococcus* sp., *E. coli, Lactobacillus* sp., *Leuconostoc* sp., *Bacillus* sp., and *Paracoccus marcusii* have demonstrated potential for β-galactosidase (lactase) production, with *Bacillus* sp. being commercialized (Nath et al. 2014). Researchers have identified that pH and incubation temperature significantly influence the rate of enzyme production. Nevertheless, the commercial production of lactase faces challenges due to intracellular localization, prompting ongoing efforts to develop cost-effective purification methods (Vasiljevic and Jelen 2001).

β-galactosidase produced by the *Kluyveromyces marxianus* NCIM 3465 from whey was successful, with the addition of dried cauliflower waste enhancing enzyme production by nearly 15% compared to whey alone (Oberoi et al. 2008). Additionally, α-amylase, a crucial enzyme, can be prepared by dairy waste through a semi-SSF method involving a mixture of food/agricultural and dairy industry water wastes. Recent study by Sar et al. (2022) reported thermostable amylase production by augmented sugarcane solid waste and whey by *Anoxybacillus beppuensis*. Furthermore, dairy industry wastes combined through orange peels have been utilized for α-amylase production (De la Torre et al. 2019).

10.4.2 Proteases

Proteases, protein hydrolyzing enzymes extensively employed in numerous food processing industries, have been produced from whey by various strains of *Serratia marcescens* and *Aspergillus* sp. and *Mucor* sp. Manganese peroxidase, a ligninolytic enzyme, can be produced from cheese whey by *Bjerkandera* sp. strain BOS55, leveraging whey's carbon and nitrogen content (Feijoo et al. 1999). Nattokinase, a serine protease with clot-dissolving abilities, was efficiently produced by *B. subtilis* utilizing cheese whey as a nutrient-rich fermentation substrate (Li et al. 2022). Supplementation with yeast extract further increased the caseinolytic activity, making it analogous to lactose- and glucose-type synthetic media and significantly lowering the production cost. Around 10% of the yeast extract supplementation to cheese whey increased the caseinolytic activity and produced the comparable yield to the lactose- and glucose-based synthetic media and reduced the substrate cost by 55–60% (Sahoo et al. 2020).

10.4.3 Other Enzymes

Penicillin acylase, a key enzyme for semisynthetic antibiotic synthesis, can be produced from whey by the recombinant strain *E. coli* W3110/pPA102 (Sambyal and Singh 2021). Penicillin amidases, which have significant biotechnological applications, including the production of 6-amino penicillanic acid (6-APA), can also be produced in this manner (Illanes and Valencia 2017). De León-Rodríguez et al. (2006) successfully produced the penicillin amidases enzyme using cheese whey as a cost-effective substrate, fermenting it with *Bacillus* sp. MARC-0103. Polygalacturonase, an important pectin enzyme, and pectin lyase are widely used in the fruit juice clarification and wine industries (Nighojkar et al. 2019). Their microbial production can be achieved by using dairy waste as an alternative source of carbon and nitrogen. In a study, *Kluyveromyces fragilis* was used to produce polygalacturonase using dehydrated sweet whey as a substrate (Karim et al. 2020). Cheese whey has also been effective in producing the cutinase-like enzyme PaE by *Pseudozyma antarctica* GB-4(1) (Watanabe et al. 2014). Inulinase production, which is essential for the creation of high-fructose syrup and fructooligosaccharides, was studied using whey powder as the carbon source for *A. niger* A42 (Dagbagh and Goksungur 2009).

10.5 Enzymes Produced from the Edible Oil Industry Waste

Oilseed crops, a significant global agricultural commodity, are rich in fats and proteins but contain fewer carbohydrates, fibers, and specific vitamins. The oils extracted from these seeds primarily consist of triglycerides and lesser amounts of free fatty acids (FFA), mono- and diacylglycerides, phytosterols, tocopherols, and tocotrienols. The oil extraction process, whether mechanical expression or solvent extraction, yields a by-product known as meal or cake. Refining edible oil involves degumming, neutralization, bleaching, deodorization, and winterization. The degumming process, which enhances oil quality in terms of color and flavor, removes phospholipids, producing phospholipid-free oil and valuable by-product gums (lecithin). The waste from edible oil production, especially de-oiled cakes, serves as an excellent substrate for producing valuable metabolites like enzymes through microbial fermentation (Thangaraju et al. 2020).

Numerous studies have demonstrated the function of cake from the oil sector as a beneficial addition to the production medium or a substrate in SSF for larger-scale enzyme synthesis. De-oiled cakes are good substrates for the synthesis of enzymes, particularly when utilizing fungal species, since they include carbon- and nitrogen-based nutrients that promote the proliferation of microorganisms. The primary enzymes produced from oil cakes include lipase, α-amylase, phytase, and glutaminase (Sivaramakrishnan and Gangadharan 2009).

10.5.1 Lipase

The earliest reported enzyme production dates to 1950 when lipase was produced by employing the *Phanerochaete chrysogenum* S1 on oilseed cakes. Subsequent studies have utilized oil cakes, including those from sesame and coconut, as preferred substrates for lipase production. In a study, Benjamin and Pandey (1997) successfully used coconut cake extract as a substrate for lipase production by *Candida rugosa. A. niger* demonstrated the ability to produce lipase enzymes using gingelly and olive oil cakes. Thermoduric fungal species, specifically *Rhizomucor pusillus,* have shown promising lipase production capacity using solid waste from the edible oil processing industry. Potential of Babassu oil cake as a substrate for lipase production by a Brazilian strain of *Penicillium restrictum* in SSF is also reported (Gombert et al. 1999). Kazeem et al. (2024) reported that *Bacillus mycoides* produced lipase using coconut oil cake as the substrate, while a *Pseudomonas* showed production of cottonseed cake. The ability of *Penicillium simplicissimum* to produce lipases using soybean oil cake as a substrate was also explored in SSF

(Greco-Duarte et al. 2023). Magdouli et al. (2017) reported that both solid and liquid wastes from the olive oil processing industry were subjected to alkaline treatment, resulting in higher lipase enzyme production in SSF when *Yarrowia lipolytica* was used. Similarly, Fickers et al. (2011) explored the lipase production capabilities of olive milling waste using *Y. lipolytica* W29 (ATCC 20255) in batch culture methods.

10.5.2 Phytase

Different types of oil cakes, including coconut, soybean, palm, peanut, cotton, and olive, have been utilized as substrates for fungal phytase production in SSF. Various strains of *Rhizopus* sp. have been confirmed to produce phytase in several studies (Suresh and Radha 2016). Rani and Ghosh (2011) recommended using a combination of different oil cakes as substrates under optimized conditions to enhance phytase production, compared to using a single type of oil cake. For example, the use of a mixture of soybean and groundnut cake as a substrate enhanced phytase production by *Mucor racemosus* NRRL 1994. Some researchers have reported that combining oil cake with other forms of FW as a substrate resulted in higher phytase yields than using oil cake alone (Ancuța and Sonia 2020). Castor oil cake has been explored for phytase production with *Aspergillus ficuum* in SSF (Buddhiwant et al. 2016). It has been observed that substrates with lower phosphorus quantities support higher enzyme production. Similarly, the addition of surfactants, such as Tween-80 and sodium oleate, has been found to enhance phytase yields and the breakdown of phytic acid (Mandviwala and Khire 2000). Buddhiwant (2017) reported on the effects of substrate moisture content, inoculum age, and homogenization intensity on phytase yields using *A. A. ficuum* NRRL 3135 in SSF. Other studies have confirmed that the optimized augmentation of coconut de-oiled cake with glucose, casein, and ammonium sulfate can significantly increase phytase activity (Soccol et al. 2014; Özbek and Ergönül 2022; Gaind and Singh 2015).

10.5.3 Protease

Protease, an important enzyme, can be produced using de-oiled cake as a substrate. For instance, *Bacillus horikoshii* has been reported to produce an extracellular alkaline protease, with higher yields achieved when protein-rich oil cakes (such as soybean) and casein are used as substrates (Joo et al. 2002). Similarly, *Bacillus clausii* has been found to produce a stable alkaline protease (resistant to oxidation and SDS). The maximum yields were reported when the substrate consisted of soybean de-oiled oil cake (15 g/l) and wheat flour. Joo et al. (2002) reported on protease production by a wild *Penicillium* sp. in SSF. In this process, soybean de-oiled meal served as the organic source of carbon and nitrogen, supporting the growth of mycelium in SSF. Various oil cakes along with FW materials, such as cereal bran and husk (wheat, rice husk) and brewery solid waste (spent brewing grain), have also been evaluated for neutral protease production by different workers (Sandhya et al. 2005; Shivakumar 2012; Mathias 2017). Recently, a significant amount of neutral metalloprotease was produced by fermenting *A. oryzae* NRRL 2217 using a substrate of coconut oil cake augmented with wheat grain bran (Sumantha et al. 2005).

10.5.4 Amylase

De-oiled cake, a by-product of biorefineries, can be effectively utilized for amylase production through microbial fermentation (Sarkar et al. 2021). Numerous studies have confirmed that various fungal strains can ferment oil cake, resulting in significant amylase production (Balakrishnan et al. 2021; Ramachandran et al. 2004; Jain and Naik 2018; Alva et al. 2007). In a study assessing several types of oil cakes for amylase production using *A. oryzae*, coconut oil cake yielded the best results in SSF (Balakrishnan et al. 2021). This finding aligns with a report by Sumantha et al. (2005), which indicated that a mixture of wheat bran and groundnut oil cake in SSF led to an increased production of enzymes. The potential of coconut oil cake in glucoamylase enzyme production by *A. niger* NCIM 1245 was also explored. It was concluded that the supplementation of inorganic nitrogen sources significantly increased enzyme yields without affecting the dry matter contents (Soccol et al. 2014). In the search of extremozymes, researchers have investigated the use of *Thermomucor indicae-seudaticae* to produce thermostable and neutral glucoamylase in SSF, using oil industry waste as the substrate (Kumar and Satyanarayana 2004). Producing thermoamylase using more cost-effective sources can significantly reduce the economic burden on industries. Additionally, it will broaden its applications in the food industry, where there is a high demand for thermostable amylase.

10.5.5 Xylanase

The escalating global demand for xylanases across various industries underscores the need for cost-effective production methods, particularly those utilizing valuable agro waste. Among the diverse range of agro-waste substrates, oil cakes have been identified as the most suitable for microbial xylanase production (Ravindran et al. 2018). As highlighted in the preceding section, filamentous fungi are the preferred candidates for xylanase production via SSF. A study by Ramachandran et al. (2007) investigated the use of castor oil cake as a substrate for xylanase production through fermentation with *T. reesei*. The study found that the yield of xylanases significantly increased when castor oil cake was used as the substrate, compared to xylan or glucose. Furthermore, the enzyme produced was found to be a mixture of enzymes, with a higher proportion of acetyl-xylan esterase, cellulase, and xylosidase. In a separate study, *A. awamori* was used to produce xylanase by fermenting palm kernel cake (PKC) and palm-pressed fiber (PPF) through SSF. This process not only yielded significant enzyme activity but also effectively decreased the lipid content and increased the nonfibrous carbohydrates (NFC) in these coproducts, thereby enhancing their nutritional value for use in animal feed (Oliveira et al. 2018).

There are limited studies available on the production of the inulinase enzyme using coconut oil cake as a substrate in SSF. In a study, fermentation using *Staphylococcus sp. RRL-1 and Kluyveromyces marxianus under SSF conditions showed inulinase production when coconut oil cake was used as the substrate. There are few reports on the production of pectin-degrading enzymes using Cryptococcus albidus* var. albidus IMAT-4735 by using wastewater from the olive oil industry as substrates, while production of extracellular glutaminase by an osmotolerant strain *Zygosaccharomyces rouxii* NRRL-Y 2547 is reported by Ramachandran et al. (2007). In their study, they utilized wheat bran and soybean oil cakes as substrates.

Furthermore, the use of palm kernel cake in SSF has been reported to produce tannase using *A. niger* ATCC 16620. Additionally, waste from the palm oil industry has been identified as a promising substrate to produce various enzymes by *A. niger* ATCC 6275. These findings highlight the potential of agro-industrial waste in enzyme production.

10.6 Enzymes Derived from Fish Waste

The industrial fish-processing sector generates a significant amount of waste, accounting for 20–60% of the initial raw material weight. This waste includes diverse residues such as whole fish and inedible parts (Välimaa et al. 2019). Both solid waste and wastewater from fish processing are rich in organic and inorganic compounds, making them excellent substrates for microbial growth and enzyme production (Sharma et al. 2020). Despite the potential of fish industry waste as an enzyme production substrate, preprocessing is necessary. This can be as simple as cooking, pressing, mincing, and drying or as complex as extracting specific components. The processed fish waste can then be used as a substrate for enzyme production by various microorganisms. Recent studies have reported the preparation of protein hydrolysates from fat-removed fish by-products using acid, alkali, and enzymatic derivation. These hydrolysates are considered excellent organic nitrogen sources for protease enzyme production (Venugopal 2021). Enzyme derived from fish waste are mainly proteases. For instance, *P. aeruginosa* MN7 and *B. subtilis* have been successfully cultured in media containing dried inedible parts of fish, resulting in significant protease enzyme production (Triki-Ellouz et al. 2003). Esakkiraj et al. (2009) found that defatted dried tuna waste powder is an optimal substrate for protease production by *B. cereus*, outperforming alkali-treated hydrolysates and commercially available peptone. However, other fermentation parameters, such as substrate composition, agitation intensity, temperature, and pH, also influence protease production (Ghorbel et al. 2005).

Various microbial strains have also been cultivated on chitinous waste materials, such as shrimp shell, squid pen, and crab shell powders, for protease production. For example, *B. subtilis* strains TKU007 and TKU004 showed high protease metabolism when fermented with different shrimp shell powders compared to other carbon and nitrogen sources (Wang and Yeh 2006). In addition, *Lacticaseibacillus paracasei* subsp. *paracasei* TKU010, *Serratia ureilytica* TKU013, and *Chryseobacterium sp.* TKU014 have been reported to produce protease in a similar manner (Wang et al. 2011). The supplementation of shrimp and crab shell powder to the microbial basal medium significantly influences the protease production when utilizing *Monascus purpureus* CCRC31499. Notably, protease produced under optimal culture conditions can be harnessed for the deproteinization of shrimp and crab shell powder wastes (Liang et al. 2006).

Wastewater from the fish-processing industry have also been used for protease production. In a study by Souissi et al. (2008), cuttlefish by-products was used as a medium for microbial growth and protease production. This approach has demonstrated enhanced protease activity in bacterial species as the nutrient requirements for protease production were met by the nutrient-rich fish waste in the wastewater.

10.7 Brewery and Distillery Waste in the Enzyme Production

The brewery industry is used in the production of the various alcoholic beverages such as beer, whisky, and wine. Barley grain is the primary raw material utilized in the production of different distilled and non-distilled alcoholic beverages. In industrial and traditional processing of cereal-based alcoholic beverages, barley grains undergo crushing and fermentation, resulting in an alcohol/water mixture that, when alcohol is intended for use in spirits, undergoes distillation to extract the alcohol. Knob et al. (2013) reported highest xylanase activity after 5 days of fermentation of brewer's waste grain at 30 °C and pH 5.0. Hooi Ling Ho (2016) utilized barley husk as the substrate for submerged fermentation by randomly mutated *B. subtilis* for xylanase synthesis.

A significant quantity of the by-product, referred to as Beverage Spent Grains (BSG), is also generated during barley processing. Constituting up to 85% of all by-products, it contributes to approximately 30–60% of the biochemical oxygen requirement and stands as a prime by-product in the entire beverage production process (Fillaudeau et al. 2006). On average, BSG represents 31% of the initial malt weight, emphasizing the need for economically processing and bio-converting this residue into value-added products (Evaristo et al. 2023). Different fungal species such as *P. ostreatus* and *Penicillium janczewskii* can successfully utilize BSG as a substrate to produce various enzymes (Zeko-Pivač et al. 2022). With superior yields, increased individual activity, lower substrate costs, and a conservative downstream process, BSG emerges as a valuable substance in enzyme production

(Moran-Aguilar et al. 2021). Currently, BSG is used in the cellulases production by cellulolytic fungus like *T. reesei*. Similarly, BSG has proven to be a best medium for the xylanase production by a *Streptomyces, Penicillium glabrum* strain from Brazil, and feruloyl esterase by *Streptomyces avermitilis* (Hassan et al. 2020).

10.8 Commercial and Households Kitchen's Waste

FW, primarily organic and rich in diverse carbon, nitrogen, and mineral sources, requires specific treatments tailored to its source and composition for effective enzyme synthesis by fermentative microbes (Chen et al. 2016). Several studies have demonstrated the potential of different types of FW in enzyme production. Hasan et al. (2017) reported amylase production by *Chryseobacterium* sp. and *Bacillus* sp. using kitchen waste. Bansal et al. (2012) reported cellulase by *A. niger* NS-2 using agricultural and kitchen-based FW, with alkali-pretreated waste and rice husk found suitable for cellulase production. Bread waste was assessed as a source for protease production by *A. awamori*, with optimal conditions identified as an airflow of 1.50 vvm, particle size around 2 cm, and fermentation duration of 144 h (Melikoglu et al. 2013, 2015). Ahmed et al. (2017) utilized lemon fruit waste peel as a substrate for pectinase production by *A. niger*, with maximum yields identified around the fifth day of fermentation. Similarly, hazelnut shell was reported as a cost-effective medium for pectinase production by *B. subtilis* (Uzuner and Cekmecelioglu 2015). Botella et al. (2007) highlighted the significance of supplementing additional carbon sources and adjusting the moisture content of grape pomace for efficient xylanase production by *A. awamori*. In another study, do Rosario Freixo et al. (2008) used tomato pomace as the sole carbon source for laccase production by *Coriolus versicolor*, with maximum quantities produced after the third day of fermentation.

10.9 Challenges in Preprocessing, Purification, and Recovery

The extraction of enzymes from fruit waste involves several critical stages such as preprocessing, purification, and recovery, each presenting its own set of challenges (Baiano 2014). Acids and alkali can disrupt the three-dimensional structure of enzymes, leading to denaturation and loss of enzymatic activity. This is particularly problematic for enzymes that are sensitive to pH changes (Baiano 2014). Furthermore, the use of strong acids and alkali can corrode processing equipment, leading to increased maintenance costs and potential contamination of the enzyme product (Ravindran and Jaiswal 2016). In addition, heat-sensitive enzymes can lose their functional properties when exposed to high temperatures, which reduces the overall yield and effectiveness of the recovered enzymes (Nadar et al. 2018). There are studies that have confirmed that excessive heat can significantly alter the structure of the fruit waste substrate thus complicating enzyme extraction and subsequent purification efforts (Galanakis 2012; Nadar et al. 2018). The complexity of fruit waste makes purification process more challenging due to presence of proteins, carbohydrates, lipids, and other organic compounds (Esparza et al. 2020). Separating the target enzymes from this mixture is challenging as during the extraction process, compounds that inhibit enzyme activity can be co-extracted (Manhongo et al. 2022). Therefore, these inhibitors must be effectively removed to obtain active enzymes. Likewise, for achieving maximum recovery of enzymes, one needs to maintain adequate balance between enzyme activity and yield (Boudrant et al. 2020). Once recovered, enzymes must remain stable during storage and transportation which often requires additional stabilizing agents or specific storage conditions (Cheng et al. 2021). Finally, ensuring that the recovered enzymes are free from microbial contamination is crucial, especially for applications in the food and pharmaceutical industries.

To overcome the abovementioned issues, milder pretreatment conditions that avoid extreme pH levels and temperatures and include the use of buffered solutions and controlled heating to preserve enzyme integrity are recommended (Shrestha et al. 2020). Using buffered solutions to maintain a stable pH environment during pretreatment helps in preserving enzyme activity (Shrestha et al. 2020). Applying lower temperatures for longer durations can be more effective in breaking down the substrate without denaturing enzymes. Further, specific enzymes like cellulases or pectinases can also be used to break down cell walls and release target enzymes from fruit waste (Kuthiala et al. 2022). Microwaves can also be used to rapidly heat the substrate, enhancing the extraction process while minimizing thermal damage to enzymes (Guzik et al. 2022). Likewise, ultrasound waves create microscopic bubbles that disrupt cell walls and release enzymes (Priya and Gogate 2021). Additionally, utilizing supercritical CO_2 provides a safe and effective solvent for enzyme extraction, avoiding the use of harmful chemicals and preserving enzyme activity (Patil et al. 2021). Immobilization is also another means to improve enzymes stability and reusability by attaching enzymes to a solid support, which allows for repeated use of enzymes in continuous processes, reducing costs and improving efficiency (Rodrigues et al. 2021). Various methods of immobilization including adsorption, covalent bonding, entrapment, and encapsulation are particularly beneficial in industrial applications. These processes can withstand harsh operational conditions and facilitate easier separation from reaction mixtures (Bié et al. 2022). Response surface methodology (RSM) is an advanced statistical technique used for optimizing the conditions for enzyme extraction and activity (Zuin et al. 2020). It involves designing experiments to evaluate the effects of multiple variables and their interactions on the outcome. Application of RSM helps determine the optimal conditions for enzyme recovery, such as pH, temperature, and substrate concentration, to maximize yield and activity (Idris et al. 2020). This method allows for a systematic and efficient approach to fine-tuning the extraction process, ultimately enhancing enzyme recovery from fruit waste (Zuin et al. 2020). Furthermore, innovative solutions, coupled with advancements in microbial strain development and fermentation technologies, are also fundamental for controlling the full potential of

FW as a sustainable source for microbial enzyme metabolism. Addressing these challenges will pave the way for eco-friendly and economically viable biotechnological processes in the enzyme industry. Production of the different enzymes from various food waste is presented in Table 10.1.

10.10 Conclusions

Using fermentative microorganisms to extract valuable enzymes from FW has emerged as a pivotal strategy for achieving industrial sustainability. This method, which is in alignment with the principles of a circular economy, transforms waste into useful bioproducts, thereby mitigating environmental pollution. The economic feasibility of this approach is underscored by the abundant availability and low cost of FW. The enzymes produced through this process are versatile, finding applications across various sectors and thereby broadening their economic impact. Despite the rapid advancement of this approach, the fermentative processes involved in enzyme extraction from FW require further refinement, particularly when dealing with the production of custom enzymes with superior performance characteristics. As industries explore sustainable alternatives, enzyme production from FW using fermentative microorganisms symbolizes both ecological responsibility and economic prudence. This method not only provides solutions to waste management challenges but also positions enzymatic biotechnology as a crucial contributor to industrial sustainability.

References

Abadulla E, Tzanov T, Costa S, Robra K-H, Cavaco-Paulo A, Gübitz GM (2000) Decolorization and detoxification of textile dyes with a laccase from *Trametes hirsuta*. Appl Environ Microbiol 66(8):3357–3362

Abdul Sattar Qureshi (2012) Production of pectinase by Bacillus subtilis EFRL 01 in a date syrup medium. Afr J Biotechnol 11(62). https://doi.org/10.5897/ajb11.4236

Adrio JL, Demain AL (2014) Microbial enzymes: tools for biotechnological processes. Biomol Ther 4(1):117–139

Aggarwal P, Kaur A, Bhise S (2017) Value-added processing and utilization of mango by-products. In: Handbook of mango fruit: production, postharvest science, processing technology and nutrition, pp 279–293

Ahmed A, Batool K, Bibi A (2017) Microbial β-glucosidase: sources, production and applications. J Appl Environ Microbiol 5(1):31–46

Ahmed H, Satheesh N, Dibaba K (2018) Functional, physical and sensory properties of cookies prepared from okara, red teff and wheat flours. Croat J Food Sci Technol 10(1):23–32

Aissaoui N, Abidi F, Mahat S, Marzouki MN (2014) Purification and biochemical characterization of a novel protease from *Penicillium digitatum*–use in bioactive peptides production. J Basic Microbiol 54(S1):S178–S189

Akpinar M, Urek RO (2014) Extracellular ligninolytic enzymes production by *Pleurotus eryngii* on agroindustrial wastes. Prep Biochem Biotechnol 44(8):772–781

Almowallad SA, Aljobair MO, Alkuraieef AN, Aljahani AH, Alsuhaibani AM, Alsayadi MM (2022) Utilization of agro industrial orange peel and sugar beet pulp wastes for fungal endo-polygalacturonase production. Saudi J Biol Sci 29(2):963–969

Alokika, Singh B (2019) Production, characteristics, and biotechnological applications of microbial xylanases. Appl Microbiol Biotechnol 103:8763–8784

Alva S, Anupama J, Savla J, Chiu YY, Vyshali P, Shruti M, Yogeetha BS, Bhavya D, Purvi J, Ruchi K, Kumudini BS (2007) Production and characterization of fungal amylase enzyme isolated from *Aspergillus sp.* JGI 12 in solid state culture. Afr J Biotechnol 6(5):576

Amoah I, Taarji N, Johnson P-NT, Barrett J, Cairncross C, Rush E (2020) Plant-based food by-products: prospects for valorisation in functional bread development. Sustain For 12(18):7785

Ananthi S, Ramasubburayan R, Palavesam A, Immanuel G (2014) Optimization and purification of lipase through solid state fermentation by *Bacillus cereus* MSU as isolated from the gut of a marine fish *Sardinella longiceps*. Int J Pharm Pharm Sci 6(5):291–298

Ancuța P, Sonia A (2020) Oil press-cakes and meals valorization through circular economy approaches: A review. Appl Sci 10(21):7432

Annamalai N, Rajeswari MV, Balasubramanian T (2014) Extraction, purification and application of thermostable and halostable alkaline protease from *Bacillus alveayuensis* CAS 5 using marine wastes. Food Bioprod Process 92(4):335–342

Anto H, Trivedi UB, Patel KC (2006) Glucoamylase production by solid-state fermentation using rice flake manufacturing waste products as substrate. Bioresour Technol 97(10):1161–1166

Anuradha K, Padma PN, Venkateshwar S, Reddy G (2014) Selection of nutrients for polygalacturonase production by *Aspergillus awamori* MTCC 9166 using Plackett-Burman design. Indian J Biotechnol 13(4):502–507

Ariffin H, Hassan MA, Shah UKM, Abdullah N, Ghazali FM, Shirai Y (2008) Production of bacterial endoglucanase from pretreated oil palm empty fruit bunch by *Bacillus pumilus* EB3. J Biosci Bioeng 106(3):231–236

Asgher M, Asad MJ, Rahman SU, Legge RL (2007) A thermostable α-amylase from a moderately thermophilic *Bacillus subtilis* strain for starch processing. J Food Eng 79(3):950–955

Aslam F, Ansari A, Aman A, Baloch G, Nisar G, Baloch AH, Rehman HU (2020) Production of commercially important enzymes from *Bacillus licheniformis* KIBGE-IB3 using date fruit wastes as substrate. J Genet Eng Biotechnol 18(1):1–7

Awad GEA, Mostafa H, Danial EN, Abdelwahed NAM, Awad HM (2015) Enhanced production of thermostable lipase from *Bacillus cereus* ASSCRC-P1 in waste frying oil based medium using statistical experimental design. J Appl Pharm Sci 5(9):7–15

Azzaz HH, Kholif AE, Murad HA, El-Bordeny NE, Ebeid HM, Hassaan NA, Anele UY (2021) A new pectinase produced from *Aspergillus terreus* compared with a commercial pectinase enhanced feed digestion, milk production and milk fatty acid profile of Damascus goats fed pectin-rich diet. Ann Anim Sci 21(2):639–656

Azzouz Z, Bettache A, Boucherba N, Prieto A, Martinez MJ, Benallaoua S, de Eugenio LI (2021) Optimization of β-1,4-Endoxylanase production by an *Aspergillus niger* strain growing on wheat straw and application in xylooligosaccharides production. Molecules 2021(26):2527. https://doi.org/10.3390/molecules26092527

Babich OO, Pokrovsky VS, Anisimova NY, Sokolov NN, Prosekov AY (2013) Recombinant l-phenylalanine ammonia lyase from *Rhodosporidium toruloides* as a potential anticancer agent. Biotechnol Appl Biochem 60(3):316–322

Baiano A (2014) Recovery of biomolecules from food wastes – a review. Molecules 19(9):14821–14842

Bajaj BK, Manhas K (2012) Production and characterization of xylanase from Bacillus licheniformis P11 (C) with potential for fruit juice and bakery industry. Biocatal Agric Biotechnol 1(4):330–337

Bajar S, Singh A, Bishnoi NR (2020) Exploration of low-cost agro-industrial waste substrate for cellulase and xylanase production using *Aspergillus heteromorphus*. Appl Water Sci 10(6):1–9

Balakrishnan M, Jeevarathinam G, Kumar SKS, Muniraj I, Uthandi S (2021) Optimization and scale-up of α-amylase production by *Aspergillus oryzae* using solid-state fermentation of edible oil cakes. BMC Biotechnol 21(1):33

Bansal N, Tewari R, Soni R, Soni SK (2012) Production of cellulases from *Aspergillus niger* NS-2 in solid state fermentation on agricultural and kitchen waste residues. Waste Manag 32(7):1341–1346

Bari MR, Alizadeh M, Farbeh F (2010) Optimizing endopectinase production from date pomace by *Aspergillus niger* PC5 using response surface methodology. Food Bioprod Process 88(1):67–72

Baysal Z, Uyar F, Aytekin Ç (2003) Solid state fermentation for production of α-amylase by a thermotolerant *Bacillus subtilis* from hot-spring water. Process Biochem 38(12):1665–1668

Benjamin S, Pandey A (1997) Coconut cake—a potent substrate for the production of lipase by *Candida rugosa* in solid state fermentation. Acta Biotechnol 17:241–251

Bhange K, Chaturvedi V, Bhatt R (2016) Ameliorating effects of chicken feathers in plant growth promotion activity by a keratinolytic strain of *Bacillus subtilis* PF1. Bioresour Bioprocess 3:1–10

Bié J, Sepodes B, Fernandes PC, Ribeiro MH (2022) Enzyme immobilization and co-immobilization: main framework, advances and some applications. PRO 10(3):494

Botella C, De Ory I, Webb C, Cantero D, Blandino A (2005) Hydrolytic enzyme production by *Aspergillus awamori* on grape pomace. Biochem Eng J 26(2–3):100–106

Botella C, Diaz A, De Ory I, Webb C, Blandino A (2007) Xylanase and pectinase production by *Aspergillus awamori* on grape pomace in solid state fermentation. Process Biochem 42(1):98–101

Boudrant J, Woodley JM, Fernandez-Lafuente R (2020) Parameters necessary to define an immobilized enzyme preparation. Process Biochem 90:66–80

Buda K, Fekete T, Ontañon OM, Campos E, Fehér C (2024) Xylanase production by *Cellulomonas phragmiteti* using lignocellulosic waste materials. Processes 12(2):258

Buddhiwant PG (2017) Phytase production under solid state fermentation and its application in feed and food, Ph.D thesis, NCIM Resource Centre, Biochemical Sciences Division, CSIR- National Chemical Laboratory, Pune, India. https://dspace.ncl.res.in/xmlui/handle/20.500.12252/4497

Buddhiwant P, Bhavsar K, Kumar VR, Khire JM (2016) Phytase production by solid-state fermentation of groundnut oil cake by *Aspergillus niger*: a bioprocess optimization study for animal feedstock applications. Prep Biochem Biotechnol 46(6):531–538

Cavalitto SF, Areas JA, Hours RA (1996) Pectinase production profile of *Aspergillus foetidus* in solid state cultures at different acidities. Biotechnol Lett 18:251–256

Chandrasekaran M, Bahkali AH (2013) Valorization of date palm (*Phoenix dactylifera*) fruit processing by-products and wastes using bioprocess technology–review. Saudi J Biol Sci 20(2):105–120

Chaudhary N, Dangi P (2022) Fruit and vegetable waste: A taste of future foods. In: Edible food packaging: applications, innovations and sustainability. Springer, pp 115–147

Chen Y, Guo R, Li Y-C, Liu H, Zhan TL (2016) A degradation model for high kitchen waste content municipal solid waste. Waste Manag 58:376–385

Chen Z, Zaky AA, Liu Y, Chen Y, Liu L, Li S, Jia Y (2019) Purification and characterization of a new xylanase with excellent stability from *Aspergillus flavus* and its application in hydrolyzing pretreated corncobs. Protein Expr Purif 154:91–97

Cheng D, Liu Y, Ngo HH, Guo W, Chang SW, Nguyen DD, Zhang S, Luo G, Bui XT (2021) Sustainable enzymatic technologies in waste animal fat and protein management. J Environ Manag 284:112040

Contato AG, Inácio FD, Brugnari T, de Araújo CAV, Maciel GM, Haminiuk CWI, Peralta RM, de Souza CGM (2020) Solid-state fermentation with orange waste: optimization of laccase production from *Pleurotus pulmonarius* CCB-20 and decolorization of synthetic dyes. Acta Sci Biol Sci 42:1–9

da Silva PO, de Alencar Guimarães NC, Serpa JDM, Masui DC, Marchetti CR, Verbisck NV, Zanoelo FF, Ruller R, Giannesi GC (2019) Application of an endo-xylanase from *Aspergillus*

japonicus in the fruit juice clarification and fruit peel waste hydrolysis. Biocatal Agric Biotechnol 21:101312
de Andrade Silva CA, Lacerda MPF, Leite RSR, Fonseca GG (2014) Physiology of *Lichtheimia ramosa* obtained by solid-state bioprocess using fruit wastes as substrate. Bioprocess Biosyst Eng 37:727–734
de Castro RJS, Sato HH (2013) Synergistic effects of agroindustrial wastes on simultaneous production of protease and α-amylase under solid state fermentation using a simplex centroid mixture design. Ind Crop Prod 49:813–821
de Lima JS, Cruz R, Fonseca JC, de Medeiros EV, de Maciel MHC, Moreira KA, de Motta CMS (2014) Production, characterization of tannase from *Penicillium montanense* URM 6286 under SSF using agroindustrial wastes, and application in the clarification of grape juice (Vitis vinifera L.). Sci World J 2014
do Rosario Freixo M, Karmali A, Frazão C, Arteiro JM (2008) Production of laccase and xylanase from *Coriolus versicolor* grown on tomato pomace and their chromatographic behaviour on immobilized metal chelates. Process Biochem 43(11):1265–1274
dos Santos TC, Gomes DPP, Bonomo RCF, Franco M (2012) Optimisation of solid state fermentation of potato peel for the production of cellulolytic enzymes. Food Chem 133(4):1299–1304
Dagbagh S, Goksungur Y (2009) Optimization of beta-galactosidase production in stirred tank bioreactor using *Kluyveromyces lactis* NRRL Y-8279. Food Sci Biotechnol 18(6):1342–1350
De la Torre I, Martin-Dominguez VI, Acedos MG, Esteban J, Santos VE, Ladero MI (2019) Utilisation/upgrading of orange peel waste from a biological biorefinery perspective. Appl Microbiol Biotechnol 103:5975–5991
De León-Rodríguez A, Rivera-Pastrana D, Medina-Rivero E, Flores-Flores JL, Estrada-Baltazar A, Ordóñez-Acevedo LG, de la Rosa APB (2006) Production of penicillin acylase by a recombinant *Escherichia coli* using cheese whey as substrate and inducer. Biomol Eng 23(6):299–305
Dhillon GS, Oberoi HS, Kaur S, Bansal S, Brar SK (2011) Value-addition of agricultural wastes for augmented cellulase and xylanase production through solid-state tray fermentation employing mixed-culture of fungi. Ind Crop Prod 34(1):1160–1167
Dhillon GS, Kaur S, Brar SK (2012) In-vitro decolorization of recalcitrant dyes through an eco-friendly approach using laccase from *Trametes versicolor* grown on brewer's spent grain. Int Biodeterior Biodegradation 72:67–75
Dhiman S, Mukherjee G, Singh AK (2018) Recent trends and advancements in microbial tannase-catalyzed biotransformation of tannins: a review. Int Microbiol 21:175–195
Dhiman S, Mukherjee G, Kumar A, Majumdar RS (2022) Purification, characterization and application study of bacterial tannase for optimization of gallic acid synthesis from fruit waste. J Sci Ind Res 81(10):1029–1036
Dobrev GT, Zhekova BY (2012) Biosynthesis, purification and characterization of endoglucanase from a xylanase producing strain *Aspergillus niger* B03. Braz J Microbiol 43:70–77
Dugmore TIJ, Clark JH, Bustamante J, Houghton JA, Matharu AS (2018) Valorisation of biowastes for the production of green materials using chemical methods. Top Curr Chem (Z) 375. https://doi.org/10.1007/s41061-017-0133-8
Elayaraja S, Velvizhi T, Maharani V, Mayavu P, Vijayalakshmi S, Balasubramanian T (2011) Thermostable α-amylase production by *Bacillus firmus* CAS 7 using potato peel as a substrate. Afr J Biotechnol 10(54):11235–11238
Erdal SERKAN, Taskin MESUT (2010) Production of α-amylase by *Penicillium expansum* MT-1 in solid-state fermentation using waste Loquat (*Eriobotrya japonica* Lindley) kernels as substrate. Rom Biotechnol Lett 15(3):5342–5350
Esakkiraj P, Immanuel G, Sowmya SM, Iyapparaj P, Palavesam A (2009) Evaluation of protease-producing ability of fish gut isolate *Bacillus cereus* for aqua feed. Food Bioprocess Technol 2:383–390
Esparza I, Jiménez-Moreno N, Bimbela F, Ancín-Azpilicueta C, Gandía LM (2020) Fruit and vegetable waste management: conventional and emerging approaches. J Environ Manag 265:110510

Esteban J, Ladero M (2018) Food waste as a source of value-added chemicals and materials: a biorefinery perspective. Int J Food Sci Technol 53(5):1095–1108

Evaristo RBW, Costa AA, Nascimento PGBD, Ghesti GF (2023) Biorefinery development based on brewers' spent grain (BSG) conversion: a forecasting technology study in the Brazilian scenario. Biomass 3(3):217–237

Faison BD, Kirk TK (1985) Factors involved in the regulation of a ligninase activity in *Phanerochaete chrysosporium*. Appl Environ Microbiol 49(2):299–304

Feijoo G, Moreira MT, Roca E, Lema JM (1999) Use of cheese whey as a substrate to produce manganese peroxidase by *Bjerkandera* sp BOS55. J Ind Microbiol Biotechnol 23:86–90

Fickers P, Marty A, Nicaud JM (2011) The lipases from *Yarrowia lipolytica*: genetics, production, regulation, biochemical characterization and biotechnological applications. Biotechnol Adv 29(6):632–644

Fillaudeau L, Blanpain-Avet P, Daufin G (2006) Water, wastewater and waste management in brewing industries. J Clean Prod 14(5):463–471

Fortkamp D, Knob A (2014) High xylanase production by *Trichoderma viride* using pineapple peel as substrate and its application in pulp biobleaching. Afr J Biotechnol 13(22):2248

Gaind S, Singh S (2015) Production, purification and characterization of neutral phytase from thermotolerant *Aspergillus flavus* ITCC 6720. Int Biodeterior Biodegradation 99:15–22

Galanakis CM (2012) Recovery of high added-value components from food wastes: conventional, emerging technologies and commercialized applications. Trends Food Sci Technol 26(2):68–87

Gao W, Lee E-J, Lee S-U, Li J, Chung C-H, Lee J-W (2012) Enhanced carboxymethylcellulase production by a newly isolated marine bacterium, *Cellulophaga lytica* LBH-14, using rice bran. J Microbiol Biotechnol 22(10):1412–1422

García Méndez MG, Morales Martínez TK, Ascacio Valdés JA, Chávez González ML, Flores Gallegos AC, Sepúlveda L (2021) Application of lactic acid bacteria in fermentation processes to obtain tannases using agro-industrial wastes. Fermentation 7:48. https://doi.org/10.3390/fermentation7020048

Gawande PV, Kamat MY (1999) Production of *Aspergillus xylanase* by lignocellulosic waste fermentation and its application. J Appl Microbiol 87(4):511–519

Georganas A, Giamouri E, Pappas AC, Papadomichelakis G, Galliou F, Manios T, Tsiplakou E, Fegeros K, Zervas G (2020) Bioactive compounds in food waste: A review on the transformation of food waste to animal feed. Food Secur 9(3):291

George SP, Ahmad A, Rao MB (2001) Studies on carboxymethyl cellulase produced by an alkalothermophilic actinomycete. Bioresour Technol 77(2):171–175

Ghorbel S, Souissi N, Triki-Ellouz Y, Dufosse L, Guerard F, Nasri M (2005) Preparation and testing of Sardinella protein hydrolysates as nitrogen source for extracellular lipase production by *Rhizopus oryzae*. World J Microbiol Biotechnol 21:33–38

Ghosh PR, Fawcett D, Sharma SB, Poinern GEJ (2016) Progress towards sustainable utilisation and management of food wastes in the global economy. Int J Food Sci 2016. https://doi.org/10.1155/2016/3563478

Gombert AK, Pinto AL, Castilho LR, Freire DM (1999) Lipase production by *Penicillium restrictum* in solid-state fermentation using babassu oil cake as substrate. Process Biochem 35(1–2):85–90

Greco-Duarte J, de Almeida FP, de Godoy MG, Lins U, Freire DMG, Gutarra MLE (2023) Simultaneous lipase production and immobilization: morphology and physiology study of *Penicillium simplicissimum* in submerged and solid-state fermentation with polypropylene as an inert support. Enzym Microb Technol 164:110173

Gupte A, Prajapati D, Bhatt A, Pandya S, Raghunathan M, Gupte S (2023) Agro-industrial residues: an eco-friendly and inexpensive substrate for fungi in the development of white biotechnology. In: Fungi and Fungal Products in Human Welfare and Biotechnology. Springer Nature Singapore, Singapore, pp 571–603

Gutierrez-Correa M, Tengerdy RP (1997) Production of cellulase on sugar cane bagasse by fungal mixed culture solid substrate fermentation. Biotechnol Lett 19:665–667

Guzik P, Kulawik P, Zając M, Migdał W (2022) Microwave applications in the food industry: an overview of recent developments. Crit Rev Food Sci Nutr 62(29):7989–8008

Haba E, Bresco O, Ferrer C, Marques A, Busquets M, Manresa A (2000) Isolation of lipase-secreting bacteria by deploying used frying oil as selective substrate. Enzym Microb Technol 26(1):40–44

Häkkinen M, Valkonen MJ, Westerholm-Parvinen A, Aro N, Arvas M, Vitikainen M, Penttilä M, Saloheimo M, Pakula TM (2014) Screening of candidate regulators for cellulase and hemicellulase production in *Trichoderma reesei* and identification of a factor essential for cellulase production. Biotechnol Biofuels 7:1–21

Hang YD, Woodams EE (1994) Production of fungal polygalacturonase from apple pomace. LWT-Food Sci Technol 27(2):194–196

Hariharan S, Nambisan P (2013) Optimization of lignin peroxidase, manganese peroxidase, and lac production from *Ganoderma lucidum* under solid state fermentation of pineapple leaf. Bioresources 8(1):250–271

Hasan MM, Marzan LW, Hosna A, Hakim A, Azad AK (2017) Optimization of some fermentation conditions for the production of extracellular amylases by using *Chryseobacterium* and *Bacillus* isolates from organic kitchen wastes. J Genet Eng Biotechnol 15(1):59–68

Hasan-Beikdashti M, Forootanfar H, Safiarian MS, Ameri A, Ghahremani MH, Khoshayand MR, Faramarzi MA (2012) Optimization of culture conditions for production of lipase by a newly isolated bacterium *Stenotrophomonas maltophilia*. J Taiwan Inst Chem Eng 43(5):670–677

Hassan SS, Tiwari BK, Williams GA, Jaiswal AK (2020) Bioprocessing of brewers' spent grain for production of xylanopectinolytic enzymes by Mucor sp. Bioresour Technol Rep 9:100371

Hatvani N, Mécs I (2001) Production of laccase and manganese peroxidase by *Lentinus edodes* on malt-containing by-product of the brewing process. Process Biochem 37(5):491–496

Heerd D, Diercks-Horn S, Fernández-Lahore M (2014) Efficient polygalacturonase production from agricultural and agro-industrial residues by solid-state culture of *Aspergillus sojae* under optimized conditions. Springerplus 3(1):1–14

Herculano PN, Porto TS, Moreira KA, Pinto GAS, Souza-Motta CM, Porto ALF (2011) Cellulase production by *Aspergillus japonicus* URM5620 using waste from castor bean (*Ricinus communis* L.) under solid-state fermentation. Appl Biochem Biotechnol 165:1057–1067

Hii K, Yeap S, Mashitah MD (2012) Cellulase production from palm oil mill effluent in Malaysia: economical and technical perspectives. Eng Life Sci 12(1):7–28

Hildebrand DC (1971) Pectate and pectin gels for differentiation of *Pseudomonas* sp. and other bacterial plant pathogens. Phytopathology 61(7):1430

Homann A, Biedendieck R, Götze S, Jahn D, Seibel J (2007) Insights into polymer versus oligosaccharide synthesis: mutagenesis and mechanistic studies of a novel levansucrase from *Bacillus megaterium*. Biochem J 407(2):189–198

Hooi Ling Ho AMC (2016) Overproduction of xylanase from mutants of *Bacillus subtilis* with barley husk as the prime carbon source under submerged fermentation after random mutagenesis using ethyl methane sulfonate (EMS) and acridine orange (AO). Br Microbiol Res J 14(1):1

Idris R, Chong CT, Asik JA, Ani FN (2020) Optimization studies of microwave-induced co-pyrolysis of empty fruit bunches/waste truck tire using response surface methodology. J Clean Prod 244:118649

Illanes A, Valencia P (2017) Industrial and therapeutic enzymes: penicillin acylase. In: Current developments in biotechnology and bioengineering. Elsevier, pp 267–305

Irfansyah AR, Koentjoro MP, Ekawati I, Prasetyo EN (2021) Utilization of *Moringa oleifera* seed waste as substrate in lipase production under different pretreatments. IOP Conf Ser Earth Environm Sci 649(1):012002

Jain R, Naik SN (2018) Adding value to the oil cake as a waste from oil processing industry: production of lipase in solid state fermentation. Biocatal Agric Biotechnol 15:181–184

Janse BJH, Pretorius IS (1995) One-step enzymatic hydrolysis of starch using a recombinant strain of *Saccharomyces cerevisiae* producing α-amylase, glucoamylase and pullulanase. Appl Microbiol Biotechnol 42:878–883

Joo HS, Kumar CG, Park GC, Kim KT, Paik SR, Chang CS (2002) Optimization of the production of an extracellular alkaline protease from *Bacillus horikoshii*. Process Biochem 38(2):155–159

Juturu V, Wu JC (2012) Microbial xylanases: engineering, production and industrial applications. Biotechnol Adv 30(6):1219–1227

Kalaichelvan P (2012) Production and optimization of pectinase from *Bacillus* sp. MFW7 using cassava waste. Asian J Plant Sci Res 2(3):369–375

Kanmani P, Kumaresan K, Aravind J (2015) Utilization of coconut oil mill waste as a substrate for optimized lipase production, oil biodegradation and enzyme purification studies in Staphylococcus pasteuri. Electron J Biotechnol 18(1):20–28

Karim A, Gerliani N, Aïder M (2020) *Kluyveromyces marxianus*: an emerging yeast cell factory for applications in food and biotechnology. Int J Food Microbiol 333:108818

Kaur SJ, Gupta VK (2017) Production of pectinolytic enzymes pectinase and pectin lyase by *Bacillus subtilis* SAV-21 in solid state fermentation. Ann Microbiol 67:333–342

Kazeem MO, Mayaki MA, Musa D (2024) Response surface optimization of lipase production by *Pseudomonas* sp. on a low-cost shea-nut cake using solid state fermentation. IFE J Sci 26(1):1–12

Khuwijitjaru P, Klinchongkon K (2020) Passion fruit. In: Valorization of fruit processing by-products. Elsevier, pp 183–201

Knob A, Beitel SM, Fortkamp D, Terrasan CRF, Almeida AFD (2013) Production, purification, and characterization of a major *Penicillium glabrum* xylanase using brewer' s spent grain as substrate. Biomed Res Int 2013(1):728735

Krishna C (1999) Production of bacterial cellulases by solid state bioprocessing of banana wastes. Bioresour Technol 69(3):231–239

Krishna C, Chandrasekaran M (1996) Banana waste as substrate for α-amylase production by *Bacillus subtilis* (CBTK 106) under solid-state fermentation. Appl Microbiol Biotechnol 46:106–111

Kumar S, Satyanarayana T (2004) Statistical optimization of a thermostable and neutral glucoamylase production by a thermophilic mold *Thermomucor indicae*-seudaticae in solid-state fermentation. World J Microbiol Biotechnol 20:895–902

Kumar D, Ashfaque M, Muthukumar M, Garg N (2012) Production and characterization of carboxymethyl cellulase from *Paenibacillus polymyxa* using mango peel as substrate Author Details. J Environ Biol 33:81–84. www.jeb.co.in

Kumar M, Singh A, Beniwal V, Salar RK (2016) Improved production of tannase by *Klebsiella pneumoniae* using Indian gooseberry leaves under submerged fermentation using Taguchi approach. AMB Express 6:1–11

Kuthiala T, Thakur K, Sharma D, Singh G, Khatri M, Arya SK (2022) The eco-friendly approach of cocktail enzyme in agricultural waste treatment: a comprehensive review. Int J Biol Macromol 209:1956–1974

Laothanachareon T, Bunterngsook B, Champreda V (2022) Profiling multi-enzyme activities of *Aspergillus niger* strains growing on various agro-industrial residues. 3 Biotech 12(1):17

Lawrence RC, Fryer TF, Reiter B (1967) The production and characterization of lipases from a *Micrococcus* and a pseudomonad. Microbiology 48(3):401–418

Lebersorger S, Schneider F (2011) Discussion on the methodology for determining food waste in household waste composition studies. Waste Manag 31(9–10):1924–1933

Leite P, Silva C, Salgado JM, Belo I (2019) Simultaneous production of lignocellulolytic enzymes and extraction of antioxidant compounds by solid-state fermentation of agro-industrial wastes. Ind Crop Prod 137:315–322

Lekshmi R, Nisha SA, Vasan PT, Kaleeswaran B (2021) A comprehensive review on tannase: microbes associated production of tannase exploiting tannin rich agro-industrial wastes with special reference to its potential environmental and industrial applications. Environ Res 201:111625

Li D, Hou L, Hu M, Gao Y, Tian Z, Fan B, Li S, Wang F (2022) Recent advances in nattokinase-enriched fermented soybean foods: a review. Food Secur 11(13):1867

Liang T-W, Lin J-J, Yen Y-H, Wang C-L, Wang S-L (2006) Purification and characterization of a protease extracellularly produced by *Monascus purpureus* CCRC31499 in a shrimp and crab shell powder medium. Enzym Microb Technol 38(1–2):74–80
Luithui Y, Baghya Nisha R, Meera MS (2019) Cereal by-products as an important functional ingredient: effect of processing. J Food Sci Technol 56:1–11
Lynd LR, Weimer PJ, Van Zyl WH, Pretorius IS (2002) Microbial cellulose utilization: fundamentals and biotechnology. Microbiol Mol Biol Rev 66(3):506–577
Maftukhah S (2019) Cellulase enzyme production using solid state fermentation method from waste–a review. Jurnal Keilmuan Dan Aplikasi Teknik 6(2):22
Magdouli S, Guedri T, Tarek R, Brar SK, Blais JF (2017) Valorization of raw glycerol and crustacean waste into value added products by Yarrowia lipolytica. Bioresour Technol 243:57–68
Mahmood AU, Greenman J, Scragg AH (1998) Orange and potato peel extracts: analysis and use as *Bacillus* substrates for the production of extracellular enzymes in continuous culture. Enzym Microb Technol 22(2):130–137
Mamma D, Kourtoglou E, Christakopoulos P (2008) Fungal multienzyme production on industrial by-products of the citrus-processing industry. Bioresour Technol 99(7):2373–2383
Mandels M, Weber J, Parizek R (1971) Enhanced cellulase production by a mutant of *Trichoderma viride*. Appl Microbiol 21(1):152–154
Mandviwala TN, Khire JM (2000) Production of high activity thermostable phytase from thermotolerant *Aspergillus niger* in solid state fermentation. J Ind Microbiol Biotechnol 24:237–243
Manhongo TT, Chimphango AFA, Thornley P, Röder M (2022) Current status and opportunities for fruit processing waste biorefineries. Renew Sust Energ Rev 155:111823
Mathias TRDS, de Fernandes Aguiar P, de Almeida e Silva JB, de Moretzsohn Mello PP, Camporese Sérvulo EF (2017) Brewery waste reuse for protease production by lactic acid fermentation. Food Technol Biotechnol 55(2):218–224
Mazhar H, Abbas N, Zamir T, Hussain Z, Ali SS (2018) Optimization study of lipolytic enzyme from *Bacillus cereus*, PCSIR NL-37. Punjab Univ J Zool 33(2):217–224
Meena P, Tripathi AD, Srivastava SK, Jha A (2013) Utilization of agro-industrial waste (wheat bran) for alkaline protease production by *Pseudomonas aeruginosa* in SSF using Taguchi (DOE) methodology. Biocatal Agric Biotechnol 2(3):210–216
Melikoglu M, Lin CSK, Webb C (2013) Stepwise optimisation of enzyme production in solid state fermentation of waste bread pieces. Food Bioprod Process 91(4):638–646
Melikoglu M, Lin CSK, Webb C (2015) Solid state fermentation of waste bread pieces by *Aspergillus awamori*: Analysing the effects of airflow rate on enzyme production in packed bed bioreactors. Food Bioprod Process 95:63–75
Milagres AMF, Santos E, Piovan T, Roberto IC (2004) Production of xylanase by *Thermoascus aurantiacus* from sugar cane bagasse in an aerated growth fermentor. Process Biochem 39(11):1387–1391
Mishra A (2006) Production of L-asparaginase, an anticancer agent, from *Aspergillus niger* using agricultural waste in solid state fermentation. Appl Biochem Biotechnol 135:33–42
Moran-Aguilar MG, Costa-Trigo I, Calderón-Santoyo M, Domínguez JM, Aguilar-Uscanga MG (2021) Production of cellulases and xylanases in solid-state fermentation by different strains of *Aspergillus niger* using sugarcane bagasse and brewery spent grain. Biochem Eng J 172:108060
Mukherjee AK, Adhikari H, Rai SK (2008) Production of alkaline protease by a thermophilic *Bacillus subtilis* under solid-state fermentation (SSF) condition using *Imperata cylindrica* grass and potato peel as low-cost medium: characterization and application of enzyme in detergent formulation. Biochem Eng J 39(2):353–361
Mulimani VH, Patil GN, R. (2000) α-Amylase production by solid state fermentation: a new practical approach to biotechnology courses. Biochem Educ 28(3):161–163
Munoz C, Guillen F, Martinez AT, Martinez MJ (1997) Induction and characterization of laccase in the ligninolytic fungus *Pleurotus eryngii*. Curr Microbiol 34:1–5
Nadar SS, Rao P, Rathod VK (2018) Enzyme assisted extraction of biomolecules as an approach to novel extraction technology: A review. Food Res Int 108:309–330

Nandini S, Nandini KE, Krishna SS (2014) Food and Agriculture Residue (FAR): A potential substrate for tannase and gallic acid production using competent microbes. J Bioprocess Biotech 5:193

Nath A, Mondal S, Chakraborty S, Bhattacharjee C, Chowdhury R (2014) Production, purification, characterization, immobilization, and application of β-galactosidase: a review. Asia Pac J Chem Eng 9(3):330–348

Nathan VK, Esther Rani M, Rathinasamy G, Dhiraviam KN, Jayavel S (2014) Process optimization and production kinetics for cellulase production by *Trichoderma viride* VKF3. Springerplus 3:1–12

Negi S, Gupta S, Banerjee R (2011) Extraction and purification of glucoamylase and protease produced by *Aspergillus awamori* in a single-stage fermentation. Food Technol Biotechnol 49(3):310–315

Nighojkar A, Patidar MK, Nighojkar S (2019) Pectinases: production and applications for fruit juice beverages. In: Processing and sustainability of beverages. Woodhead Publishing, pp 235–273

Nthangeni MB, Patterton H-G, van Tonder A, Vergeer WP, Litthauer D (2001) Over-expression and properties of a purified recombinant *Bacillus licheniformis* lipase: a comparative report on Bacillus lipases. Enzym Microb Technol 28(7–8):705–712

O'Connor J, Hoang SA, Bradney L, Dutta S, Xiong X, Tsang DCW, Ramadass K, Vinu A, Kirkham MB, Bolan NS (2021) A review on the valorisation of food waste as a nutrient source and soil amendment. Environ Pollut 272:115985

Oberoi HS, Bansal S, Dhillon GS (2008) Enhanced β-galactosidase production by supplementing whey with cauliflower waste. Int J Food Sci Technol 43(8):1499–1504

Oberoi HS, Chavan Y, Bansal S, Dhillon GS (2010) Production of cellulases through solid state fermentation using kinnow pulp as a major substrate. Food Bioprocess Technol 3:528–536

Olajuyigbe FM, Fatokun CO, Oyelere OM (2018) Biodelignification of some agro-residues by *Stenotrophomonas* sp. CFB-09 and enhanced production of ligninolytic enzymes. Biocatal Agric Biotechnol 15:120–130

Oliveira AC, Amorim GM, Azevêdo JAG, Godoy MG, Freire DM (2018) Solid-state fermentation of co-products from palm oil processing: production of lipase and xylanase and effects on chemical composition. Biocatal Biotransformation 36(5):381–388

Ong KL, Kaur G, Pensupa N, Uisan K, Lin CSK (2018) Trends in food waste valorization for the production of chemicals, materials and fuels: case study South and Southeast Asia. Bioresour Technol 248:100–112

Özbek ZA, Ergönül PG (2022) Utilisation of oilseed meals in food industry. In: Waste to food: returning nutrients to the food chain. Academic Publishers, Wageningen, pp 79–84

Padma PN, Anuradha K, Nagaraju B, Kumar VS, Reddy G (2012) Use of pectin rich fruit wastes for polygalacturonase production by *Aspergillus awamori* MTCC 9166 in solid state fermentation. J Bioprocess Biotech 2(2):1000116

Palaniveloo K, Amran MA, Norhashim NA, Mohamad-Fauzi N, Peng-Hui F, Hui-Wen L, Kai-Lin Y, Jiale L, Chian-Yee MG, Jing-Yi L (2020) Food waste composting and microbial community structure profiling. Processes 8(6):723

Papageorgiou M, Skendi A (2018) Introduction to cereal processing and by-products. In: Sustainable recovery and reutilization of cereal processing by-products. Woodhead Publishing, pp 1–25

Patil NP, Chaudhari BL (2010) Production and purification of pectinase by soil isolate *Penicillium* sp. and search for better agro-residue for its SSF. Recent Res Technol 2(7):36–42

Patil PD, Patil SP, Kelkar RK, Patil NP, Pise PV, Nadar SS (2021) Enzyme-assisted supercritical fluid extraction: an integral approach to extract bioactive compounds. Trends Food Sci Technol 116:357–369

Patil PJ, Usman M, Mehmood A, Ahmad S, Haider J, Zhang C, Teng C, Li X (2022) Cereal grain tea beverages and their potential health properties. In: Functional cereals and cereal foods: properties, functionality and applications. Springer, pp 289–333

Patra F, Duary RK (2020) Waste from dairy processing industries and its sustainable utilization. In: Sustainable food waste management: concepts and innovations, pp 127–154
Peer RL, Thorneloe SA, Epperson DL (1993) A comparison of methods for estimating global methane emissions from landfills. Chemosphere 26(1–4):387–400
Picart P, Diaz P, Pastor FIJ (2007) Cellulases from two *Penicillium* sp. strains isolated from subtropical forest soil: production and characterization. Lett Appl Microbiol 45(1):108–113
Poondla V, Bandikari R, Subramanyam R, Obulam VSR (2015) Low temperature active pectinases production by *Saccharomyces cerevisiae* isolate and their characterization. Biocatal Agric Biotechnol 4(1):70–76
Prajapati J, Dudhagara P, Patel K (2021) Production of thermal and acid-stable pectinase from *Bacillus subtilis* strain BK-3: optimization, characterization, and application for fruit juice clarification. Biocatal Agric Biotechnol 35:102063
Prakasham RS, Rao CS, Sarma PN (2006) Green gram husk – an inexpensive substrate for alkaline protease production by *Bacillus* sp. in solid-state fermentation. Bioresour Technol 97(13):1449–1454
Priya, Gogate PR (2021) Ultrasound-assisted intensification of activity of free and immobilized enzymes: a review. Ind Eng Chem Res 60(27):9650–9668
Prueckler M, Siebenhandl-Ehn S, Apprich S, Hoeltinger S, Haas C, Schmid E, Kneifel W (2014) Wheat bran-based biorefinery 1: composition of wheat bran and strategies of functionalization. LWT-Food Sci Technol 56(2):211–221
Ramachandran S, Patel AK, Nampoothiri KM, Chandran S, Szakacs G, Soccol CR, Pandey A (2004) Alpha amylase from a fungal culture grown on oil cakes and its properties. Braz Arch Biol Technol 47:309–317
Ramachandran S, Singh SK, Larroche C, Soccol CR, Pandey A (2007) Oil cakes and their biotechnological applications–a review. Bioresour Technol 98(10):2000–2009
Ramesh A, Harani Devi P, Chattopadhyay S, Kavitha M (2020) Commercial applications of microbial enzymes. In: Microbial enzymes: roles and applications in industries, pp 137–184
Rani R, Ghosh S (2011) Production of phytase under solid-state fermentation using *Rhizopus oryzae*: novel strain improvement approach and studies on purification and characterization. Bioresour Technol 102(22):10641–10649
Rao MB, Tanksale AM, Ghatge MS, Deshpande VV (1998) Molecular and biotechnological aspects of microbial proteases. Microbiol Mol Biol Rev 62(3):597–635
Ravindran R, Jaiswal AK (2016) Microbial enzyme production using lignocellulosic food industry wastes as feedstock: a review. Bioengineering 3(4):30
Ravindran R, Hassan SS, Williams GA, Jaiswal AK (2018) A review on bioconversion of agro-industrial wastes to industrially important enzymes. Bioengineering 5(4):93
Rebhun M, Hadar Y (2005) Use of agro-industrial waste for production of laccase and manganese peroxidase from white-rot basidiomycetes. Int J Med Mushrooms 7:459
Reddy GV, Babu PR, Komaraiah P, Roy K, Kothari IL (2003) Utilization of banana waste for the production of lignolytic and cellulolytic enzymes by solid substrate fermentation using two *Pleurotus* species (*P. ostreatus* and *P. sajor-caju*). Process Biochem 38(10):1457–1462
Risdianto H, Sofianti E, Suhardi SH, Setiadi T (2012) Optimisation of laccase production using white rot fungi and agriculture wastes in solid state fermentation. ITB J Eng Sci 44(2):93–105
Rodrigues RC, Berenguer-Murcia Á, Carballares D, Morellon-Sterling R, Fernandez-Lafuente R (2021) Stabilization of enzymes via immobilization: multipoint covalent attachment and other stabilization strategies. Biotechnol Adv 52:107821
Sabater C, Ruiz L, Delgado S, Ruas-Madiedo P, Margolles A (2020) Valorization of vegetable food waste and by-products through fermentation processes. Front Microbiol 11:581997
Sahoo A, Mahanty B, Daverey A, Dutta K (2020) Nattokinase production from *Bacillus subtilis* using cheese whey: effect of nitrogen supplementation and dynamic modelling. J Water Process Eng 38:101533

Salemdeeb R, Zu Ermgassen EKHJ, Kim MH, Balmford A, Al-Tabbaa A (2017) Environmental and health impacts of using food waste as animal feed: a comparative analysis of food waste management options. J Clean Prod 140:871–880

Salimi E, Saragas K, Taheri ME et al (2019) The role of enzyme loading on starch and cellulose hydrolysis of food waste. Waste Biomass Valoriz 10:3753–3762. https://doi.org/10.1007/s12649-019-00826-3

Sambyal K, Singh RV (2021) Exploitation of *E. coli* for the production of penicillin G amidase: a tool for the synthesis of semisynthetic β-lactam antibiotics. J Genet Eng Biotechnol 19(1):156

Sandhya C, Sumantha A, Szakacs G, Pandey A (2005) Comparative evaluation of neutral protease production by *Aspergillus oryzae* in submerged and solid-state fermentation. Process Biochem 40(8):2689–2694

Sandhya R, Kurup G (2013) Screening and isolation of pectinase from fruit and vegetable wastes and the use of orange waste as a substrate for pectinase production. Int Res J Biol Sci 2(9):34–39

Sangkharak K, Khaithongkaeo P, Chuaikhunupakarn T, Choonut A, Prasertsan P (2021) The production of polyhydroxyalkanoate from waste cooking oil and its application in biofuel production. Biomass Convers Biorefinery 11:1651–1664

Sar T, Harirchi S, Ramezani M, Bulkan G, Akbas MY, Pandey A, Taherzadeh MJ (2022) Potential utilization of dairy industries by-products and wastes through microbial processes: A critical review. Sci Total Environ 810:152253

Sarangi PK, Singh AK, Sonkar S, Shadangi KP, Srivastava RK, Gupta VK, Parikh J, Sahoo UK, Govarthanan M (2023) Biorefinery solutions for food processing wastes: A sustainable bioeconomic perspective. Ind Crop Prod 205:117488

Saravanan P, Muthuvelayudham R, Viruthagiri T (2012) Application of statistical design for the production of cellulase by *Trichoderma reesei* using mango peel. Enzyme Res 2012(1):157643

Sarkar N, Chakraborty D, Dutta R, Agrahari P, Bharathi SD, Singh AA, Jacob S (2021) A comprehensive review on oilseed cakes and their potential as a feedstock for integrated biorefinery. J Adv Biotechnol Exp Ther 4(3):376

Sawaya WN (2017) Impact of food losses and waste on food security. In: Water, Energy & Food Sustainability in the Middle East: the Sustainability Triangle, Springe Cham pp 361–388

Schalchli H, Hormazábal E, Rubilar O, Briceño G, Mutis A, Zocolo GJ, Diez MC (2017) Production of ligninolytic enzymes and some diffusible antifungal compounds by white-rot fungi using potato solid wastes as the sole nutrient source. J Appl Microbiol 123(4):886–895

Selvam K, Selvankumar T, Rajiniganth R, Srinivasan P, Sudhakar C, Senthilkumar B, Govarthanan M (2016) Enhanced production of amylase from *Bacillus* sp. using groundnut shell and cassava waste as a substrate under process optimization: waste to wealth approach. Biocatal Agric Biotechnol 7:250–256

Seyis I, Aksoz N (2005) Effect of carbon and nitrogen sources on xylanase production by *Trichoderma harzianum* 1073 D3. Int Biodeterior Biodegradation 55(2):115–119

Sharma R, Oberoi HS, Dhillon GS (2016) Fruit and vegetable processing waste: renewable feed stocks for enzyme production. In: Agro-industrial wastes as feedstock for enzyme production. Elsevier, pp 23–59

Sharma P, Gaur VK, Kim S-H, Pandey A (2020) Microbial strategies for bio-transforming food waste into resources. Bioresour Technol 299:122580

Sharma V, Tsai ML, Nargotra P, Chen CW, Kuo CH, Sun PP, Dong CD (2022) Agro-industrial food waste as a low-cost substrate for sustainable production of industrial enzymes: a critical review. Catalysts 12(11):1373

Sharma V, Singh A, Grenier M, Singh V, Thakur M (2024) Waste valorization in food industries: a review of sustainable approaches. In: Thakur M (ed) Sustainable food systems (Volume II), World Sustainability Series. Springer, Cham. https://doi.org/10.1007/978-3-031-46046-3_9

Shivakumar S (2012) Production and characterization of an acid protease from a local *Aspergillus* sp. by solid substrate fermentation. Arch Appl Sci Res 4(1):188–199

Shrestha L, Kulig B, Moscetti R, Massantini R, Pawelzik E, Hensel O, Sturm B (2020) Optimisation of physical and chemical treatments to control browning development and enzymatic activity on fresh-cut apple slices. Food Secur 9(1):76

Shukla J, Kar R (2006) Potato peel as a solid state substrate for thermostable α-amylase production by thermophilic *Bacillus* isolates. World J Microbiol Biotechnol 22:417–422

Sinjaroonsak S, Chaiyaso T, H-Kittikun, A. (2020) Optimization of cellulase and xylanase productions by *Streptomyces thermocoprophilus* TC13W using low cost pretreated oil palm empty fruit bunch. Waste Biomass Valoriz 11:3925–3936

Sivaramakrishnan S, Gangadharan D (2009) Edible oil cakes. In: Biotechnology for Agro-industrial Residues Utilisation: Utilisation of Agro-residues. Springer, pp 253–271

Soccol CR, de Souza Vandenberghe LP, Spier MR, Karp SG, Weingartner V, Montibeller DNXS, Goelzer F, Batista MCT, Soccol VT (2014) Enzymes production during value-addition of agro-industrial wastes. In: Enzymes in Value-addition of Wastes, Nova Science, p 1

Sodhi HK, Sharma K, Gupta JK, Soni SK (2005) Production of a thermostable α-amylase from *Bacillus* sp. PS-7 by solid state fermentation and its synergistic use in the hydrolysis of malt starch for alcohol production. Process Biochem 40(2):525–534

Sohail M, Siddiqi R, Ahmad A, Khan SA (2009) Cellulase production from *Aspergillus niger* MS82: effect of temperature and pH. New Biotechnol 25(6):437–441

Sondhi S, Kaur R, Madan J (2021) Purification and characterization of a novel white highly thermostable laccase from a novel *Bacillus* sp. MSK-01 having potential to be used as anticancer agent. Int J Biol Macromol 170:232–238

Soni SK, Kaur A, Gupta JK (2003) A solid-state fermentation based bacterial α-amylase and fungal glucoamylase system and its suitability for the hydrolysis of wheat starch. Process Biochem 39(2):185–192

Souissi N, Ellouz-Triki Y, Bougatef A, Blibech M, Nasri M (2008) Preparation and use of media for protease-producing bacterial strains based on by-products from Cuttlefish (*Sepia officinalis*) and wastewaters from marine-products processing factories. Microbiol Res 163(4):473–480

Sumantha A, Sandhya C, Szakacs G, Soccol CR, Pandey A (2005) Production and partial purification of a neutral metalloprotease by fungal mixed substrate fermentation. Food Technol Biotechnol 43(4):313–319

Suresh S, Radha KV (2016) Statistical optimization and mutagenesis for high level of phytase production by *Rhizopus oligosporus* MTCC 556 under solid state fermentation. J Environ Biol 37(2):253

Sutaoney P, Akhand A, Meshram M, Sinha S, Joshi V, Shahadat M (2023) Tannase production using green biotechnology and its applications: a review. Biochem Eng J 202:109163

Sztajer H, Lünsdorf H, Erdmann H, Menge U, Schmid R (1992) Purification and properties of lipase from *Penicillium simplicissimum*. Biochim Biophys Acta (BBA)-Lipids Lipid Metab 1124(3):253–261

Tai WY, Tan JS, Lim V, Lee CK (2019) Comprehensive studies on optimization of cellulase and xylanase production by a local indigenous fungus strain via solid state fermentation using oil palm frond as substrate. Biotechnol Prog 35(3):e2781

Tang B, Pan H, Tang W, Zhang Q, Ding L, Zhang F (2012) Fermentation and purification of cellulase from a novel strain *Rhizopus stolonifer* var. reflexus TP-02. Biomass Bioenergy 36:366–372

Tanruean K, Penkhrue W, Kumla J, Suwannarach N, Lumyong S (2021) Valorization of lignocellulosic wastes to produce phytase and cellulolytic enzymes from a thermophilic fungus, *Thermoascus aurantiacus* sl16w, under semi-solid state fermentation. J Fungi 7(4):286

Thangaraju S, Pulivarthi MK, Natarajan V (2020) Waste from oil-seed industry: a sustainable approach. In: Sustainable Food Waste Management: Concepts and Innovations, Springer Singapore, pp 177–190

Thi NBD, Kumar G, Lin C-Y (2015) An overview of food waste management in developing countries: current status and future perspective. J Environ Manag 157:220–229

Triki-Ellouz Y, Ghorbel B, Souissi N, Kammoun S, Nasri M (2003) Biosynthesis of protease by *Pseudomonas aeruginosa* MN7 grown on fish substrate. World J Microbiol Biotechnol 19:41–45

Uma C, Gomathi D, Muthulakshmi C, Gopalakrishnan VK (2010) Production, purification and characterization of invertase by *Aspergillus flavus* using fruit peel waste as substrate. Adv Biol Res 4(1):31–36

Umsza-Guez MA, Díaz AB, Ory ID, Blandino A, Gomes E, Caro I (2011) Xylanase production by *Aspergillus awamori* under solid state fermentation conditions on tomato pomace. Braz J Microbiol 42:1585–1597

Unban K, Kodchasee P, Shetty K, Khanongnuch C (2020) Tannin-tolerant and extracellular tannase producing Bacillus isolated from traditional fermented tea leaves and their probiotic functional properties. Food Secur 9(4):490

Uyar F, Baysal Z (2004) Production and optimization of process parameters for alkaline protease production by a newly isolated *Bacillus* sp. under solid state fermentation. Process Biochem 39(12):1893–1898

Uzuner S, Cekmecelioglu D (2015) Enhanced pectinase production by optimizing fermentation conditions of *Bacillus subtilis* growing on hazelnut shell hydrolyzate. J Mol Catal B Enzym 113:62–67

Vachher M, Sen A, Kapila R, Nigam A (2021) Microbial therapeutic enzymes: a promising area of biopharmaceuticals. Curr Res Biotechnol 3:195–208

Välimaa A-L, Mäkinen S, Mattila P, Marnila P, Pihlanto A, Mäki M, Hiidenhovi J (2019) Fish and fish side streams are valuable sources of high-value components. Food Qual Saf 3(4):209–226

Vasiljevic T, Jelen P (2001) Production of β-galactosidase for lactose hydrolysis in milk and dairy products using thermophilic lactic acid bacteria. Innovative Food Sci Emerg Technol 2(2):75–85

Veerabhadrappa MB, Shivakumar SB, Devappa S (2014) Solid-state fermentation of Jatropha seed cake for optimization of lipase, protease and detoxification of anti-nutrients in Jatropha seed cake using *Aspergillus versicolor* CJS-98. J Biosci Bioeng 117(2):208–214

Venugopal V (2021) Valorization of seafood processing discards: bioconversion and bio-refinery approaches. Front Sustain Food Syst 5:611835

Vera Zambrano M, Dutta B, Mercer DG, MacLean HL, Touchie MF (2019) Assessment of moisture content measurement methods of dried food products in small-scale operations in developing countries: A review. Trends Food Sci Technol 88:484–496. https://doi.org/10.1016/j.tifs.2019.04.006

Verni M, Rizzello CG, Coda R (2019) Fermentation biotechnology applied to cereal industry by-products: nutritional and functional insights. Front Nutr 6:42

Virupakshi S, Babu KG, Gaikwad SR, Naik GR (2005) Production of a xylanolytic enzyme by a thermoalkaliphilic *Bacillus* sp. JB-99 in solid state fermentation. Process Biochem 40(1):431–435

Villas-Bôas SG, Esposito E, de Mendonça MM (2003) Bioconversion of apple pomace into a nutritionally enriched substrate by Candida utilis and Pleurotus ostreatus. World J Microbiol Biotechnol 19:461–467

Vite-Vallejo O, Palomares LA, Dantán-González E, Ayala-Castro HG, Martínez-Anaya C, Valderrama B, Folch-Mallol J (2009) The role of N-glycosylation on the enzymatic activity of a *Pycnoporus sanguineus* laccase. Enzym Microb Technol 45(3):233–239

Volkov PV, Rozhkova AM, Zorov IN, Sinitsyn AP (2020) Cloning, purification and study of recombinant GH3 family β-glucosidase from *Penicillium verruculosum*. Biochimie 168:231–240

Wadhwa M, Bakshi MPS, Makkar HPS (2016) Wastes to worth: value added products from fruit and vegetable wastes. CABI Rev 2015:1–25

Waghmare PR, Kshirsagar SD, Saratale RG, Govindwar SP, Saratale GD (2014) Production and characterization of cellulolytic enzymes by isolated *Klebsiella* sp. PRW-1 using agricultural waste biomass. Emirates J Food Agric 26:44–59

Wang SL, Yeh PY (2006) Production of a surfactant-and solvent-stable alkaliphilic protease by bioconversion of shrimp shell wastes fermented by *Bacillus subtilis* TKU007. Process Biochem 41(7):1545–1552

Wang Q, Wang X, Wang X, Ma H (2008) Glucoamylase production from food waste by *Aspergillus niger* under submerged fermentation. Process Biochem 43(3):280–286

Wang S-L, Liang T-W, Yen Y-H (2011) Bioconversion of chitin-containing wastes for the production of enzymes and bioactive materials. Carbohydr Polym 84(2):732–742

Wang F, Hu J-H, Guo C, Liu C-Z (2014) Enhanced laccase production by *Trametes versicolor* using corn steep liquor as both nitrogen source and inducer. Bioresour Technol 166:602–605

Watanabe T, Shinozaki Y, Suzuki K, Koitabashi M, Yoshida S, Sameshima-Yamashita Y, Kitamoto HK (2014) Production of a biodegradable plastic-degrading enzyme from cheese whey by the phyllosphere yeast *Pseudozyma antarctica* GB-4 (1) W. J Biosci Bioeng 118(2):183–187

Yadav DN, Tushir S, Guru PN, Yadav DK, Vishwakarma RK (2021) Technological advancements in processing of legumes and pulses. In: Advances in cereals processing technologies. CRC Press, pp 79–107

Zaier H, Maktouf S, Roussos S, Rhouma A (2021) Filamentous fungi isolated from Tunisian olive mill wastes: use of solid-state fermentation for enzyme production. Notulae Botanicae Horti Agrobotanica Cluj-Napoca 49(1):12125

Zambare V (2010) Solid state fermentation of *Aspergillus oryzae* for glucoamylase production on agro residues. http://ijolsci.org/content/2010/4/16-25.pdf

Zeko-Pivač A, Tišma M, Žnidaršič-Plazl P, Kulisic B, Sakellaris G, Hao J, Planinić M (2022) The potential of brewer's spent grain in the circular bioeconomy: state of the art and future perspectives. Front Bioeng Biotechnol 10:870744

Zhang F, Wang T, Wang X, Lü X (2021) Apple pomace as a potential valuable resource for full-components utilization: a review. J Clean Prod 329:129676

Zhu Y, Luan Y, Zhao Y, Liu J, Duan Z, Ruan R (2023) Current technologies and uses for fruit and vegetable wastes in a sustainable system: a review. Food Secur 12(10):1949

Zilly A, dos Santos Bazanella GC, Helm CV, Araújo CAV, de Souza CGM, Bracht A, Peralta RM (2012) Solid-state bioconversion of passion fruit waste by white-rot fungi for production of oxidative and hydrolytic enzymes. Food Bioprocess Technol 5:1573–1580

Zuin VG, Segatto ML, Zanotti K (2020) Towards a green and sustainable fruit waste valorisation model in Brazil: optimisation of homogenizer-assisted extraction of bioactive compounds from mango waste using a response surface methodology. Pure Appl Chem 92(4):617–629

Chapter 11
Pyomelanin Biosynthesis by *Pseudomonas*: A Sustainable Resource for Cosmetic Formulations

Harleen Bahri, Krishna Kumar Rai, Shiv Kumar Dwivedi, and Jitendra Mishra

Abstract The cosmetic industry's growing shift toward sustainable alternatives to synthetic dyes has spotlighted microbial pigments, with pyomelanin emerging as a highly promising multifunctional candidate. Synthesized through the oxidative polymerization of homogentisic acid (HGA) by *Pseudomonas* and related species, pyomelanin is a nitrogen-free melanin that exhibits unique physicochemical properties, including water solubility, broad-spectrum ultra-violet (UV) absorbance, and robust antioxidant activity. This chapter elucidates the biosynthetic pathway of pyomelanin via HGA accumulation, highlighting key genetic regulations—such as *hmgA* knockout and *hppD* overexpression—and environmental factors like light exposure and oxidative stress that enhance production. Industrial-scale fermentation approaches, utilizing engineered microbial strains to achieve high yields, are also discussed. Owing to its dual function as a photostable pigment and redox-active metal chelator, pyomelanin offers solutions to formulation challenges in sunscreens and antioxidant-rich cosmetics, outperforming hydrophobic counterparts such as violacein and carotenoids. Furthermore, regulatory considerations and CRISPR-mediated pathway optimization strategies emphasized its potential for sustainable, high-performance cosmetic applications. By integrating microbial biosynthesis with precision engineering, pyomelanin represents a paradigm shift toward eco-friendly, multifunctional colorants in the modern skincare industry.

Keywords Pyomelanin · Melanin · *Pseudomonas* · Cosmetics · Homogentisic acid

---

H. Bahri · K. K. Rai · J. Mishra (✉)
Amity School of Biological Sciences, Amity University Punjab, Mohali, Punjab, India
e-mail: jmishra@pb.amity.edu; mishrajitendra57@gmail.com

S. K. Dwivedi
Department of Environmental Science, School of Earth and Environmental Sciences, Babasaheb Bhimrao Ambedkar University, Lucknow, Uttar Pradesh, India

N. K. Arora et al. (eds.), *Environmental Sustainability and Biotechnology: Opportunities and Challenges*, Microorganisms for Sustainability 46,
https://doi.org/10.1007/978-981-96-9025-1_11

11.1 Introduction

Cosmetic and skincare industries have traditionally relied on synthetic colorants to add vibrant hues to different skincare products. These synthetic dyes, often derived from petroleum-based precursors, offer color diversity but lead to growing concerns due to their environmental persistence, toxicological risk, and nonrenewable origin (Nambela et al. 2020). The production and disposal of synthetic colorants contribute significantly to industrial effluent loads, particularly through heavy metal contamination and azo dye residues (Slama et al. 2021). According to Virtue Market Research,[1] the global natural pigments market was estimated at USD 4.9 billion in 2022 and is projected to reach USD 7.41 billion by 2030, growing at a CAGR of 5.3% during the forecast period. This trend is increasing due to various regulatory restrictions, such as REACH[2] (Registration, Evaluation, Authorisation and Restriction of Chemicals) in the EU, which now mandates the removal of several synthetic pigments from consumer-use categories.

Therefore, microbial pigments have emerged as one of the most promising alternatives (Ramesh et al. 2019). Unlike plant-based dyes, which depend on seasonality, land usage, and crop variation, microbes offer fast growth rates, batch reproducibility, and independent scalability (Abdi et al. 2024). Additionally, many microbial pigments exhibit antioxidants, antimicrobial, anti-inflammatory, or ultraviolet (UV)-absorbing properties, adding therapeutic value to cosmetic formulations (Venil et al. 2013).

Among microbial pigments, pyomelanin has gained notable attention in the cosmetic industry (Mavridi-Printezi et al. 2020). This reddish-brown, water-soluble melanin pigment is a product of tyrosine catabolism and is produced by a variety of Gram-negative bacteria, particularly *Pseudomonas aeruginosa*, *Burkholderia cenocepacia*, and other related species (Lorquin et al. 2022; Rodríguez-Rojas et al. 2009). In 1994, researchers discovered that the enzyme 4-hydroxyphenylpyruvate dioxygenase produced this pigment from homogentisate in *Streptomyces avermitilis*. This pigment was structurally like one identified decade earlier in *P. aeruginosa*, which had already been named "pyomelanin" in 1972.

Pyomelanin is known for its high antioxidant activity and photoprotective behavior—which makes it suitable for skincare formulations (El-Naggar and Saber 2022). In microbial physiology, pyomelanin plays an important role by neutralizing reactive oxygen species (ROS), enabling biofilm formation, and increasing bacterial resilience under environmental stress (Keith et al. 2007). These bacterial straits contribute toward its robustness and hold strong potential toward developing high-performance skincare. This chapter contributes to the understanding of pyomelanin, its biosynthesis, physicochemical properties, production methods, and cosmetic

[1] https://virtuemarketresearch.com/report/natural-pigments-market

[2] https://environment.ec.europa.eu/topics/chemicals/reach-regulation_en

applications. It aims to present pyomelanin as a next-generation, sustainable, and multifunctional colorant for the modern skincare industry.

11.2 Physicochemical Properties of Pyomelanin

Pyomelanin is a negatively charged, reddish-brown pigment produced extracellularly by various bacteria, particularly *Pseudomonas* species (Ketelboeter and Bardy 2017). Pyomelanin is classified within the diverse group of allomelanins, a category of melanins synthesized across all biological kingdoms from precursors such as 1,8-dihydroxynaphthalene (1,8-DHN), 1,3,6,8-tetrahydroxynaphthalene (1,3,6,8-THN), catechols, γ-glutaminyl-4-hydroxybenzene (GHB), and homogentisic acid (HGA) (Lorquin et al. 2022). Pyomelanin, a nitrogen-free melanin variant, is synthesized through the oxidative polymerization of pyrrole-2,3,5-tricarboxylic acid, forming repeating 2-acetyl-1,4-benzoquinone subunits with a molecular mass of 10–14 kDa—smaller than most natural pigments (Qin and Xia 2024). During polymerization, the $-CH_{2COOH}$ group remains intact, while oxidation at the phenol's ortho position generates quinone structures (Turick et al. 2002). These quinones confer unique electrochemical properties, including robust electron transfer efficiency and reversible redox activity, making pyomelanin as valuable biomaterial.

Unlike most melanins that are hydrophobic and insoluble, pyomelanin exhibits strong hydrophilicity and is soluble in water due to the presence of phenolic and carboxyl functional groups in its polymeric structure (Mavridi-Printezi et al. 2023). This water solubility is an advantage in skincare formulations where aqueous media are predominant, allowing for uniform dispersion and stable coloring without the need for organic solvents (Lorquin et al. 2022). Pyomelanin demonstrates broad-spectrum UV absorbance, peaking around 250–280 nm, which aligns with the UV B range and some UV A overlap (Bayram 2021). This property makes it a potent natural photoprotective agent.

Pyomelanin has shown better antioxidant activity when compared to other bacterial pigments such as violacein or prodigiosin (Choi et al. 2021). The antioxidant potential is primarily attributed to its phenolic groups, which function as electron donors in redox reactions. This activity not only protects the producing bacteria from oxidative stress but also provides protection to free radicals and oxidative pollutants (Lorquin et al. 2022). Due to its carboxylate and phenolate groups, pyomelanin exhibits high-affinity metal chelation properties, particularly toward transition metals such as Fe^{3+}, Cu^{2+}, and Zn^{2+} (Kolhe et al. 2022). Pyomelanin-forming bacteria generally grow at pH 6–7, while the HGA autoxidation is optimal at pH 8–9 (Lorquin et al. 2022).

11.3 Pyomelanin Biosynthetic Pathway

The biosynthesis of pyomelanin (distinct from animal pheomelanin) in *Pseudomonas* involves a secondary metabolic pathway that diverges from canonical L-tyrosine catabolism, commonly referred to as the homogentisate pathway (Fig. 11.1) through specific enzymatic and regulatory mechanisms (Pavan et al. 2020). This pathway is conserved across various bacteria, notably in *P. aeruginosa*, *Acinetobacter baumannii*, and *Legionella pneumophila* (Fonseca et al. 2020; Panda et al. 2025; Pavan et al. 2020). During biosynthesis, L-tyrosine initially undergoes transamination via the pyridoxal phosphate (PLP)-dependent enzyme tyrosine aminotransferase (TyrB), yielding 4-hydroxyphenylpyruvate (4-HPP) while transferring the amino group to an α-ketoacid acceptor (Yabuuchi and Ohyama 1972). Subsequently, 4-HPP is oxidatively cleaved by 4-hydroxyphenylpyruvate dioxygenase (HppD), a nonheme Fe^{2+}-dependent enzyme that catalyzes the incorporation of molecular oxygen, leading to the formation of homogentisic acid via decarboxylation and hydroxylation (Moran 2005).

Under wild-type conditions, HGA enters the homogentisate pathway, where homogentisate 1,2-dioxygenase (HmgA) cleaves the aromatic ring to produce maleylacetoacetate, a precursor for entry into the tricarboxylic acid (TCA) cycle (Arias-Barrau et al. 2004). However, in *hmgA*-mutant or knockout strains (e.g., *P. aeruginosa* PAO1 ΔhmgA), this catabolic step is disrupted, leading to the periplasmic accumulation of HGA due to impaired metabolic flux and active or passive efflux mechanisms (Lorquin et al. 2022). The redox-active HGA undergoes spontaneous auto-oxidation in the extracellular milieu, driven by ambient oxygen and alkaline pH, producing semiquinone radicals and reactive oxygen species (ROS) such as superoxide (O_2^{-}•) and hydrogen peroxide ($H_{2O2)}$ (Galeb et al. 2022; Li et al. 2021). These reactive intermediates facilitate radical-mediated polymerization, resulting in the formation of pyomelanin's heterogeneous, carboxylated polymer structure (Galeb et al. 2022). This oxygen-dependent process is further potentiated by extracellular oxidases and redox-cycling agents, highlighting the interplay between abiotic oxidation and enzymatic priming in microbial melanogenesis (Al Khatib et al. 2021). The resulting pyomelanin, a nitrogen-free, carboxylated polymer distinct from eumelanin, confers multiple adaptive advantages, including UV protection and iron chelation and serving as a redox buffer under oxidative stress conditions (Motovilov and Mostert 2024). Genetic and proteomic studies have revealed regulatory cross-talk between pyomelanin synthesis and central carbon metabolism, emphasizing its role as a metabolic sink under conditions of HmgA dysfunction (Schmaler-Ripcke et al. 2009).

Pyomelanin synthesis is also well documented in *Aspergillus fumigatus*. While both *Pseudomonas* and *Aspergillus* synthesize pyomelanin via HGA polymerization, *Aspergillus* employs a tightly regulated, stress-responsive pathway integrated with developmental processes, whereas *Pseudomonas* relies on metabolic "leakage" resulting from disruption of the canonical catabolic pathway (Perez-Cuesta et al. 2020).

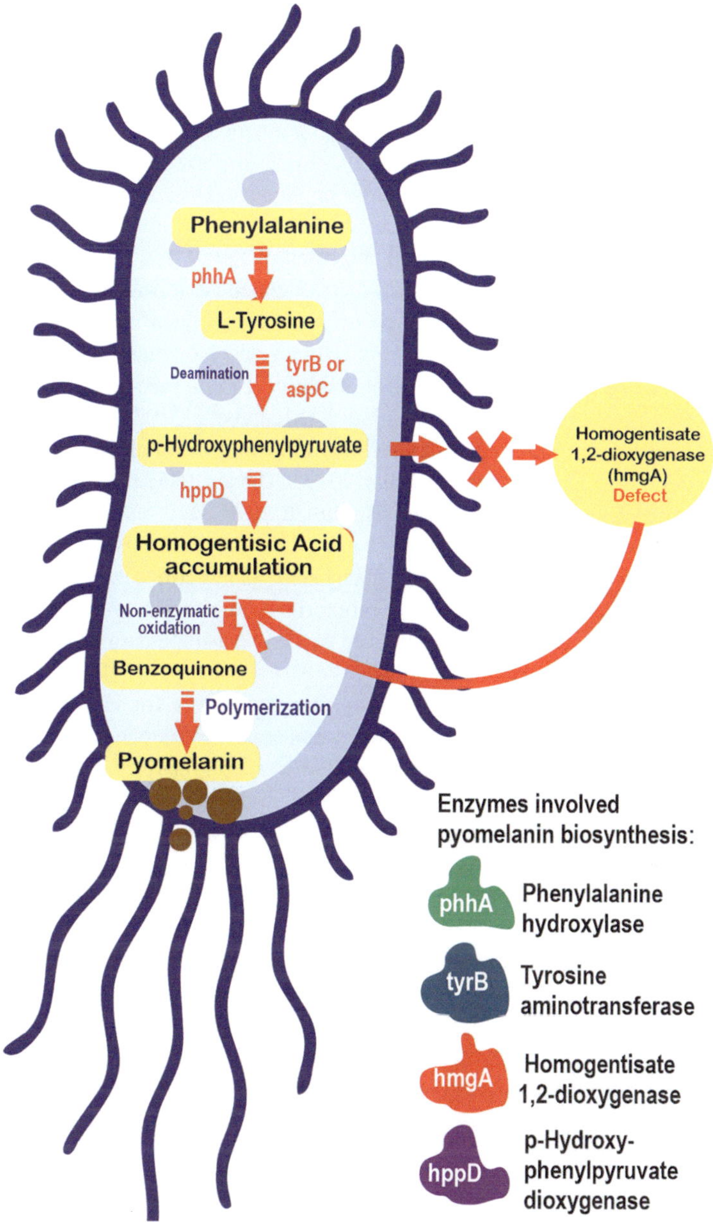

Fig. 11.1 Biosynthesis of pyomelanin in bacteria

11.4 Biosynthetic Gene Cluster for Pyomelanin in *Pseudomonas*

In prokaryotes, pyomelanin is a reddish-brown pigment synthesized from the oxidative polymerization of homogentisic acid and is transported into the extracellular milieu, conferring a distinct coloration to organisms (Díaz Appella et al. 2024). While pyomelanin's biosynthesis is linked to tyrosine degradation, much remains unknown about its processing beyond the roles of *hmgA* and *hpd* (Kosuru et al. 2021). Plethora of studies have uncovered novel genetic determinants and transport mechanisms critical for pyomelanin production, revealing a complex regulatory network and unknown molecular players (Hunter and Newman 2010; Styczynski et al. 2022). Classically, pyomelanin production in *P. aeruginosa* arises from mutations in *hmgA*, leading to HGA accumulation (Menon et al. 2022). However, transposon mutagenesis in *P. aeruginosa* have identified 26 additional genes contributing to melanogenesis (Hunter and Newman 2010). These include genes involved in aromatic amino acid catabolism, such as *phhA/B*, *phhC*, and *phhR*, which mediate the conversion of chorismate to HGA via phenylalanine and tyrosine (Arias-Barrau et al. 2004). Disruption of this pathway at any step via mutations in *phhA/B* (phenylalanine hydroxylase), *phhC* (tetrahydrobiopterin synthesis), or *hpd* abolishes pyomelanin production, indicating the key role of the chorismate pathway (Hunter and Newman 2010; Styczynski et al. 2022).

The *hmgABC* cluster encodes the enzymes responsible for HGA degradation (Arias-Barrau et al. 2004). *hmgA* encodes a nonheme Fe^{2+}-dependent dioxygenase with a conserved cupin domain. Structural studies reveal that Fe^{2+} coordinates with His183, His228, and Glu202 in the active site, facilitating HGA cleavage (Moustafa et al. 2024). A single amino acid substitution (G378R) destabilizes this structure, abolishing enzymatic activity and triggering pyomelanin production (Moustafa et al. 2024). *hmgB*, part of the GST superfamily, possesses a glutathione-binding motif critical for isomerizing maleylacetoacetate (Donoso et al. 2021). Though not directly involved in pigment synthesis, *hmgB* dysfunction disrupts metabolic flux, indirectly promoting HGA accumulation. *hmgC*, encoding a metallohydrolase, completes the pathway by hydrolyzing fumarylacetoacetate (Donoso et al. 2021).

hmgA expression is tightly regulated by environmental cues. Oxidative stressors like hydrogen peroxide downregulate *hmgA* via OxyR, a redox-sensitive transcription factor, redirecting HGA toward pyomelanin synthesis. Metal ions (Fe^{3+}, Cu^{2+}) modulate transcription through Fur (ferric uptake regulator), which represses *hmgA* under iron-replete conditions (Moustafa et al. 2024). Recent studies highlight gold nanoparticles as novel modulators, enhancing *hmgA* expression and pigment production, potentially through oxidative stress induction (Elnegery et al. 2021; Moustafa et al. 2024). Quorum sensing (QS) systems, such as LasR and RhlR, may also influence *hmg* genes, linking pyomelanin production to population density and biofilm formation (Elnegery et al. 2021).

Homology modeling and molecular dynamics simulations have elucidated mechanistic details of HmgA inactivation (Díaz Appella et al. 2024). Sequence alignment of HmgA proteins across species reveals conserved iron-binding residues (Fig. 11.2), including His183, His228, and Glu202, critical for catalytic activity (Zhang et al. 2023). Predicted structures of HmgB show conserved glutathione-binding motifs (Fig. 11.2b), while HmgC models align with class I amidohydrolase folds (Fig. 11.2c). These computational tools have helped confirm that mutations in hmgA near metal-coordinating residues severely reduce enzymatic activity, providing molecular insight into pyomelanin overproduction (Zhang et al. 2023). Mutations near Fe^{2+}-binding residues (e.g., G378R) distort the active site, preventing HGA cleavage (Díaz Appella et al. 2024). Cryo-EM and X-ray crystallography of homologous enzymes in *A. fumigatus* corroborate these findings, highlighting evolutionary conservation. Comparative genomics reveals *hmgA* orthologs in *B.cepacia* and *Shewanella oneidensis*, though regulatory mechanisms differ, reflecting niche-specific adaptations (Fig. 11.2).

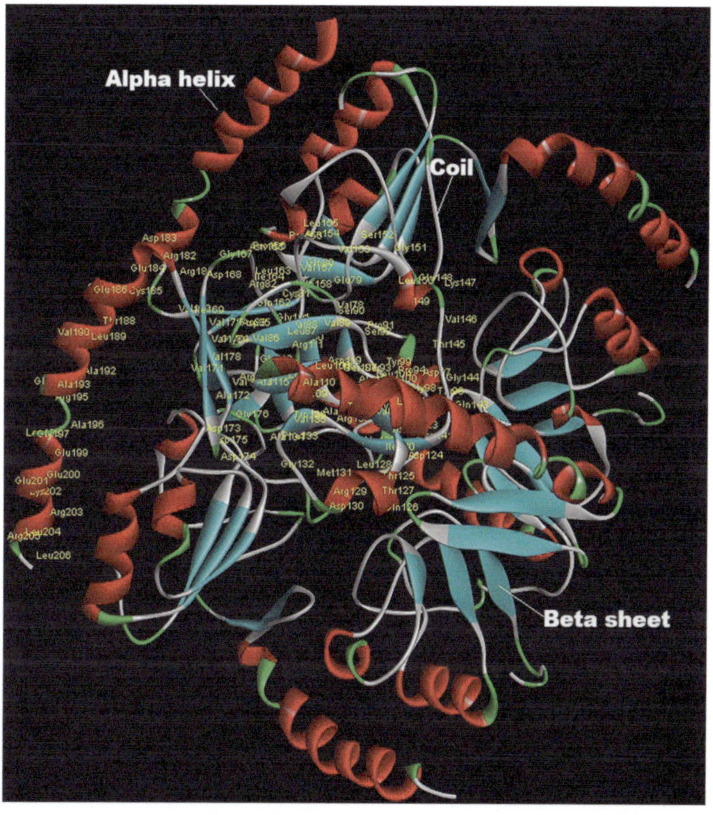

Fig. 11.2 The 3D structure of the hmgA protein highlights the amino acid sequences involved in ligand binding. The structural elements include alpha helices (red), beta sheets (cyan), and coils (green). Specific amino acids involved in ligand interactions are annotated in yellow, showcasing their positions and functional roles within the protein's structure

11.5 Fermentation Process for Pyomelanin Production

Industrial-scale melanin production primarily utilizes submerged batch or fed-batch fermentation systems, employing wild types of strains or genetically engineered *Escherichia coli* (Lee et al. 2022). Production is optimized by modulating environmental stressors and enhancing tyrosine metabolism, as pyomelanin is synthesized under nutrient-limited or oxidative conditions (Pavan et al. 2020). For laboratory-scale fermentation, bacterial cultures are typically grown in minimal media supplemented with L-tyrosine (0.5–3.0 g/L) at 30–37 °C with shaking at 180–220 rpm overnight (Singh et al. 2018). In large-scale batch or fed-batch fermentations, volumes range from 2 to 100 L for pilot scales and up to 1000 L for commercial production. Cost-effective media for commercial-scale production often incorporate agro-industrial waste (e.g., molasses, whey, corn steep liquor) supplemented with L-tyrosine as the carbon precursor, and occasionally glucose or glycerol as co-substrates (Lin et al. 2024; Tsouko et al. 2023). Optimal fermentation conditions include temperatures of 30–37 °C, pH maintained between 6.8 and 7.2 (controlled using NaOH or HCl), aeration rates of 1–2 vvm (volumes of air per volume of media per minute), and agitation speeds of 180–400 rpm, ammonium salt addition, and H_2O_2 stimulation (Zhou et al. 2024; Zhu et al. 2025). Key induction strategies to enhance pyomelanin production involve the knockout or repression of the hmgA gene, which prevents the degradation of HGA, leading to its accumulation and subsequent pyomelanin synthesis (Ketelboeter and Bardy 2015). Iron limitation has also been reported to induce pigment production, as low Fe^{3+} levels stimulate siderophore and pyomelanin synthesis (Styczynski et al. 2022). After 24–72 h of fermentation, the culture typically exhibits a dark brown coloration indicative of pyomelanin production. Cells are removed by centrifugation at 8000× g for 10 min. The supernatant containing soluble pyomelanin is acidified to pH ~2.0 using HCl to precipitate the pigment. The precipitated pigment is then collected by centrifugation or filtration, washed, and dried under vacuum or freeze-dried for further applications. Typical pyomelanin yields range from 200 to 500 mg/L in wild-type *Pseudomonas* strains and can reach up to 1.8 g/L in engineered *E. coli* strains (Bolognese et al. 2019; Martínez et al. 2019).

11.6 Influence of Light on Pyomelanin Production

Recent research has begun to explore light as a potential modulator of pyomelanin biosynthesis in bacteria. Although the detailed molecular mechanisms remain under investigation, several bacterial species have demonstrated altered pigment yields in response to light exposure, particularly under UV and blue light wavelengths (Cordero and Casadevall 2020).

A study by Guo et al. (2024) reported that *Pseudoalteromonas lipolytica* strain P3, a marine bacterium, exhibited increase in pyomelanin production when exposed to high-intensity visible light. The light acted as a physical stressor, prompting oxidative stress and enhancing the release and polymerization of HGA into pyomelanin. In another study, Zhao et al. (2022a, b) demonstrated that n carbapenem-resistant *Acinetobacter baumannii* strain (DETAB-R21) harboring a natural *hmgA* deletion exhibited higher resistance to UV irradiation (Zhao et al. 2022a, b). Although pyomelanin production in this strain was constitutive due to the gene disruption, UV exposure further emphasized the protective role of pyomelanin by improving survival against DNA damage and oxidative stress (Zhao et al. 2022a, b). This suggests that while UV does not directly upregulate pigment genes in hmgA mutants, pyomelanin offers a significant selective advantage under UV stress. In *P. aeruginosa*, despite the absence of dedicated light sensors such as phytochromes, redox-sensitive transcriptional regulators like Sox, OhR, and OxyR respond indirectly to oxidative stress caused by light-induced reactive oxygen species (ROS) (Sevilla et al. 2019). These regulators orchestrate antioxidant defenses, thereby influencing conditions favoring homogentisic acid accumulation and subsequent pyomelanin synthesis (Chen et al. 2011).

Emerging strategies are investigating light-controlled fermentation setups (photobioreactors) to modulate pigment production (Mohd Said 2023). By simulating natural light cycles or applying UV-A/B pulses, it might be possible to nongenetically regulate oxidative stress pathways and stimulate pyomelanin biosynthesis. Moreover, coupling light exposure with iron limitation or oxidative triggers may further enhance pigment yields (Styczynski et al. 2022) although systematic experimental validation is needed.

Future research employing RNA-Seq, CRISPR interference (CRISPRi), and transcriptome profiling could help unravel the precise regulatory links between light exposure, oxidative stress responses, and pyomelanin biosynthesis in *Pseudomonas* and related species.

11.7 Scope of Pyomelanin as a Pigment in Cosmetics

The multifunctionality of pyomelanin makes it an advanced bio-derived colorant for modern skincare and cosmetic applications. Its physicochemical traits—such as water solubility, photoprotective, and redox activity—make it suitable for a variety of product formulations while aligning with regulatory, environmental, and consumer safety expectations.

11.7.1 UV Absorbance and Photoprotection

Pyomelanin is widely recognized for its broad UV-visible absorption spectrum (200–700 nm), a characteristic it shares with eumelanin (Lorquin et al. 2022). Remarkably, even under extreme UVA-visible irradiation—far exceeding doses used in standard photostability testing—purified pyomelanin demonstrates exceptional resistance to degradation. Its practical utility is further highlighted by studies showing that bacterial-derived pyomelanin can substantially enhance the sun protection factor (SPF) of commercial sunscreens (Styczynski et al. 2022). In contrast, other natural pigments, such as violacein and prodigiosin, exhibit limited UV stability despite their vivid coloration. These compounds display narrow absorption peaks (violacein at ~585 nm; prodigiosin in the visible red spectrum) and undergo rapid photodegradation under UV exposure, rendering them ineffective for long-term UV protection (Tang et al. 2025). Similarly, carotenoids like β-carotene and astaxanthin absorb primarily in the visible and UVA ranges but are highly susceptible to photodegradation and oxidative bleaching under light and oxygen exposure, further limiting their protective efficacy (Amorim et al. 2022).

11.7.2 Antioxidant Capacity and Stability

Pyomelanin is recognized for its robust antioxidant functionality. Unlike many conventional antioxidants, pyomelanin scavenges ROS without being readily consumed in the process, a property attributed to its stable redox-active structure (Turick et al. 2010). In vitro assays, such as DPPH free radical scavenging, confirm its potent radical-quenching capacity, showing its potential as a durable antioxidant agent (Xiao and Li 2023). Critically, pyomelanin does not generate ROS upon UV irradiation—a notable advantage over some human melanins—and maintains both its chemical integrity and antioxidant activity even under prolonged UV A exposure (Lorquin et al. 2022). Its "hyperstable" antioxidant behavior arises from a resilient polyquinone-based architecture, which prevents the structural breakdown common in other pigments (Lorquin et al. 2021). This dual role as a photostable UV absorber and persistent ROS scavenger positions pyomelanin as a uniquely versatile biomaterial for photoprotection (Roy and Rhim 2022). Enzymatically synthesized pyomelanin have demonstrated thermal stability, resistance to photodegradation, and effective ROS scavenging in human skin cell models, reinforcing its suitability for dermatological and cosmetic use (Lorquin et al. 2021). In contrast, carotenoids like β-carotene and astaxanthin, though potent antioxidants in dark conditions, suffer from inherent chemical fragility. Their highly unsaturated structures render them prone to oxidation and degradation under light, heat, and oxygen exposure, leading to rapid loss of color, antioxidant efficacy, and even the formation of prooxidant byproducts under certain conditions (Mordi et al. 2020). Similarly, violacein and prodigiosin—pigments with narrow absorption maxima—exhibit limited UV stability, undergoing rapid photodegradation that compromises their utility in long-term UV protection (Tang et al. 2025).

By combining broad-spectrum UV absorption, resistance to photochemical breakdown, and sustained antioxidant activity, pyomelanin addresses critical limitations of both traditional and emerging pigment-based photoprotective agents. Its stability under extreme conditions, coupled with its multifunctional role, makes it a compelling candidate for next-generation sunscreens and antioxidant-infused materials.

11.7.3 Metal Ion Chelation and Formulation Stability

Pyomelanin's ability to chelate metal ions adds a critical dimension to its utility in cosmetic formulations, enhancing both stability and functionality. Structurally, pyomelanin contains redox-active catechol and quinone moieties that enable robust interactions with metal ions. Early work by Turick et al. (2010) demonstrated its capacity to bind Fe (III), even reducing it to Fe(II), a process suggesting dynamic redox cycling. Recent studies reveal its exceptional chelation efficiency: Shewanella-derived pyomelanin exhibited ~300-fold and ~950-fold increases in manganese (Mn) and iron (Fe) binding, respectively, compared to controls when exposed to 2 mM $MnSO_4$ and FeSO4 solutions (Kuttan et al. 2023). FT-IR spectral analysis confirmed that metal coordination occurs via hydroxyl, carboxylic acid, and amine functional groups within the pigment's architecture, elucidating the molecular basis of its binding behavior (Kuttan et al. 2023).

Beyond environmental applications, pyomelanin's metal-binding properties have biological and industrial relevance. In bacterial systems, its ferric-reductase activity facilitates iron uptake, while in vitro assays show its iron-chelation efficacy rivals that of the siderophore deferoxamine at micromolar concentrations (Chatfield and Cianciotto 2007). This trait is particularly advantageous in cosmetics, where trace metals like iron and copper can catalyze oxidative degradation of formulations via Fenton reactions. Pyomelanin's ability to sequester these prooxidant metals not only mitigates spoilage but also enhances product shelf life. Supporting this, studies in yeast demonstrate that pyomelanin-mediated Cu^{2+} chelation confers heavy-metal tolerance, directly linking its binding capacity to cellular protection (Larroude et al. 2021).

By integrating metal chelation with its inherent antioxidant and UV-protective roles, pyomelanin emerges as a multifunctional additive. Its polyquinone structure ensures stability under thermal and photochemical stress, preventing the degradation pathways seen in traditional chelators like EDTA. This dual functionality—acting as both a pigment and an in situ preservative—positions pyomelanin as a groundbreaking ingredient for sustainable, multi-action cosmetic formulations.

11.7.4 Solubility and Formulation Incorporation

Bacterial pyomelanin exists in two distinct forms with divergent formulation implications: water-soluble pyomelanin (Pyosol) and water-insoluble pyomelanin *(Pyoinsol)* (Urbaniak et al. 2023a, b). While both variants demonstrate biocompatibility and antioxidant activity, *Pyosol* stands out for cosmetic applications due to its unique physicochemical profile. Its high aqueous solubility, hygroscopic nature, and enhanced thermal stability enable seamless integration into water-based formulations (e.g., serums, lotions) without requiring solubilizing agents. Furthermore, Pyosol exhibits strong biological activity, including immune pathway activation, making it particularly suited for antiaging creams, skin-protective products, and antioxidant-rich cosmetics (Urbaniak et al. 2023a, b). In contrast, *Pyoinsol*, though stable, suffers from limited solubility and reduced bioactivity, constraining its utility in complex formulations (Urbaniak et al. 2023a, b). Pyomelanin's inherent water compatibility offers a stark advantage over other natural pigments. For instance, violacein, a violet bisindole pigment, is entirely water insoluble and requires organic solvents (ethanol, acetone) or lipid carriers for incorporation into cosmetics (Arif et al. 2017). Prodigiosin, a red pigment, displays negligible solubility in water and alcohols, often necessitating harsh solvents like chloroform or DMSO for formulation (Lu et al. 2024), while carotenoids (e.g., β-carotene, astaxanthin), are lipid-soluble terpenoids that demand emulsifiers, encapsulation, or nanoparticle delivery systems to disperse in aqueous products (Mezzomo and Ferreira 2016).

These solubility challenges force manufacturers to rely on additives that may compromise product stability, safety, or sustainability. Pyosol circumvents these issues entirely: Its water-soluble nature allows direct incorporation into hydrophilic formulations, eliminating the need for solvents or complex delivery systems. This simplifies production, reduces costs, and aligns with clean-label trends in cosmetics. Moreover, its hygroscopic properties enhance skin hydration, while its thermal stability ensures long-term performance in diverse formulations. By combining solubility, multifunctionality, and formulation flexibility, pyomelanin—particularly its *Pyosol* variant—represents a paradigm shift in pigment-based cosmetic design, outperforming conventional hydrophobic alternatives in both practicality and efficacy.

11.7.5 Biocompatibility: Noncytotoxicity and Hypoallergenicity

Current evidence strongly supports pyomelanin's noncytotoxicity, with multiple in vitro studies confirming its safety in mammalian cell models. For example, pyomelanin derived from *Yarrowia lipolytica* exhibited no cytotoxic effects on human keratinocytes at concentrations relevant to cosmetic formulations (Larroude et al. 2021). Similarly, Lorquin et al. (2021) demonstrated that pyomelanin remained nontoxic to skin cells even after prolonged UV irradiation, confirming that it does not generate harmful photoproducts under light stress. Structurally, pyomelanin shares key features with human melanin—a biocompatible polymer with a long history of safe use in food and cosmetics (e.g., sepia melanin from cephalopod ink). This biochemical similarity, coupled with the absence of reported sensitization or allergenic responses in the literature, suggests a low likelihood of adverse immune reactions.

However, while existing data imply hypoallergenic potential, direct experimental validation of pyomelanin's allergenicity is currently lacking. For instance, no peer-reviewed studies have yet conducted standardized human tests (e.g., repeated insult patch trials) to confirm its non-sensitizing properties. Thus, claims of hypoallergenicity remain provisional, relying on indirect evidence such as structural analogy to natural melanins and the absence of reported irritation in limited usage contexts.

11.7.6 Regulatory Status and Certification Potential

Pyomelanin's natural microbial origin and demonstrated low toxicity provide a foundational rationale for its safety in consumer applications. In the food sector, the FDA's Generally Recognized as Safe (GRAS) designation is attainable for substances with established safety under intended use. While pyomelanin itself lacks GRAS certification, its structural analogs—such as cuttlefish melanin (used for centuries in culinary applications like squid ink)—have a documented history of safe consumption. Recent studies propose that pyomelanin could meet GRAS criteria with rigorous purity testing and production in GRAS-certified microbial hosts (e.g., *Aspergillus oryzae* or *Bacillus subtilis*), minimizing risks of toxin contamination (Sen et al. 2019).

In cosmetics, pyomelanin may align with ECOCERT/COSMOS standards for natural and sustainable ingredients. It is synthesized via fermentation of plant-derived amino acids (e.g., tyrosine), meeting criteria for natural origin (Ruf et al. 2024). Precedent exists for melanin-based pigments in certified cosmetics such as catechol melanin from black olives, and plant-derived "bio-melanin" is used in UV-protective formulations. However, pyomelanin has yet to appear in commercial cosmetic inventories, largely due to unindustrialized production (Kiran et al. 2017). This gap presents significant intellectual property (IP) opportunities. Innovations in cost-effective biosynthesis (e.g., optimized fermentation) or novel applications (e.g., hybrid sunscreens with pyomelanin-infused emulsifiers) could be patentable, as prior art focuses predominantly on synthetic melanins (e.g., DOPA-melanin) or melanin-coated minerals rather than microbial pyomelanin pathways.

11.8 Prospects for Enhancing Pyomelanin Production

The biosynthesis of pyomelanin initiates with the conversion of *L*-tyrosine to homogentisic acid, which subsequently undergoes spontaneous polymerization into pyomelanin (Singh et al. 2018). Recent genetic strategies have validated ways to enhance metabolic flux toward HGA accumulation. Notably, heterologous overexpression of *hppD* in *E. coli* has been shown to redirect metabolic flux from 4-hydroxyphenylpyruvate specifically toward the HGA-melanin pathway, bypassing native competing routes of tyrosine degradation (Zhang et al. 2023). This modification significantly increased pyomelanin production, achieving 804.4 mg/L within 48 h, by streamlining substrate availability and minimizing metabolic losses—an advantage for industrial fermentation processes.

Additionally, the *sawR* gene (PA3133) has been identified as a novel regulatory element influencing pyomelanin biosynthesis. Repression of *sawR* upregulates HGA accumulation by downregulating *hmgA* expression, thereby blocking HGA degradation and promoting enhanced pyomelanin formation (Ben-David et al. 2018). Engineering *sawR*-based regulatory circuits could offer innovative avenues for controlled, on-demand melanin production, with potential applications in biomedical, cosmetic, and anti-biofilm technologies. Interestingly, external stimuli such as acoustic waves have been shown to influence *sawR*-mediated pathways, introducing nongenetic triggers for dynamic pyomelanin synthesis.

Furthermore, studies in fungi suggest similar strategies may be applied across biological domains. In *A. fumigatus*, deletion of *hmgA* prevented HGA degradation, leading to elevated pyomelanin accumulation (Schmaler-Ripcke et al. 2009), highlighting the potential for fungal systems to be exploited for pyomelanin bioproduction.

Advanced genome-editing technologies, particularly CRISPR-Cas9 and CRISPR/nCas9 systems, have further refined the optimization of pyomelanin pathways for cosmetic industry. For example, *hmgA*-knockout mutants generated in *P. aeruginosa* strains PAO1 and PA14 via CRISPR/nCas9 exhibited pyomelanin pigments with distinct structural and functional properties compared to naturally melanogenic strains, such as PAM and PexM (Díaz Appella et al. 2024). These findings emphasize the power of precision genome editing to tailor pigment structure and functionality for specialized industrial applications (Fig. 11.3).

11.9 Conclusion

Pyomelanin, synthesized by *Pseudomonas* species through the oxidative polymerization of homogentisic acid, emerges as a transformative, eco-friendly alternative to synthetic pigments in the cosmetics industry. Its unique physicochemical properties—including water solubility, broad-spectrum UV absorption, antioxidant efficacy, and metal-chelating capabilities—address critical formulation challenges in sunscreens and antioxidant-rich skincare products. Genetic engineering strategies, such as *hmgA* knockout and *hppD* overexpression, coupled with optimized fermentation processes using agro-industrial waste, enable scalable and cost-effective production. Pyomelanin outperforms traditional pigments like violacein and carotenoids in photostability, solubility, and multifunctionality, aligning with regulatory demands for sustainable, nontoxic ingredients. Future advancements in CRISPR-mediated pathway optimization and light-regulated biosynthesis promise further enhancements in yield and functionality. By bridging microbial biotechnology with cosmetic innovation, pyomelanin exemplifies a paradigm shift toward high-performance, environmentally responsible colorants, poised to redefine the standards of modern skincare formulations.

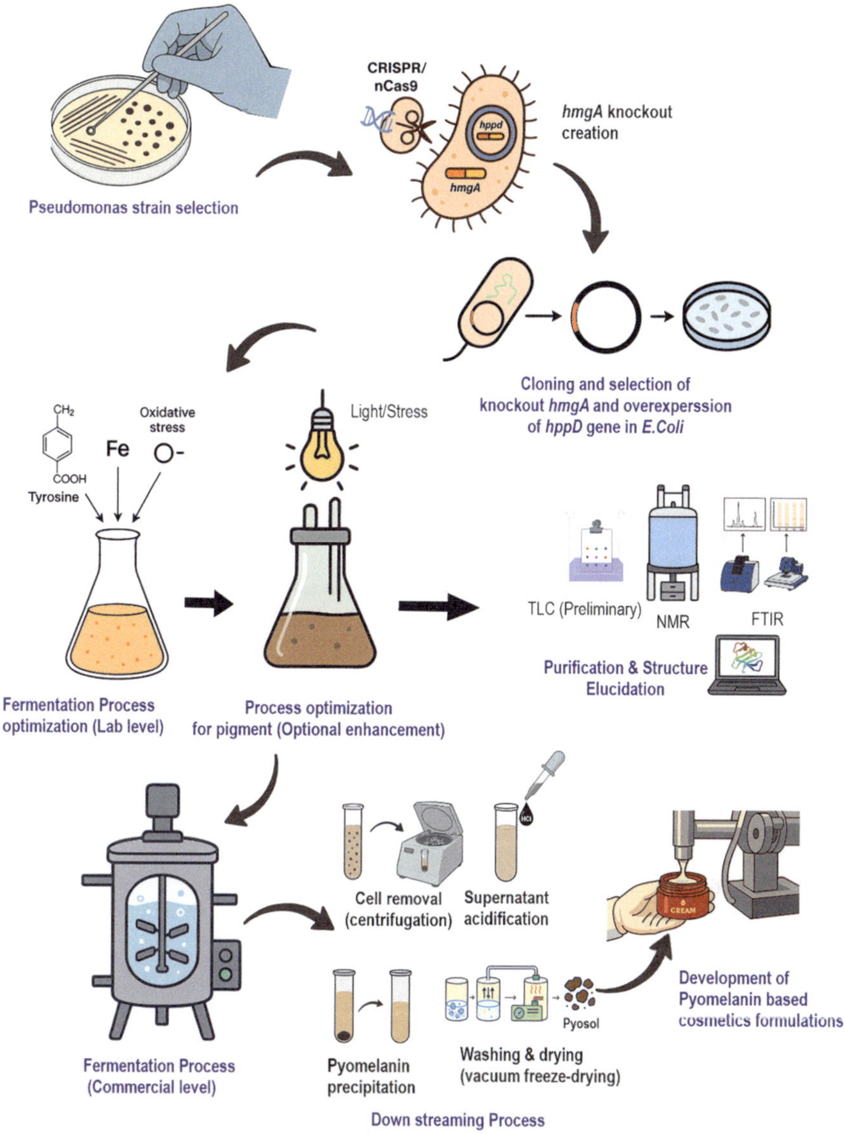

Fig. 11.3 Schematic overview of the genetic engineering and bioprocess pipeline for *Pseudomonas* strain optimization and pyomelanin production. Key steps include the following: Strain Selection: CRISPR/nCas9-mediated knockout (e.g., *hmgA*) and overexpression (e.g., *hppD*) strategies for enhancing pyomelanin synthesis. Cloning in *E. coli*: Validation of a structural characterization of pyomelanin using thin-layer chromatography (TLC) (preliminary), nuclear magnetic resonance (NMR), and Fourier-transform infrared spectroscopy (FTIR). Lab-Level Fermentation: Process optimization under tyrosine-rich media, iron limitation, oxidative stress, and optional light/stress induction to boost pigment yield. Downstream Processing: Commercial-scale fermentation, cell removal via centrifugation, supernatant acidification, pyomelanin precipitation, and vacuum freeze-drying for purification. Product Development: Formulation of water-soluble Pyosol and integration into cosmetic applications (e.g., sunscreens, antioxidant creams, antiaging products)

References

Abdi G, Tendulkar R, Thatte C, Mishra S, Desai B, Surve S, Tarighat MA (2024) Scaling up nature's chemistry: A guide to industrial production of valuable metabolites. In: Advances in metabolomics. Springer Nature Singapore, pp 331–375

Al Khatib M, Costa J, Spinelli D, Capecchi E, Saladino R, Baratto MC, Pogni R (2021) Homogentisic acid and gentisic acid biosynthesized pyomelanin mimics: structural characterization and antioxidant activity. Int J Mol Sci 22(4):1739. https://doi.org/10.3390/ijms22041739

Amorim IS, Almeida MCS, Chaves RPF, Chisté RC (2022) Technological applications and color stability of carotenoids extracted from selected Amazonian fruits. Food Sci Technol 42:e01922. https://doi.org/10.1590/fst.01922

Arias-Barrau E, Olivera ER, Luengo JM, Fernández C, Galán B, García JL, Díaz E, Miñambres B (2004) The homogentisate pathway: A central catabolic pathway involved in the degradation of L-phenylalanine, L-tyrosine, and 3-hydroxyphenylacetate in *Pseudomonas putida*. J Bacteriol 186(15):5062–5077. https://doi.org/10.1128/JB.186.15.5062-5077.2004

Arif S, Batool A, Khalid N, Ahmed I, Janjua HA (2017) Comparative analysis of stability and biological activities of violacein and starch-capped silver nanoparticles. RSC Adv 7(8):4468–4478. https://doi.org/10.1039/C6RA25806A

Bayram S (2021) Production, purification, and characterization of *Streptomyces* sp. strain MPPS2 extracellular pyomelanin pigment. Arch Microbiol 203:2129–2140. https://doi.org/10.1007/s00203-021-02437-w

Ben-David Y, Zlotnik E, Zander I, Yerushalmi G, Shoshani S, Banin E (2018) SawR, a new regulator controlling pyomelanin synthesis in *Pseudomonas aeruginosa*. Microbiol Res 206:91–98. https://doi.org/10.1016/j.micres.2017.10.004

Bolognese F, Scanferla C, Caruso E, Orlandi VT (2019) Bacterial melanin production by heterologous expression of 4-hydroxyphenylpyruvate dioxygenase from *Pseudomonas aeruginosa*. Int J Biol Macromol 133:1072–1080. https://doi.org/10.1016/j.ijbiomac.2019.04.061

Chatfield CH, Cianciotto NP (2007) The secreted pyomelanin pigment of *Legionella pneumophila* confers ferric reductase activity. Infect Immun 75(8):4062–4070. https://doi.org/10.1128/IAI.00489-07

Chen PR, Brugarolas P, He C (2011) Redox signaling in human pathogens. Antioxid Redox Signal 14(6):1107–1118. https://doi.org/10.1089/ars.2010.3377

Choi SY, Lim S, Yoon KH, Lee JI, Mitchell RJ (2021) Biotechnological activities and applications of bacterial pigments violacein and prodigiosin. J Biol Eng 15(1):1–16. https://doi.org/10.1186/s13036-021-00262-9

Cordero RJB, Casadevall A (2020) Functions of fungal melanin beyond virulence. Fungal Biol Rev 34(3):109–118. https://doi.org/10.1016/j.fbr.2020.07.002

Díaz Appella MN, Kolender A, Oppezzo OJ, López NI, Tribelli PM (2024) The structural complexity of pyomelanin impacts UV shielding in *Pseudomonas* species with different lifestyles. FEBS Lett 598(21):2702–2716. https://doi.org/10.1002/1873-3468.15000

Donoso RA, Ruiz D, Gárate-Castro C, Villegas P, González-Pastor JE, de Lorenzo V, Gonzalez B, Pérez-Pantoja D (2021) Identification of a self-sufficient cytochrome P450 monooxygenase from *Cupriavidus pinatubonensis* JMP134 involved in 2-hydroxyphenylacetic acid catabolism via the homogentisate pathway. Microb Biotechnol 14(5):1944–1960. https://doi.org/10.1111/1751-7915.13865

El-Naggar NE, Saber WIA (2022) Natural melanin: current trends, and future approaches, with especial reference to microbial source. Polymers 14(7):1339. https://doi.org/10.3390/polym14071339

Elnegery AA, Mowafy WK, Zahra TA, Abou El-Khier NT (2021) Study of quorum-sensing LasR and RhlR genes and their dependent virulence factors in *Pseudomonas aeruginosa* isolates from infected burn wounds. Access Microbiol 3(1):000211. https://doi.org/10.1099/acmi.0.000211

Fonseca É, Freitas F, Caldart R, Morgado S, Vicente AC (2020) Pyomelanin biosynthetic pathway in pigment-producer strains from the pandemic *Acinetobacter baumannii* IC-5. Mem Inst Oswaldo Cruz 115:e200371. https://doi.org/10.1590/0074-02760200371

Galeb HA, Lamantia A, Robson A, König K, Eichhorn J, Baldock SJ, Ashton MD, Baum JV, Mort RL, Robinson BJ, Schacher FH, Chechik V, Taylor AM, Hardy JG (2022) The polymerization of homogentisic acid in vitro as a model for pyomelanin formation. Macromol Chem Phys 223(6):2100489. https://doi.org/10.1002/macp.202100489

Guo Z, Feng Q, Mao X, Guo N, Yin Y, Liu T (2024) Increased secretion of bacterial pyomelanin caused by light accelerates corrosion of low alloy steel. Chemosphere 359:142353. https://doi.org/10.1016/j.chemosphere.2024.142353

Hunter RC, Newman DK (2010) A putative ABC transporter, hatABCDE, is among molecular determinants of pyomelanin production in *Pseudomonas aeruginosa*. J Bacteriol 192(22):5962–5971. https://doi.org/10.1128/JB.01021-10

Keith KE, Killip L, He P, Moran GR, Valvano MA (2007) *Burkholderia cenocepacia* C5424 produces a pigment with antioxidant properties using a homogentisate intermediate. J Bacteriol 189(24):9057–9065. https://doi.org/10.1128/JB.00436-07

Ketelboeter LM, Bardy SL (2015) Methods to inhibit bacterial pyomelanin production and determine the corresponding increase in sensitivity to oxidative stress. J Vis Exp 102:e53105. https://doi.org/10.3791/53105

Ketelboeter LM, Bardy SL (2017) Characterization of 2-(2-nitro-4-trifluoromethylbenzoyl)-1,3-cyclohexanedione resistance in pyomelanogenic *Pseudomonas aeruginosa* DKN343. PLoS One 12(6):e0178084. https://doi.org/10.1371/journal.pone.0178084

Kiran GS, Jackson SA, Priyadharsini S, Dobson ADW, Selvin J (2017) Synthesis of Nm-PHB (nanomelanin-polyhydroxy butyrate) nanocomposite film and its protective effect against biofilm-forming multi-drug resistant *Staphylococcus aureus*. Sci Rep 7(1):1–13. https://doi.org/10.1038/s41598-017-08816-y

Kolhe N, Damle E, Pradhan A, Zinjarde S (2022) A comprehensive assessment of *Yarrowia lipolytica* and its interactions with metals: current updates and future prospective. Biotechnol Adv 59:107967. https://doi.org/10.1016/j.biotechadv.2022.107967

Kosuru RY, Roy A, Bera S (2021) Antagonistic roles of gallates and ascorbic acid in pyomelanin biosynthesis of *Pseudomonas aeruginosa* biofilms. Curr Microbiol 78:3843–3852. https://doi.org/10.1007/s00284-021-02655-x

Kuttan SP, Abdulaziz A, Chekidhenkuzhiyil J et al (2023) Characterization of pyomelanin secreted by *Shewanella* sp. and their application in metal recovery. Environ Sci Pollut Res 30:6705–6715. https://doi.org/10.1007/s11356-022-22686-6

Larroude M, Onésime D, Rué O, Nicaud J-M, Rossignol T (2021) A *Yarrowia lipolytica* strain engineered for pyomelanin production. Microorganisms 9(4):838. https://doi.org/10.3390/microorganisms9040838

Lee HS, Choi JY, Kwon SJ, Park ES, Oh BM, Kim JH, Lee PC (2022) Melanin biopolymer synthesis using a new melanogenic strain of *Flavobacterium kingsejongi* and a recombinant strain of *Escherichia coli* expressing 4-hydroxyphenylpyruvate dioxygenase from F. kingsejongi. Microb Cell Factories 21(1):75. https://doi.org/10.1186/s12934-022-01800-w

Li Y, Ye Z, Lu P, Lu L (2021) Pyomelanin produced by *Streptomyces* sp. ZL-24 and its protective effects against SH-SY5Y cells injury induced by hydrogen peroxide. Sci Rep 11(1):16649. https://doi.org/10.1038/s41598-021-94598-3

Lin C, Lu P, Ma J, Kan T, Han X, Liu S, Ji Z, Mao J (2024) Investigation into the production of melanin from by-products of Huangjiu brewing. Food Secur 13(19):3063. https://doi.org/10.3390/foods13193063

Lorquin F, Ziarelli F, Amouric A et al (2021) Production and properties of non-cytotoxic pyomelanin by laccase and comparison to bacterial and synthetic pigments. Sci Rep 11:8538. https://doi.org/10.1038/s41598-021-87328-2

Lorquin F, Piccerelle P, Orneto C, Robin M, Lorquin J (2022) New insights and advances on pyomelanin production: from microbial synthesis to applications. J Ind Microbiol Biotechnol 49(4):kuac013. https://doi.org/10.1093/jimb/kuac013

Lu Y, Liu D, Jiang R, Li Z, Gao X (2024) Prodigiosin: unveiling the crimson wonder – a comprehensive journey from diverse bioactivity to synthesis and yield enhancement. Front Microbiol 15:1412776. https://doi.org/10.3389/fmicb.2024.1412776

Martínez LM, Martinez A, Gosset G (2019) Production of melanins with recombinant microorganisms. Front Bioeng Biotechnol 7:285. https://doi.org/10.3389/fbioe.2019.00285

Mavridi-Printezi A, Guernelli M, Menichetti A, Montalti M (2020) Bio-applications of multifunctional melanin nanoparticles: from nanomedicine to nanocosmetics. Nanomaterials 10(11):2276. https://doi.org/10.3390/nano10112276

Mavridi-Printezi A, Menichetti A, Mordini D, Montalti M (2023) Functionalization of and through melanin: strategies and bio-applications. Int J Mol Sci 24(11):9689. https://doi.org/10.3390/ijms24119689

Menon ND, Penziner S, Montaño ET, Zurich R, Pride DT, Nair BG, Nizet V (2022) Increased innate immune susceptibility in hyperpigmented bacteriophage-resistant mutants of *Pseudomonas aeruginosa*. Antimicrob Agents Chemother 66(8):e00239–e00222. https://doi.org/10.1128/aac.00239-22

Mezzomo N, Ferreira SRS (2016) Carotenoids functionality, sources, and processing by supercritical technology: A review. J Chem 2016:3164312. https://doi.org/10.1155/2016/3164312

Mohd Said F (2023) Production of microbial pigment in a bioreactor. In: Plant specialized metabolites: phytochemistry, ecology and biotechnology. Springer Nature, pp 1–19. https://doi.org/10.1007/978-3-031-30037-0_40-1

Moran GR (2005) 4-Hydroxyphenylpyruvate dioxygenase. Arch Biochem Biophys 433(1):117–128. https://doi.org/10.1016/j.abb.2004.08.015

Mordi RC, Ademosun OT, Ajanaku CO, Olanrewaju IO, Walton JC (2020) Free radical mediated oxidative degradation of carotenes and xanthophylls. Molecules 25(5):1038. https://doi.org/10.3390/molecules25051038

Motovilov KA, Mostert AB (2024) Melanin: nature's fourth bioorganic polymer. Soft Matter 20(29):5635–5651. https://doi.org/10.1039/D4SM00491D

Moustafa DA, Wu L, Ivey M, Fankhauser SC, Goldberg JB (2024) Mutation of hmgA, encoding homogentisate 1,2-dioxygenase, is responsible for pyomelanin production but does not impact the virulence of *Burkholderia cenocepacia* in a chronic granulomatous disease mouse lung infection. Microbiol Spectrum 12(7):e00410–e00424. https://doi.org/10.1128/spectrum.00410-24

Nambela L, Haule LV, Mgani Q (2020) A review on source, chemistry, green synthesis and application of textile colorants. J Clean Prod 246:119036. https://doi.org/10.1016/j.jclepro.2019.119036

Panda L, Krishnapriya GK, Sawant AR, Singh SK, Gupta M, Shashikala P, Prashanth K (2025) Genomic insights into drug resistance and virulence determinants in rare pyomelanin-producing clinical isolates of *Acinetobacter baumannii*. Eur J Clin Microbiol Infect Dis 44(3):533–547. https://doi.org/10.1007/s10096-024-05008-1

Pavan ME, López NI, Pettinari MJ (2020) Melanin biosynthesis in bacteria, regulation and production perspectives. Appl Microbiol Biotechnol 104(4):1357–1370. https://doi.org/10.1007/s00253-019-10245-y

Perez-Cuesta U, Aparicio-Fernandez L, Guruceaga X, Martin-Souto L, Abad-Diaz-de-Cerio A, Antoran A, Buldain I, Hernando FL, Ramirez-Garcia A, Rementeria A (2020) Melanin and pyomelanin in *Aspergillus fumigatus*: from its genetics to host interaction. Int Microbiol 23(1):55–63. https://doi.org/10.1007/s10123-019-00078-0

Qin Y, Xia Y (2024) Melanin in fungi: advances in structure, biosynthesis, regulation, and metabolic engineering. Microb Cell Factories 23(1):334. https://doi.org/10.1186/s12934-024-02614-8

Ramesh C, Vinithkumar NV, Kirubagaran R, Venil CK, Dufossé L (2019) Multifaceted applications of microbial pigments: current knowledge, challenges and future directions for public health implications. Microorganisms 7(7):186. https://doi.org/10.3390/microorganisms7070186

Rodríguez-Rojas A, Mena A, Martín S, Borrell N, Oliver A, Blázquez J (2009) Inactivation of the hmgA gene of *Pseudomonas aeruginosa* leads to pyomelanin hyperproduction, stress resistance and increased persistence in chronic lung infection. Microbiology 155(Pt 4):1050–1057. https://doi.org/10.1099/mic.0.024745-0

Roy S, Rhim JW (2022) New insight into melanin for food packaging and biotechnology applications. Crit Rev Food Sci Nutr 62(17):4629–4655. https://doi.org/10.1080/10408398.2021.1878097

Ruf J, Menrad K, Emberger-Klein A (2024) The local range of natural cosmetics: consumer perceptions and assessments in Germany. Bus Strateg Environ 33(7):7462–7478. https://doi.org/10.1002/bse.3862

Schmaler-Ripcke J, Sugareva V, Gebhardt P, Winkler R, Kniemeyer O, Heinekamp T, Brakhage AA (2009) Production of pyomelanin, a second type of melanin, via the tyrosine degradation pathway in Aspergillus fumigatus. Appl Environ Microbiol 75(2):493–503. https://doi.org/10.1128/AEM.02077-08

Sen T, Barrow CJ, Deshmukh SK (2019) Microbial pigments in the food industry – challenges and the way forward. Front Nutr 6:7. https://doi.org/10.3389/fnut.2019.00007

Sevilla E, Bes MT, González A, Peleato ML, Fillat MF (2019) Redox-based transcriptional regulation in prokaryotes: revisiting model mechanisms. Antioxid Redox Signal 30(13):1651–1696. https://doi.org/10.1089/ars.2017.7442

Singh D, Kumar J, Kumar A (2018) Isolation of pyomelanin from bacteria and evidences showing its synthesis by 4-hydroxyphenylpyruvate dioxygenase enzyme encoded by hppD gene. Int J Biol Macromol 119:864–873. https://doi.org/10.1016/j.ijbiomac.2018.08.003

Slama HB, Chenari Bouket A, Pourhassan Z, Alenezi FN, Silini A, Cherif-Silini H, Belbahri L (2021) Diversity of synthetic dyes from textile industries, discharge impacts and treatment methods. Appl Sci 11(14):6255. https://doi.org/10.3390/app11146255

Styczynski M, Rogowska A, Nyabayo C, Decewicz P, Romaniuk F, Pączkowski C, Dziewit L (2022) Heterologous production and characterization of a pyomelanin of Antarctic *Pseudomonas* sp. ANT_H4: A metabolite protecting against UV and free radicals, interacting with iron from minerals and exhibiting priming properties toward plant hairy roots. Microb Cell Factories 21:261. https://doi.org/10.1186/s12934-022-01990-3

Tang Q, Li Z, Chen N, Luo X, Zhao Q (2025) Natural pigments derived from plants and microorganisms: classification, biosynthesis, and applications. Plant Biotechnol J 23(2):592–614. https://doi.org/10.1111/pbi.14522

Tsouko E, Tolia E, Sarris D (2023) Microbial melanin: renewable feedstock and emerging applications in food-related systems. Sustain For 15(9):7516. https://doi.org/10.3390/su15097516

Turick CE, Tisa LS, Caccavo F Jr (2002) Melanin production and use as a soluble electron shuttle for Fe(III) oxide reduction and as a terminal electron acceptor by Shewanella algae BrY. Appl Environ Microbiol 68(5):2436–2444. https://doi.org/10.1128/AEM.68.5.2436-2444.2002

Turick CE, Knox AS, Becnel JM, Ekechukwu AA, Milliken CE (2010) Properties and function of pyomelanin. Biopolymers 449(72):10–5772. https://doi.org/10.5772/10273

Urbaniak MM, Gazińska M, Rudnicka K, Płociński P, Nowak M, Chmiela M (2023a) In vitro and in vivo biocompatibility of natural and synthetic *Pseudomonas aeruginosa* pyomelanin for potential biomedical applications. Int J Mol Sci 24(9):7846. https://doi.org/10.3390/ijms24097846

Urbaniak MM, Rudnicka K, Gościniak G, Chmiela M (2023b) Can pyomelanin produced by *Pseudomonas aeruginosa* promote the regeneration of gastric epithelial cells and enhance *Helicobacter pylori* phagocytosis? Int J Mol Sci 24(18):13911. https://doi.org/10.3390/ijms241813911

Venil CK, Zakaria ZA, Ahmad WA (2013) Bacterial pigments and their applications. Process Biochem 48(7):1065–1079. https://doi.org/10.1016/j.procbio.2013.06.006

Xiao M, Li W (2023) Fundamentals and applications of optically active melanin-based materials. In: Gosset G (ed) Melanins: functions, biotechnological production, and applications. Springer, Cham, pp 157–188. https://doi.org/10.1007/978-3-031-27799-3_7

Yabuuchi E, Ohyama A (1972) Characterization of "Pyomelanin"-producing strains of Pseudomonas aeruginosa. Int J Syst Evol Microbiol 22(2):53–64. https://doi.org/10.1099/00207713-22-2-53

Zhang Q, Yang X, Lin L, Wu S, Wang P, Wei W, Wei D (2023) Pyomelanin production via heterologous expression of 4-hydroxyphenylpyruvate dioxygenase (HPPD) and construction of HPPD inhibitor screening model. J Biosci Bioeng 135(2):93–101. https://doi.org/10.1016/j.jbiosc.2022.10.005

Zhao F, Liu H, Yao Y, Zhang L, Zhou Z, Leptihn S, Yu Y, Hua X, Fu Y (2022a) Description of a rare pyomelanin-producing carbapenem-resistant *Acinetobacter baumannii* strain coharboring chromosomal OXA-23 and NDM-1. Microbiol Spectrum 10(4):e02144–e02122. https://doi.org/10.1128/spectrum.02144-22

Zhao Y, Sun J, Song Y, Wang M, Song X, Li H (2022b) Spontaneous inactivation of hmgA in *Acinetobacter baumannii* leads to pyomelanin production and UV protection. Front Microbiol 13:893684. https://doi.org/10.3389/fmicb.2022.893684

Zhou R, Ma L, Qin X, Zhu H, Chen G, Liang Z, Zeng W (2024) Efficient production of melanin by *Aureobasidium melanogenum* using a simplified medium and pH-controlled fermentation strategy with the cell morphology analysis. Appl Biochem Biotechnol 196(2):1122–1141. https://doi.org/10.1007/s12010-023-04594-8

Zhu Q, Zhao B, Ji N, Zhu Y, Shi X (2025) Effects of ORF14 gene on melanin expression, fermentation conditions and properties of melanin production in modified strains. J Biotechnol 398:224–234. https://doi.org/10.1016/j.jbiotec.2025.01.001

Chapter 12
Proteomics in Unravelling Secondary Metabolite Synthesis in Plants: Exploring the Microbial Connection

Pritha Pal and Sibashish Baksi

Abstract Metabolism in plants and their secondary products play vital role in various signaling pathways governing ecological interactions including defense. Plant secondary metabolites are progressively used as food additives or bio-preservative agents and used for aromatic and culinary purposes. At molecular level, the MYB, bHLH, AP2/ERF, SPL, and NAC are some crucial transcription factors that regulate gene expression against biotic or abiotic stressors. The mevalonate pathway deals with HMG-CoA reductase, mevalonate-5-kinase, mevalonate-5-phosphate decarboxylase, and phosphomevalonate kinase, whereas DHAP synthase, chorismite synthase, DHQ dehydratase, and shikimate kinase are some of the most important enzymes of the shikimate pathway. Proteomics help to perceive a clear picture of the regulation of plant genome expression in response to different biotic and abiotic stressors that eventually couple with secondary metabolite production, and the same can also be used as complementary tools for functional analysis and biomarker discovery research. Plant secondary metabolites are essential for microbial interactions, quorum sensing, and plant defense. Genes such as CHS (chalcone synthase) and F3H (flavonoid 3-hydroxylase) regulate flavonoids, which influence nodulation and mycorrhizal symbiosis. Microbes such as *Rhizobium*, *Pseudomonas*, and arbuscular mycorrhizal fungi (AMF) influence these pathways, improving nutrient acquisition and resistance to diseases. These plant-microbe interactions highlight the possibility of sustainable agricultural and biotechnological uses. Alteration in the expression of differential proteins, variation in cellular protein localization and their accumulation patterns, and posttranscriptional and posttranslational modifications are often found associated with the presence of stressors. This chapter focuses on

P. Pal
School of Life Sciences, Swami Vivekananda University, Kolkata, India

S. Baksi (✉)
School of Life Sciences, Swami Vivekananda University, Kolkata, India

Department of Biotechnology and Biosciences, University of Milan-Bicocca, Milan, Italy
e-mail: sibashish.baksi@gmail.com

N. K. Arora et al. (eds.), *Environmental Sustainability and Biotechnology: Opportunities and Challenges*, Microorganisms for Sustainability 46,
https://doi.org/10.1007/978-981-96-9025-1_12

the understanding of protein profiling coupled with secondary metabolomics to study the cellular pathways associated with stress responses.

Keywords Alkaloids · Flavonoids · Rhizosphere · Rhizomicrobiome · Proteomics · Biosynthesis

12.1 Introduction

Numerous chemical substances created by the plant cell through metabolic pathways descended from the basic metabolic pathways are known as secondary plant metabolites. The biological effects of secondary metabolites have been demonstrated to vary, which gives many ancient civilizations' usage of herbs in traditional medicine a scientific foundation. They can defend plants from pathogens since they have been described as antibiotic, antifungal, and antiviral. Additionally, they contain significant UV absorption chemicals, protecting leaves from severe light damage. Secondary metabolites are often considered as platform chemicals to synthesize food, drugs, agricultural products, and biofuels. In crop breeding and metabolic engineering research, secondary metabolites play a significant role. However, especially in non-model plants, our knowledge of secondary metabolism is far from complete. Understanding secondary metabolism is now possible, thanks to functional genomics, which also encompasses transcriptomics, proteomics, and metabolomics. A potent tool for understanding secondary metabolism is provided by functional genomics, which includes transcriptomics, proteomics, and metabolomics. At the mRNA level, differentially expressed transcripts are found using transcriptomics. Transcriptomic technology is frequently implemented using DNA microarray systems. It is appropriate for model plants and requires genomic data. There are few genomic data for most plants and cDNA. For non-model plants including *Panax ginseng*, *Medicago truncatula*, and *Catharanthus roseus*, AFLP provides an alternative functional genomic method (Vuylsteke et al. 2007). RNA sequencing, which is suited for both model and non-model plants, is growing in popularity as a method for transcriptomics due to the drop in sequencing costs. In secondary metabolic pathways, proteins act as catalysts for chemical processes. Elicitation or inhibition of some of the pathway's enzymes may cause an increase in beneficial metabolites and a decrease in toxic ones. A high-throughput method for simultaneously determining numerous proteins is provided by proteomics. Only a few publications on secondary metabolism have been published due to challenges with sample preparation and low protein concentrations, which mainly focus on the identification of novel proteins (Jacobs et al. 2005; Lei et al. 2005; Nam et al. 2005). Secondary plant metabolites eventually operate on many molecular targets depending on the situation. Enzymes, mediators, transcription factors, and even nucleic acids may be among these targets. As a result, it is feasible to infer the knowledge of the resulting pharmacological and toxicological effects as well as any potential synergistic or antagonistic effects brought on by the usage of herbal formulas with numerous components. Due to this, despite its age, the isolation and structural elucidation of secondary plant metabolites is still a vast and rapidly expanding field, and techniques for separation and analysis are constantly improving.

The present chapter focuses on the application of proteomics to study the variation of gene expression resulting in the upregulation and downregulation of different secondary metabolites. The approaches will involve the respective enzyme systems acting behind the implementation of the expression of secondary metabolites, impacting various stress conditions in plants. The importance of proteomic studies along with transcriptomic and genomic profiling has already proven successful in protein expression profiling of secondary metabolites, which will mainly be highlighted in this chapter.

12.2 Plants Utilized in Routine Production of Secondary Metabolites

Different environmental factors like temperature, moisture, and light intensity exhibit significant effects on plant growth rate and secondary metabolite production. Moreover, the environmental factors also significantly affect the microbiome population and rhizosphere health, which eventually take part in crosstalk among the plant species. Species development, diversity, and distribution are therefore significantly affected by the ecology and environment and the same is represented in Table 12.1. However, tea, thyme, cannabis, sage, and citrus fruits are common herbs and/or plants that are produced worldwide in different types of environments. Different hybrids and species of those aforementioned plants have also been developed with higher productivity and enhanced adaptability to environmental stress (Maraveas et al. 2021).

12.3 Secondary Plant Metabolites: Biosynthetic Pathways

Plants have sophisticated metabolic pathways associated with the biosynthesis of different primary metabolites like organic acids, flavonoids, phenolics, amino acids, and plant hormones and secondary metabolites like alkaloids, amines, nonprotein amino acids, amygdalin, dhurrin, and polyketides. Even though secondary metabolites house a wide array of phytochemicals having different physiochemical properties, most of them are synthesized from some basic platform chemicals through some distinct and restricted metabolic pathways including glycolysis, TCA cycle, and shikimate pathway (Crozier et al. 2006; Wink 2010).

Table 12.1 Plant species: A brief description, ecology, and associated secondary metabolites

Popular plant species	Diversity, ecology, and distribution	Secondary metabolites	Reference
Artemisia sp.	Range from woody shrubs to herbaceous perennial Range from herbs to subshrubs and shrubs Distributed in steppe, semidesert, and desert steppe Distributed mainly on uncultivated hillsides in lower diversity in coastal plains or ranges	Sesquiterpene Lactone Artemisinin	Davis et al. (2012), Guerriero et al. (2018), Liu et al. (2019), Olsen (1921)
Urtica dioica	Herbaceous Perennial Found in many cooler temperate parts of the world—in Africa, the America, Asia, Australia, and Europe In North America, it is widely distributed in Canada and the USA and also can be found in northernmost Mexico. It grows in abundance in the Pacific Northwest, especially in places where the annual rainfall is high.	Phenolic compounds like flavonoids and anthocyanins Hydroxycinnamic acid derivatives	
Coffea arabica L.	Large shrub Can be found in any part of Africa, Latin America, Southeast Asia, India, China, and assorted islands in the Caribbean and in the Pacific	Caffeine (1,3,7-*N*-trimethylxanthine)	

Orchis militaris	Perennial herb in the family orchids Can be found in Indochina, Southwest Australia, Europe, Northern Asia, and North America In Europe, their biodiversity center is in Southern Europe, mainly in the Mediterranean region	Phytoalexins (9,10-dihydrophenanthrenes): Orchinol	Averyanov et al. (2016), Bruelheide and Heinemeyer (2002), Evans et al. (2020), Fisch et al. (1973), Teoh (2016), Waite and Farrell (1998)
Loroglossum hircinum		Loroglossol Hircinol	
Digitalis purpurea	Herbaceous biennial or short-lived perennial plant Widespread in most of temperate Europe, also naturalized in parts of North America and some other temperate regions	Triterpenes and steroids	
Vitis sp.	A genus of 79 accepted species of vining plants in the flowering plant family Vitaceae Found in the temperate regions of the Northern Hemisphere, especially in North America and eastern Asia	Stilbenoids (tamoxifen and diethylstilbenoids)	
Dendrobium draconis	A high ranging, differently sized, ranging from warm to cool growing epiphytic orchid Found in Assam India, Myanmar, Thailand, Cambodia, Laos, and Vietnam	Bibenzyls (gigantol and batatasin III, erianin)	
Dendrobium chrysotoxum	Huge cultivated species of orchid Native to Southeast Asia, growing naturally in Myanmar, Bhutan, Yunnan (China), Manipur, Assam (India), Bangladesh, Andaman Islands, Laos, Nepal, Thailand, and Vietnam		
Capsicum annuum L.	Fruit vegetable that is a member of the Solanaceae family Tropical America is the natural home of the genus Capsicum.	Phenolic compound Proline	Ashraf et al. (2018), Orobiyi et al. (2017), Dossou-Aminon et al. (2016)
Sorghum bicolor L.	Cereal crops Africa and India are the main producers. Sorghum can be cultivated in many dry and semiarid parts of the world.	Polyphenols: ferulic acid. Caffeic acid, apigeninidin, and luteolinidin	

(continued)

Table 12.1 (continued)

Popular plant species	Diversity, ecology, and distribution	Secondary metabolites	Reference
Gramineae	Monocotyledonous plants Pacific and eastern Asia	Benzoxazinoids	Niemeyer (1988), Whyte (1974)
Papaver somniferum	Annual herb Native to the western Mediterranean region and now grown throughout Europe and Asia	Codeine Morphine	Velu et al. (2018), Lu et al. (2010), Rozov-Ung et al. (2014), Peralta et al. (2021), Tétényi (2010)
Solanum tuberosum	Herbaceous plant with hairy stem (sometimes nearly hairless) Native to South America and now cultivated worldwide	Solanine	
Scutellaria baicalensis	Perennial herbaceous plant Cultivated in Asia, Siberia, and parts of Russia	Baicalin	Pant et al. (2021), Billah et al. (2019), He et al. (2021), Yu et al. (2021)
Erigeron breviscapus	Annual to perennial herb under the Asteraceae family Grown in middle and high altitudes of southwestern part of China	Scutellarin	
Rauvolfia serpentina	Herb/shrub/small tree Native to tropical and subtropical regions, like Asia	Ajmaline	Isah (2019), Kumar et al. (2022), Semwal et al. (2021), Kean et al. (2017), Pandey et al. (2022)
Dioscorea deltoidea	Tuber crop Found in hilly regions or tropical and subtropical regions of the world	Diosgenin	
Bacopa monnieri	Perennial herb Wetlands and muddy shores	Bacosides	
Nicotiana tabacum	Perennial Native tropical and subtropical America	Alkaloids	Ahlawat et al. (2024), Rawat et al. (2013), Watt (1987), Lim (n.d.)
Senecio jacobaea	Biennial or perennial plant Several parts of the world		
Glycine max	Bushy, hairy, annual legume Native to East Asia		

12.3.1 Flavonoid Biosynthetic Pathway: Metabolomic and Transcriptomic Analyses

Among the secondary metabolites, flavonoids are of prime interest as they possess different kinds of medicinal benefits, including anticancer, antioxidant, anti-inflammatory, and antiviral properties (Ullah et al. 2020). The chemical structure of flavonoids can be expressed as C6-C3-C6 and known as chalcones where C6 and C3 refer to the benzene ring and a 3-carbon long linking chain, respectively. The two benzene rings are known as ring A and ring B, whereas an additional heterocyclic pyran or pyrone ring, termed ring C, can also be found in some flavonoids (Corradini et al. 2011; FRS and Baxter 1999).

The acetate pathway and shikimate pathway are the two key processes responsible for flavonoid biosynthesis (Fig. 12.1). During the acetate pathway, glucose is transformed into malonyl-CoA molecules and three such molecules subsequently synthesize ring A whereas the shikimate pathway depends on synthesizing ring B. Phenylalanine, an end product of the shikimate pathway, synthesizes 4-coumaroyl-CoA that further synthesize ring B of flavonoids. Rings A and B are further condensates and generate chalcone that subsequently transformed into flavanone through an isomerase-catalyzed cyclization. This flavanone serves as the precursor of other 7000 flavonoids. However, all known flavonoids share this common biosynthetic pathway where 4-coumaroyl-CoA also acts as a central intermediate of different metabolic pathways involved in synthesizing different secondary plant metabolites including lignols, flavonoids, isoflavonoids, coumarins, aurones, stilbenes, catechin, and others (Fig. 12.2) (Khoo et al. 2017; Nabavi et al. 2020).

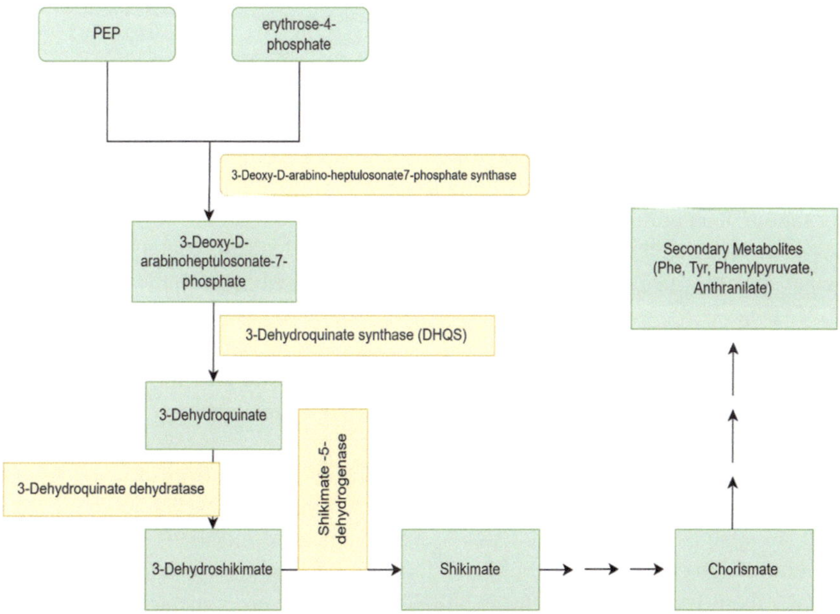

Fig. 12.1 Shikimate pathway. [Adopted from Tiwari et al. (2019)]

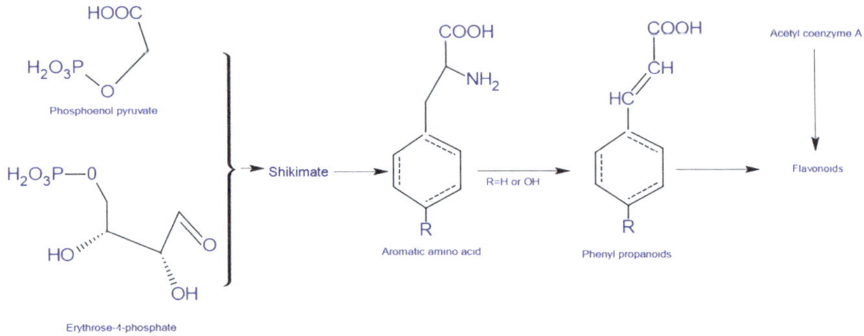

Fig. 12.2 Structural representation of flavonoid biosynthetic pathway. [Adopted from Nabavi et al. (2020)]

2-phenylchroman (flavan) is the main structural unit of the ring C containing flavonoids whereas their structural diversity critically depends on the oxidation of this ring and its degree of unsaturation. Based on the basic structure, flavonoids can be categorized into limited groups where differences among individual compounds are developed by hydroxylation, methylation, glycosylation, and acylation that substitute different chemical groups at various ring positions (Khoo et al. 2017; Nabavi et al. 2020).

12.3.2 Alkaloid Biosynthesis: Biochemistry and Molecular Regulation

Alkaloids, nitrogenous alkaline metabolites, on the other hand, have the power to harm and heal living organisms and have been used for centuries across countries. They were among the first pharmaceuticals developed for therapeutic purposes. However, the biosynthetic pathway of the alkaloids is comprised of some complex phenomena and divided into (i) accumulation of amine and aldehyde precursor, (ii) formation of an iminium cation, and (iii) a Mannich-like reaction. However, the first step is considered the most crucial one (Lichman 2021). This Mannich-like step is considered the first step committed to the alkaloid synthesis pathway (Lichman 2021) (Fig. 12.3).

12.3.2.1 Accumulation of Amine and Aldehyde Precursor

Compositional nitrogen in complex alkaloids often derives from primary amine derivatives, instead of incorporating amino acids directly. However, amine derivatives are also essential intermediates of other important metabolic pathways, and therefore, high concentrations of amine derivatives are necessary to accumulate and engage them for alkaloid synthesis without disturbing other metabolic pathways.

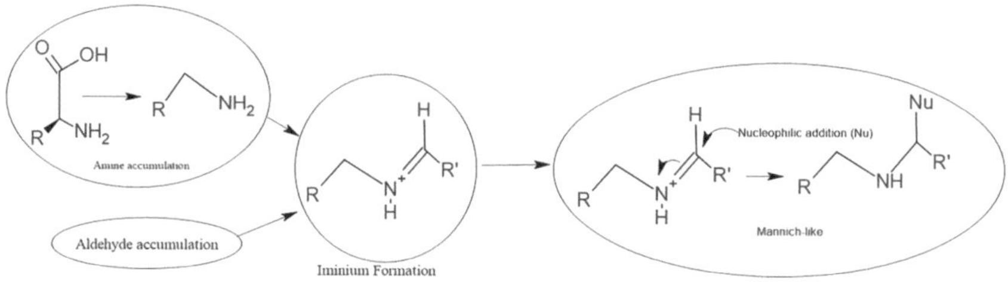

Fig. 12.3 Structural representation of alkaloid biosynthesis: a generalized view. [Adopted from Lichman (2021)]

Duplication of responsible genes and upregulation of genetic expression are some of the approaches that can be used for high amine synthesis. Polyamines are commonly generated from lysine, arginine, or ornithine, while aromatic amines are formed from tryptophan and tyrosine (Fig. 12.4) (Chen et al. 2019; Ishiai et al. 2016; Maeda 2019; Moghe and Last 2015; Serotonin et al. 2018).

Plant alkaloids contain pyrrolinium or piperideine and are derived from simple polyamines putrescine and cadaverine, respectively. On the other hand, pyrrolizidine alkaloids are developed by dimerization of triamine putrescine and/or homospermidine. Cadaverine, however, acts as a precursor of piperideine moiety of alkaloids and belongs to granatane and quinolizidine families (Lichman 2021).

Tyramine and tryptamine, which are amine derivatives of the aromatic amino acids tyrosine and tryptophan, are also used to make alkaloids. Tyrosine decarboxylase and tryptophan decarboxylase act on amino acids to produce the amine derivatives. Moreover, dopamine, a chemical intermediate derived from tyrosine through hydroxylation and decarboxylation, also acts as a precursor of different plant alkaloids (Fig. 12.5) (Graça 2015; Hagel and Facchini 2013; Lichman 2021).

However, unlike conventional alkaloids, betalain-producing plants derive the pigments directly by incorporating tyrosine without decarboxylation (DeLoache et al. 2015).

LDOPA and cyclo-DOPA are nonproteinogenic amino acids that function as precursors of betalains. They are produced by hydroxylating tyrosine and then oxidative cycling. Tryptamine, an amine derivative of tryptophan, is a precursor to β-carboline alkaloids such as harmala alkaloids, nigakinone, and monoterpene indoles (DeLoache et al. 2015; O'Connor and Maresh 2006).

Accumulation of aldehyde precursors at levels almost comparable to amine precursors is required to continue to the next step of alkaloid biosynthesis, which involves the creation of a reactive iminium intermediate. Polyamine-derived alkaloids need the formation of amino aldehydes by oxidative deamination of terminal polyamines (Lichman 2021). When oxidative deamination occurs, amine oxidases or FAD-dependent polyamine oxidases catalyze the spontaneous formation of a cyclic iminium intermediate (Wang et al. 2019).

Surprisingly, aldehydes that are included in alkaloids, produced from amino acids, accumulate separately from the amine precursor (Lichman 2021).

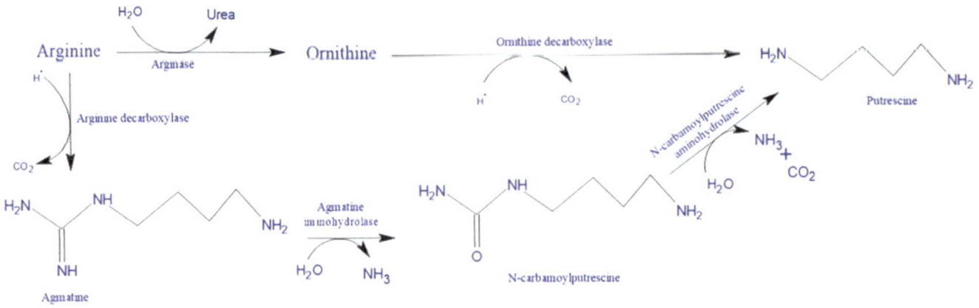

Fig. 12.4 Structural biology of putrescine biosynthesis from ornithine or arginine. [Adopted from Lichman (2021)]

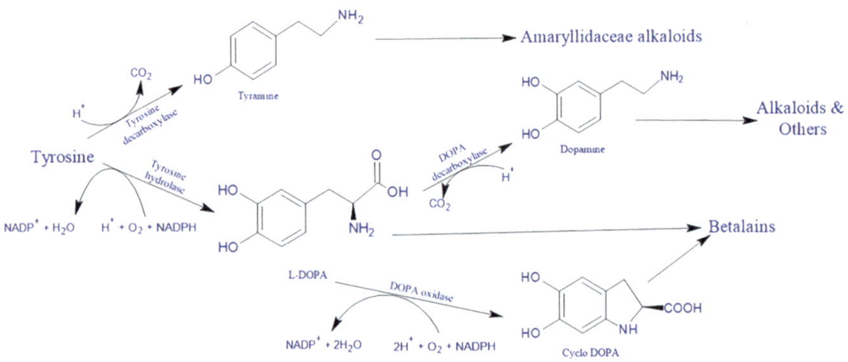

Fig. 12.5 Molecular basis of tyrosine-derived alkaloids biosynthesis. [Adopted from Lichman (2021)]

Benzylisoquinolines, another type of alkaloids, are synthesized from tyrosine-derivative dopamine and 4-hydroxyphenylacetaldehyde (Fig. 12.6). This 4-hydroxyphenylacetaldehyde molecule is derived from tyrosine in a two-step procedure, viz., transamination followed by decarboxylation. However, another alternative of this route has been identified in *Macleaya cordata* where benzylisoquinolines are derived from 3,4-dihydroxyphenylacetaldehyde and not from 4-hydroxyphenylacetaldehyde (Lee and Facchini 2011; Liu et al. 2017; Singh et al. 2019).

Benzaldehydes, specifically 3,4-dihydrobenzaldehyde, are another group of aldehydes that contribute to the synthesis of amaryllidaceae alkaloids. These substituted aldehydes are synthesized as a product of the phenylalanine-derived phenylpropanoid pathway (Gallage et al. 2014; Kilgore and Kutchan 2016; Lichman 2021).

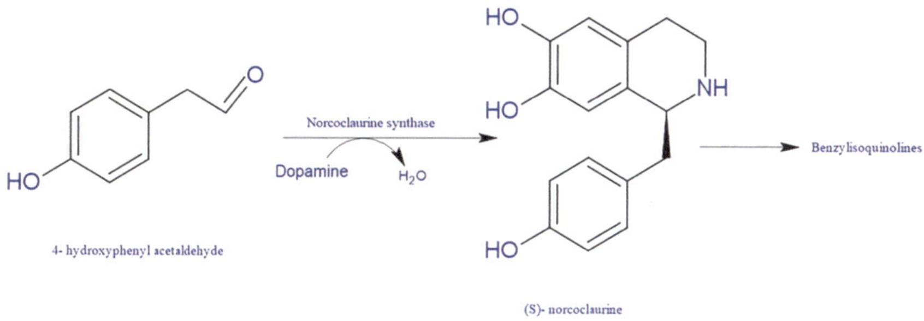

Fig. 12.6 4-hydroxyphenylacetaldehyde biosynthesis and its incorporation into alkaloids of benzylisoquinolines: structural elucidation. [Adopted from Lichman (2021)]

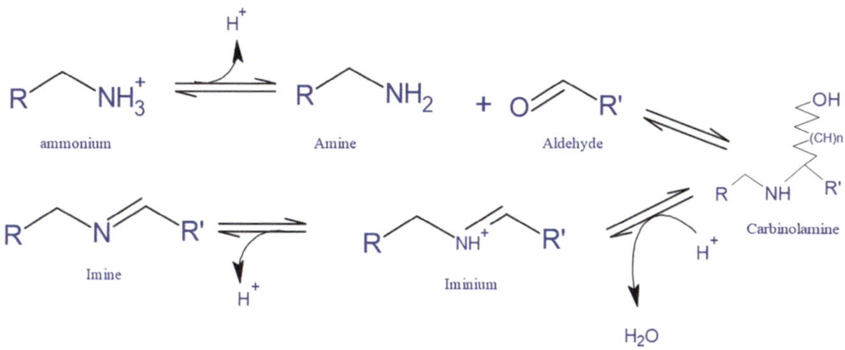

Fig. 12.7 Iminium formation: molecular mechanism. [Adopted from Lichman (2021)]

12.3.2.2 Iminium Formation

Electrophilic iminium cation formation crucially depends on reversible condensation of primary amine along with an aldehyde (Fig. 12.7). However, the concentration of iminium is often found low in an aqueous environment whereas an elevated concentration can be achieved with cyclic or conjugated iminium ion (Atherton et al. 2004; Eldin et al. 1995; Eldin and Jencks 1995; Hupe 1984; Scheller et al. 2015).

Surprisingly, under the range of physiological pH, a plentiful amount of reactive iminium species are found rather than imine species. Cyclic iminiums such as N-methylpyrollinium and D1-piperideine are found as intermediates of the polyamine alkaloid biosynthesis pathway and are formed through intramolecular condensation of linear amino aldehydes. On the contrary, intermolecular condensation takes place during the synthesis of aromatic alkaloids where noncyclic iminium is synthesized as the intermediate species (Hupe 1984; Lichman 2021; Rouchaud and Braekman 2009).

12.3.2.3 Mannich-Like Reaction

Mannich-like reaction is often considered the key step in alkaloid biosynthesis where iminium reacts irreversibly with a nucleophile. Intermolecular Mannich-like reaction can be witnessed during the synthesis of polyamine-derived alkaloids where the nucleophiles are essentially needed to be accumulated in the same subcellular location as the iminium (Lichman 2021). However, biosynthesis of tropane alkaloids in the *Solanaceae* family of plants is considered as the most well-understood Mannich-like reaction involved in synthesizing polyamine-derived alkaloids (Lichman 2021). In this case, 3-oxoglutaric acid's enol form attacks N-methylpyrollinium as a nucleophile, whereas pyrrolidine ketide synthase, a type III polyketide synthase, catalyzes the conversion of two malonyl-CoA units into 3-oxoglutaric acid. However, N-methylpyrollinium and malonyl-CoA produce 4-(1-methyl-2-pyrrolidinyl)-3-oxobutanoic acid during incubation with pyrrolidine ketide synthase. This important intermediate is then transformed into tropinone by tropinone synthase through a spontaneous intramolecular decarboxylative Mannich condensation (Bedewitz et al. 2018; Huang et al. 2019; Ping et al. 2019).

12.3.3 Plant Terpenoids: Biosynthesis and Regulation

Terpenoids are a group of chemicals that play a significant role in signal transduction and plant defense mechanism. They possess pharmaceutical properties and also attract insects for pollination purposes (Ashour et al. 2010). The mevalonate pathway is a well-known pathway committed to terpene biosynthesis. Isopentenyl diphosphate and dimethylallyl diphosphate are two intermediates in the mevalonate pathway that acts also as precursors of terpenoids. While the mevalonic acid process is carried out in the cytosol, peroxisomes, and endoplasmic reticulum, they are generated by two compartmentalized mechanisms (Carrie et al. 2007; Singh and Sharma 2015). The 3-hydroxy-3-methylglutaryl CoA, developed following condensation of three units of acetyl CoA, is the precursor of mevalonic acid, which is subsequently converted to isopentenyl diphosphate through phosphorylation and decarboxylation. Finally, mevalonic acid is synthesized following the reduction of 3-hydroxy3-methylglutaryl CoA catalyzed by 3-hydroxy-3-methylglutaryl CoA reductase (Basson et al. 1988; Igual et al. 1992; Rodwell et al. 2000). On the contrary, *Arabidopsis thaliana* phosphorylates mevalonic acid with the help of mevalonate kinase and phosphomevalonate kinase and produces mevalonate-5-diphosphate (Lluch et al. 2000; Tsay and Robinson 1991). Finally, mevalonate-5-diphosphate is decarboxylated to provide isopentenyl diphosphate. Isopentenyl diphosphate is the end product of the mevalonate pathway dedicated to terpenoid biosynthesis (Dhe-Paganon et al. 1994) (Fig. 12.8).

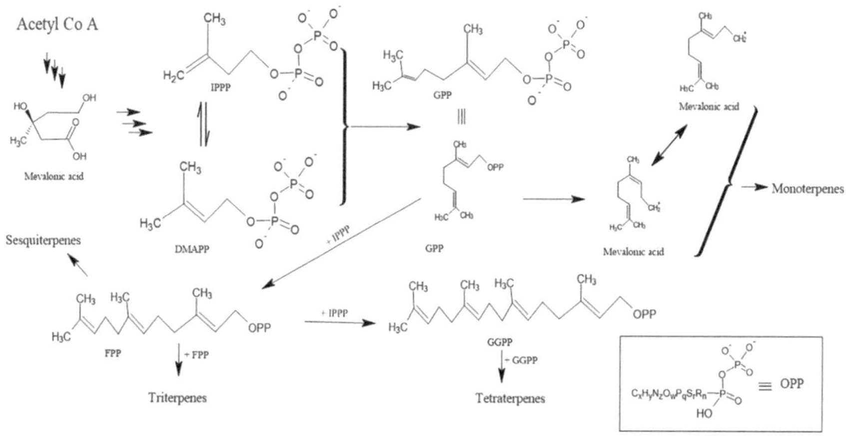

Fig. 12.8 Terpenoid biosynthesis: molecular insight. [Adopted from Habtemariam (2019)]

Plastids also take part in terpenoid biosynthesis where pyruvic acid and glyceraldehydes-3-phosphate are condensed by 1-deoxy-D-xylulose 5-phosphate synthase and produce 1-deoxy-D-xylulose 5-phosphate (Sprenger et al. 1997). 1-deoxy-D-xylulose 5-phosphate reductoisomerase subsequently reduces 1-deoxy-D-xylulose 5-phosphate to 2-C-methyl-D-erythritol 4-phosphate. The conjugation of 2-C-methyl-derythritol 4-phosphate and 4-cytidine 5-phosphate results in the formation of 4-cytidine 5-phospho-2-C-methyl erythritol, which is catalyzed by 2-C-methyl D-erythritol 4-phosphate cytidyltransferase. Next, 4-cytidine-5-phospho-2C-methyl erythritol is converted to 2-C-methyl erythritol 2,4-cyclodiphosphate by the enzyme 2-C-methyl erythritol 2,4-cyclodiphosphate synthase (Calisto et al. 2007; Rohdich et al. 2000; Steinbacher et al. 2003; Takahashi et al. 1998). Mevalonate, however, is furthermore converted to IPP, which subsequently forms geranyl geranyl diphosphate by the catalysis of geranyl geranyl diphosphate synthase and finally synthesizes diterpenes including sterols and triterpenoids (Sparg et al. 2004).

12.4 Metabolic Engineering: A Step Toward the Production of Enhanced and Novel Secondary Metabolites

The biosynthetic pathway of plant species is regulated by several proteins and most of them are transcription factors. The development of engineered plant proteins with different substrate specificity, the introduction of new genes, and the upregulation of genetic expression are some of the promising approaches that have been practiced in recent years for metabolic engineering purposes.

12.4.1 Metabolic Engineering for Enhanced Production of Flavonoids and Anthocyanins

The first focus for genetic engineering was the upregulation of flavonoid and anthocyanin production since their biosynthetic pathway was well established and the effect can be easily accessed through visual changes in flower color. However, upregulation of the expression of flavonoids has been considered as a potential approach to increase its traces in food, and chalcone isomerase has been found as the key regulatory enzyme for the same (Davies 2000; Dixon and Steele 1999; Forkmann and Martens 2001; Muir et al. 2001). A 78-fold increase in flavonoid production has been achieved following the upregulation of Petunia CHI gene expression in tomatoes. Improved isoflavans production in heterologous plant species, on the other hand, depends on the modification of the phenylpropanoid pathway and transcription factors are used for the same. However, constitutive transcription factor expression in wild-type or heterologous plant species has an effect on phenylpropanoid biosynthesis pathways. For example, induction of anthocyanin biosynthesis in maize kernels has been achieved by overexpressing transcription factors R and C1 that ultimately upregulate the complete flavonoid biosynthesis pathway in undifferentiated maize cells, cultured in vitro (Grotewold et al. 1998). Moreover, chalcone synthase along with overexpressed C1 and R transcription factors is used to develop fungal-resistant rice by activating anthocyanin biosynthesis. In *Arabidopsis* also, overexpression of PAP1, a single MYB-type transcription factor, is coupled with the synthesis of purple pigment in the plant during development (Borevitz et al. 2000; Gandikota et al. 2001). However, repression of transcription factors was also found responsible for increasing certain types of secondary metabolites in plants. For example, *Arabidopsis* with knocked-out MYB4 produce an enhanced level of sinapate esters, which eventually accumulate in leaves and increase the tolerance of UV-B irradiation. Similarly, overexpressed FaMYB1, an MYB-type protein from strawberry, in tobacco reduced pigment deposition in flowers along with the concentration of flavonols and anthocyanin. This phenomenon suggests that FaMYB1 acts as a repressor protein and represses certain stages of the flavonoid biosynthesis (Aharoni et al. 2001; Jin et al. 2000).

12.4.2 Application of Metabolic Engineering to the Production of Alkaloids

Different plants synthesize different species-specific alkaloids and most of them share partially similar biosynthetic pathways with common intermediates. Pathways for terpenoid in suspension culture of *C. roseus* are critically regulated by the expression of involved genes. Overexpression of the genes is associated with enhanced tryptamine production whereas the amount of indole alkaloids surprisingly remains constant (Verpoorte et al. 2000). Strictosidine is a common intermediate of antitumor alkaloids vinblastine, vincristine, and camptothecin from which different species-specific alkaloid biosynthetic pathways diverge (Verpoorte et al. 1998). Tryptamine, another crucial intermediate, can be overexpressed by upregulating the gene tryptophan decarboxylase. However, this phenomenon does not increase the number of downstream alkaloids, whereas upregulated strictosidine synthase exhibits a contradictory result (Canel et al. 1998). Regulation of transcriptional factors ORCA2 and ORCA3 have also been carried out to control different pathways associated with alkaloid biosynthesis in *C. roseus*. However, upregulated ORCA3 expression did not exhibit any increment in alkaloid biosynthesis, and the same established the fact that it doesn't affect the regulation of the G10H gene encoding cytochrome P450 enzyme, which is responsible for synthesizing secologanin, an intermediate of the mevalonate pathway and precursor of indole alkaloids. However, a threefold increase in the production of secologanin has been achieved by the direct introduction of loganin, the precursor of secologanin (Memelink et al. 2001; Menke et al. 1999; van der Fits and Memelink 2000). The synthesis of isoquinoline, another pharmaceutically important alkaloid, can also be upregulated by modifying the expression of cellular enzymes. For example, the production of berberine is regulated by the enzyme (S)-scoulerine 9-O-methyltransferase, and overexpression of the responsible gene can increase berberine production from 79% to 91%. This example proves the fact that by increasing the expression of any enzymes, flux of any intermediate can be regulated (Sato et al. 2001). Hyoscyamine-6β-hydroxylase catalyzes the hyoscyamine bioconversion to scopolamine, a value-added product, and overexpression of the encoding genes increases the scopolamine levels by 100 times in the hairy root culture of *Hyoscyamus muticus* (Jouhikainen et al. 1999). Biosynthesis of tropane and pyrrolidine in *Atropa belladonna* and *Nicotiana sylvestris*, respectively, involved methylputrescine biosynthesis, and the first dedicated step and overproduction of alkaloids has been achieved by overexpressing the tobacco putrescine N-methyltransferase gene (Verpoorte and Memelink 2002).

12.4.3 Metabolic Engineering in Terpenoid Biosynthesis: Monoterpenes and Sesquiterpenes

Biosynthesis of different plastidial terpenoids, including carotenoids, monoterpenes, and diterpenes, has been assessed in the recent past and the role of the 2-C-methyl-D-erythritol-4-phosphate (MEP) pathway in the biosynthesis of plastidial terpenoids has been discovered. Modification or cloning of the genes involved in this pathway has been considered as a potential strategy for metabolic engineering. Knocking out the cytochrome P450 enzyme from tobacco trichome glands by using co-suppression or antisense strategy increased the production of diterpenoid cembratriene-ol and decreased its oxidized product cembratriene-diol, which eventually increases the resistance of tobacco plants against aphids. Moreover, an increased flux in the sesquiterpenoid biosynthetic pathway following the upregulation of chimeric farnesyl diphosphate synthase gene in *Artemisia annua* showcased a twofold to threefold increase in the antimalaria drug. Even in tomatoes, overexpression of S-linalool synthase transgene coupled with increased S-linalool production, which is basically a monoterpenoid flavor compound and surprisingly no modification has been witnessed in the levels of other terpenoids. Researchers achieved a drastic elevation in 1-deoxy-d-xylulose 5-phosphate reductoisomerase activities in mint by overexpressing the involved genes, and the same also exhibited a 50% increase in essential oil production, which is monoterpenoid.

Enrichment of mint oil with an increased relative percentage of methanol can also be accomplished through blocking any cellular reaction(s) dedicated to the production of menthofuran using a monoterpenoid metabolic network (Broun and Somerville 2001; Lange and Croteau 1999; Lange et al. 2000; Lewinsohn et al. 2001; Li et al. 2010; Mahmoud and Croteau 2001; Wang et al. 2001; Ye et al. 2000).

The β-carotene production in rice was a major achievement in metabolic engineering and was achieved by introducing phytoene synthase, phytoene desaturase, and lycopene β-cyclase genes and overexpressing them (Fraser et al. 2002; Mann et al. 2000; Römer et al. 2000; Rosati et al. 2000; Shewmaker et al. 1999; Ye et al. 2000).

Following a course of research on metabolic engineering and secondary metabolite production, metabolite production has been considered as an orchestrated phenomenon made up of several signaling networks that are structured like neutral networks. Boolean networks provide a simple and logical framework for understanding crosstalk among plant cells and developing models to depict information flow.

12.5 Production of New Compounds: Synergies Between Synthetic Biology and Metabolic Engineering

Combinatorial biochemistry is a promising approach for the production of a new type of secondary metabolites where enzymes with new substrate specificity are introduced. *Thalictrum tuberosum* O-methyltransferase subunits recombine to produce heterodimeric enzymes with novel substrate specificity, which are employed to generate new alkaloids in plants (Frick et al. 2001). However, limited information about the genetic crosstalk and metabolic genes set boundaries for the application of recombinatorial biochemistry in plants. The formation of new compounds in plants was first successfully carried out by introducing gene encoding dihydroflavonol reductase into petunia following isolation from maize, and the same experiments developed a new brick red flower color (Meyer et al. 1987). Production of benzoic acid derivatives from chorismate in plant cells has been achieved by introducing microbial genes into plant cells. Biosynthesis of salicylic acid in tobacco has also been successfully achieved by introducing two microbial genes into tobacco, and they were regulated by the CaMV 35S promoter. Moreover, the constitutive expression of the genes was also induced by coupling their expression with the Rubisco signal peptide sequences. Verberne and colleagues conducted the studies and generated plants with enhanced resistance to viruses and fungi by using the entC gene, which codes for isochorismate synthase and was isolated from *Escherichia coli*, and the orfD gene, which codes for isochorismate pyruvate lyase and was isolated from *Pseudomonas fluorescence* (Verpoorte et al. 2000). Hain and colleagues generated transgenic tobacco, tomato, potato, and oil seed rape plants with improved fungal resistance, and the major mechanism for this feature is constitutive production of the stilbene phytoalexin resveratrol. Overexpressing the stilbene synthase gene from grapevine resulted in constitutive synthesis of stilbene phytoalexin resveratrol (Hain and Grimmig 2000).

35S promoter and its enhancer sequence play a crucial role in gene regulation and *Agrobacterium tumefaciens* assisted insertion of the enhancer, and in tetrameric form, 1.5 kb upstream of the STS transcription start site upregulates the expression in tobacco by 14 times (Verpoorte et al. 2000).

12.6 Alkaloids and Flavonoids: Roles in Quorum Signaling and Microbial Interactions

Plant secondary metabolites have a wide range of pharmacological activities and serve key biological functions. They serve as both a defensive mechanism against pathogen assaults and a valuable therapeutic resource. The role of microorganisms in the control of plant secondary metabolism is widely established. The presence of certain microorganisms can boost secondary metabolite production, and their positive effects are influenced by ambient factors and plant-related bacteria.

Flavonoids in legumes have an important role in the onset of nodule development in the nitrogen-fixing symbiosis by triggering rhizobial nodulation and nodulation-related genes. Legume hosts continually release flavonoids, but their concentrations in the rhizosphere rise dramatically in the presence of suitable rhizobial strains (Liu and Murray 2016; Shaw et al. 2006). Most terrestrial plants have mycorrhizal fungi as vital symbionts, which are activated when the soil lacks phosphorus. In addition, mycorrhizal fungus can improve rhizobia symbiosis and shield plants from diseases. One theory is that the mycorrhizal fungus-induced flavonoids also promote the production of Nod factors. For instance, mycorrhizal fungi in soybeans produce daidzein, which also functions as a nod gene inducer for *Bradyrhizobium japonicum*. Conversely, mycorrhizal symbionts in *M. truncatula* generate cumestrol, which may function as a nod gene inhibitor in *S. meliloti*. Flavonoid structures interact with rhizobial NodD proteins to induce transcription of nodulation genes producing lipo-chitooligosaccharide Nod factors, which cause root hair deformation and aid rhizobial entrance through infection threads. Growth of successful infection thread is most likely depending on rhizobial synthesis of extracellular polysaccharides and proteins, which can be stimulated by flavonoid compounds (Chen et al. 2024; Khaliq ct al. 2022; Shaw et al. 2006; Wahab et al. 2023).

Flavonols also help to resist infections. One of the most commonly dispersed flavonols, quercetin, has potent antibacterial effects. Quercetin binds to the GyrB component of *E. coli* DNA gyrase, inhibiting its ATPase activity. However, the cleavage in DNA was only produced at concentrations greater than 80μM. Quercetin also reduced the development of the fungus *Neurospora crassa*. Carnation (*Dianthus caryophyllus*) defends against *Fusarium* infections by raising the content of kaemferide, a fungitoxic flavonol triglycoside, to as low as 50 nM. Global gene expression investigations have also revealed that when *M. truncatula* plants are attacked by *Phymatotrichopsis omnivora*, the pathogen responsible for cotton root rot disease, flavonoid production increases. To effectively colonize the plant, pathogens such as *P. omnivorous*, *Nectria haematococca*, and *Rhizoctonia solani* have "learned" to avoid many of these flavonoids (Chagas et al. 2022; Mahmud et al. 2023; Rodríguez et al. 2023; Shamsudin et al. 2022; Shaw et al. 2006; Yan et al. 2024).

It is becoming more widely recognized that flavonoids are also a component of a group of signals that are transmitted between arbuscular mycorrhizal fungus (AMF) and their host plants throughout the presymbiotic, colonization, and symbiotic phases of the symbiotic relationship. Flavonoids have been found to have an impact on hyphae development, spore germination, and colonization extent. For instance, the flavonol quercetin presymbiotically promotes AMF spore germ tube growth and hyphal branching.

Apart from pterocarpan phytoalexins' traditional antibacterial functions, other flavonoid structures—like catechin—have been shown to be microbially toxic and may exist in exudates at harmful amounts (Chagas et al. 2022; Mahmud et al. 2023; Rodríguez et al. 2023; Shamsudin et al. 2022; Shaw et al. 2006; Yan et al. 2024).

Other extensive and chemically variable natural molecules found in plants are terpenoids, which are found in many different plant tissues. The effect of microbes on terpenoids has been the subject of several investigations so far.

Nearly all plants' internal tissues are asymptomatically colonized by microorganisms like endophytic bacteria. According to research, adding endophytic bacteria, such as *P. fluorescens*, *Bacillus pumilus*, *Acinetobacter pittii*, and others, to medicinal plants greatly boost their terpene content. Furthermore, the buildup of terpenes is intimately associated with plant growth promoting rhizobacteria (PGPR) inoculation. When PGPR (*Mucilaginibacter* sp. and *Pseudomonas* sp.) were injected into cannabis, the terpene concentration rose by almost 23% and 18%, respectively, in comparison to the control. It is abundantly evident that endophytic bacteria and PGPR, particularly *Bacillus*, *Pseudomonas*, *Acinetobacter*, and *Mucilaginibacter*, may promote plant development and generate terpene chemicals (Lv et al. 2024; Pandey et al. 2023).

The buildup of plant terpenoids is likewise intimately related to the microbiota. For instance, when *A. annua* plants were infected with four bacterial endophytic combinations, the output of artemisinin rose by almost 658% in comparison to the uninoculated reference. Citrus monoterpene synthesis is influenced by both endophytic and rhizosphere soil microbes. The seed microbiota of *Salvia miltiorrhiza* has a wide variety of terpenoid metabolic pathways, suggesting that these microbes play a key role in the accumulation of secondary metabolites. These investigations showed that the rhizosphere soil microbiome, seed microbiome, and synthetic microbial community may all encourage the production of terpenoids in a variety of plants (Chen et al. 2021; Lv et al. 2024; Pandey et al. 2023; Vandana et al. 2021).

Alkaloids, which have a major impact on plant development, are mostly metabolites of amino acids that include nitrogen heterocycles. The *norank_f_ Anaerolineaceae* of *Lycium barbarum* considerably increased the accumulation of betaine in the rhizosphere, whereas the rhizosphere bacteria N5.18, *Stenotrophomonas maltophilia*, enhanced the opium poppy's overall alkaloid output. Treatment with a low quantity of exogenous PGPR (*P. fluorescens*) was observed to increase the amount of alkaloids in *C. roseus*, as well as endophytic and rhizosphere bacteria (Herbert 2003; Lv et al. 2024).

Phytohormones are produced by certain microbes to encourage photosynthesis and plant development. *P. fluorescens* ALEB7B synthesizes indole acetic acid (IAA) to boost photosynthesis, raise terpene precursor levels, and improve plant root growth and vigor by promoting the conversion of anaerobic sesquiterpenes into oxygenated sesquiterpenes. Numerous natural compounds can be produced by plant photosynthesis, but the NADPH generated by this process affects the rate at which secondary metabolites are synthesized because the endoplasmic reticulum's P450 monooxygenase is involved in the synthesis of bioactive substances and NADPH electron transport provides energy (Donoso et al. 2017; Egamberdieva et al. 2017; Lv et al. 2024; Orozco-Mosqueda et al. 2023).

By changing the amounts of endogenous plant hormones that control secondary metabolism, microorganisms promote the expression of a number of significant genes and transcription factors. For instance, the signaling process triggered by endophytic fungi involves the interaction of jasmonic acid (JA) signaling with phytohormones (e.g., GA and ET). This alters the transcription factors involved in the biosynthesis of Taxol, which either directly or indirectly controls the expression of genes or transcription factors related to synthesis, thereby influencing the accumulation of Taxol in *Taxus chinensis* (Eichmann et al. 2021; Lv et al. 2024; Shi et al. 2024; Wang et al. 2023).

It has been discovered that mycorrhizalization increases abscisic acid (ABA) and JA levels. Transcription factor MYC2, which controls the production of secondary metabolites including phenolic acids and anthocyanins, can be activated by these hormones. In addition to inducing the reactions of ET and JA, inoculation with *P. fluorescens* may increase the synthesis of GA and CK. In turn, these reactions might raise the transcription factor CrWRKT2 expression as well as the levels of the TIA biosynthesis pathway's tabersonine 16-hydroxylase (T16H), *C. roseus* peroxidase (CrPRX), geraniol-10-hydroxylase (G10H), and deacetylvindoline-4-O-acetyltransferase. Phytohormones are key signaling molecules that help plants produce secondary metabolites. Additionally, JA significantly upregulated the 3-hydroxy-3-methylglutaryl-CoA reductase (HMGR) genes, which led to an increase in sesquiterpene components. AMF has been shown to improve the phenylalanine route and boost the synthesis of phenolic compounds in roots by increasing the buildup of H_2O_2, SA, and NO, which sets off a signaling cascade. Increased amounts of SA, H_2O_2, and NO signaling molecules in the roots during AMF colonization of the root system support the phenylalanine metabolic pathway, which raises the synthesis of phenylalanine ammonia lyase (PAL) and adds to the rise in phenolic compounds (Lv et al. 2024; Pozo et al. 2008; Song et al. 2022).

Flavonoids and alkaloids are important in microbial interactions and quorum signaling, affecting both symbiotic and defensive plant-microbe connections. Flavonoids regulate nodulation, improve mycorrhizal connections, and have antibacterial characteristics, whereas alkaloids aid plant defense and development by altering secondary metabolism. Microorganisms have a substantial impact on the manufacture of these chemicals by changing phytohormone levels and activating important metabolic pathways. The interaction between plant secondary metabolites with microbial communities is critical for plant health, nutrient acquisition, and disease resistance. Understanding these dynamics can pave the way for more sustainable farming methods and innovative biotechnology uses.

12.7 Future Prospective

Intensive agriculture methods on restricted land with low fertilizer and agrochemical inputs provide a major worldwide concern. Despite continual advances in omics-based techniques, experimental and computational validation processes are not standardized. The discovery and characterization of metabolite-protein interactions provides a valuable viewpoint for studying individual metabolite bioactivities and, more broadly, understanding biological regulatory mechanisms. As a result, newly unbiased and systematic techniques are revealing new insights into metabolite-protein interactomes. Challenges and new directions for plant proteomics during the next decade may be defined at the biology, proteomics, experimental, and analytical, wet and dry approach levels (Diwan et al. 2022; Jorrin Novo 2021).

The comparison of metabolic profiling of different developmental stages, with a focus on seed maturation and germination, will aid in understanding developmental processes, the regulatory effects of various metabolites, and how this knowledge can be applied to develop methods for metabolite regulation, protein engineering, and the development of engineered species. We should move beyond the analysis of roots, leaves, and seeds to other organs such as xylem, phloem, and the extracellular fraction in order to understand how a plant functions as a whole, identify the core proteome, and understand organ communications, sink-source relationships, metabolite synthesis, and their regulation by the biotic environment (e.g., mycorrhiza). In this aspect, in vitro cultivation techniques will be beneficial, despite the fact that they have removed us from natural environments (Jorrin Novo 2021).

Plant responses to stress are the primary issue in plant proteomics and secondary metabolite synthesis, but most research have focused on individual stressors. As in nature, combination stresses occur; thus, the experimental design should be modified in this direction. When researching secondary metabolite synthesis in various organs and tissues, it is important to remember that the sample is a mix of diverse cell signals, each having its own signature and responding differently to environmental variables depending on location. Metabolite formation is triggered by several metabolic and signal transduction pathways, as well as gene expression control, and is mediated by permanent or transitory protein or molecular complexes whose knowledge and comprehension are based on interactomic techniques (Bontinck et al. 2018; Iacobucci et al. 2021; Jorrin Novo 2021; Kivivirta et al. 2019; Yang et al. 2020).

Validating proteomics data and combining it with other molecular, genetic, physiological, cell biology, and phenotype-based methods is imperative. There are now more options for comparative and functional omics research, thanks to the multiplexing technique and Activity-Based Proteome Profiling (ABPP) (Latha et al. 2021; Rey et al. 2020; Veenstra 2021; Yu et al. 2020). There is still opportunity for improvement at the analytical level, particularly in identifying low and very low abundant or resistant proteins or metabolites, even though the MS upstream steps of the proteomics workflow are founded on tried-and-true protocols of protein extraction, fractionation, and separation (Jorrin-Novo 2014, 2020).

Protein-mediated molecular events are known to cause not only changes in protein abundance or metabolite profile but also posttranslational modifications (PTMs), macro- and micromolecular interactions, cleavage, or conformational changes, processes that may happen concurrently but are not feasible to experimentally afford on a proteome-wide scale. Native mass spectrometry, cross-linking mass spectrometry, and affinity purification-MS are techniques intended particularly for researching PTMs and/or cross-protein interactions. A novel approach known as "limited proteolysis-coupled mass spectrometry" (LiP-MS) has been developed to assess changed protein structure at the -omics level. This can be regarded as the most recent advancement in the proteomics field (Allison and Bechara 2019; Feng et al. 2014; Gemperline et al. 2019; Zhu et al. 2016).

The innovative PROMIS technique enables systematic research that allows reporting on complete biological networks, giving an effective means to find novel metabolites as well as novel metabolite functions simultaneously. Modern technical advancements enhance our analytical capabilities, such as ligand-detected NMR created in bacteria, which gives great resolution of direct binding of weak contacts. However, not all possible methodologies have been effectively applied to plants (Diether et al. 2019; Venegas-Molina et al. 2021).

In the future, certain key criteria can be addressed in terms of the accuracy of existing methodologies to studying metabolite-protein interactions. First, sufficient sample preparation techniques should be promoted to confirm accurate protein structure, such as selecting appropriate buffers, pH levels, cofactor content, and so on. Second, sophisticated new algorithms are being created to cope with more complicated data, such as the LiP-Quant technique, which evaluates positive interactions using a machine learning-based composite score. Third, confirming and analyzing biological in vivo functioning is essential. Unfortunately, it is a significant bottleneck that necessitates costly and low-throughput testing. Nonetheless, such investigation is required to separate true interactions from false positives or accidental physical interactions that lack biological activity (Piazza et al. 2019; Venegas-Molina et al. 2021).

Furthermore, identifying complex microbial communities in varied habitats remains a considerable problem, exacerbated by the fact that only a tiny proportion of microorganisms have been adequately described to yet. The key challenge is determining the best technology and approaches for efficiently addressing specific issues. Furthermore, limits in accessible databases provide another obstacle that must be solved in order to develop microbial research fully. Despite these developments, the industry continues to face challenges in developing analytical tools to combine diverse datasets. To address this, there is an increasing need for multi-omics research that combines traditional methodologies such as metabolomics, transcriptomics, proteomics, and metagenomics with new and emerging techniques such as genomes, epigenomics, and lipidomics (Kimotho and Maina 2024; Kumar et al. 2023). Furthermore, understanding the precise processes and kinetics of molecular signal identification, transmission, and response in various plant-microbe interactions necessitates advanced experimental and analytical methodologies.

The intricacy and variety of biological systems provide a number of difficulties, including the potential for problems like unclear data interpretation and trouble differentiating between correlations and causal variables. Furthermore, omics methods can produce enormous volumes of data, requiring advanced computing tools as well as proficiency in processing and interpretation. Furthermore, it might be difficult to fully capture the range of interactions since plant-microbe interactions are dynamic across many environmental circumstances and developmental phases. The reliance of omics methods on high-quality reference genomes and databases, which might not be available for non-model species or poorly defined microbial taxa, is another drawback (Jain et al. 2024).

12.8 Conclusion

Understanding the proteomic profile of plants is crucial to understanding the effects of different biotic and abiotic stresses on the expression of various proteins. This understanding is also essential for studying their role in the synthesis and regulation of secondary metabolites. It has been found that tyrosine decarboxylase (TDC) and N-methyltransferase (NMT) are two of the most important enzymes for alkaloid biosynthesis, whereas terpene synthases (TPS) and geranyl diphosphate synthase (GPPS) are key components of terpenoid biosynthesis. Similarly, chalcone synthase (CHS), dihydroflavonol 4-reductase (DFR), and flavonoid 3′-hydroxylase (F3′H) are among the many enzymes that play key roles in flavonoid regulation.

By understanding these cascades, the concentration of various secondary metabolites of commercial value can be upregulated or downregulated by controlling the expression of key intermediates. Additionally, the regulation of metabolites can also affect quorum sensing, which eventually shapes the soil atmosphere in ways that benefit plants. This approach also enables the selective growth of specific plants. Thus, by deciphering the omics of metabolite synthesis and regulation, the artificial implementation of "survival of the fittest" can be mimicked, which can be useful for conservation, research, and purposes beneficial to humans.

Funding Statement No funding to declare.

Conflict of Interest The authors declare that they have no competing interests.

References

Aharoni A, De Vos CR, Wein M, Sun Z, Greco R, Kroon A, Mol JN, O'Connell AP (2001) The strawberry FaMYB1 transcription factor suppresses anthocyanin and flavonol accumulation in transgenic tobacco. Plant J 28(3):319–332

Ahlawat YK, Singh M, Manorama K, Lakra N, Zaid A, Zulfiqar F (2024) Plant phenolics: neglected secondary metabolites in plant stress tolerance Brazilian. J Bot 47(3):703–721. https://doi.org/10.1007/s40415-023-00949-x

Allison TM, Bechara C (2019) Structural mass spectrometry comes of age: new insight into protein structure, function and interactions. Biochem Soc Trans 47(1):317–327

Ashour M, Wink M, Gershenzon J (2010) Biochemistry of terpenoids: monoterpenes, sesquiterpenes and diterpenes. In: Annual plant reviews volume 40: biochemistry of plant secondary metabolism, pp 258–303

Ashraf MA, Iqbal M, Rasheed R, Hussain I, Riaz M, Arif MS (2018) Environmental stress and secondary metabolites in plants. In: Plant metabolites and regulation under environmental stress. Elsevier, pp 153–167

Atherton JH, Blacker J, Crampton MR, Grosjean C (2004) The Strecker reaction: kinetic and equilibrium studies of cyanide addition to iminium ions. Org Biomol Chem 2(18):2567–2571

Averyanov LV, Ponert J, Nguyen PT, Van Duy N, Khang NS, Nguyen VC (2016) A survey of *Dendrobium* Sw. sect. Formosae (Benth. & Hook. f.) Hook. f. in Cambodia, Laos and Vietnam. Adansonia 38(2):199–217

Basson ME, Thorsness M, Finer-Moore J, Stroud RM, Rine J (1988) Structural and functional conservation between yeast and human 3-hydroxy-3-methylglutaryl coenzyme A reductases, the rate-limiting enzyme of sterol biosynthesis. Mol Cell Biol 8(9):3797–3808

Bedewitz MA, Jones AD, D'Auria JC, Barry CS (2018) Tropinone synthesis via an atypical polyketide synthase and P450-mediated cyclization. Nat Commun 9(1):5281

Billah MM, Khan F, Niaz K (2019) *Scutellaria baicalensis* Georgi. In: Nonvitamin and nonmineral nutritional supplements. Elsevier, pp 403–408

Bontinck M, Van Leene J, Gadeyne A, De Rybel B, Eeckhout D, Nelissen H, De Jaeger G (2018) Recent trends in plant protein complex analysis in a developmental context. Front Plant Sci 9:640

Borevitz JO, Xia Y, Blount J, Dixon RA, Lamb C (2000) Activation tagging identifies a conserved MYB regulator of phenylpropanoid biosynthesis. Plant Cell 12(12):2383–2393

Broun P, Somerville C (2001) Progress in plant metabolic engineering. Proc Natl Acad Sci 98(16):8925–8927

Bruelheide H, Heinemeyer A (2002) Climatic factors controlling the eastern and altitudinal distribution boundary of *Digitalis purpurea* L. in Germany. Flora - Morphol Distrib Funct Ecol Plants 197(6):475–490. https://doi.org/10.1078/0367-2530-00064

Calisto BM, Perez-Gil J, Bergua M, Querol-Audi J, Fita I, Imperial S (2007) Biosynthesis of isoprenoids in plants: Structure of the 2C-methyl-d-erithrytol 2, 4-cyclodiphosphate synthase from Arabidopsis thaliana. Comparison with the bacterial enzymes. Protein Sci 16(9):2082–2088

Canel C, Lopes-Cardoso MI, Whitmer S, van der Fits L, Pasquali G, van der Heijden R, Hoge JHC, Verpoorte R (1998) Effects of over-expression of strictosidine synthase and tryptophan decarboxylase on alkaloid production by cell cultures of *Catharanthus roseus*. Planta 205:414–419

Carrie C, Murcha MW, Millar AH, Smith SM, Whelan J (2007) Nine 3-ketoacyl-CoA thiolases (KATs) and acetoacetyl-CoA thiolases (ACATs) encoded by five genes in *Arabidopsis thaliana* are targeted either to peroxisomes or cytosol but not to mitochondria. Plant Mol Biol 63:97–108

Chagas MSS, Behrens MD, Moragas-Tellis CJ, Penedo GX, Silva AR, Gonçalves-de-Albuquerque CF (2022) Flavonols and flavones as potential anti-inflammatory, antioxidant, and antibacterial compounds. Oxidative Med Cell Longev 2022(1):9966750

Chen D, Shao Q, Yin L, Younis A, Zheng B (2019) Polyamine function in plants: metabolism, regulation on development, and roles in abiotic stress responses. Front Plant Sci 9:1945

Chen L, Zhang X, Li Q, Yang X, Huang Y, Zhang B, Ye L, Li X (2024) Phosphatases: decoding the role of mycorrhizal fungi in plant disease resistance. Int J Mol Sci 25(17):9491

Chen Y, Hu B, Xing J, Li C (2021) Endophytes: the novel sources for plant terpenoid biosynthesis. Appl Microbiol Biotechnol 105(11):4501–4513

Corradini E, Foglia P, Giansanti P, Gubbiotti R, Samperi R, Laganà A (2011) Flavonoids: chemical properties and analytical methodologies of identification and quantitation in foods and plants. Nat Prod Res 25(5):469–495

Crozier A, Clifford MN, Ashihara H (2006) Plant secondary metabolites. In: Occurrence, structure and role in the human diet. Blackwell Publishers

Davies K (2000) Plant colour and fragrance. In: Metabolic engineering of plant secondary metabolism. Springer, pp 127–163

Davis AP, Gole TW, Baena S, Justin M (2012) The impact of climate change on Indigenous Arabica Coffee (*Coffea arabica*): predicting future trends and identifying priorities. PloS One 7(11):e47981. https://doi.org/10.1371/journal.pone.0047981

DeLoache WC, Russ ZN, Narcross L, Gonzales AM, Martin VJ, Dueber JE (2015) An enzyme-coupled biosensor enables (S)-reticuline production in yeast from glucose. Nat Chem Biol 11(7):465–471

Dhe-Paganon S, Magrath J, Abeles RH (1994) Mechanism of mevalonate pyrophosphate decarboxylase: evidence for a carbocationic transition state. Biochemistry 33(45):13355–13362

Diether M, Nikolaev Y, Allain FH, Sauer U (2019) Systematic mapping of protein-metabolite interactions in central metabolism of *Escherichia coli*. Mol Syst Biol 15(8):e9008

Diwan D, Rashid MM, Vaishnav A (2022) Current understanding of plant-microbe interaction through the lenses of multi-omics approaches and their benefits in sustainable agriculture. Microbiol Res 265:127180

Dixon RA, Steele CL (1999) Flavonoids and isoflavonoids—a gold mine for metabolic engineering. Trends Plant Sci 4(10):394–400

Donoso R, Leiva-Novoa P, Zúñiga A, Timmermann T, Recabarren-Gajardo G, González B (2017) Biochemical and genetic bases of indole-3-acetic acid (auxin phytohormone) degradation by the plant-growth-promoting rhizobacterium *Paraburkholderia phytofirmans* PsJN. Appl Environ Microbiol 83(1):e01991–e01916

Dossou-Aminon I, Dansi A, Ahissou H, Cissé N, Vodouhè R, Sanni A (2016) Climate variability and status of the production and diversity of sorghum (*Sorghum bicolor* (L.) Moench) in the arid zone of northwest Benin. Genet Resour Crop Evol 63(7):1181–1201. https://doi.org/10.1007/s10722-015-0310-y

Egamberdieva D, Wirth SJ, Alqarawi AA, Abd-Allah EF, Hashem A (2017) Phytohormones and beneficial microbes: essential components for plants to balance stress and fitness. Front Microbiol 8:2104

Eichmann R, Richards L, Schäfer P (2021) Hormones as go-betweens in plant microbiome assembly. Plant J 105(2):518–541
Eldin S, Digits JA, Huang S-T, Jencks WP (1995) Lifetime of an aliphatic iminium ion in aqueous solution. J Am Chem Soc 117(24):6631–6632
Eldin S, Jencks WP (1995) Lifetimes of iminium ions in aqueous solution. J Am Chem Soc 117(17):4851–4857
Evans A, Janssens S, Jacquemyn H (2020) Impact of climate change on the distribution of four closely related Orchis (Orchidaceae). Spec Divers 12(8):312. https://doi.org/10.3390/d12080312
Feng Y, De Franceschi G, Kahraman A, Soste M, Melnik A, Boersema PJ, De Laureto PP, Nikolaev Y, Oliveira AP, Picotti P (2014) Global analysis of protein structural changes in complex proteomes. Nat Biotechnol 32(10):1036–1044
Forkmann G, Martens S (2001) Metabolic engineering and applications of flavonoids. Curr Opin Biotechnol 12(2):155–160
Fraser PD, Romer S, Shipton CA, Mills PB, Kiano JW, Misawa N, Drake RG, Schuch W, Bramley PM (2002) Evaluation of transgenic tomato plants expressing an additional phytoene synthase in a fruit-specific manner. Proc Natl Acad Sci 99(2):1092–1097
Frick S, Ounaroon A, Kutchan TM (2001) Combinatorial biochemistry in plants: the case of O-methyltransferases. Phytochemistry 56(1):1–4
FRS JH, Baxter H (1999) The handbook of natural flavonoids. John Wiley & Sons, New York
Fisch MH, Flick BH, Arditti J (1973) Structure and antifungal activity of hircinol loroglossol and orchinol. Phytochemistry 12(2):437–441. https://doi.org/10.1016/0031-9422(73)80036-6
Gallage NJ, Hansen EH, Kannangara R, Olsen CE, Motawia MS, Jørgensen K, Holme I, Hebelstrup K, Grisoni M, Møller BL (2014) Vanillin formation from ferulic acid in *Vanilla planifolia* is catalysed by a single enzyme. Nat Commun 5(1):4037
Gandikota M, De Kochko A, Chen L, Ithal N, Fauquet C, Reddy AR (2001) Development of transgenic rice plants expressing maize anthocyanin genes and increased blast resistance. Mol Breed 7:73–83
Gemperline DC, Marshall RS, Lee K-H, Zhao Q, Hu W, McLoughlin F, Scalf M, Smith LM, Vierstra RD (2019) Proteomic analysis of affinity-purified 26S proteasomes identifies a suite of assembly chaperones in *Arabidopsis*. J Biol Chem 294(46):17570–17592
Graça J (2015) Suberin: the biopolyester at the frontier of plants. Front Chem 3:62
Grotewold E, Chamberlin M, Snook M, Siame B, Butler L, Swenson J, Maddock S, St. Clair G, Bowen B (1998) Engineering secondary metabolism in maize cells by ectopic expression of transcription factors. Plant Cell 10(5):721–740
Guerriero G, Roberto Berni J, Muñoz-Sanchez A, Apone F, Abdel-Salam EM, Qahtan AA, Alatar AA, Cantini C, Cai G, Hausman J-F, Sohail K, Teresa Hernández-Sotomayor SSM, Faisal M (2018) Production of plant secondary metabolites: examples tips and suggestions for biotechnologists. Gene 9(6):309. https://doi.org/10.3390/genes9060309
Habtemariam S (2019) Introduction to plant secondary metabolites—from biosynthesis to chemistry and antidiabetic action. In: Medicinal foods as potential therapies for type-2 diabetes and associated diseases, pp 109–132
Hagel JM, Facchini PJ (2013) Benzylisoquinoline alkaloid metabolism: a century of discovery and a brave new world. Plant Cell Physiol 54(5):647–672
Hain R, Grimmig B (2000) Modification of plant secondary metabolism by genetic engineering. In: Metabolic engineering of plant secondary metabolism. Springer, pp 217–231
He S, Dong X, Zhang G, Fan W, Duan S, Shi H, Li D, Li R, Chen G, Long G, Zhao Y, Chen M, Yan M, Yang J, Lu Y, Zhou Y, Chen W, Dong Y, Yang S (2021) High quality genome of Erigeron breviscapus provides a reference for herbal plants in Asteraceae. Abstr Mol Ecol Resour 21(1):153–169. https://doi.org/10.1111/1755-0998.13257
Herbert RB (2003) The biosynthesis of plant alkaloids and nitrogenous microbial metabolites. Nat Prod Rep 20(5):494–508

Huang J-P, Fang C, Ma X, Wang L, Yang J, Luo J, Yan Y, Zhang Y, Huang S-X (2019) Tropane alkaloids biosynthesis involves an unusual type III polyketide synthase and non-enzymatic condensation. Nat Commun 10(1):4036

Hupe DJ (1984) Enzyme reactions involving imine formation. New Compr Biochem 6:271–301

Iacobucci I, Monaco V, Cozzolino F, Monti M (2021) From classical to new generation approaches: an excursus of-omics methods for investigation of protein-protein interaction networks. J Proteome 230:103990

Igual J, Gonzalez-Bosch C, Dopazo J, Pérez-Ortín JE (1992) Phylogenetic analysis of the thiolase family. Implications for the evolutionary origin of peroxisomes. J Mol Evol 35:147–155

Isah T (2019) Stress and defense responses in plant secondary metabolites production. Biol Res 52(1):39. https://doi.org/10.1186/s40659-019-0246-3.

Ishiai S, Kondo H, Hattori T, Mikami M, Aoki Y, Enoki S, Suzuki S (2016) Hordenine is responsible for plant defense response through jasmonate-dependent defense pathway. Physiol Mol Plant Pathol 96:94–100

Jacobs DI, Gaspari M, van der Greef J, van der Heijden R, Verpoorte R (2005) Proteome analysis of the medicinal plant *Catharanthus roseus*. Planta 221:690–704

Jain A, Sarsaiya S, Singh R, Gong Q, Wu Q, Shi J (2024) Omics approaches in understanding the benefits of plant-microbe interactions. Front Microbiol 15:1391059

Jin H, Cominelli E, Bailey P, Parr A, Mehrtens F, Jones J, Tonelli C, Weisshaar B, Martin C (2000) Transcriptional repression by AtMYB4 controls production of UV-protecting sunscreens in *Arabidopsis*. EMBO J 19(22):6150–6161

Jorrin-Novo JV (2014) Plant proteomics methods and protocols. Springer

Jorrin-Novo JV (2020) What is new in (plant) proteomics methods and protocols: the 2015–2019 Quinquennium. Plant Proteomics: Methods and Protocols:1–10

Jorrin Novo J (2021) Proteomics and plant biology: contributions to date and a look towards the next decade. Expert Rev Proteomics 18(2):93–103

Jouhikainen K, Lindgren L, Jokelainen T, Hiltunen R, Teeri TH, Oksman-Caldentey K-M (1999) Enhancement of scopolamine production in *Hyoscyamus muticus* L. hairy root cultures by genetic engineering. Planta 208:545–551

Kean J, Downey L, Stough C (2017) Systematic overview of Bacopa monnieri (L.) Wettst. Dominant poly-herbal formulas in children and adolescents. Medicines 4(4):86. https://doi.org/10.3390/medicines4040086

Khaliq A, Perveen S, Alamer KH, Zia Ul Haq M, Rafique Z, Alsudays IM, Althobaiti AT, Saleh MA, Hussain S, Attia H (2022) Arbuscular mycorrhizal fungi symbiosis to enhance plant–soil interaction. Sustain For 14(13):7840

Khoo HE, Azlan A, Tang ST, Lim SM (2017) Anthocyanidins and anthocyanins: colored pigments as food, pharmaceutical ingredients, and the potential health benefits. Food Nutr Res 61:1361779

Kilgore MB, Kutchan TM (2016) The Amaryllidaceae alkaloids: biosynthesis and methods for enzyme discovery. Phytochem Rev 15:317–337

Kimotho RN, Maina S (2024) Unraveling plant–microbe interactions: can integrated omics approaches offer concrete answers? J Exp Bot 75(5):1289–1313

Kivivirta K, Herbert D, Lange M, Beuerlein K, Altmüller J, Becker A (2019) A protocol for laser microdissection (LMD) followed by transcriptome analysis of plant reproductive tissue in phylogenetically distant angiosperms. Plant Methods 15:1–11

Kumar U, Raj S, Sreenikethanam A, Maddheshiya R, Kumari S, Han S, Kapoor KK, Bhaskar R, Bajhaiya AK, Gahlot DK (2023) Multi-omics approaches in plant–microbe interactions hold enormous promise for sustainable agriculture. Agronomy 13(7):1804

Kumar S, Kumari D, Singh B (2022) Genus Rauvolfia: a review of its ethnopharmacology phytochemistry quality control/quality assurance pharmacological activities and clinical evidence. J Ethnopharmacol 295115327. https://doi.org/10.1016/j.jep.2022.115327

Lange BM, Croteau R (1999) Genetic engineering of essential oil production in mint. Curr Opin Plant Biol 2(2):139–144

Lange BM, Rujan T, Martin W, Croteau R (2000) Isoprenoid biosynthesis: the evolution of two ancient and distinct pathways across genomes. Proc Natl Acad Sci 97(24):13172–13177

Latha M, Dolui AK, Vijayaraj P (2021) Proteoform of Arabidopsis seed storage protein identified by functional proteomics approach exhibits acyl hydrolase activity during germination. Int J Biol Macromol 172:452–463

Lee E-J, Facchini PJ (2011) Tyrosine aminotransferase contributes to benzylisoquinoline alkaloid biosynthesis in opium poppy. Plant Physiol 157(3):1067–1078

Lei Z, Elmer AM, Watson BS, Dixon RA, Mendes PJ, Sumner LW (2005) A two-dimensional electrophoresis proteomic reference map and systematic identification of 1367 proteins from a cell suspension culture of the model legume *Medicago truncatula** S. Mol Cell Proteomics 4(11):1812–1825

Lewinsohn E, Schalechet F, Wilkinson J, Matsui K, Tadmor Y, Nam K-H, Amar O, Lastochkin E, Larkov O, Ravid U (2001) Enhanced levels of the aroma and flavor compound S-linalool by metabolic engineering of the terpenoid pathway in tomato fruits. Plant Physiol 127(3):1256–1265

Li Y, Wang C, Wang F, Dong H, Guo S, Yang J, Xiao P (2010) Chemical constituents of *Dendrobium candidum. Zhongguo Zhong yao za zhi= Zhongguo zhongyao zazhi*=China journal of Chinese materia medica 35(13):1715–1719

Lichman BR (2021) The scaffold-forming steps of plant alkaloid biosynthesis. Nat Prod Rep 38(1):103–129

Lim TK (n.d.) Edible medicinal and non medicinal plants volume 9 modified stems roots bulbs. Springer Netherlands, Dordrecht

Liu C-W, Murray JD (2016) The role of flavonoids in nodulation host-range specificity: an update. Plan Theory 5(3):33

Liu X, Liu Y, Huang P, Ma Y, Qing Z, Tang Q, Cao H, Cheng P, Zheng Y, Yuan Z (2017) The genome of medicinal plant Macleaya cordata provides new insights into benzylisoquinoline alkaloids metabolism. Mol Plant 10(7):975–989

Liu Z, Chen Y, Lian B, Zhang Z, Zhao Y, Ji Z, Lv Y, Li H (2019) Comparative study on population ecological distribution and extracellular enzyme activities of endophytic fungi in *Artemisia annua*. J Biosci Med 07(08):94–105. https://doi.org/10.4236/jbm.2019.78008

Lluch MA, Masferrer A, Arró M, Boronat A, Ferrer A (2000) Molecular cloning and expression analysis of the mevalonate kinase gene from *Arabidopsis thaliana*. Plant Mol Biol 42:365–376

Lu M-K, Shih Y-W, Chien T-TC, Fang L-H, Huang H-C, Chen P-S (2010) ALPHA.-Solanine inhibits human melanoma cell migration and invasion by reducing Matrix Metalloproteinase-2/9 activities. Biol Pharm Bullet 33(10):1685–1691. https://doi.org/10.1248/bpb.33.1685

Lv J, Yang S, Zhou W, Liu Z, Tan J, Wei M (2024) Microbial regulation of plant secondary metabolites: Impact, mechanisms and prospects. Microbiol Res 12:127688

Maeda HA (2019) Evolutionary diversification of primary metabolism and its contribution to plant chemical diversity. Front Plant Sci 10:881

Mahmoud SS, Croteau RB (2001) Metabolic engineering of essential oil yield and composition in mint by altering expression of deoxyxylulose phosphate reductoisomerase and menthofuran synthase. Proc Natl Acad Sci 98(15):8915–8920

Mahmud AR, Ema TI, Siddiquee MF-R, Shahriar A, Ahmed H, Mosfeq-Ul-Hasan M, Rahman N, Islam R, Uddin MR, Mizan MFR (2023) Natural flavonols: actions, mechanisms, and potential therapeutic utility for various diseases. Beni-Suef Univ J Basic Appl Sci 12(1):47

Mann V, Harker M, Pecker I, Hirschberg J (2000) Metabolic engineering of astaxanthin production in tobacco flowers. Nat Biotechnol 18(8):888–892

Maraveas C, Bayer IS, Bartzanas T (2021) Recent advances in antioxidant polymers: From sustainable and natural monomers to synthesis and applications. Polymers 13(15):2465

Memelink J, Verpoorte R, Kijne JW (2001) ORCAnization of jasmonate-responsive gene expression in alkaloid metabolism. Trends Plant Sci 6(5):212–219

Menke FL, Champion A, Kijne JW, Memelink J (1999) A novel jasmonate-and elicitor-responsive element in the periwinkle secondary metabolite biosynthetic gene Str interacts with a jasmonate-and elicitor-inducible AP2-domain transcription factor, ORCA2. EMBO J 18:4455

Meyer P, Heidmann I, Forkmann G, Saedler H (1987) A new petunia flower colour generated by transformation of a mutant with a maize gene. Nature 330(6149):677–678

Moghe GD, Last RL (2015) Something old, something new: conserved enzymes and the evolution of novelty in plant specialized metabolism. Plant Physiol 169(3):1512–1523

Muir SR, Collins GJ, Robinson S, Hughes S, Bovy A, Ric De Vos CHV, van Tunen AJ, Verhoeyen ME (2001) Overexpression of petunia chalcone isomerase in tomato results in fruit containing increased levels of flavonols. Nat Biotechnol 19(5):470–474

Nabavi SM, Šamec D, Tomczyk M, Milella L, Russo D, Habtemariam S, Suntar I, Rastrelli L, Daglia M, Xiao J (2020) Flavonoid biosynthetic pathways in plants: Versatile targets for metabolic engineering. Biotechnol Adv 38:107316

Nam MH, Kim SI, Liu JR, Yang DC, Lim YP, Kwon K-H, Yoo JS, Park YM (2005) Proteomic analysis of Korean ginseng (*Panax ginseng* CA Meyer). J Chromatogr B 815(1–2):147–155

Niemeyer HM (1988) Hydroxamic acids (4-hydroxy-14-benzoxazin-3-ones) defence chemicals in the gramineae. Phytochemistry 27(11):3349–3358. https://doi.org/10.1016/0031-9422(88)80731-3

O'Connor SE, Maresh JJ (2006) Chemistry and biology of monoterpene indole alkaloid biosynthesis. Nat Prod Rep 23(4):532–547

Orozco-Mosqueda MDC, Santoyo G, Glick BR (2023) Recent advances in the bacterial phytohormone modulation of plant growth. Plan Theory 12(3):606

Orobiyi A, Loko YL, Sanoussi F, Adjatin A, Gbaguidi A, Sanni DA (2017) Horticultural practices and varietal diversity of chili pepper (*Capsicum annuum* L.) in Central and Northern Benin. Genet Resour Crop Evol 64(2):419–436. https://doi.org/10.1007/s10722-016-0370-7

Olsen C (1921) The ecology of Urtica Dioica. J Ecol 9(1):1. https://doi.org/10.2307/2255757

Pant P, Pandey S, Dall'Acqua S (2021) The influence of environmental conditions on secondary metabolites in medicinal plants: a literature review. Chem Biodivers 18(11):e2100345

Pandey SA, Kumari M, Ruchi S, Akansha S, Ashmita R, Naveen B, Gaurav (2022) Comparison of antioxidant phytochemical profiling of *Bacopa monnieri* (Brahmi). Sci Temp 13(2):285–292. https://doi.org/10.58414/SCIENTIFICTEMPER.13.2.2022.285-292

Pandey P, Tripathi A, Dwivedi S, Lal K, Jhang T (2023) Deciphering the mechanisms, hormonal signaling, and potential applications of endophytic microbes to mediate stress tolerance in medicinal plants. Front Plant Sci 14:1250020

Peralta IE, Clausen AM, Zorrilla C, Ames M, Digilio A, Rodriguez F, Domenico C, Riccardo AM, Raffaella E (2021) Wild and cultivated potato species diversity taxonomy and conservation. In: The wild Solanums genomes. Springer International Publishing, Cham, pp 51–94

Piazza I, Beaton N, Bruderer R, Knobloch T, Barbisan C, Siepe I, Rinner O, de Souza N, Picotti P, Reiter L (2019) LiP-Quant, an automated chemoproteomic approach to identify drug targets in complex proteomes. bioRxiv:860072

Ping Y, Li X, You W, Li G, Yang M, Wei W, Zhou Z, Xiao Y (2019) De novo production of the plant-derived tropine and pseudotropine in yeast. ACS Synth Biol 8(6):1257–1262

Pozo MJ, Van Der Ent S, Van Loon L, Pieterse CM (2008) Transcription factor MYC2 is involved in priming for enhanced defense during rhizobacteria-induced systemic resistance in *Arabidopsis thaliana*. New Phytol 180(2):511–523

Rawat A, Mali RR (2013) Phytochemical properties and pharmcological activities of *Nicotiana tabacum*: a review. Indian J Pharm Biol Res 1(02):74–82. https://doi.org/10.30750/ijpbr.1.2.9

Rey M-D, Valledor L, Castillejo MÁ, Sánchez-Lucas R, López-Hidalgo C, Guerrero-Sanchez VM, Colina FJ, Escandón M, Maldonado-Alconada AM, Jorrín-Novo JV (2020) Recent advances in MS-based plant proteomics: proteomics data validation through integration with other classic and-omics approaches. Progress in Botany 81:77–101

Rodríguez B, Pacheco L, Bernal I, Piña M (2023) Mechanisms of action of flavonoids: antioxidant, antibacterial and antifungal properties. Ciencia, Ambiente y Clima 6(2):33–66

Rodwell VW, Beach MJ, Bischoff KM, Bochar DA, Darnay BG, Friesen JA, Gill JF, Hedl M, Jordan-Starck T, Kennelly PJ (2000) 3-Hydroxy-3-methylglutaryl-CoA reductase. Methods Enzymol 324:259–280

Rohdich F, Wungsintaweekul J, Eisenreich W, Richter G, Schuhr CA, Hecht S, Zenk MH, Bacher A (2000) Biosynthesis of terpenoids: 4-diphosphocytidyl-2C-methyl-D-erythritol synthase of *Arabidopsis thaliana*. Proc Natl Acad Sci 97(12):6451–6456

Römer S, Fraser PD, Kiano JW, Shipton CA, Misawa N, Schuch W, Bramley PM (2000) Elevation of the provitamin A content of transgenic tomato plants. Nat Biotechnol 18(6):666–669

Rosati C, Aquilani R, Dharmapuri S, Pallara P, Marusic C, Tavazza R, Bouvier F, Camara B, Giuliano G (2000) Metabolic engineering of beta-carotene and lycopene content in tomato fruit. Plant J 24(3):413–420

Rouchaud A, Braekman JC (2009) Synthesis of new analogues of the tetraponerines, vol 2009. Wiley Online Library, p 2666

Rozov-Ung AI, John M, Moore Gregory E, Elias W, Khawam Jeffrey M, Lackner John R, Semler Michael D, Sitrin (2014) Detection of drug effects on gastric emptying and contractility using a wireless motility capsule. Abstr BMC Gastroenterol 14(1). https://doi.org/10.1186/1471-230X-14-2

Sato F, Hashimoto T, Hachiya A, Tamura K-I, Choi K-B, Morishige T, Fujimoto H, Yamada Y (2001) Metabolic engineering of plant alkaloid biosynthesis. Proc Natl Acad Sci 98(1):367–372

Scheller PN, Lenz M, Hammer SC, Hauer B, Nestl BM (2015) Imine reductase-catalyzed intermolecular reductive amination of aldehydes and ketones. ChemCatChem 7(20):3239–3242

Semwal P, Sakshi P, Natália C-M (2021) Dioscorea deltoidea wall. Ex Griseb: a review of traditional uses bioactive compounds and biological activities. Food Biosci:41100969. https://doi.org/10.1016/j.fbio.2021.100969

Serotonin IP, Serotonergic NIV, Invertebrate MO (2018) Serotonin: the mediator that spans evolution. Regulation 203(238):1

Shamsudin NF, Ahmed QU, Mahmood S, Ali Shah SA, Khatib A, Mukhtar S, Alsharif MA, Parveen H, Zakaria ZA (2022) Antibacterial effects of flavonoids and their structure-activity relationship study: A comparative interpretation. Molecules 27(4):1149

Shaw LJ, Morris P, Hooker JE (2006) Perception and modification of plant flavonoid signals by rhizosphere microorganisms. Environ Microbiol 8(11):1867–1880

Shewmaker CK, Sheehy JA, Daley M, Colburn S, Ke DY (1999) Seed-specific overexpression of phytoene synthase: increase in carotenoids and other metabolic effects. Plant J 20(4):401–412

Shi X, Zhao Y, Xu M, Ma L, Adams JM, Shi Y (2024) Insights into plant–microbe interactions in the rhizosphere to promote sustainable agriculture in the new crops era. New Crops 1:100004

Singh A, Menéndez-Perdomo IM, Facchini PJ (2019) Benzylisoquinoline alkaloid biosynthesis in opium poppy: an update. Phytochem Rev 18(6):1457–1482

Singh B, Sharma RA (2015) Plant terpenes: defense responses, phylogenetic analysis, regulation and clinical applications. *3*. Biotech 5(2):129–151

Song C, Cao Y, Dai J, Li G, Manzoor MA, Chen C, Deng H (2022) The multifaceted roles of MYC2 in plants: toward transcriptional reprogramming and stress tolerance by jasmonate signaling. Front Plant Sci 13:868874

Sparg, S., Light, M., & Van Staden, J. (2004). Biological activities and distribution of plant saponins. J Ethnopharmacol, 94(2–3), 219–243.

Sprenger GA, Schörken U, Wiegert T, Grolle S, De Graaf AA, Taylor SV, Begley TP, Bringer-Meyer S, Sahm H (1997) Identification of a thiamin-dependent synthase in *Escherichia coli* required for the formation of the 1-deoxy-D-xylulose 5-phosphate precursor to isoprenoids, thiamin, and pyridoxol. Proc Natl Acad Sci 94(24):12857–12862

Steinbacher S, Kaiser J, Eisenreich W, Huber R, Bacher A, Rohdich F (2003) Structural basis of fosmidomycin action revealed by the complex with 2-C-methyl-D-erythritol 4-phosphate synthase (IspC): implications for the catalytic mechanism and anti-malaria drug development. J Biol Chem 278(20):18401–18407

Takahashi S, Kuzuyama T, Watanabe H, Seto H (1998) A 1-deoxy-D-xylulose 5-phosphate reductoisomerase catalyzing the formation of 2-C-methyl-D-erythritol 4-phosphate in an alternative nonmevalonate pathway for terpenoid biosynthesis. Proc Natl Acad Sci 95(17):9879–9884

Tétényi P (2010) Opium poppy (*Papaver somniferum*): botany and horticulture. In: Horticultural reviews. Wiley, pp 373–408

Tiwari B, Kharwar S, Tiwari D (2019) Pesticides and rice agriculture. In: Cyanobacteria. Elsevier, pp 303–325

Tsay YH, Robinson GW (1991) Cloning and characterization of ERG8 an essential gene of *Saccharomyces cerevisiae* that encodes phosphomevalonate kinase. Mol Cell Biol 11(2):620–631

Ullah A, Munir S, Badshah SL, Khan N, Ghani L, Poulson BG, Emwas A-H, Jaremko M (2020) Important flavonoids and their role as a therapeutic agent. Molecules 25(22):5243

van der Fits L, Memelink J (2000) ORCA3, a jasmonate-responsive transcriptional regulator of plant primary and secondary metabolism. Science 289(5477):295–297

Vandana UK, Rajkumari J, Singha LP, Satish L, Alavilli H, Sudheer PD, Chauhan S, Ratnala R, Satturu V, Mazumder PB (2021) The endophytic microbiome as a hotspot of synergistic interactions, with prospects of plant growth promotion. Biology 10(2):101

Velu G, Palanichamy V, Rajan AP (2018) Phytochemical and pharmacological importance of plant secondary metabolites in modern medicine. In: Roopan S, Madhumitha G (eds) Bioorganic phase in natural food: an overview. Springer International Publishing, Cham, pp 135–156

Veenstra TD (2021) Omics in systems biology: current progress and future outlook. Proteomics 21(3–4):2000235

Venegas-Molina J, Molina-Hidalgo FJ, Clicque E, Goossens A (2021) Why and how to dig into plant metabolite–protein interactions. Trends Plant Sci 26(5):472–483

Verpoorte R, Memelink J (2002) Engineering secondary metabolite production in plants. Curr Opin Biotechnol 13(2):181–187

Verpoorte R, van der Heijden R, Memelink J (1998) Plant biotechnology and the production of alkaloids: prospects of metabolic engineering. In: The alkaloids: chemistry and biology, vol 50. Elsevier, pp 453–508

Verpoorte R, van der Heijden R, Memelink J (2000) Engineering the plant cell factory for secondary metabolite production. Transgenic Res 9:323–343

Vuylsteke M, Peleman JD, Van Eijk MJ (2007) AFLP-based transcript profiling (cDNA-AFLP) for genome-wide expression analysis. Nat Protoc 2(6):1399–1413

Wahab A, Muhammad M, Munir A, Abdi G, Zaman W, Ayaz A, Khizar C, Reddy SPP (2023) Role of arbuscular mycorrhizal fungi in regulating growth, enhancing productivity, and potentially influencing ecosystems under abiotic and biotic stresses. Plan Theory 12(17):3102

Waite S, Farrell L (1998) Population biology of the rare military orchid (Orchis militaris L.) at an established site in Suffolk England. Botan J Linnean Soc 126(1-2):109–121. https://doi.org/10.1111/j.1095-8339.1998.tb02520.x

Wang E, Wang R, DeParasis J, Loughrin JH, Gan S, Wagner GJ (2001) Suppression of a P450 hydroxylase gene in plant trichome glands enhances natural-product-based aphid resistance. Nat Biotechnol 19(4):371–374

Wang W, Paschalidis K, Feng J-C, Song J, Liu J-H (2019) Polyamine catabolism in plants: a universal process with diverse functions. Front Plant Sci 10:561

Wang Y, Li Z, Xu M, Xiao Z, Liu C, Du B, Xu D, Li L (2023) Signal molecules regulate the synthesis of secondary metabolites in the interaction between endophytes and medicinal plants. PRO 11(3):849

Watt TA (1987) The biology and toxicity of ragwort (*Senecio jacobaea* L.) and its herbicidal and biological control. Herbage Abstr 57(1):1–16

Whyte RO (1974) Tropical grazing lands. Springer Netherlands, Dordrecht

Wink M (2010) Introduction: biochemistry, physiology and ecological functions of secondary metabolites. In: Annual plant reviews volume 40: Biochemistry of plant secondary metabolism, pp 1–19

Yan Y, Xia X, Fatima A, Zhang L, Yuan G, Lian F, Wang Y (2024) Antibacterial activity and mechanisms of plant flavonoids against gram-negative bacteria based on the antibacterial statistical model. Pharmaceuticals 17(3):292

Yang X, Yang S, Qi H, Wang T, Li H, Zhang Z (2020) PlaPPISite: a comprehensive resource for plant protein-protein interaction sites. BMC Plant Biol 20:1–11

Ye X, Al-Babili S, Kloti A, Zhang J, Lucca P, Beyer P, Potrykus I (2000) Engineering the provitamin A (β-carotene) biosynthetic pathway into (carotenoid-free) rice endosperm. Science 287(5451):303–305

Yu Q, Xiao H, Jedrychowski MP, Schweppe DK, Navarrete-Perea J, Knott J, Rogers J, Chouchani ET, Gygi SP (2020) Sample multiplexing for targeted pathway proteomics in aging mice. Proc Natl Acad Sci 117(18):9723–9732

Yu Y, Ouyang Z, Guo J, Zeng W, Zhao Y, Huang L (2021) Complete chloroplast genome sequence of *Erigeron breviscapus* and characterization of chloroplast regulatory elements. Front Plant Sci 12. https://doi.org/10.3389/fpls.2021.758290

Zhu X, Yu F, Yang Z, Liu S, Dai C, Lu X, Liu C, Yu W, Li N (2016) In planta chemical cross-linking and mass spectrometry analysis of protein structure and interaction in *Arabidopsis*. Proteomics 16(13):1915–1927

MIX
Papier aus verantwortungsvollen Quellen
Paper from responsible sources
FSC® C105338

FSC
www.fsc.org

If you have any concerns about our products,
you can contact us on
ProductSafety@springernature.com

In case Publisher is established outside the EU,
the EU authorized representative is:
Springer Nature Customer Service Center GmbH
Europaplatz 3, 69115 Heidelberg, Germany

Printed by Libri Plureos GmbH
in Hamburg, Germany